T0331636

Optimal Control and Geometry: Integrable Systems

The synthesis of symplectic geometry, the calculus of variations, and control theory offered in this book provides a crucial foundation for the understanding of many problems in applied mathematics.

Focusing on the theory of integrable systems, this book introduces a class of optimal control problems on Lie groups whose Hamiltonians, obtained through the Maximum Principle of optimality, shed new light on the theory of integrable systems. These Hamiltonians provide an original and unified account of the existing theory of integrable systems. The book particularly explains much of the mystery surrounding the Kepler problem, the Jacobi problem, and the Kowalewski Top. It also reveals the ubiquitous presence of elastic curves in integrable systems up to the soliton solutions of the non-linear Schroedinger's equation.

Containing a useful blend of theory and applications, this is an indispensable guide for graduates and researchers, in many fields from mathematical physics to space control.

PROFESSOR JURDJEVIC is one of the founders of geometric control theory. His pioneering work with H. J. Sussmann was deemed to be among the most influential papers of the century and his book, *Geometric Control Theory*, revealed the geometric origins of the subject and uncovered important connections to physics and geometry. It remains a major reference on non-linear control. Professor Jurdjevic's expertise also extends to differential geometry, mechanics and integrable systems. His publications cover a wide range of topics including stability theory, Hamiltonian systems on Lie groups, and integrable systems. He has spent most of his professional career at the University of Toronto.

CAMBRIDGE TRACTS IN MATHEMATICS

GENERAL EDITORS

B. BOLLOBÁS, W. FULTON, A. KATOK, F. KIRWAN, P. SARNAK, B. SIMON, B. TOTARO

All the titles listed below can be obtained from good booksellers or from Cambridge University Press. For a complete series listing visit: http://www.cambridge.org/mathematics

Optimal Control and Geometry: Integrable Systems

VELIMIR JURDJEVIC
University of Toronto

CAMBRIDGE
UNIVERSITY PRESS

Shaftesbury Road, Cambridge CB2 8EA, United Kingdom

One Liberty Plaza, 20th Floor, New York, NY 10006, USA

477 Williamstown Road, Port Melbourne, VIC 3207, Australia

314–321, 3rd Floor, Plot 3, Splendor Forum, Jasola District Centre, New Delhi – 110025, India

103 Penang Road, #05–06/07, Visioncrest Commercial, Singapore 238467

Cambridge University Press is part of Cambridge University Press & Assessment,
a department of the University of Cambridge.

We share the University's mission to contribute to society through the pursuit of
education, learning and research at the highest international levels of excellence.

www.cambridge.org
Information on this title: www.cambridge.org/9781107113886

© Cambridge University Press & Assessment 2016

First published 2016

A catalogue record for this publication is available from the British Library

Library of Congress Cataloging-in-Publication data
Names: Jurdjevic, Velimir.
Title: Optimal control and geometry : integrable systems / Velimir Jurdjevic,
University of Toronto.
Description: Cambridge : Cambridge University Press, 2016. |
Series: Cambridge studies in advanced mathematics ; 154 |
Includes bibliographical references and index.
Identifiers: LCCN 2015046457 | ISBN 9781107113886 (Hardback : alk. paper)
Subjects: LCSH: Control theory. | Geometry, Differential.
| Hamiltonian systems. | Lie groups. | Manifolds (Mathematics)
Classification: LCC QA402.3 .J88 2016 | DDC 515/.642–dc23
LC record available at http://lccn.loc.gov/2015046457

ISBN   978-1-107-11388-6   Hardback

Contents

Acknowledgments

This book grew out of the lecture notes written during the graduate courses that I gave at the University of Toronto in the mid 2000s. These courses made me aware of the need to bridge the gap between mainstream mathematics, differential geometry, and integrable systems, and control theory, and this realization motivated the initial conception of the book.

I am grateful to the University of Toronto for imposing the mandatory retirement that freed my time to carry out the necessary research required for the completion of this project.

There are several conferences and workshops which have had an impact on this work. In particular I would like to single out the Conference of Geometry, Dynamics and Integrable systems, first held in 2010 in Belgrade, Serbia and second in Sintra – Lisbon, Portugal in 2011, the INDAM meeting on Geometric Control and sub-Riemannian Geometry, held in Cortona, Italy in 2012, the IV Ibaronamerican Meeting on Geometric Mechanics and Control held in Rio de Janeiro in 2014, and the IMECC/Unicamp Fourth School and Workshop on Lie theory held in Campinas, Brazil in 2015. I thank the organizers for giving me a chance to present some of the material in the book and benefit from the interaction with the scientific community.

I also would like to thank the editorial staff of Cambridge University Press for providing an invaluable assistance in transforming the original manuscript to its present form. In particular, my thanks go to David Hemsley for his painstaking work in unraveling the inconsistencies and ambiguities in my original submission.

I feel especially indebted to my wife Ann Bia for taking care of the life around us during my preoccupation with this project. Thank you Ann.

Introduction

Upon the completion of my book on geometric control theory, I realized that this subject matter, which was traditionally regarded as a domain of applied mathematics connected with the problems of engineering, made important contributions to mathematics beyond the boundaries of its original intent. The fundamental questions of space control, starting with the possibility of navigating a dynamical system from an initial state to a given final state, all the way to finding the best path of transfer, inspired an original theory of differential systems based on Lie theoretic methods, and the quest for the best path led to the Maximum Principle of optimality. This theory, apart from its relevance for the subject within which it was conceived, infuses the calculus of variations with new and fresh insights: controllability theory provides information about the existence of optimal solutions and the Maximum Principle leads to the solutions via the appropriate Hamiltonians. The new subject, a synthesis of the calculus of variations, modern symplectic geometry and control theory, provides a rich foundation indispensable for problems of applied mathematics.

This recognition forms the philosophical underpinning for the book. The bias towards control theoretic interpretations of variational problems provides a direct path to Hamiltonian systems and reorients our understanding of Hamiltonian systems inherited from the classical calculus of variations in which the Euler–Lagrange equation was the focal point of the subject. This bias also reveals a much wider relevance of Hamiltonian systems for problems of geometry and applied mathematics than previously understood, and, at the same time, it offers a distinctive look at the theory of integrable Hamiltonian systems.

This book is inspired by several mathematical discoveries in the theory of integrable systems. The starting point was the discovery that the mathematical formalism initiated by G. Kirchhoff to model the equilibrium configurations of a thin elastic bar subjected to twisting and bending torques at its ends can be reformulated as an optimal control problem on the orthonormal frame bundle of $\mathbb{R}^3$, with obvious generalizations to any Riemannian manifold. On three-dimensional spaces of constant curvature, where the orthonormal frame bundle coincides with the isometry group, this generalization of Kirchhoff's elastic model led to a left-invariant Hamiltonian H on a six-dimensional Lie group whose Hamiltonian equations on the Lie algebra showed remarkable similarity with the equations of motion for the heavy top (a rigid body fixed at a point and free to move around this point under the gravitational force). Further study revealed an even more astonishing fact, that the associated control Hamiltonian system is integrable precisely in three cases under the same conditions as the the heavy top [JA; Jm].

This discovery showed that the equations of the heavy top form an invariant subsystem of the above Hamiltonian system and that the solvability of the equations for the top is subordinate to the integrability of this Hamiltonian system (and not the other way around as suggested by Kirchhoff and his "kinetic analogue" metaphor [Lv]. More importantly, this discovery suggested that integrability of mechanical tops is better understood throgh certain left-invariant Hamiltonians on Lie groups, rather than through conventional methods within the confines of Newtonian physics.

The above discovery drew attention to a larger class of optimal control problems on Lie groups G whose Lie algebra $\mathfrak{g}$ admits a Cartan decomposition $\mathfrak{g} = \mathfrak{p} \oplus \mathfrak{k}$ subject to

$$[\mathfrak{p}, \mathfrak{p}] \subseteq \mathfrak{k}, [\mathfrak{p}, \mathfrak{k}] = \mathfrak{p}, [\mathfrak{k}, \mathfrak{k}] \subseteq \mathfrak{k}. \tag{I.1}$$

These optimal problems, defined by an element $A \in \mathfrak{p}$ and a positive definite quadratic form $\langle \, , \rangle$ on $\mathfrak{k}$, consist of finding the solutions of the affine control system

$$\frac{dg}{dt} = g(A + U(t)), U(t) \in \mathfrak{k}, \tag{I.2}$$

that conform to the given boundary conditions $g(0) = g_0$ and $g(T) = g_1$, for which the integral $\frac{1}{2} \int_0^T \langle U(t), U(t) \rangle \, dt$ is minimal. This class of optimal control problems is called *affine-quadratic*. We show that any affine-quadratic problem is well defined for any regular element A in $\mathfrak{p}$ in the sense that for any any pair of points g_0 and g_1 in G, there exists a time $T > 0$, and a control $U(t)$ on $[0, T]$ that generates a solution $g(t)$ in (2) with $g(0) = g_0$ and $g(T) = g_1$, and attains the minimum of $\frac{1}{2} \int_0^T \langle U(t), U(t) \rangle \, dt$.

Remarkably, the Hamiltonians associated with these optimal problems reveal profound connections with integrable systems. Not only do they link mechanical tops with geodesic and elastic problems, but also reveal the hidden symmetries, even for the most enigmatic systems such as Jacobi's geodesic problem on the ellipsoid, and the top of Kowalewski.

This book lays out the mathematical foundation from which these phenomena can be seen in a unified manner. As L. C. Young notes in his classical book on the calculus of variations and optimal control [Yg], problems of optimality are not the problems to tackle with bare hands, but only when one is properly equipped. In the process of preparing ourselves for the tasks ahead it became necessary to amalgamate symplectic and Poisson geometry with control theory. This synthesis forms the theoretic background for problems of optimality. Along the way, however, we discovered that this theoretic foundation also applies to the classic theory of Lie groups and symmetric spaces as well. As a result, the book turned out to be as much about Lie groups and homogeneous spaces, as is about the problems of the calculus of variations and optimal control.

The subject matter is introduced through the basic notions of differential geometry, manifolds, vector fields, differential forms and Lie brackets. The first two chapters deal with the accessibility theory based on Lie theoretic methods, an abridged version of the material presented earlier in [Jc]. The orbit theorem of this chapter makes a natural segue to the chapters on Lie groups and Poisson manifolds, where it is used to prove that a closed subgroup of a Lie group is a Lie group and that a Poisson manifold is foliated by symplectic manifolds. The latter result is then used to show that the dual of a Lie algebra is a Poisson manifold, with its Poisson structure inherited from the symplectic structure of the cotangent bundle, in which the symplectic leaves are the coadjoint orbits. This material ends with a discussion of left-invariant Hamiltonians, a prelude to the Maximum Principle and differential systems with symmetries.

The chapter on the Maximum Principle explains the role of optimal control for problems of the calculus of variations and provides a natural transition to the second part of the book on integrable systems. The Maximum Principle is presented through its natural topological property as a necessary condition for a trajectory to be on the boundary of the reachable set. The topological view of this principle allows for its strong formulation over an enlarged system, called the Lie saturate, that includes all the symmetries of the system. This version of the Maximum Principle is called the Saturated Maximum Principle. It is then shown that Noether's theorem and the related Moment map associated with the symmetries are natural consequences of the Saturated Maximum Principle.

This material forms the theoretic background for the second part of the book, which, for the most part, deals with specific problems. This material begins with a presentation of the non-Euclidean geometry from the Hamiltonian point of view. This choice of presentation illustrates the relevance of the above formalism for the problems of geometry and also serves as the natural segue to the chapter on Lie groups G with an involutive automorphism σ and to the geometric problems on G induced by the associated Cartan decomposition $\mathfrak{g} = \mathfrak{p} \oplus \mathfrak{k}$ of the Lie algebra $\mathfrak{g}$ of G. In these situations the Cartan decomposition then yields a splitting $\mathcal{F}_{\mathfrak{p}} \oplus \mathcal{F}_{\mathfrak{k}}$ of the tangent bundle TG, with $\mathcal{F}_{\mathfrak{p}}$ and $\mathcal{F}_{\mathfrak{k}}$ the families of left-invariant vector fields on G that take values in $\mathfrak{p}$, respectively $\mathfrak{k}$, at the group identity e. The distributions defined by these families of vector fields, called vertical and horizontal form a basis for the class of variational problems on G described below.

Vertical distribution $\mathcal{F}_{\mathfrak{k}}$ is involutive and its orbit through the group identity e is a connected Lie subgroup K of G whose Lie algebra is $\mathfrak{k}$. This subgroup is contained in the set of fixed points of σ and is the smallest Lie subgroup of G with Lie algebra equal to $\mathfrak{k}$, and can be regarded as the structure group for the homogeneous space $M = G/K$.

Horizontal family $\mathcal{F}_{\mathfrak{p}}$ is in general not involutive. We then use the Orbit theorem to show that on semi-simple Lie algebras the orbit of $\mathcal{F}_{\mathfrak{p}}$ through the group identity is equal to G if and only if $[\mathcal{F}_{\mathfrak{p}}, \mathcal{F}_{\mathfrak{p}}] = \mathcal{F}_{\mathfrak{k}}$, or, equivalently, if and only if $[\mathfrak{p}, \mathfrak{p}] = \mathfrak{k}$. This controllability condition, translated to the language of the principal bundles, says that $[\mathfrak{p}, \mathfrak{p}] = \mathfrak{k}$ is a necessary and sufficient condition that any two points in G can be connected by a horizontal curve in G, where a horizontal curve is a curve that is tangent to $\mathcal{F}_{\mathfrak{p}}$.

The aforementioned class of problems on G is divided into two classes each treated somewhat separately. The first class of problems, inspired by the Riemannian problem on $M = G/K$ defined by a positive-definite, Ad_K-invariant quadratic form $\langle\,,\,\rangle$ on $\mathfrak{p}$ is treated in Chapter 8. In contrast to the existing literature on symmetric spaces, which introduces this subject matter through the geodesic symmetries of the underlying symmetric space [Eb; Hl], the present exposition is based on the pioneering work of R. W. Brockett [Br1; Br2] and begins with the sub-Riemannian problem of finding a horizontal curve $g(t)$ in G of minimal length $\int_0^T \sqrt{\left\langle g^{-1}(t)\frac{dg}{dt}, g^{-1}(t)\frac{dg}{dt}\right\rangle}\, dt$ that connects given points g_0 and g_1 under the assumption that $[\mathfrak{p}, \mathfrak{p}] = \mathfrak{k}$.

We demonstrate that this intrinsic sub-Riemannian problem is fundamental for the geometry of the underlying Riemannian symmetric space G/K, in the sense that all of its geometric properties can be extracted from $\mathfrak{g}$, without ever descending onto the quotient space G/K. We show that the associated Hamiltonian system is completely integrable and that its solutions can be written in closed form as

$$g(t) = g_0 \exp t(A + B) \exp(-tB), A \in \mathfrak{p}, B \in \mathfrak{h}. \tag{I.3}$$

The projection of these curves on the underlying manifold G/K coincides with the curves of constant geodesic curvature, with $B = 0$ resulting in the geodesics. We then extract the Riemannian curvature tensor

$$\kappa(A, B) = \langle [[A, B], A], B \rangle, A \in \mathfrak{p}, B \in \mathfrak{p}. \tag{I.4}$$

from the associated Jacobi equation. The chapter ends with a detailed analysis of the Lie algebras associated with symmetric spaces of constant curvature, the setting frequently used in the rest of the text.

The second aforementioned class of problems, called affine-quadratic, presented in Chapter 9, in a sense is complementary to the sub-Riemannian case mentioned above, and is most naturally introduced in the language of control theory as an optimal control problem over an affine distribution $\mathcal{D}(g) = \{g(A+U) : U \in \mathfrak{k}$ defined by an element A in $\mathfrak{p}$ and a positive-definite quadratic form $Q(u, v)$ defined on $\mathfrak{k}$. The first part deals with controllability, as a first step to the well-posedness of the problem. We first note a remarkable fact that any semi-simple Lie algebra $\mathfrak{g}$, as a vector space, carries two Lie bracket structures: the semi-simple Lie algebra and the semi-direct product Lie algebra induced by the adjoint action of K on $\mathfrak{p}$. This means that the affine-quadratic problem on a semi-simple Lie group G then admits analogous formulation on the semi-direct product $G_s = \mathfrak{p} \rtimes K$. Hence, the semi-direct affine-quadratic problem is always present behind every semi-simple affine problem. We refer to this semi-direct affine problem as *the shadow problem*. We then show that every affine system is controllable whenever A is a regular element in $\mathfrak{p}$. This fact implies that the corresponding affine-quadratic problem is well posed for any positive-definite quadratic form on $\mathfrak{k}$.

On semi-simple Lie groups G with K compact and with a finite center, the Killing form is negative-definite on $\mathfrak{k}$ and can be used to define an Ad_K invariant, positive-definite bilinear form $\langle \, , \, \rangle$ on $\mathfrak{k}$. The corresponding optimal control system is Ad_K-invariant and hence can be regarded as the canonical affine-quadratic problem on G. It is then natural to consider the departures from the canonical case defined by a quadratic form $\langle Q(u), v \rangle$ for some linear transformation Q on $\mathfrak{k}$ which is positive-definite relative to $\langle \, , \, \rangle$.

Any such affine-quadratic problem induces a left-invariant affine Hamiltonian

$$H = \frac{1}{2} \langle Q^{-1}(L_\mathfrak{k}), L_\mathfrak{k} \rangle + \langle A, L_\mathfrak{p} \rangle \tag{I.5}$$

on the Lie algebra $\mathfrak{g} = \mathfrak{p} \oplus \mathfrak{k}$ obtained by the Maximum Principle, where $L_\mathfrak{k}$ and $L_\mathfrak{p}$ denote the projections of an element $L \in \mathfrak{g}$ on the factors $\mathfrak{k}$ and $\mathfrak{p}$. The Hamiltonians which admit a spectral representation of the form

$$\frac{dL_\lambda}{dt} = [M_\lambda, L_\lambda] \text{ with}$$

$$M_\lambda = Q^{-1}(L_{\mathfrak{k}}) - \lambda A, \text{ and } L_\lambda = -L_{\mathfrak{p}} + \lambda L_{\mathfrak{h}} + (\lambda^2 - s)B \tag{I.6}$$

for some matrix B, are called *isospectral*. In this notation s is a parameter, equal to zero in the semi-simple case and equal to one in the semi-direct case.

The spectral invariants of $L_\lambda = L_{\mathfrak{p}} - \lambda L_{\mathfrak{k}} + (\lambda^2 - s)B$ are constants of motion and are in involution with each other relative to the Poisson structure induced by either the semi-simple Lie algebra $\mathfrak{g}$ or by the semi-direct product $\mathfrak{g}_s = \mathfrak{p} \ltimes \mathfrak{k}$ (see [Rm], also [Bv; RT]).

We show that an affine Hamiltonian H is isospectral if and only

$$[Q^{-1}(L_{\mathfrak{k}}), A] = [L_{\mathfrak{k}}, B] \tag{I.7}$$

for some matrix $B \in \mathfrak{p}$ that commutes with A. In the isospectral case every solution of the homogeneous part

$$\frac{dL_{\mathfrak{k}}}{dt} = [Q^{-1}(L_{\mathfrak{k}}), L_{\mathfrak{k}}] \tag{I.8}$$

is the projection of a solution $L_{\mathfrak{p}} = sB$ of the affine Hamiltonian system (I.6) and hence admits a spectral representation

$$\frac{dL_{\mathfrak{k}}}{dt} = [Q^{-1}(L_{\mathfrak{k}}) - \lambda A, L_{\mathfrak{k}} - \lambda B]. \tag{I.9}$$

The above shows that the fundamental results of A. T. Fomenko and V. V. Trofimov [Fa] based on Manakov's seminal work on the n-dimensional Euler's top are subordinate to the isospectral properties of affine Hamitonian systems on $\mathfrak{g}$, in the sense that the spectral invariants of $L_{\mathfrak{k}} - \lambda B$ are always in involution with a larger family of functions generated by the spectral invariants of $L_\lambda = -L_{\mathfrak{p}} + \lambda L_{\mathfrak{h}} + (\lambda^2 - s)B$ on $\mathfrak{g}$.

The spectral invariants of L_λ belong to a larger family of functions on the dual of the Lie algebra whose members are in involution with each other, and are sufficiently numerous to guarantee integrability in the sense of Liouville on each coadjoint orbit in $\mathfrak{g}^*$ [Bv].

We then show that the cotangent bundles of space forms, as well as the cotangent bundles of oriented Stiefel and oriented Grassmannian manifolds can be realized as the coadjoint orbits in the space of matrices having zero trace, in which case the restriction of isospectral Hamiltonians to these orbits results in integrable Hamiltonians on the underlying manifolds. In particular, we show that the restriction of the canonical affine Hamiltonian to the cotangent bundles of non-Euclidean space forms (spheres and hyperboloids) is given by

$$H = \frac{1}{2}||x||_\epsilon^2 ||y||_\epsilon^2 - \frac{1}{2}(Ax, x)_\epsilon, \epsilon = \pm 1. \tag{I.10}$$

This Hamiltonian governs the motion of a particle on the space form under a quadratic potential $V = \frac{1}{2}(Ax, x)_\epsilon$. We then show that all of these mechanical systems are completely integrable by computing the integrals of motion generated by the spectral invariants of the matrix L_λ. These integrals of motion coincide with the ones presented by J. Moser in [Ms2] in the case of C. Newmann's system on the sphere.

Remarkably, the degenerate case $A = 0$ provides a natural explanation for the enigmatic discovery of V.A. Fock that the solutions of Kepler's problem move along the geodesics of the space forms [Fk; Ms1; O1; O2]. We show that the stereographic projections from the sphere, respectively the hyperboloid, can be extended to the entire coadjoint orbit in such a way that the extended map is a symplectomorphism from the coadjoint orbit onto the cotangent bundle of $\mathbb{R}^n/\{0\}$ such that $H = \frac{1}{2}(x, x)_\epsilon (y, y)_\epsilon$ is mapped onto $E = \frac{1}{2}||p||^2 - \frac{1}{||q||}$ and the energy level $H = \frac{\epsilon}{2h^2}$ is mapped onto the energy level $E = -\frac{1}{2}\epsilon h^2$. Therefore $E < 0$ in the spherical case and $E > 0$ in the hyperbolic case. The Euclidean case $E = 0$ is obtained by a limiting argument when ϵ is regarded as a continuous parameter which tends to zero. This correspondence also identifies the angular momentum and the Runge–Lenz vector associated with the problem of Kepler with the moment map associated with the Hamiltonian H.

The chapter on the matrices in $sl_{n+1}(R)$ also includes a discussion of a left-invariant geodesic problem on the group of upper triangular matrices that is relevant for the solutions of a Toda lattice system. Our exposition then turns to Jacobi's geodesic problem on the ellipsoid and the origins of its integrals of motion. We show that there is a surprising and beautiful connection between this classical problem and isospectral affine Hamiltonians on $sl_n(R)$ that sheds much light on the symmetries that account for the integrals of motion. The path is somewhat indirect: rather than starting with Jacobi's problem on the ellipsoid $x \cdot D^{-1}x = 1$, we begin instead with a geodesic problem on the sphere in which the length is given by the elliptic metric $\int_0^T \sqrt{(Dx(t) \cdot x(t))}\, dt$. It turns out that the Hamiltonian system corresponding to the elliptic problem on the sphere is symplectomorphic to the Hamiltonian system associated with the geodesic problem of Jacobi on the ellipsoid, but in contrast to Jacobi's problem, the Hamiltonian system on the sphere can be represented as a coadjoint orbit. It turns out that the Hamiltonian system associated with the elliptic problem on the sphere is equal to the restriction of an isospectral affine Hamiltonian system H on $sl_{n+1}(R)$, and hence inherits the integrals of motion from the spectral matrix L_λ. In fact, the Hamiltonian is given by

$$H = \frac{1}{2}\langle D^{-1}L_{\mathfrak{k}}D^{-1}, L_{\mathfrak{k}}\rangle + \langle D^{-1}, L_{\mathfrak{p}},\rangle \tag{I.11}$$

and its spectral matrix by $L_\lambda = L_{\mathfrak{p}} - \lambda L_{\mathfrak{k}} + (\lambda^2 - s)D$. This observation reveals that the mechanical problem of Newmann and the elliptic problem on the sphere share the same integrals of motion. This discovery implies not only that all three problems – the mechanical problem of Newmann, Jacobi's problem on the ellipsoid and the elliptic problem on the sphere – are integrable, but it also identifies the symmetries that account for their integrals of motion. These findings validate Moser's speculation that the symmetries that account for these integrals of motion are hidden in the Lie algebra $sl_{n+1}(R)$ [Ms3].

The material then shifts to the rigid body and the seminal work of S. V. Manakov mentioned earlier. We interpret Manakov's integrability results in the realm of isospectral affine Hamiltonians, and provide natural explanations for the integrability of the equations of motion for a rigid body in the presence of a quadratic Newtonian field (originally discovered by O. Bogoyavlensky in 1984 [Bg1].

We then consider the Hamiltonians associated with the affine-quadratic problems on the isometry groups $SE_3(R)$, $SO_4(R)$ and $SO(1,3)$. These Hamiltonians contain six parameters: three induced by the left-invariant metric and another three corresponding to the coordinates of the drift vector. The drift vector reflects how the tangent of the curve is related to the orthonormal frame along the curve. To make parallels with a heavy top, we associate the metric parameters with the principal moments of inertia and the coordinates of the drift vector with the coordinates of the center of gravity. Then we show that these Hamiltonians are integrable precisely under the same conditions as the heavy tops, with exactly three integrable cases analogous to the top of Euler, top of Lagrange and the top of Kowalewski.

The fact that the Lie algebras $so_4(R)$ and $so(1,3)$ are real forms for the complex Lie algebra $so_4(\mathbb{C})$ suggests that the Hamiltonian equations associated with Kirchhoff's problem should be complexified and studied on $so_4(\mathbb{C})$ rather than on the real Lie algebras. This observation seems particularly relevant for the Kowalewski case. We show that the Hamiltonian system that corresponds to her case admits four holomorphic integrals of motion, one of which is of the form

$$I_4 = \left(\frac{1}{2\lambda}z_1^2 - bw_1 + s\frac{\lambda}{2}b^2\right)\left(\frac{1}{2\lambda}z_2^2 - \bar{b}w_2 + s\frac{\lambda}{2}\bar{b}^2\right),$$

where $s = 0$ corresponds to the semi-direct case and $s = 1$ to the semi-simple case. For $s = 0$ and $\lambda = 1$ this integral of motion coincides with the one obtained by S. Kowalewski in her famous paper of 1889 [Kw]. The passage to complex Lie algebras validates Kowalewska's mysterious use of complex variables and also improves the integration procedure reported in [JA].

Our treatment of the above Hamiltonians reveals ubiquitous presence of elastic curves in these Hamitonians. Elastic curves are the projections of extremal curves associated with the functional $\frac{1}{2} \int_0^T \kappa^2(s)\, ds$. In Chapter 16 we consider this problem in its own right as the curvature problem. Parallel to the curvature problem we also consider the problem of finding a curve of shortest length among the curves that satisfy fixed tangential directions at their ends and whose curvature is bounded by a given constant c. This problem is referred to as the Dubins–Delauney problem. Our interest in Delauney–Dubins problem is inspired by a remarkable paper of L. Dubins of 1957 [Db] in which he showed that optimal solutions exist in the class of continuosly differentiable curves having Lebesgue integrable second derivatives, and characterized optimal solutions in the plane as the concatenations of arcs of circles and straight line segments with the number of switchings from one arc to another equal to at most two.

We will show that the solutions of n-dimensional Dubins' problem on space forms are essentially three dimensional and are characterized by two integrals of motion I_1 and I_2. Dubins' planar solutions persist on the level $I_2 = 0$, while on $I_2 \neq 0$ the solutions are given by elliptic functions obtained exactly as in the paper of J. von Schwarz of 1934 in her treatment of the problem of Delaunay [VS]. Our solutions also clarify Caratheodory's fundamental formula for the problem of Delauney at the end of his book on the calculus of variations. [Cr, p. 378].

This chapter also includes a derivation of the Hamiltonian equation associated with the curvature problem on a general symmetric space G/K corresponding to the Riemannian symmetric pair (G, K). The corresponding formulas show clear dependence of this problem on the Riemannian curvature of the underlying space. We then recover the known integrability results on the space forms explained earlier in the book, and show the connections with rolling sphere problems discovered in [JZ].

The book ends with with a brief treatment of infinite-dimensional Hamiltonian systems and their relevance for the solutions of the non-linear Schroedinger equation, the Korteveg–de Vries equation and Heisenberg's magnetic equation. This material is largely inspired by another spectacular property of the elastic curves – they appear as the soliton solutions in the non-linear Schroedinger equation. We will be able to demonstrate this fact by introducing a symplectic structure on an infinite-dimensional Fréchet manifold of framed curves of fixed length over a three-dimensional space form. We will then use the symplectic form to identify some partial differential equations of mathematical physics with the Hamiltonian flows generated by the functionals defined by the geometric invariants of the underlying curves, such as the curvature and the torsion functionals.

Keeping in mind the reader who may not be familiar with all aspects of this theory we have made every effort to keep the exposition self-contained and integrated in a way that minimizes the gap between different fields. Unavoidably, some aspects of the theory have to be taken for granted such as the basic knowledge of manifolds and differential equations.

1

The Orbit Theorem and Lie determined systems

Let us begin with the basic concepts and notations required to set the text in motion starting from differentiable manifolds as the point of departure. Throughout the text, manifolds will be generally designated by the capital letters $M, N, O, \ldots$ and their points by the lower case letters $x, y, z, \ldots$. Unless otherwise stated, all manifolds will be finite dimensional, smooth, and second countable, that is, can be covered by countably many coordinate charts.

Local charts on M will be denoted by (U, ϕ) with U a coordinate neighborhood and ϕ a coordinate map on U. For each point $x \in U$ the coordinates $\phi(x)$ in $\mathbb{R}^n$ will be denoted by $(x_1, x_2, \ldots, x_n)$. For notational simplicity, we will often write $x = (x_1, \ldots, x_n)$, meaning that $(x_1, \ldots, x_n)$ is the coordinate representation of a point x in some coordinate chart (ϕ, U).

The set of smooth functions f defined on open subsets of M will be denoted by $C^\infty(M)$. It is a ring with respect to pointwise addition $(f + g)(x) = f(x) + g(x)$ and pointwise multiplication $(fg)(x) = f(x)g(x)$, both defined on the intersection of their domains.

On smooth manifolds tangent vectors v can be regarded both as the equivalence classes of curves and as the derivations. The first case corresponds to the notion of an "arrow": a tangent vector v at x is defined as the equivalence class of parametrized curves $\sigma(t)$ defined in some open interval I containing 0 such that $\sigma(0) = x$ with $\sigma_1 \sim \sigma_2$ if and only if in each coordinate chart (U, ϕ) $\frac{d}{dt} \phi \circ \sigma_1|_{t=0} = \frac{d}{dt} \phi \circ \sigma_2|_{t=0}$. We shall follow the usual custom and write

$$v = \frac{d}{dt} \phi \circ x(t)|_{t=0} = \left(\frac{dx_1}{dt}, \ldots, \frac{dx_n}{dt} \right),$$

where $x(t)$ is any representative in the equivalence class of curves that defines v.

In the second case, tangent vector are defined by their action on functions resulting in directional derivatives. As such, tangent vectors v at a point x are linear mappings from $C^\infty(M)$ into $\mathbb{R}$ that satisfy the Leibnitz formula,

$v(fg) = f(x)v(g) + g(x)v(f)$, for any functions f and g. In this context, tangent vectors in local coordinates will be written as $v = \sum_{i=1}^{n} v_i \frac{\partial}{\partial x_i}$, where $\frac{\partial}{\partial x_1}, \ldots \frac{\partial}{\partial x_n}$ denotes the usual basis of tangent vectors defined by $\frac{\partial}{\partial x_i} f = \frac{\partial f}{\partial x_i}, f \in C^{\infty}(M)$.

These two notions of tangent vectors are reconciled through the pairing

$$v(f) = \frac{d}{dt} f \circ x(t)|_{t=0} \sum_{i=1}^{n} v_i \frac{\partial f}{\partial x_i}.$$

We will use $T_x M$ to denote the tangent space at x, and use TM to denote the tangent bundle of M; TM is a smooth manifold whose dimension is equal to $2dim(M)$. Each coordinate chart (U, ϕ) in M induces a coordinate chart $(TU, \tilde{\phi})$ in TM with $\tilde{\phi}(v) = (x_1, \ldots, x_n, v_1, \ldots, v_n)$ for every $v \in T_x(U)$ and $x \in U$.

Recall that cotangent vectors at x are the equivalence classes of functions in $C^{\infty}(M)$ that vanish at x, with f and g in the same equivalence class if and only if f and g coincide in some open neighborhood of x and $\frac{d}{dt} f \circ \sigma(t)|_{t=0} = \frac{d}{dt} g \circ \sigma(t)|_{t=0}$ for every curve σ on M such that $\sigma(0) = x$. The pairing $\langle [f], [\sigma] \rangle = \frac{d}{dt} f \circ \sigma(t)|_{t=0}$ reflects the duality between tangent and cotangent vectors and identifies cotangent vectors as linear functions on $T_x(M)$.

In local coordinates, cotangent vectors will be written as the sums $\sum_{i=1}^{n} \frac{\partial f}{\partial x_i} dx^i$, where $dx_, \ldots, dx_n$ denotes the dual basis relative to the basis $\frac{\partial}{\partial x_1}, \ldots, \frac{\partial}{\partial x_n}$. The cotangent space at x will be denoted by $T_x^* M$ and the cotangent bundle $\cup \{T^*(M) : x \in M\}$ will be denoted by $T^* M$. The cotangent bundle is also a smooth manifold whose dimension is twice that of the underlying manifold M.

1.1 Vector fields and differential forms

Since these objects are fundamental for this study, it is essential to be precise about their meanings. Recall that a vector field X on M is a smooth mapping from M into TM such that $\pi \circ X = I$, where π denotes the natural projection from TM onto M. Thus $X(x)$ belongs to $T_x(M)$ for each point x in M. In the language of vector bundles, vector fields are sections of the tangent bundle. The space of smooth vector fields will be denoted by $V^{\infty}(M)$. In local coordinates vector fields X will be represented either by the arrow vector $(X_1(x_1, \ldots, x_n), \ldots, X_n(x_1, \ldots, x_n))$, or by the expression

$$X(x_1, \ldots, x_n) = \sum_{i=1}^{n} X_i(x_1, \ldots, x_n) \frac{\partial}{\partial x_i},$$

depending on the context. The function $\sum_{i=1}^{n} X_i(x_1, \ldots, x_n)\frac{\partial f}{\partial x_i}(x_1, \ldots, x_n)$ will be denoted by Xf. This action of vector fields on functions identifies vector fields with derivations on M, that is, it identifies $V^\infty(M)$ with the linear mappings D on $C^\infty(M)$ that satisfy

$$D(fg)(x) = f(x)(D(g)(x)) + g(x)(D(f)(x)) \tag{1.1}$$

for all functions f and g in $C^\infty(M)$.

The space $V^\infty(M)$ has a rich mathematical structure. To begin with, it is a module over the ring of smooth functions under the operations:

(i) $((fX)g)(x) = f(x)(Xg)(x)$ for any function g and all x in M.
(ii) $(X + Y)f = Xf + Yf$ for all functions f.

Secondly, $V^\infty(M)$ is a Lie algebra under the addition defined by (ii) above and the Lie bracket $[X, Y] = Y \circ X - X \circ Y$, where $[X, Y]$ means that $[X, Y]f = Y(Xf) - X(Yf)$ for every $f \in C^\infty(M)$. The reader can easily show that in local coordinates $[X, Y]$ is given by $Z = \sum_{i=1}^{n} Z_i \frac{\partial}{\partial x_i}$ with

$$Z_i = \sum_{j=1}^{n} \frac{\partial X_i}{\partial x_j} Y_j - \frac{\partial Y_i}{\partial x_j} X_j. \tag{1.2}$$

There seems to be no established convention about the sign of the Lie bracket. In some books the Lie bracket is taken as the negative of the one defined above (for instance, [AM] or [Hl]).

Differential forms are geometric objects dual to vector fields. They are defined analogously, as the smooth mappings ω from M into T^*M such that $\pi \circ \omega = I$, where now π is the natural projection from T^*M onto M. In local coordinates, ω will be written as $\omega(x_1, \ldots, x_n) = \sum_{i=1}^{n} \omega_i(x_1, \ldots, x_n)dx_i$ for some some smooth functions $\omega_1, \ldots, \omega_n$. Differential forms act on vector fields to produce functions $\omega(X)$ given by $\omega(X) = \sum_{i=1}^{n} \omega_i X_i$ in each chart (U, ϕ).

Differential forms are contained in the complex of exterior differential forms in which functions in $C^\infty(M)$ are considered as the forms of degree 0, and the differential forms defined above as the forms of degree 1. Differential forms of degree k can be defined in several ways [BT; Ar]. For our purposes, it will be convenient to define them through the action on vector fields. A differential form ω of degree k is any mapping $\omega : \underbrace{V^\infty(M) \times \cdots \times V^\infty(M)}_{k} \to C^\infty(M)$

that satisifies

$$\omega_x(X_1, \ldots X_{i-1}, f X_i + g W_i, X_{i+1}, \ldots, X_k)$$
$$= f\omega(X_1, \ldots, X_n) + g(\omega(X_1, \ldots, W_i, \ldots X_k),$$

for each $i \in \{1, \ldots, k\}$ and each function f and g, and

$$\omega(X_1, \ldots X_i, \ldots, X_j, \ldots, X_k)) = -\omega((X_1, \ldots X_j, \ldots, X_i, \ldots, X_k)),$$

for each index i and j.

Forms of degree k will be denoted by $\Omega^k(M)$. It follows that forms of degree k are k-multilinear and skew-symmetric mappings over $V^\infty(M)$. The skew-symmetry property implies that $\Omega^k(M) = 0$, for $k > dim(M)$.

Alternatively, differential forms can be defined through the wedge products. The wedge product $\omega_1 \wedge \omega_2$ of 1-forms ω_1 and ω_2 is a 2-form defined by

$$(\omega_1 \wedge \omega_2)(X, Y) = \omega_1(X)\omega_2(Y) - \omega_1(Y)\omega_2(X).$$

Any 2-form ω van be expressed as a wedge product of 1-forms. To demonstrate, let $X = \sum_{i=1}^{n} X^i \frac{\partial}{\partial x_i}$ and $Y = \sum_{i=1}^{n} Y^i \frac{\partial}{\partial x_i}$. It follows that

$$\omega(X, Y) = \sum_{i,j}^{n} X^i Y^j \omega\left(\frac{\partial}{\partial x_i}, \frac{\partial}{\partial x_j}\right) = \sum_{i>j} (X^i Y^j - X^j Y^i)\omega\left(\frac{\partial}{\partial x_i}, \frac{\partial}{\partial x_j}\right).$$

But then $(dx_i \wedge dx_j)(X, Y) = \omega\left(\frac{\partial}{\partial x_i}, \frac{\partial}{\partial x_j}\right)(X^i Y^j - X^i Y^i)$. Hence,

$$\omega = \sum_{i,j}^{n} \omega_{ij}(dx_i \wedge dx_j),$$

where ω_{ij} are the functions $\omega\left(\frac{\partial}{\partial x_i}, \frac{\partial}{\partial x_j}\right)$.

We now come to another indispensable theoretic ingredient, the exterior derivative.

Definition 1.1 The exterior derivative d is a mapping from $\Omega^k(M)$ into $\Omega^{k+1}(M)$ defined by

$$df(X) = X(f), \text{ when } k = 0, d\omega(X_1, \ldots, X_{k+1})$$

$$= \sum_{i=1}^{k+1} (-1)^{i+1} X_i \omega(X_1, \ldots, \hat{X}_i, \ldots, X_{k+1})$$

$$- \sum_{i<j} (-1)^{i+j} \omega([X_i, X_j], \ldots, \hat{X}_i, \ldots, X_{k+1})$$

for $k > 0$, where the hat above an entry indicates the absence of that entry from the expression. For instance, $\omega(X_1, \hat{X}_2, X_3) = \omega(X_1, X_3)$, $\omega(X_1, X_2, \hat{X}_3) = \omega(X_1, X_2)$.

In particular, the exterior derivative of a 1-form ω is given by

$$d\omega(X_1, X_2) = X_1 \omega(X_2) - X_2 \omega(X_1) + \omega([X_1, X_2]). \tag{1.3}$$

To show the exterior derivative in more familiar terms [BT], let $X_1 = \frac{\partial}{\partial x_1}, \ldots, X_n = \frac{\partial}{\partial x_n}$ denote the standard basis relative to a system of coordinates $x_1, \ldots, x_n$. If f is a function, then $df = \sum_{i=1}^{n} \omega_i dx_i$ for some functions $\omega_1, \ldots, \omega_n$. It follows that $\omega_i = df(X_i) = X_i(f) = \frac{\partial f}{\partial x_i}$, and $df = \sum_{i=1}^{n} \frac{\partial f}{\partial x_i} dx_i$. Therefore, the exterior derivative of f coincides with the directional derivative.

Consider now the exterior derivative of a 1-form $\omega = \sum_{i=1}^{n} \omega_i(x) dx_i$. It follows that

$$d\omega(X_i, X_j) = X_i\omega(X_j) - X_j\omega(X_i) = \frac{\partial \omega_i}{\partial x_j} - \frac{\partial \omega_j}{\partial x_i},$$

since $[X_i, X_j] = 0$. Hence,

$$d\omega = \sum_{i=1,j=1}^{n} \left(\frac{\partial \omega_i}{\partial x_j} - \frac{\partial \omega_j}{\partial x_i} \right) dx_i \wedge dx_j.$$

The exterior derivative of a 2-form $\omega = \omega_1 dx_2 \wedge dx_3 + \omega_2 dx_3 \wedge dx_1 + \omega_3 dx_1 \wedge dx_2$ in $\mathbb{R}^3$ is given by

$$d\omega(X, Y, Z) - \left(\frac{\partial \omega_1}{\partial x_1} + \frac{\partial \omega_2}{\partial x_2} + \frac{\partial \omega_3}{\partial x_3} \right) (X \cdot (Y \wedge Z),$$

where $X \cdot (Y \wedge Z)$ denotes the signed volume defined by the vectors $X = (X_1, X_2, X_3), Y = (Y_1, Y_2, Y_3)$ and $Z = (Z_1, Z_2, Z_3)$.

Differential forms in $\mathbb{R}^3$ are intercheangeably identified with vector fields via the following identification:

$$(w \in \Omega^1(\mathbb{R}^3) \iff W \in V^\infty(\mathbb{R}^3)) \iff w(X) = (W \cdot X), X \in V^\infty(\mathbb{R}^3).$$

In particular, if $w = df$ then the corresponding vector field is called the gradient of f and is usually denoted by $grad(f)$.

The expressions $\left(\frac{\partial \omega_2}{\partial x_3} - \frac{\partial \omega_3}{\partial x_2}, \frac{\partial \omega_3}{\partial x_1} - \frac{\partial \omega_1}{\partial x_3}, \frac{\partial \omega_1}{\partial x_2} - \frac{\partial \omega_2}{\partial x_1} \right)$ and $\left(\frac{\partial \omega_1}{\partial x_1} + \frac{\partial \omega_2}{\partial x_2} + \frac{\partial \omega_3}{\partial x_3} \right)$ are known as the curl and the divergence of a vector field $\omega_1 \frac{\partial}{\partial x_1} + \omega_2 \frac{\partial}{\partial x_2} + \omega_3 \frac{\partial}{\partial x_3}$. In this terminology, $d^2 = 0$ coincides with the well-known formulas of vector calculus

$$curl(grad) = 0 \quad \text{and} \quad div(curl) = 0.$$

Differential forms ω for which the exterior derivative is equal to 0 are called closed. The forms ω for which $\omega = d\gamma$ for some form γ are called exact. It can be shown that $d^2 = 0$, therefore exact forms are automatically closed. The quotient of k-closed forms over the exact k-forms is called the kth de Rham cohomology of M.

1.2 Flows and diffeomorphisms

Let us now consider differential equations

$$\frac{d\sigma}{dt}(t) = X(\sigma(t)), \tag{1.4}$$

defined by a vector field X in a manifold M.

Solution curves $\sigma(t)$ are called integral curves of X. In local coordinates, integral curves are the solutions of a system of ordinary differential equations

$$\frac{d\sigma_i}{dt}(t) = X_i(\sigma_1(t), \ldots, \sigma_n(t)), i = 1, \ldots, n.$$

It then follows from the basic theory of differential equations that for each initial point x there exists integral curves $\sigma(t)$ of X defined on an open interval $I = (-\epsilon, \epsilon)$ such that $\sigma(0) = x$. Any such curve can be extended to a maximal open interval $I_x = (e^-(x), e^+(x))$ whose end points are called the negative and the positive escape time. All integral curves of X that pass through a common point x have the same negative and positive escape time. The solution curve $\sigma(t), t \in (e^-(x), e^+(x))$ is called the integral curve of X through x.

Definition 1.2 Let X be a vector field and let $\Delta = \{(x, t) : x \in M, t \in (e^-(x), e^+(x))\}$. The mapping $\phi : \Delta \to M$ defined by $\phi(x, t) = \sigma(t)$ will be called the flow, or a dynamical system induced by X.

The theory of ordinary differential equations concerning the existence and uniqueness of solutions and their smooth dependence on the initial conditions can be summarized by the following essential properties:

1. $\phi(x, 0) = x$ for each $x \in M$.
2. $\phi(x, s + t) = \phi(\phi(x, s), t) = \phi(\phi(x, t), s)$ for all $(x, s), (x, t)$ and $(x, s + t)$ in Δ.
3. ϕ is smooth.
4. $\frac{\partial}{\partial t}\phi(x, t) = X \circ \phi(x, t)$.

Conversely, any smooth mapping $\phi : \Delta \to M$ with Δ an open subset of $M \times \mathbb{R}$, and a neighborhood of $M \times \{0\}$ that satisfies properties $(1), (2), (3)$, necessarily satisfies property (4) with $X = \frac{\partial \phi}{\partial t}(x, t)|_{t=0}$.

Vector field X is called the infinitesimal generator of the flow. The set $\{\phi(x, t) : t \in \mathbb{R}\}$ is called the trajectory through x, or the motion through x.

Definition 1.3 A mapping F from a manifold M onto a manifold N is called a diffeomorphism if F is invertible with both F and its inverse smooth. Manifolds M and N are said to be diffeomorphic if there is a diffeomorphism between them.

Flows of vector fields induce diffeomorphisms in the following sense. Let U be an open set in M whose closure is compact. Then there is an open interval $I = (-a, a)$ such that $U \times I$ is contained in Δ. The mapping $\Phi_t(x) = \phi(x, t)$ is a diffeomorphism from U onto $\Phi_t(U)$. Indeed, $\Phi_t^{-1} = \Phi_{-t}$.

A vector field X is said to be complete if $\Delta = M \times \mathbb{R}$, i.e., if each integral curve of X is defined for all $t \in \mathbb{R}$. Complete vector fields induce global flows $\phi : M \times \mathbb{R} \to M$. If X is a complete vector field then the corresponding family of diffeomorphisms $\{\Phi_t : t \in \mathbb{R}\}$ is called the one-parameter group of diffeomorphisms induced by X. Indeed, $\{\Phi_t : t \in \mathbb{R}\}$ is a group under the composition with $\Phi_t \circ \Phi_s = \Phi_{t+s}$ and $\Phi_t^{-1} = \Phi_{-t}$. If a vector field is not complete then its flow defines a local group of diffeomorphisms in some neighborhood of each point in M. It is known that all vector fields on compact manifolds are complete.

The shift in perspective from dynamical systems to groups of diffeomorphisms suggests another name for the trajectories. The trajectory through x becomes the orbit through x under the one-parameter group of diffeomorphisms. Each name evokes its own orientation, hence both will be used depending on the context.

1.2.1 Duality between points and linear functionals

The two ways of seeing vector fields, as arrows or as derivations, calls for further notational distinctions that elucidate the calculations with their flows. If X is a vector field then it is natural to write $X(q)$ for the induced tangent vector at q, seen as the arrow with its base at q and the direction $X(q)$, We will use a different notation for the same tangent vector seen as a derivation; $\hat{q} \circ X$ will denote the same tangent vector defined by $(\hat{q} \circ X)(f) = X(f)(q)$ for any function f, where now $\hat{q} : C^\infty(M) \to \mathbb{R}$ denotes the evaluation of f at q, i.e., $\hat{q}(f) = f(q)$.

It is easy to verify that for each point $q \in M$, $\hat{q}$ is a homomorphism from $C^\infty(M)$ into $\mathbb{R}$, the latter viewed as the ring under multiplication and addition, that is, $\hat{q}$ is a mapping from $C^\infty(M)$ into $\mathbb{R}$ that satisfies:

1. $\hat{q}(\alpha f + \beta g) = \alpha \hat{q}(f) + \beta \hat{q}(g)$ for all real numbers α and β and all functions f and g, and
2. $\hat{q}(fg) = \hat{q}(f)\hat{q}(g)$ for all functions f and g.

Conversely, for any non-trivial homomorphism ϕ from $C^\infty(M)$ into $\mathbb{R}$ there exists a unique point q in M such that $\hat{q} = \phi$ ([AS]). Therefore, the correspondence $q \to \hat{q}$ identifies points in M with linear functionals on $C^\infty(M)$ that satisfy (1) and (2) above.

The dualism between points and linear functionals carries over to diffeomorphisms. If F is any diffeomorphism on M then $\hat{F}$ will denote the pull back on functions in $C^\infty(M)$ defined by $\hat{F}(f) = f \circ F$. It follows that $\hat{F}$ is a ring automorphism on $C^\infty(M)$. Conversely, any ring automorphism on $C^\infty(M)$ is of the form $\hat{F}$ for some diffeomorphism F, as can be easily shown by the same proof as above.

We will now extend this notation to the flows Φ_t induced by vector fields on M, and let $\exp tX$ denote $\hat{\Phi}_t$ for the flow $\{\Phi_t : t \in R\}$ induced by X. It follows that

$$\exp(t+s)X = \exp tX \circ \exp sX = \exp sX \circ \exp tX, s, t \in R. \qquad (1.5)$$

The above implies that $\exp tX|_{t=0} = I$ and $\exp -tX = (\exp tX)^{-1}$. Moreover,

$$\frac{d}{dt}\exp tX = X \circ \exp tX = \exp tX \circ X. \qquad (1.6)$$

Let us now note an important fact that will be useful in the calculations below. Suppose that $\{\Phi_t : t \in \mathbb{R}\}$ is the flow induced by a vector field X. Then, $\{F \circ \Phi_t \circ F^{-1} : t \in \mathbb{R}\}$ is a one-parameter group of diffeomorphisms on M for any diffeomorphism F, and hence is generated by some vector field Y. It follows that $Y = F_*X \circ F^{-1}$, where F_* denotes the tangent map induced by F. Recall that $F_*(v) = w$, where $v = \frac{d\sigma}{dt}|_{t=0}, \sigma(0) = q$, and $w = \frac{d}{dt}(F(\sigma(t))|_{t=0}$. Since $F \circ \Phi_t \circ F^{-1}$ acts on points, $Y = F_*X \circ F^{-1}$ is the arrow representation of the infinitesimal generator of $\{F \circ \Phi_t \circ F^{-1} : t \in \mathbb{R}\}$.

As a derivation, $Y = \hat{F}^{-1} \circ \exp tX \circ \hat{F}$ by the following calculation:

$$Yf = \frac{d}{dt}f \circ (F \circ \Phi_t \circ F^{-1}) = X(\hat{F}(f)) \circ F^{-1} = (\hat{F}^{-1} \circ X \circ \hat{F})(f),$$

and therefore

$$\exp tY = \hat{F}^{-1} \circ \exp tX \circ \hat{F}. \qquad (1.7)$$

Equation (1.6) yields the following asymptotic formula:

$$\exp tX \approx I + tX + \frac{t^2}{2}X^2 + \cdots + \frac{t^n}{n!} + \cdots . \qquad (1.8)$$

Then,

$$\exp tY \circ \exp tX \circ \exp -tY \circ \exp -tX$$

$$= \left(I + tY + \frac{t^2}{2}Y^2 + \cdots\right) \circ \left(I + tX + \frac{t^2}{2}X^2 + \cdots\right)$$

$$\circ \left(I - Yt + \frac{t^2}{2}Y^2 + \cdots\right) \circ \left(I - tX + \frac{t^2}{2}X^2 + \cdots\right)$$

$$= \left(I + t(X + Y) + \frac{t^2}{2}(X^2 + 2X \circ Y + Y^2) + \cdots \right)$$

$$\circ \left((I - t(X + Y) + \frac{t^2}{2}(X^2 + 2X \circ Y + Y^2) + \cdots \right)$$

$$= I + t^2(Y \circ X - X \circ Y) + \cdots ,$$

which in turn yields an important formula

$$\frac{d}{dt} \exp \sqrt{t}Y \circ \exp \sqrt{t}X \circ \exp -\sqrt{t}Y \circ \exp -\sqrt{t}X|_{t=0} = [X, Y]. \qquad (1.9)$$

Similar calculations show that the Lie bracket $[X, Y]$ can alternatively be defined by the formula

$$\frac{\partial^2}{\partial t \partial s} \exp -tX \circ \exp sY \circ \exp tX|_{t=s=0} = [X, Y]. \qquad (1.10)$$

The preceeding formula can be seen in slightly more general terms according to the following definitions.

Definition 1.4 If X is a vector field then $adX : V^\infty(M) \to V^\infty(M)$ denotes the mapping $adX(Y) = [X, Y], Y \in V^\infty(M)$. If F is a diffeomorphism on M, then $Ad_F : V^\infty(M) \to V^\infty(M)$ is defined as

$$Ad_F(X) = \hat{F}^{-1} \circ X \circ \hat{F}$$

for all X in $V^\infty(M)$.

It then follows from (1.10) that

$$\frac{d}{dt}Ad_{\exp tX} = Ad_{\exp tX} \circ adX = adX \circ Ad_{\exp tX}. \qquad (1.11)$$

1.3 Orbits of families of vector fields: the Orbit theorem

It is well known that each orbit of a one-parameter group of diffeomorphisms $\{\Phi_t\}$ generated by a vector field X is a submanifold of the ambient manifold M; the orbit through a critical point of X is zero dimensional, otherwise an orbit is one dimensional. These orbits are often referred to as the leafs of X, in which case M is said to be foliated by the leaves of X. There are two pertinent observations about these orbits that are relevant for the text below:

1. The orbits are not of the same dimension whenever X has critical points.
2. It may happen that an orbit is an immersed rather than an embedded submanifold of M. Recall that a submanifold is called embedded if its

topology coincides with the relative topology induced by the topology of the ambient manifold. For immersed submanifolds, all relatively open sets are open in the submanifold topology, but there may be other open sets which are not in this class.

For instance, each orbit of the flow $\Phi_t(z, w) = \{z \exp t\theta, w \exp t\phi, z \in \mathbb{C}, w \in \mathbb{C}, |z|^2 = 1, |w|^2 = 1\}$ is dense on the torus $T^2 = \{z \in \mathbb{C} : |z|^2 = 1\} \times \{w \in \mathbb{C} : |w|^2 = 1\}$ whenever the ratio $\frac{\theta}{\phi}$ is irrational. The sets $\{\Phi_t(z, w) : t \in (t_0, t_1)\}$ are open in the orbit topology, but are not equal to the intersections of open sets in T^2 with the orbit.

Remarkably, the manifold structure of orbits generated by one vector field extends to arbitrary families of vector fields, and that is the content of the Orbit theorem. To be more precise, let $\mathcal{F}$ be an arbitrary family of vector fields (finite or infinite) which, for simplicity of exposition only, will be assumed to consist of complete vector fields.

For each $X \in \mathcal{F}$, Φ_t^X will denote the one-parameter group of diffeomorphisms on M generated by X, and $G(\mathcal{F})$ will denote the group of diffeomorphisms generated by $\cup\{\Phi_t^X : X \in \mathcal{F}, t \in \mathbb{R}\}$. A typical element in $G(\mathcal{F})$ is of the form

$$g = \Phi_{t_p}^{X_p} \circ \Phi_{t_{p-1}}^{X_{p-1}} \circ \cdots \Phi_{t_1}^{X_1} \tag{1.12}$$

for a subset $\{X_1, \ldots, X_p\}$ of $\mathcal{F}$ and some numbers $t_1, t_2, \ldots, t_p$, or

$$\hat{g} = \exp t_1 X_1 \circ \exp t_2 X_2 \circ \exp t_{p-1} X_{p-1} \circ \exp t_p X_p. \tag{1.13}$$

Definition 1.5 The set $\{g(x) : g \in G(\mathcal{F})\}$ will be called the orbit of $\mathcal{F}$ through x and will be denoted by $\mathcal{O}_{\mathcal{F}}(x)$.

The orbit of $\mathcal{F}$ through a point x can be defined analogously on the space of functions as the set of automorphisms ϕ on $C^\infty(M)$ of the form

$$\phi = \hat{x} \circ \exp t_1 X_1 \circ \exp t_2 X_2 \circ \exp t_{p-1} X_{p-1} \circ \exp t_p X_p.$$

with the understanding that the automorphism $\hat{x} \circ F$ is identified with the point $F(x)$.

Proposition 1.6 The Orbit theorem   *Each orbit $\mathcal{O}_{\mathcal{F}}(x)$ is a connected (possibly immersed) submanifold of M.*

This theorem, well known in the control community, has not yet found its proper place in the literature on geometry and therefore may not be so familiar to the general reader. Partly for that reason, but mostly because of the importance for the subsequent applications, we will outline the most important features of its proof (for more detailed proofs the reader can either consult the original sources [Sf; Ss] or the books [AS; Jc].

The prerequisite for the proof is a manifold version of the implicit function theorem, known as the constant rank theorem.

The constant rank theorem   *Let N and M be manifolds and let $F : N \to M$ be a smooth mapping whose rank of the tangent map $F_*(x)$ is constant as x varies over the points of N. Let k denote this rank and let n denote the dimension of M. Then,*

(a) $F^{-1}(y)$ is an $(n - k)$-dimensional embedded submanifold of N for each y in the range of F.

(b) Each point x in N has a neighborhood U such that $F(U)$ is an embedded submanifold of M of dimension k.

Sketch of the proof of the Orbit theorem   Let N denote the orbit of $\mathcal{F}$ through a point x in M. We then have the following:

(i) **The orbit topology** The topology on N as the strongest topology under which all mappings

$$\{t_1, t_2, \ldots, t_p\} \to \Phi_{t_p}^{X_p} \circ \Phi_{t_{p-1}}^{X_{p-1}} \circ \cdots \Phi_{t_1}^{X_1}(y) \qquad (1.14)$$

are continuous as the mappings from $\mathbb{R}^p$ into N, where y is an arbitrary point of N, and $\{X_1, \ldots, X_p\}$ an arbitrary finite subset of $\mathcal{F}$. Both the choice of the vector fields in $\mathcal{F}$ and their number is arbitrary. Since all such mappings are continuous, the topology of N if finer than the relative topology induced by the topology of the ambient manifold. In particular, N is Hausdorff because M is Hausdorff.

(ii) **Local charts** To define local charts at a point z in N, let F denote a mapping of the form (1.14) that satisfies:

1. $F(\hat{s}) = z$ for some point $\hat{s}$ in $\mathbb{R}^p$, and
2. The rank of the tangent map $F_*(\hat{s})$ is maximal among all the mappings given by (1.14).

Let $Rk(z)$ denote the rank of $F_*(\hat{s})$ defined by (1) above. It turns out that $Rk(z)$ is constant as z varies over the points of N. Let k denote the common value of $Rk(z)$. For each $\hat{z}$ in N let F denote any mapping of the form (1.14) that satisfies conditions (1) and (2) above. Then the rank of F is constant in some neighborhood of $\hat{s}$ in $\mathbb{R}^p$. By the constant rank theorem there exists a neighborhood U of $\hat{t}$ in $\mathbb{R}^p$ such that $F(U)$ is an embedded k-dimensional submanifold of M. Any such set $F(U)$ will be referred to as a local integral manifold of $\mathcal{F}$ at $\hat{z}$. Then it can be shown that each local integral manifold is open in the orbit topology of N.

(iii) **N is second countable** We will supply a complete proof of this assertion since it was omitted in the proof in [Jc]. The fact that each point of N is contained in a coordinate neighborhood that is an embedded submanifold of M implies that N is second countable by the following topological arguments based on the notion of paracompactness.

Recall that a topological space is *paracompact* if every open cover of the space has a locally finite refinement. A collection of subsets of a topological space is said to be *locally finite* if every point of the space has a neighborhood that meets only finitely many sets of the collection and an open covering $\mathcal{V}$ of a topological space is a *refinement* of a covering $\mathcal{U}$ if every $V \in \mathcal{V}$ is a subset of an element of $\mathcal{U}$.

The following fact is crucial for our proof: a locally compact topological space is paracompact if and only if it is second countable. The proof of this assertion can be found in Royden [Ro]. So it suffices to show that each orbit is paracompact.

To show that N is paracompact let $\mathcal{U}$ be an open cover of N and let z be a point of N. Designate by U an open neighborhood of z that is an embedded submanifold of N. Let $\{U_\alpha\}$ denote the open sets in $\mathcal{U}$ that intersect U. Since U is embedded manifold, each $U_\alpha \cap U$ is relatively open, and therefore $U_\alpha \cap U = O_\alpha \cap U$ for some open set O_α.

Since $\cup O_\alpha$ is an open submanifold of M, it is second countable and hence paracompact. Let $\{V_\alpha\}$ denote a locally finite refinement of $\{O_\alpha\}$, and let V be an open neighborhood of z that meets only finitely many V_α. Then $V \cap U$ is a neighborhood of z that meets only finitely many $V_\alpha \cap U_\alpha$, and $\{V_\alpha \cap U_\alpha\}$ is an open refinement of $\{U_\alpha\}$. Therefore, N is paracompact.

(iv) **Tangency properties of the orbits** A vector field X on M is said to be *tangent* to a submanifold N if $X(x)$ belongs to $T_x(N)$ for each point x in N. This notion extends to families of vector fields. We will say that a family $\mathcal{F}$ is tangent to N if every vector field X in $\mathcal{F}$ is tangent to N. An integral curve $\sigma(t)$ of a vector field X that is tangent to N remains in N for t in some open interval $(-\epsilon, \epsilon)$ whenever $\sigma(0) \in N$ (this fact is an immediate consequence of the existence of solutions to the Cauchy problem in differential equations). Conversely, if each integral curve $\sigma(t)$ of X that initiates on N remains in N for t in an open neighborhood of 0 then X is tangent to N.

Evidently $\mathcal{F}$ is tangent to each of its orbits. Moreover, for each y in an orbit N of $\mathcal{F}$ the curve $\sigma_t(s) = \hat{y} \circ \exp -tX \circ \exp sY \circ \exp tX$ is contained in N and satisfies $\sigma_t(0) = y$. Therefore, $\frac{d\sigma}{ds}|_{s=0} = \hat{y} \circ \exp -tX \circ Y \circ \exp tX$ belongs to T_yN for all t. Since each tangent space $T_y(N)$ is closed, $\frac{d}{dt}(\hat{y} \circ \exp -tX \circ Y \circ \exp tX)|_{t=0} = [X, Y](y)$ belongs to $T_y(N)$. Therefore, each Lie bracket $[X, Y]$ is tangent to the orbits of $\mathcal{F}$. The same applies to the brackets of higher orders. To take full advantage of these observations we need a few definitions.

If $\mathcal{F}$ is a family of vector fields then and $x \in M$, then $\mathcal{F}_x$ will denote the set of tangent vectors $\{X(x) : X \in \mathcal{F}\}$. The set $\mathcal{F}_x$ will be called the evaluation of $\mathcal{F}$ at x.

Definition 1.7 If $\mathcal{F}$ is any family of vector fields, then the Lie algebra generated by $\mathcal{F}$ is the smallest Lie algebra, in the sense of set inclusion, that contains $\mathcal{F}$. It will be denoted by $Lie(\mathcal{F})$. Its evaluation at any point x will be denoted by $Lie_x(\mathcal{F})$.

Typical elements in $Lie(\mathcal{F})$ are linear combinations of the Lie brackets of the form

$$[X_m, [X_{m-1}, \ldots, [X_2, X_1]] \ldots],$$

with $\{X_1, X_2, \ldots, X_m\} \subseteq \mathcal{F}$. An easy induction on the number of iterated Lie brackets shows that $Lie(\mathcal{F})$ is tangent to the orbits of $\mathcal{F}$. Hence $Lie_y(\mathcal{F}) \subseteq T_y(\mathcal{O}_{\mathcal{F}}(x))$ for all y in an orbit $\mathcal{O}_{\mathcal{F}}(x)$. In general, however, the tangent space of an orbit of $\mathcal{F}$ at a point y is generated by the tangent maps associated with to the linear span of tangent vectors of the form

$$\hat{y} \cap Ad_{\exp t_p X_p} \circ Ad_{\exp t_{p-1} X_{p-1}} \circ \cdots \circ Ad_{\exp t_1 X_1}(X) \tag{1.15}$$

associated with the mappings $\hat{y} \circ \exp t_p X_p \circ \exp t_{p-1} X_{p-1} \circ \exp t_1 X_1 \circ \exp tX$. In fact, vectors in (1.15) are tangent to the curve

$$\sigma(t) = \hat{q} \circ \exp t_p X_p \circ \cdots \circ \exp t_1 X_1 \exp tX$$

at $t = 0$. The example below shows that $\hat{y} \circ Ad_{\exp tY}(X)$ need not be in $Lie_y(\mathcal{F})$ and that the dimension of N exceeds the dimension of $Lie_y(\mathcal{F})$.

Example 1.8 Let $M = \mathbb{R}^2$ and let $\mathcal{F} = \{X, Y\}$ with

$$X(x, y) = \frac{\partial}{\partial x}, \text{ and } Y(x, y) = \frac{\partial}{\partial x} + \phi(x)\frac{\partial}{\partial y},$$

where ϕ is a smooth function that satisfies $\phi(x) = 0$ for $x \leq 0$ and $\phi(x) > 0$ for $x > 0$. Then $ad^k X(Y) = \phi^{(k)}\frac{\partial}{\partial y}$ and $[Y, adX(Y)] = (\phi^{(1)})^2 - \phi\phi^{(2)})\frac{\partial}{\partial y}$. Thus $Lie(\mathcal{F})$ is an infinite-dimensional Lie algebra whenever the ring of functions generated by $\phi^{(k)}, k = 0, 1 \ldots$ is infinite dimensional.

It follows that $Lie_q(\mathcal{F})$ is one-dimensional subspace of the tangent space at q for points $q = (x, y)$ with $x \le 0$. For all other points $Lie_q(\mathcal{F}) = \mathbb{R}^2$. The reader can easily show that there is only one orbit of $\mathcal{F}$ equal to the entire plane $\mathbb{R}^2$.

1.4 Distributions and Lie determined systems

The Orbit theorem can be recast as a theorem in the theory of distributions [Sh; St]. Loosely speaking, a k-dimensional distribution is a collection of k-dimensional subspaces $\mathcal{D}(x)$ of each tangent space T_xM which varies smoothly with the base point x. One-dimensional distributions reduce to line bundles in TM.

Distributions can be seen as generalized differential equations in which the ordinary differential equation $\frac{d\sigma}{dt} = X(\sigma(t))$ is replaced by a more general equation $\frac{d\sigma}{dt} \in \mathcal{D}(\sigma(t))$. A curve $\sigma(t)$ that is a solution of the above inclusion equation is called an integral curve of $\mathcal{D}$. The problem is to find conditions on $\mathcal{D}$ that guarantee the existence of integral curves through each initial point x.

A submanifold N that contains the point x and also contains every integral curve of $\mathcal{D}$ that passes through x is called an integral manifold of $\mathcal{D}$ through x. Distribution $\mathcal{D}$ is called integrable if integral manifolds exist through each point $x \in M$ and are of the same dimension as $\mathcal{D}$.

Below we shall consider slight generalizations of the distributions that appear in the literature on differential geometry. For our purposes:

Definition 1.9 A distribution $\mathcal{D}$ is a subset of the tangent bundle TM consisting of linear subspaces $\mathcal{D}(q)$ of $T_q(M)$ for $q \in M$. A distribution is said to be smooth if for each q there exist smooth vector fields $X_1, \ldots, X_k$ in a neighborhood U of q such that:

(a) $X_1(q), \ldots, X_k(q)$ is a basis for $\mathcal{D}(q)$, and
(b) $\{X_1(x), X_2(x), \ldots, X_k(x)\} \subseteq \mathcal{D}(x)$ for all points x in U.

In what follows all distributions are assumed to be smooth.

A vector field X is said to be tangent to a distribution $\mathcal{D}$ if $X(x) \in \mathcal{D}(x)$ for all $x \in M$. Alternatively, smooth distributions could have been defined in terms of vector fields that are tangent to the distribution for the following reasons. If $X_1, \ldots, X_k$ and U are as in Definition 1.9, then let U_0 be an open neighborhood of q such that its closure is a compact subset of U. Then there exists a smooth function α on M such that $\alpha = 1$ on U_0, and $\alpha = 0$ outside U [Hl]. This implies that the modified vector fields $\tilde{X}_1 = \alpha X_1, \ldots, \tilde{X}_k = \alpha X_k$ are tangent to $\mathcal{D}$ and are a basis for $\mathcal{D}(q)$.

The reader should note that the above definition allows for the possibility that the dimension of the distribution may vary with the base point q. We will also make another small departure from the existing literature on this subject, and extend admissible solutions to absolutely continuous curves, in which case we will say that an absolutely continuous curve $x(t)$ is an integral curve of a distribution $\mathcal{D}$ if $\frac{dx}{dt} \in \mathcal{D}(\sigma(t))$ for almost all points t in the domain of x. Recall that a parametrized curve $x(t)$, $t \in [t_0, t_1]$ in $\mathbb{R}^n$ is said to be absolutely continuous if $\frac{dx}{dt}$ exists almost everywhere in $[t_0, t_1]$ and is integrable, and, moreover, $x(t) - x(s) = \int_s^t \frac{dx}{d\tau} d\tau$ holds for all s, t in $[t_0, t_1]$. Then a curve $x(t)$ on a manifold M is said to be absolutely continuous if it is absolutely continuous in every system of coordinates.

It follows that an absolutely continuous curve $x(t)$, $t \in [t_0, t_1]$ is an integral curve of a distribution $\mathcal{D}$ whenever there exist vector fields $X_1, \ldots, X_m$ tangent to $\mathcal{D}(x(t))$ and measurable and bounded functions $u_1(t), \ldots, u_m(t)$ such that

$$\frac{dx}{dt} = \sum_{i=1}^m u_i(t) X_i(x(t)) \tag{1.16}$$

for each point t at which x is differentiable.

The existence theory of differential equations then guarantees that for each choice of vector fields $X_1, \ldots, X_m$ that are tangent to a distribution $\mathcal{D}$, and each choice of bounded and measurable functions $u_1(t), \ldots, u_m(t)$ on an interval $[0, T]$ there exist solution curves $x(t)$ of (1.16) defined on some interval $[0, s)$, $s > 0$. If the interval $[0, s)$ is maximal, then $x(t)$ is a unique integral curve that satisfies the above differential equation and passes through a fixed point at $t = 0$.

A distribution is said to be *involutive* if the Lie bracket of vector fields tangent to the distribution is also tangent to the distribution. A distribution $\mathcal{D}$ is said to be *integrable* if for each x in M there exists a submanifold N_x that contains x and in addition satisfies $T_y(N_x) = \mathcal{D}(y)$ for all points $y \in N_x$. Such a manifold is called an *integral manifold of $\mathcal{D}$ through x*. An integral manifold that is connected and is not a subset of any other connected integral manifold is said to be *maximal*. The example below shows that the distributions need not be of constant rank.

Example 1.10 Let $M_n(R)$ denote the space of all $n \times n$ matrices and let $\mathcal{D}(x) = \{Ax : A^T = -A, A \in M_n(R)\}$, for each $x \in \mathbb{R}^n$. Then $\mathcal{D}(0) = \{0\}$ and $\mathcal{D}(x) = \{y : x \cdot y = 0\}$ for each $x \neq 0$, because any vector y which is orthogonal to x can be written as $y = \sum_{i=1}^n \frac{y_i}{x_j}(e_i \wedge e_j)x$ (here x_j denotes the coordinate of x that is not equal to zero). Evidently, $\mathcal{D}$ is smooth, since it is defined by linear vector fields $X(x) = Ax$.

It is easy to see that the distribution $\mathcal{D}$ in Example 1.8 is integrable. The maximal integral manifold through each point $x_0 \neq 0$ is the sphere $||x|| = ||x_0||$.

Each distribution $\mathcal{D}$ defines a family of vector fields $\vec{\mathcal{D}}$ that are tangent to $\mathcal{D}$. The following proposition relates integrable manifolds of distributions to the orbits of $\vec{\mathcal{D}}$.

Proposition 1.11 *If a distribution $\mathcal{D}$ is integrable then it is necessarily involutive and the orbits of $\vec{\mathcal{D}}$ coincide with the maximal integral manifolds of $\mathcal{D}$.*

Proposition 1.12 The Frobenius theorem *Suppose that $\mathcal{D}$ is an involutive distribution such that the rank of $\mathcal{D}(x)$ is constant for all $x \in M$. Then $\mathcal{D}$ is integrable and the orbits of $\vec{\mathcal{D}}$ coincide with the maximal integral manifolds of $\mathcal{D}$.*

Let us first prove the Frobenius theorem.

Proof Let N denote an orbit of $\vec{\mathcal{D}}$ through a point q. We will show that $dim(N) = dim\mathcal{D}(q)$. Suppose that $\mathcal{F} = \{X_1, \ldots, X_k\}$ is an involutive family of vector fields in $\vec{\mathcal{D}}$ such that $X_1(q), \ldots, X_k(q)$ are linearly independent on an open set U in M. It suffices to show that each tangent vector $\dot{y} \circ Ad_{\exp tX_j}(X_i)$ belongs to the linear span of vectors in $\mathcal{F}_y$ for each $y \in U$, because then the tangent space T_yN is the linear span of $\dot{y} \circ Ad_{\exp tX_j}(X_i)$ (expression (1.15)).

Since $\mathcal{F}$ is involutive there exist smooth functions $\alpha_{ij}^{(m)}$ on U such that

$$[X_i, X_j](y) = \sum_{m=1}^{k} \alpha_{ij}^{(m)}(y)X_m(y)$$

for all y in U. If Φ_t denote the flow of X_i, then let $\sigma_j(t) = Ad_{\Phi_t}(X_j(y))$. Then,

$$\frac{d}{dt}\sigma_j(t) = \Phi_{t*}[X_i, X_j]\Phi_{-t}(y) = \sum_{m=1}^{k}\alpha_{ij}^{(m)}(\Phi_{-t}y)\Phi_{t*}X_m\Phi_{-t}(y) = \sum_{m=1}^{k}\alpha_{ij}^{(m)}(\Phi_{-t}y)\sigma_m.$$

This system is a linear system of equations

$$\frac{d\sigma_j}{dt} = \sum_{m=1}^{k} A_{jm}(t)\sigma_m(t), \qquad (1.17)$$

where $A(t)$ is the matrix with entries $A_{jm}(t) = \alpha_{ij}^{(m)}(\Phi_{-t}y)$. Let now $\phi_j(t) = \sum_{m=1}^{k} x_{jm}(t)X_m(y)$, where $x_{ij}(t)$ satisfy the following system of linear differential equations:

$$\frac{dx_{jm}}{dt} = \sum_{i=1}^{k} A_{ji}(t)x_{im}(t), x_{jm}(0) = \delta_{jm}, j = 1, \ldots, k.$$

These solutions exist and are unique. Then

$$\frac{d\phi_j}{dt} = \sum_{m=1}^{k} \frac{dx_{jm}}{dt} X_m(y) = \sum_{m=1}^{k} \sum_{i=1}^{k} A_{ji}(t) x_{im}(t) X_m(y) = \sum_{i=1}^{k} A_{ji}(t) \phi_i(t).$$

So both $\phi_1(t), \dots, \phi_k(t)$ and $\sigma_1(t), \dots, \sigma_k(t)$ are the solutions of the same differential equation with the same initial conditions at $t = 0$, hence, must be equal to each other. This shows that each curve $\sigma_j(t)$ belongs to the linear span of $\mathcal{F}(y)$. $\qquad\square$

Let us now turn to the proof of Proposition 1.11.

Proof Suppose that N is a connected integral manifold of $\mathcal{D}$. Let $\vec{\mathcal{D}}_N$ denote the restriction of $\vec{\mathcal{D}}$ to N. The orbits of $\vec{\mathcal{D}}_N$ partition N. Since $T_y(N) = \mathcal{D}(y) = \vec{\mathcal{D}}(y)$ for all points y in N it follows that each orbit of $\vec{\mathcal{D}}_N$ is open in N. Since N is connected there is only one orbit and therefore, N is equal to the orbit of $\vec{\mathcal{D}}_N$. This argument shows that $N \subseteq \mathcal{O}_{\vec{\mathcal{D}}}(x)$. The same argument shows that $\mathcal{D}$ is involutive because the Lie algebra generated by $\vec{\mathcal{D}}_N$ is tangent to its orbit, hence is tangent to N.

It remains to show that the orbit of $\vec{\mathcal{D}}$ through x is of the same dimension as N. Let q be any point of N. Since N is an integral manifold of $\mathcal{D}$ there exists a basis of vector fields $X_1, \dots, X_k$ in $\vec{\mathcal{D}}$ and a neighborhood U of q such that $X_1(y) \dots, X_k(y)$ are linearly independent at each point $y \in U$. Let $\mathcal{F}(y)$ be the linear span of $(X_1(y), \dots, X_k(y))$ at each point $y \in U$. Since $\mathcal{F}$ is involutive, the theorem of Frobenius (Proposition 1.11) applies. Therefore, N and the orbit through any point of N are of the same dimension and consequently the orbit through any point of N is the maximal integral manifold of $\mathcal{D}$. $\qquad\square$

Definition 1.13 A family of vector fields $\mathcal{F}$ is said to be Lie determined if the distribution defined by $Lie(\mathcal{F})$ is integrable.

Corollary 1.14 *Any family $\mathcal{F}$ of vector fields, such that $Lie_x\mathcal{F}$ is of constant rank at all points of an orbit is Lie determined.*

In view of the preceding theorem, $\mathcal{F}$ is Lie determined if $Lie_x(\mathcal{F}) = T_x(\mathcal{O}_x(\mathcal{F}))$ for all x in M, that is, if the tangent spaces of the orbits of $\mathcal{F}$ coincide with the evaluation of $Lie(\mathcal{F})$ at the points of each orbit. The following cases of Lie determined systems are well known.

Example 1.15 The distribution $\mathcal{D} = Lie(\mathcal{F})$ defined by the family of vector fields in Example 1.8 is involutive, but not of constant rank. It does not have integral manifolds anywhere along the y axis, and hence, is not integrable.

On the other hand, for analytic systems the structure of orbits is simple, thanks to this theorem.

Proposition 1.16 The Hermann–Nagano theorem *Suppose that M is an analytic manifold. Then any analytic family of vector fields $\mathcal{F}$ is Lie determined.*

Proof When X and Y are analytic then

$$x \circ Ad_{\exp tX}(Y) = \sum_{k=1}^{\infty} \frac{t^k}{k!} x \circ ad^k X(Y)$$

for small t, and therefore, $Ad_{\exp tX}(Y)$ belongs to $Lie(\mathcal{F})$. Consequently, the tangent spaces of an orbit of $\mathcal{F}$ coincide with the evaluation of $Lie(\mathcal{F})$ at the points of the orbit.   □

Example 1.17 Let $\mathcal{F} = \{X, Y\}$ be the family in $\mathbb{R}^3$ defined by two linear vector fields $X(q) = Aq$ and $Y(q) = Bq$ with A and B the following matrices:

$$A = \begin{pmatrix} 0 & -1 & 0 \\ 1 & 0 & 0 \\ 0 & 0 & 0 \end{pmatrix}, B = \begin{pmatrix} 1 & 0 & 0 \\ 0 & -1 & 0 \\ 0 & 0 & 1 \end{pmatrix}.$$

The Lie bracket of linear fields $X(q) = Aq$ and $Y(q) = Bq$ is linear and is given by the matrix $C = AB - BA$.

It follows that $Lie(\mathcal{F})$ is equal to the linear span of the matrices: A, B, C, D, where A and B are as defined above, and

$$C = \begin{pmatrix} 0 & 1 & 0 \\ 1 & 0 & 0 \\ 0 & 0 & 0 \end{pmatrix}, D = \begin{pmatrix} 1 & 0 & 0 \\ 0 & -1 & 0 \\ 0 & 0 & 0 \end{pmatrix}.$$

An easy calculation shows that

$$Lie_q(\mathcal{F}) = \begin{cases} \mathbb{R}^3 & \text{for } x^2 + y^2 \neq 0 \text{ and } z \neq 0, \\ \mathbb{R}^2 & \text{for } x^2 + y^2 \neq 0 \text{ and } z = 0, \\ \mathbb{R} & \text{for } x^2 + y^2 = 0 \text{ and } z \neq 0, \\ 0 & \text{for } x^2 + y^2 + z^2 = 0, \end{cases}$$

at a point $q = \begin{pmatrix} x \\ y \\ z \end{pmatrix}$. Since the vector fields are analytic, $\mathcal{F}$ is Lie determined by the Hermann–Nagano theorem. Indeed, the orbits of $\mathcal{F}$ are zero dimensional through the origin, one dimensional through non-zero points along the z-axis, two dimensional in the punctured plane $z = 0$ and $x^2 + y^2 \neq 0$, and three-dimensional at all points $x^2 + y^2 \neq 0$ and $z \neq 0$.

2

Control systems: accessibility and controllability

2.1 Control systems and families of vector fields

A control system on a manifold M is any differential system on M of the form

$$\frac{dx}{dt} = F(x(t), u(t)), \tag{2.1}$$

where $u(t) = (u_1(t), \ldots, u_m(t))$. Functions $u_1(t), \ldots, u_m(t)$ are called controls. They are usually assumed to be bounded and measurable on compact intervals $[t_0, t_1, T]$ and take values in a prescribed subset U of $\mathbb{R}^m$. A trajectory is any absolutely continuous curve $x(t)$ defined on some interval $I = [t_0, t_1]$ which satisfies (2.1) almost everywhere in $[t_0, t_1]$ for some control function $u(t)$.

We will assume that F is sufficiently regular that the Cauchy problem $x(\tau) = x_0$ associated with the time-varying vector field $X_u(t) = F(x, u(t))$ admits a unique solution $x(t)$ on some interval $[t_0, t_1]$ that varies smoothly relative to the initial point x_0. It is well known in the theory of differential equations [CL] that these properties will be fulfilled under the following conditions:

1. The vector field $X_u(x) = F(x, u)$ is smooth for each u in U.
2. The mapping $(x, u) \rightarrow F(x, u)$ from $M \times \bar{U}$ into TM is continuous.
3. The mapping $(x, u) \in M \times \bar{U} \rightarrow \frac{\partial F}{\partial x}(x, u)$ is continuous in any choice of local coordinates.

Under these conditions, each initial point x_0 and each control function $u(t)$ give rise to a unique trajectory $x(t)$ that emanates from x_0 at the initial time $t = 0$. Control theory is fundamentally concerned with the following questions:

1. Controllability: Given two states x_0 and x_1 in M is there a control $u(t)$ that steers x_0 to x_1 in some (or a priori fixed) positive time T?

19

2. Motion planning: If the answer to the first question is affirmative, then what is the trajectory that provides the given transfer?
3. Optimality: What is an optimal way of steering x_0 to x_1?

Alternatively, these questions could be reinterpreted as questions about the nature of points reachable by the trajectories of the system, and this shift in emphasis identifies the reachable sets as the basic objects of study associated with any control system. This chapter is devoted to the qualitative properties of the reachable sets.

There are essentially three kinds of reachable sets from a given initial point x_0: points in M reachable in exactly T units of time, points reachable in at most T units of time, and points reachable in any positive time. These sets will be denoted respectively by $\mathcal{A}(x_0, T)$, $\mathcal{A}(x_0, \leq T)$, and $\mathcal{A}(x_0)$. Evidently,

$$\mathcal{A}(x_0, T) \subseteq \mathcal{A}(x_0, \leq T) \subseteq \mathcal{A}(x_0).$$

The following notion is basic.

Definition 2.1 System (2.1) is said to be controllable if $\mathcal{A}(x_0) = M$ for each initial point x_0. It is said to be strongly controllable if $\mathcal{A}(x_0, T) = M$ for each x_0 in M and all $T > 0$.

This subject matter originated with linear systems in $\mathbb{R}^n$,

$$\frac{dx}{dt} = Ax + \sum_{i=1}^{m} u_i(t) b_i, \tag{2.2}$$

where A is an $n \times n$ matrix and $b_1, \ldots, b_m$ are vectors in $\mathbb{R}^n$, and the following theorem:

Proposition 2.2 *If $U = \mathbb{R}^m$, then (2.2) is strongly controllable if and only if the linear span of $\cup_{k \geq 0} \{A^k b_i, i = 1, \ldots, m\}$ is equal to $\mathbb{R}^n$, that is, if and only if the rank of the matrix*

$$C = (B \, AB \cdots A^{n-1} B), \tag{2.3}$$

where B denotes the matrix with columns $b_1, \ldots, b_m$, is equal to n.

This proposition is remarkable in the sense that it characterizes the nature of the reachable sets in terms of an algebraic condition that completely bypasses the need to solve the differential equation. It turns out that this controllability criterion lends itself to Lie algebraic interpretations applicable to non-linear situations as well.

Before elaborating on this point further, let us first note that linear systems belong to a particular class of systems

$$\frac{dx}{dt} = X_0(x) + \sum_{i=1}^{m} u_i(t)X_i(x) \qquad (2.4)$$

for some choice of vector fields $X_0, \ldots, X_m$ on M. Such systems are known as *control affine systems*, where X_0 is called the drift. The remaining vector fields $X_1, \ldots, X_m$ are called controlled vector fields.

Apart from their interest for control theory, affine systems figure prominently in mechanics and geometry. For instance, variational problems involving geometric invariants of curves, such as problems involving curvature and torsion, are naturally formulated on the orthonormal frame bundle via the Serret–Frenet differential systems. Then the Serret–Frenet system can be regarded as an affine control system on the group of motions of $\mathbb{R}^n$ with the curvatures $\kappa_1, \kappa_2, \ldots, \kappa_{n-1}$ playing the role of controls. We shall see later that affine systems also play an important part in the theory of mechanical tops.

Any attempt to bridge control theory with mechanics and geometry, however, requires some preliminary remarks justifying the passage to absolutely continuous trajectories, a generalization that might seem totally alien to both a geometer and a physicist. For a control practitioner, the need for measurable controls is dictated by the solutions of optimal problems involving inequality constraints. Even the simplest optimal problems with bounds on controls result in chattering controls that take the solutions outside the realm of the usual Euler–Lagrange equation. For such a person measurable controls are indispensable.

This bias towards measurable controls, however, resulted in an important side effect: it drew attention to piecewise constant controls and led to a new paradigm in which a control system was replaced by a family of vector fields. This paradigm shift identified a control system with a polysystem, a generalization of a dynamical system consisting of a single vector field, and uncovered the Lie bracket as a basic tool for studying its geometric properties.

With these remarks in mind let us return to control system (2.1) and its family of vector fields

$$\mathcal{F} = \{X : X(x) = F(x, u), u \in U\} \qquad (2.5)$$

generated by the constant controls. For each X in $\mathcal{F}$, the semi-orbit $\{x : x = \Phi_t^X(x_0), t \geq 0\}$ is contained in the reachable set $\mathcal{A}(x_0)$. The set of points reachable by the piecewise constant controls from x_0 at some positive time

T consist of points x which can be written as

$$x = \Phi_{t_p}^{X_p} \circ \Phi_{t_{p-1}}^{X_{p-1}} \cdots \circ \Phi_{t_1}^{X_1}(x_0), , t_1 \geq 0, \ldots, t_p \geq 0 \qquad (2.6)$$

for some vector fields $X_1, \ldots, X_p$ in $\mathcal{F}$, or dually as functions

$$\hat{x} = \hat{x}_0 \circ \exp t_1 X_1 \circ \exp t_2 X_2 \circ \cdot \circ \exp t_p X_p, t_1 \geq 0, t_2 \geq 0, \ldots, , t_p \geq 0. \quad (2.7)$$

The concatenation of flows by the elements of $\mathcal{F}$ admits a geometric interpretation in the group of diffeomorphisms of M through the following objects.

Definition 2.3 If $\mathcal{F}$ is a family of vector fields then $G(\mathcal{F})$ denotes the group of diffeomorphisms generated by $\{\exp tX : t \in \mathbb{R}\}$, $X \in \mathcal{F}$, and $S(\mathcal{F})$ will denote the semigroup generated by $\{\exp tX : t \geq 0\}$, $X \in \mathcal{F}$.

Any diffeomorphism Φ in $G(\mathcal{F})$ is of the form

$$\Phi = \exp t_1 X_1 \circ \exp t_2 X_2 \circ \cdots \circ \exp t_p X_p$$

for some vector fields $X_1, \ldots, X_k$ in $\mathcal{F}$ and numbers $t_1, \ldots, t_k$, while a diffeomorphism Φ in $S(\mathcal{F})$ is of the same form except that the numbers $t_1, \ldots, t_p$ are all non-negative.

The reachable sets $A(x_0, T), \mathcal{A}_{\mathcal{F}}(x_0, \leq T)$ and $\mathcal{A}(x_0)$ of $\mathcal{F}$ are defined completely analogously to the reachable sets of a control system, with obvious extensions to arbitrary families of vector fields and not just the family induced by (2.1). Of course, the reachable sets defined by $\mathcal{F}$ correspond to the points reachable by piecewise controls, only when $\mathcal{F}$ is induced by a system (2.1). The fact that the piecewise constant controls are dense in the class of bounded and measurable controls implies that the topological closure of the reachable sets of $\mathcal{F}$ and the control system F are the same.

It follows that the reachable sets of $\mathcal{F}$ can be interpreted either as the orbits of the semi-group $S(\mathcal{F})$, or as the semi-orbits of the group $G(\mathcal{F})$ through the point x_0. The theory of accessiblity emanates from the following:

Proposition 2.4 The Accessibility theorem *Let $\mathcal{F}$ be any Lie determined family of vector fields on a manifold M, and let N denote an orbit of $\mathcal{F}$ through a point x_0. Then the interior of $\mathcal{A}_{\mathcal{F}}(x_0, \leq T)$ relative to the orbit topology of N is dense in $\mathcal{A}_{\mathcal{F}}(x_0, \leq T)$ for any $T > 0$.*

For a proof see [Jc].

Definition 2.5 A family of vector fields $\mathcal{F}$ on M is said to have the accessibility property at x if $\mathcal{A}_{\mathcal{F}}(x, \leq T)$ has a non-empty interior in M.

Definition 2.6 A family of vector fields on M is controllable if $\mathcal{A}_{\mathcal{F}}(x) = M$ for each $x \in M$. It is strongly controllable if $\mathcal{A}_{\mathcal{F}}(x, \leq T) = M$ for all $T > 0$ and all $x \in M$.

Corollary 2.7 *A Lie determined family of vector fields $\mathcal{F}$ on a manifold M has the accessibility property at x if and only if $Lie_x\mathcal{F} = T_xM$. Moreover, the accessibility property at a single point implies the accessibility property at all points of M whenever M is connected.*

Proof If $\mathcal{F}$ has the accessibility property at x then the orbit of $\mathcal{F}$ through x is of the same dimension as the ambient manifold M. Since $\mathcal{F}$ is Lie determined, $Lie_x\mathcal{F} = T_xM$. Then $\mathcal{F}$ has the accessibility property at each point of the orbit of $\mathcal{F}$ by the Accessibility theorem (Proposition 2.4). The orbit through x is both open and closed, hence it is equal to M, whenever M is connected. $\qquad\square$

Proposition 2.8 *Let $\mathcal{F}$ be a family of vector fields on M such that $Lie_x(\mathcal{F}) = T_xM$ for all $x \in M$. If $\mathcal{A}_{\mathcal{F}}(x)$ is dense in M for some x, then $\mathcal{A}_{\mathcal{F}}(x) = M$.*

Proof The fact that $Lie_x(\mathcal{F}) = T_xM$ for each x implies that each orbit of $\mathcal{F}$ is open in M. Since $\mathcal{A}_{\mathcal{F}}(x)$ is connected and dense implies that there is only one orbit of $\mathcal{F}$.

Let $-\mathcal{F} = \{-X : X \in \mathcal{F}\}$. The orbits of $\mathcal{F}$ and $-\mathcal{F}$ are the same. Then $\mathcal{A}_{-\mathcal{F}}(y, \leq T)$ contains an open set in M for each $y \in M$ and each $T > 0$ by the Accessibility theorem. To say that $z \in \mathcal{A}_{-\mathcal{F}}(y, \leq T)$ is the same as saying that $y \in \mathcal{A}_{\mathcal{F}}(z, \leq T)$.

To show that each $y \in M$ is reachable from x by $\mathcal{F}$, let O be an open set contained in $\mathcal{A}_{-\mathcal{F}}(y, \leq T)$ and let $z \in O \cap \mathcal{A}_{\mathcal{F}}(x)$. Then $y \in \mathcal{A}_{\mathcal{F}}(z, \leq T)$ and $\mathcal{A}_{\mathcal{F}}(z, \leq T) \subseteq \mathcal{A}_{\mathcal{F}}(x)$ imply that $y \in \mathcal{A}_{\mathcal{F}}(x)$. $\qquad\square$

Example 2.9 Positive semi-orbits of the flow $(e^{it\theta}z, e^{it\phi}w)$ are dense on the torus $T^2 = \{(z, w) : |z| = |w| = 1\}$ whenever the ratio $\frac{\theta}{\phi}$ is irrational. So it may happen that the reachable set is dense without being equal to the entire space.

Let us now return to the linear systems and give a Lie theoretic proof for Proposition 2.2.

Proof We want to show that (2.2) is strongly controllable if and only if the linear span of $\{b_i, Ab_i, \ldots, A^{n-1}b_i, i = 1, \ldots, m\}$ is equal to $\mathbb{R}^n$ [Kl]. This control system induces an affine distribution

$$\mathcal{F} = \left\{ X_0 + \sum_{i=1}^{m} u_iX_i, u = (u_1, \ldots, u_m) \in \mathbb{R}^m \right\}$$

in $\mathbb{R}^n$ with X_0 a linear field $X_0(x) = Ax$ and each controlled vector field X_i constant and equal to b_i.

Then, $ad^k X_0(X_i)(x) = A^k b_i$ and $[X_i, X_j] = 0$ for any $i, j \geq 1$. It follows that $Lie_x(\mathcal{F})$ is equal to the linear span of Ax and vectors $A^k b_i$, $i = 1, \ldots, m$, $k = 0, 1, \ldots$. In particular, the evaluation of $Lie(\mathcal{F})$ at the origin is equal to the linear span of $A^k b_i$, $i = 1, \ldots, m$, $k = 0, 1, \ldots$. Since $A^k b$ for $k \geq n$ is linearly dependent on $\{b, Ab, A^2 b, \ldots, A^{n_1} b\}$, it suffices to consider powers $A^k b_i$, $i = 1, \ldots, m$, $k = 0, 1, \ldots, n - 1$. Therefore, $Lie_0(\mathcal{F})$ is equal to the range space of the controllability matrix C.

If (2.2) is strongly controllable, then the orbit of $\mathcal{F}$ through the origin is equal to $\mathbb{R}^n$, and by the Hermann–Nagano theorem, the dimension of each orbit is the same as the rank of the Lie algebra. Hence, the rank of the controllability matrix must be n.

Conversely, if the dimension of $Lie_0(\mathcal{F})$ is equal to n then the reachable sets $\mathcal{A}_{\mathcal{F}}(0, \leq T)$ and $\mathcal{A}_{-\mathcal{F}}(0, \leq T)$ contain open sets in $\mathbb{R}^n$ for each $T > 0$ by Proposition 2.4. But each of these reachable sets are also linear subspaces of $\mathbb{R}^n$, hence each of them is equal to $\mathbb{R}^n$. So any point x_0 can be steered to the origin in an arbitrarily short amount of time by a trajectory of (2.2) and also the origin can be steered to any point x in arbitrarily short amount of time by a trajectory of (2.2). Hence (2.2) is strongly controllable.   □

Let us now consider the affine distributions

$$\mathcal{F} = \left\{ X_0 + \sum_{i=1}^{m} u_i X_i, \ u = (u_1, \ldots, u_m) \in \mathbb{R}^m \right\} \tag{2.8}$$

associated with a control affine system (2.3) consisting of analytic vector fields $X_0, \ldots, X_m$. All analytic systems are Lie determined as a consequence of the Hermann–Nagano theorem. When $X_0 = 0$ then $\mathcal{F}$ reduces to the distribution defined in Chapter 1. In that case,

Proposition 2.10 *Control system $\frac{dx}{dt} = \sum_{i-1}^{m} u_i X_i(x)$ with $u = (u_1, \ldots, u_m) \in \mathbb{R}^m$ is strongly controllable if and only if $Lie_x \mathcal{F} = T_x M$, for all $x \in M$.*

Proof This proposition is a paraphrase of the Orbit theorem (Proposition 1.6).
                   □

For affine systems, with non-zero drift, there are no natural conditions that guarantee controllability, even when $Lie(\mathcal{F}$ is of full rank at all points of M. For instance, consider:

Example 2.11

$$\frac{dx}{dt} = Ax + uBx \tag{2.9}$$

in $M = \mathbb{R}^2/(0)$ with A and B matrices and $u(t)$ a scalar control. Let us first take

$$A = \begin{pmatrix} 0 & 1 \\ 1 & 0 \end{pmatrix}, B = \begin{pmatrix} 1 & 0 \\ 0 & -1 \end{pmatrix}.$$

It is easy to verify that $Lie(\mathcal{F})$ consist of linear fields Cx with C an arbitrary 2×2 matrix with zero trace. Hence, $Lie_x(\mathcal{F}) = \mathbb{R}^2$ for each $x \neq 0$.

However, (2.9) is not controllable because

$$\frac{d}{dt}((x_1(t)x_2(t)) = x_1^2(t) + x_2^2(t),$$

and $x_1(t)x_2(t) \geq 0$ whenever $x_1(0)x_2(0) \geq 0$. On the other hand, (2.9) becomes controllable on $\mathbb{R}^2/(0)$ when A is replaced by $A = \begin{pmatrix} 0 & 1 \\ -1 & 0 \end{pmatrix}$. For then, the flow of $X_0 = Ax$ is periodic. Therefore, $\{\pm X_0 + uX_1\}$ and $\{X_0 + uX_1\}$ have the same reachable sets, and hence, the reachable sets coincide with the orbits. Since $Lie_x\mathcal{F} = \mathbb{R}^2$ at all points $x \neq 0$ there is only one orbit of $\mathcal{F}$, and therefore, (2.9) is controllable.

2.2 The Lie saturate

Even though there is no general theory that guarantees controllability, there are some controllability criteria that can be applied in certain situations to obtain positive results. One such criterion is based on the notion of the Lie saturate, It is defined as follows.

Definition 2.12 The Lie saturate of a family of vector fields $\mathcal{F}$ is the largest set $\hat{\mathcal{F}}$ in $Lie(\mathcal{F})$, in the sense of set inclusion, such that

$$\bar{\mathcal{A}}_\mathcal{F}(x) = \bar{\mathcal{A}}_{\hat{\mathcal{F}}}(x)$$

for each $x \in M$. The Lie saturate will be denoted by $LS(\mathcal{F})$.

Its significance is described by the following:

Proposition 2.13 The controllability criterion *A Lie determined family $\mathcal{F}$ is controllable on M if and only if $LS(\mathcal{F}) = Lie(\mathcal{F})$.*

Proof If $\mathcal{LS}(\mathcal{F}) = Lie(\mathcal{F})$ then $\mathcal{LS}(\mathcal{F})$ is controllable. Since $\mathcal{A}_{\mathcal{LS}(\mathcal{F})}(x)$ is contained in the closure of $\mathcal{A}_{\mathcal{F}}(x)$, $\mathcal{A}_{\mathcal{F}}(x)$ is dense in M. But then $\mathcal{F}$ is controllable by Proposition 3.   □

In general, there is no constructive procedure to go from $\mathcal{F}$ to its Lie saturate. Nevertheless, there are some constructive steps that can be taken to enlarge a given family of vector fields without altering the closure of its reachable sets. The first step involves a topological closure of the family, which, in turn, requires a topology on the space of vector fields.

The most convenient topology $V^\infty(M)$ is that defined by the smooth uniform convergence on compact subsets of M. This means that a sequence of vector fields $X^{(m)}$ converges to a vector field X if for any chart (U, ϕ) and any compact set C in M

$$\frac{\partial^k X_i^{(m)}}{\partial x_1^{i_1} \cdots \partial x_n^{i_n}} \rightarrow \frac{\partial^k X_i}{\partial x_1^{i_1} \cdots \partial x_n^{i_n}}$$

uniformly on $\phi(C)$ in $\mathbb{R}^n$ for each k and each multi-index $i_1 + \cdots + i_k = n$. Here, $(X_1^{(m)}, \ldots, X_n^{(m)})$ and $(X_1, \ldots, X_n)$ denote the coordinate vectors of X^m and X. The following theorem is a paraphrase of well-known facts from the theory of differential equations (see also [Jc])

Proposition 2.14 *Suppose that a sequence of complete vector fields X^m converges to a vector field X and suppose that $x(t)$ is an integral curve of X defined on an interval $[0, T]$. Let y_k be any sequence of points in M that converges to $x(0)$ and let $x_k^{(m)}(t)$ denote the integral curves of X^m with $x_k^m(0) = y_k$. Then there exists an integer k_0 such that for $k > k_0$ each curve $x_k^m(t)$ is defined on $[0, T]$ and converges uniformly to $x(t)$ on $[0, T]$.*

This proposition easily implies

Corollary 2.15 *Let $\bar{\mathcal{F}}$ denote the topological closure of $\mathcal{F}$. Then,*

$$\bar{\mathcal{A}}_{\bar{\mathcal{F}}}(x) = \bar{\mathcal{A}}_{\mathcal{F}}(x)$$

for each $x \in M$, where $\bar{\mathcal{A}}$ denotes the topological closure of the set $\mathcal{A}$.

Corollary 2.16 *The Lie saturate is a closed family of vector fields.*

The next step makes use of weak limits in the space of controls according to the following:

Proposition 2.17 *Let $u^{(k)}$ be a sequence in $L^\infty([0, T], \mathbb{R}^m)$ that converges weakly to u^∞ in $L^\infty([0, T], \mathbb{R}^m)$, and let $Z_k(t) = X_0 + \sum_{i=1}^m u_i^{(k)}(t)X_i$ and $Z(t) = X_0 + \sum_{i=1}^m u_i^\infty(t)X_i$. Suppose that $x(t)$ is an integral curve of $Z(t)$*

defined on the interval $[0, T]$. *Then for all sufficiently large* k, *the integral curves* $x^{(k)}(t)$ *of* $Z_k(t)$ *with* $x^{(k)}(0) = x(0)$ *are defined on* $[0, T]$, *and converge uniformly to* $x(t)$ *on* $[0, T]$.

For a proof see [Jc, p. 118]. We just remind the reader that a sequence of functions $u^{(k)}$ in a Banach space B converges weakly to a point u in B if $L(u^{(k)})$ converges to $L(u)$ for every linear function L in the dual of B. In the case that B is equal to $L^1([t_0, t_1], \mathbb{R}^m)$, then $L^\infty([t_0, t_1], \mathbb{R}^m)$ is its dual, and weak convergence means that $\sum_{i=1}^{m} \int_{t_0}^{t_1} u_i^{(k)}(t) f_i(t)\, dt$ converges to $\sum_{i=1}^{m} \int_{t_0}^{t_1} u_i(t) f_i(t)\, dt$ for every choice of functions $f = (f_1, \ldots, f_m)$ in $L^\infty([t_0, t_1], \mathbb{R}^m)$.

Proposition 2.18 *Let* $\mathcal{F} = \{X, Y\}$ *where* X *and* Y *are smooth vector fields. Let* $x(t)$ *denote an integral curve of the convex combination* $\lambda X + (1 - \lambda) Y$. *If* $x(t)$ *is defined on an interval* $[0, T]$, *then* $x(T) \in cl(A_{\mathcal{F}}(x(0), T))$. *Consequently,*

$$\hat{x} \circ \exp(\lambda X + (1 - \lambda) Y)(T) \in cl(\mathcal{A}(x, T))$$

for each $x \in M$ *for which* $\hat{x} \circ \exp(\lambda X + (1 - \lambda) Y)(T)$ *is defined.*

Proof Let $0 = t_0 < t_1 < \cdots < t_{2n} = t$ be an equidistant partition of the interval $[0, T]$. If I_k denotes the interval $(t_{k-1}, t_k]$, then let u_k denote the characteristic function of the set $I_2 \cup I_4 \cup \cdots \cup I_{2n}$ and let v_k denote the characteristic function of the set $I_1 \cup I_3 \cup \cdots \cup I_{2n-1}$. Then

$$\lim_{k \to \infty} \int_0^T u_k(\tau) f(\tau)\, d\tau = \frac{1}{2} \int_0^T f(\tau)\, d\tau \text{ and } \lim_{k \to \infty} \int_0^T v_k(\tau) f(\tau)\, d\tau = \frac{1}{2} \int_0^T f(\tau)\, d\tau$$

for every function f in $L^\infty([0, t], \mathbb{R})$. Therefore, both sequences of functions converge weakly to the constant function $u = v = \frac{1}{2}$.

Let z_n denotes the integral curve of $Z_n(t) = 2\lambda u_n(t) X + 2(1 - \lambda)(1 - u_n(t)) Y$ that satisfies $z_n(0) = x(0)$. It follows that $z_n(T) \in cl(A_{\mathcal{F}}(x(0), T)$ for each n, and that Z_n converges weakly to $Z = \lambda X + (1 - \lambda) Y$. But then z_n converges uniformly to $x(t)$ on the interval $[0, T]$. Hence, $x(T) \in cl(A_{\mathcal{F}}(x(0), T))$. $\square$

Corollary 2.19 $cl(A_{\mathcal{F}}(x, T)) = cl(A_{ch(\mathcal{F})}(x, T))$, *where* $ch((\mathcal{F})$ *denotes the convex hull of* $\mathcal{F}$.

Corollary 2.20 $cl(A_{\mathcal{F}}(x) = cl(A_{co(\mathcal{F})}(x)$, *where* $co(\mathcal{F})$ *denotes the the positive convex cone generated by* $\mathcal{F}$.

Proof The reachable sets $A_{\mathcal{F}}(x)$ are invariant under positive reparametrizations of vector fields in $\mathcal{F}$. $\square$

In addition, the reachable sets are unaltered by symmetries. The following notion is basic.

Definition 2.21 A diffeomorphism Φ is a normalizer for $\mathcal{F}$ if

$$\Phi^{-1}(\mathcal{A}_{\mathcal{F}}(\Phi(x)) \subseteq \bar{\mathcal{A}}_{\mathcal{F}}(x) \text{ and } (\Phi_*^{-1}\mathcal{F}\Phi)(x) \subseteq Lie_x(\mathcal{F})$$

for all $x \in M$.

Definition 2.22 For each diffeomorphism Φ and each vector field X, $\Phi_\sharp(X)$ will denote the vector field $\Phi_*(X) \circ \Phi^{-1}$, and $\Phi_\sharp(\mathcal{F}) = \{\Phi_\sharp(X) : X \in \mathcal{F}\}$.

Then:

Proposition 2.23 *Suppose that $\mathcal{N}(\mathcal{F})$ denotes the set of normalizers for a family $\mathcal{F}$. Then*

$$\{\Phi_\sharp(X) : X \in \mathcal{F}, \Phi \in \mathcal{N}(\mathcal{F})\} \subseteq LS(\mathcal{F}).$$

The proof is obvious. It will be convenient for future reference to assemble all these facts into the following:

Proposition 2.24 *The Lie saturate $\mathcal{LS}(\mathcal{F})$ is a closed set of vector fields, invariant under its normalizer, and also invariant under the following enlargements:*

1. *If $Y_1, Y_2, \ldots, Y_p$ are any set of vector fields in $\mathcal{LS}(\mathcal{F})$ then the positive affine hull $\{\alpha_1 Y_1 + \cdots + \alpha_p Y_p : \alpha_i \geq 0, i = 1, \ldots, p\}$ is also contained in $\mathcal{LS}(\mathcal{F})$.*
2. *If $\mathcal{V}$ is a vector space of vector fields in $\mathcal{LS}(\mathcal{F})$, then $Lie(\mathcal{V})$ is in $\mathcal{LS}(\mathcal{F})$.*
3. *If $\pm Y$ is in $\mathcal{LS}(\mathcal{F})$, then $(\Phi_\lambda^Y)_* X \Phi_{-\lambda}^Y$ is in $\mathcal{LS}(\mathcal{F})$ for any $\lambda \in \mathbb{R}$, and any $X \in \mathcal{LS}(\mathcal{F})$. Here, $\Phi_{\lambda *}^Y$ denotes its tangent map of the flow $\{\Phi_\lambda^Y : \lambda \in \mathbb{R}\}$ induced by Y.*

Let us now return to the affine control systems (2.4) with these symmetry tools at our disposal. It follows that $\alpha(X_0 + u_i X_i) \in \mathcal{LS}(\mathcal{F})$ for each $\alpha > 0$ by 1 in Proposition 2.24. Then, $\lambda X_i = lim_{\alpha \to 0}(\alpha(X_0 + \lambda\frac{u}{\alpha}X_i)$ is in $\mathcal{LS}(\mathcal{F})$ for each number λ. But then the vector space generated by the controlled vector fields $X_1, \ldots, X_m$ is in the Lie saturate of $\mathcal{F}$ by 2 in Proposition 2.24.

Hence, $\{Ad_{\exp \lambda_i X_i}(X_0)$ is in the Lie saturate of $\mathcal{F}$ for each λ_i. We will be able to show that in some situations $-X_0$ is in the positive convex cone spanned $\{Ad_{\exp \lambda_i X_i}(X_0), (\lambda_1, \ldots, \lambda_m) \in \mathbb{R}^m, i = 1, \ldots, m\}$, which then implies controllability whenever $Lie(\mathcal{F})$ is of full rank on M.

We will come back to this theory when discussing the Maximum Principle. In the meantime, we will first integrate this material with other geometric structures, with Lie groups at the core.

3

Lie groups and homogeneous spaces

The mathematical formalism developed so far has natural applications to the theory of Lie groups. To highlight these contributions, and also to better orient the subject for further use, it seems best to start at the beginning and develop the relevant concepts in a self contained manner.

Definition 3.1 A group G is called a real Lie group if G is a real analytic manifold and the group operations

$$(x, y) \to xy \text{ and } x \to x^{-1}$$

are real analytic, the first as a mapping from $G \times G$ into G and the second as a mapping from G into G. A group G is called a complex Lie group if G is a complex manifold and the group operations are holomorphic.

We will assume that all Lie groups are real unless explicitly stated otherwise.

Prototypical Example: the general linear group The group $GL(E)$ of all linear automorphisms of a finite-dimensional real vector space E is a Lie group.

Proof Each basis $e_1, \ldots, e_n$ in E identifies automorphisms T in $GL(E)$ with $n \times n$ matrices $X(T) = (x_{ij}(T))$ defined by $Te_i = \sum_{j=1}^{n} x_{ji}e_j$. Then the entries of the matrix $X(T)$ provide a correspondence between points of $\mathbb{R}^{n^2}$ and elements of $GL(E)$. If ϕ denotes this correspondence, then $GL(E)$ is topologized by the finest topology such that ϕ is a homeomorphism. Under this topology the coordinate neighborhoods U are the inverse images of open sets in $\mathbb{R}^{n^2}$ defined by $Det(X(T)) \neq 0$. Matrices that correspond to different bases are conjugate to each other, hence are defined by analytic (rational) functions.

The composition of elements in $GL(E)$ corresponds to the products of matrices. Since the entries of the product depend polynomially on the entries of the matrices, the group multiplication is analytic. Similarly the entries of the inverse of a matrix are rational functions of the entries of the matrix, hence analytic.   □

It is a common practice to denote $GL(\mathbb{R}^n)$ by $GL_n(R)$, a convention that will be adopted in this text.

Any Lie group G has two distinguished groups of diffeomorphisms, the group of left translations $\{L_g : g \in G\}$ and the group of right translations $\{R_g : g \in G\}$. The left translations are defined by $L_g(x) = gx$ and the right translations by $R_g(x) = xg$.

We shall follow the convention established earlier and use $\hat{F}$ to denote the automorphism on $C^\infty(M)$ induced by a diffeomorphism F.

Definition 3.2 Vector field X on a Lie group G is called left-invariant if $\hat{L}_g^{-1} \circ X \circ \hat{L}_g = X$ for any $g \in G$. Right-invariant vector fields are defined by $\hat{R}_g^{-1} \circ X \circ \hat{R}_g = X$.

It follows that the left (right)-invariant vector fields satisfy

$$(L_g)_* X(x) = X(L_g(x)) = X(gx) \ ((R_g)_* X(x) = X(R_g(x)) = X(xg) \ ), \text{ for all } g \in G.$$

Since $X(g) = (L_g)_* X(e)$, both left- and right-invariant vector fields are determined by their values at the group identity e. Moreover, each tangent vector A in $T_e(G)$ determines a unique left (right)-invariant vector field X_A defined by $X_A(g) = (L_g)_* A$ (respectively, $X_A(g) = (R_g)_* A$).

Proposition 3.3 *The Lie bracket of left (right) invariant vector fields is left (right) invariant vector field.*

Proof

$$[\hat{F}^{-1} \circ X \circ \hat{F}, \hat{F}^{-1} \circ Y \circ \hat{F}] = \hat{F}^{-1} \circ [X, Y] \circ \hat{F}.$$

□

We will use $\mathcal{F}_l$ and $\mathcal{F}_r$ to designate the Lie algebras of left (right) invariant vector fields on G.

Proposition 3.4 *Let $F : G \to G$ denote the diffeomorphism $F(x) = x^{-1}$ Then,*

$$X_l \to \hat{F}^{-1} \circ X_r \circ \hat{F}$$

is an isomorphism from $\mathcal{F}_r$ onto $\mathcal{F}_l$.
Moreover, if $A = X_r(e)$ then $(F \circ X_r \circ F^{-1})(e) = -A$.

Proof Let $Y = \hat{F}^{-1} \circ X_r \circ \hat{F}$. Then,

$$\hat{L}_g^{-1} \circ Y \circ \hat{L}_g = \hat{L}_g^{-1} \hat{F}^{-1} \circ X_r \circ \hat{F} \hat{L}_g = \hat{F}^{-1} \hat{R}_g \circ X_r \circ \hat{R}_g \hat{F} = \hat{F}^{-1} \circ X_r \circ \hat{F} = Y$$

because $\hat{L}_g^{-1} \hat{F}^{-1} = \hat{F}^{-1} \hat{R}_g$. Therefore Y is left-invariant.

To prove the second statement let $\sigma(\epsilon)$ denote a curve in G such that $\sigma(0) = e$ and $\frac{d\sigma}{d\epsilon}(0) = A$. Then $\frac{d}{d\epsilon}\sigma^{-1}(\epsilon)|_{\epsilon=0} = -A$, and therefore $F_*(A) = -A$. □

It follows from the preceeding proposition that both of these algebras are finite dimensional and that their common dimension is equal to the dimension of G.

3.1 The Lie algebra and the exponential map

Let $\mathfrak{g}$ denote the tangent space of G at the group identity e. Invariant vector fields on G induce a Lie bracket on $\mathfrak{g}$ by the following rule.

Definition 3.5 If A and B are elements of the Lie algebra $\mathfrak{g}$ then their Lie bracket $[A, B]$ is defined by

$$[A, B] = [X_A, X_B](e)$$

where X_A and X_B denote the left-invariant vector fields such that $X_A(e) = A$ and $X_B(e) = B$.

Remark 3.6 The Lie bracket could be defined also by right-invariant vector fields in which case the Lie bracket $[A, B]$ would be the negative of the one defined by the left-invariant vector fields.

The Jacobi identity on $\mathfrak{g}$ follows from the Jacobi identity in the Lie algebra of vector fields. In fact,

$$[X_A, [X_B, X_C]](e) + [X_C, [X_A, X_B]](e) + [X_B, [X_C, X_A]](e) = 0$$

reduces to

$$[A, [B, C]] + [C, [A, B]] + [B, [C, A]] = 0.$$

It follows that $\mathfrak{g}$ is a Lie algebra under the vector addition and the Lie bracket defined above. It is called the Lie algebra of G and will be denoted by $\mathfrak{g}$.

Proposition 3.7 *The Lie algebra $gl_n(R)$ of $GL_n(R)$ is equal to $M_n(R)$, the vector space of all $n \times n$ matrices with real entries, with the Lie bracket given by $[A, B] = BA - AB$.*

Proof Since $GL_n(R) \subset M_n(R)$, $gl_n(R) \subseteq M_n(R)$. Any matrix A in $M_n(R)$ defines a one-parameter group of left translations $\Phi_t(g) = ge^{tA}$. Its infinitesimal generator is the left-invariant vector field $X(g) = gA$. This shows that $gl_n(R) = M_n(R)$.

We showed in (1.10) in Chapter 1 that the Lie bracket conforms to the following formula:

$$\frac{\partial^2}{\partial t \partial s} \exp -tX \circ \exp sY \circ \exp tX|_{t=s=0} = [X, Y].$$

So if $X(g) = gA$ and $Y(g) = gB$ are any left-invariant fields then according to the above formula

$$\frac{\partial^2}{\partial t \partial s} e^{-tA} e^{sB} e^{tA}|_{t=s=0} = [X, Y](e) = [A, B]$$

The above yields $[A, B] = (BA - AB)$.   □

Remark 3.8 The commutator $[A, B]$ of matrices A and B is usually denoted by $AB - BA$ rather than its negative. Our choice is dictated by the choice of the sign in the definition of the Lie bracket. Each choice is natural in some situations and less natural in others. For instance, in the case of linear fields $X(x) = Ax$ and $Y(x) = Bx$ in R^n the choice $]X, Y] = Y \circ X - X \circ Y$ implies that $[X, Y](x) = (AB - BA)x$, which seems more natural than its negative. It is imperative to maintain the conventions initially chosen, otherwise confusion is bound to follow.

Proposition 3.9 *Each left (right)-invariant vector field X is analytic and complete.*

The flow $\Phi : G \times R \to G$ of a left-invariant vector field $X(g)$ is given by $\Phi(g, t) = ge(t)$, where $e(t)$ is a one-parameter abelian subgroup of G.

Proof The fact that invariant vector fields are analytic follows directly from the analyticity of the group multiplication. To show completeness, let X be a left-invariant vector field and let $\xi(t)$ be the left-translate $L_g(\sigma(t))$ of an integral curve $\sigma(t)$ of X. Then,

$$\frac{d}{dt}\xi(t) = (L_g)_* \frac{d\sigma}{dt} = (L_g)_* X \circ \sigma(t) = X \circ L_g(\sigma(t)) = X \circ \xi(t).$$

Therefore $\xi(t)$ is also an integral curve of X.

The fact that the integral curves of X are invariant under the left translations has several direct consequences. Firstly, it implies that the integral curve through any point g is equal to the left-multiple $ge(t)$ of the integral curve $e(t)$ through the group identity e. Secondly, it implies that the integral curves are defined for all t for the following reasons:

Let $I = [-a, a]$ denote an interval such that $e(t)$ is defined for all $t \in I$. Such an interval exists as a consequence of the local existence of solutions of differential equations. Let e^+ denote the positive escape time for $e(t)$. Suppose that $e^+ < \infty$. Let T be any number such that $0 < e^+ - T < \frac{a}{2}$. Then,

$$\sigma(t) = \begin{cases} e(t), t \leq T \\ e(T)e(t), t \in I \end{cases}$$

is an integral curve of X that is defined on the interval $[0, e^+ + \frac{a}{2}]$. By the uniqueness of solutions of differential equations, $e(t) = \sigma(t)$, contradicting the finiteness of e^+. The last part follows from the general property $\Phi_{t+s}(e) = \Phi_t \circ \Phi_s(e) = \Phi_s \circ \Phi_t(e)$. □

If X is a left-invariant vector field $X(g) = (L_g)_* A$ then its integral curve through the group identity will be denoted by e^{At}. Since $\{e^{At} : t \in \mathbb{R}\}$ is a one-parameter abelian subgroup of G, e^{At} agrees with the exponential of a matrix when G is a linear group of matrices, that is,

$$e^{At} = I + At + \frac{t^2}{2}A^2 + \cdots + \frac{t^n}{n!}A^n + \cdots . \tag{3.1}$$

For a general Lie group G the above formula is to be interpreted functionally in terms of the convergent series

$$\hat{e} \circ \exp tX = e + e \circ tX + \ldots e \circ \frac{t^n}{n!}X^n + \cdots .$$

Since $e \circ X^n f = e \circ A^n(f)$ for all n we may write

$$e \circ \exp tX = e^{tA} = I + At + \frac{t^2}{2}A^2 + \cdots + \frac{t^n}{n!}A^n + \cdots ,$$

with the understanding that both sides of the formula act on functions.

Definition 3.10 If $\mathfrak{h}$ is a Lie subalgebra of $\mathfrak{g}$ then $\exp \mathfrak{h} = \{e^A : A \in \mathfrak{h}\}$ is called the exponential of $\mathfrak{h}$.

Proposition 3.11

(a) *If $\mathfrak{g} = \mathfrak{k} \oplus \mathfrak{p}$ then $(\exp \mathfrak{k})(\exp \mathfrak{p})$ contains a neighborhood U of the identity in G.*
(b) *There is an open neighborhood U of e such that $U \subseteq \exp \mathfrak{g}$.*
(c) *If G is compact and connected then $\exp \mathfrak{g} = G$.*

Proof Let $A_1, \ldots, A_k$ denote a basis in $\mathfrak{k}$, and let $B_1, \ldots, B_p$ denote a basis in $\mathfrak{p}$. The mapping

$$F(s_1, \ldots, s_p, t_1, \ldots, t_k) = e^{A(t_1, \ldots, t_k)} e^{B(s_1, \ldots, s_p)}$$

where $A(t_1, \ldots, t_k) = \sum_{i=1}^{k} t_i A_i$ and $B(s_1 \ldots, s_p) = \sum_{i=1}^{p} s_i B_i$ satisfies

$$\left.\frac{\partial F}{\partial t_i}\right|_{s=t=0} = A_i, i = 1, \ldots, k \text{ and } \left.\frac{\partial F}{\partial s_i}\right|_{s=t=0} = B_i, i = 1, \ldots, p$$

It follows from the inverse function theorem that the range of F covers a neighborhood U of $F(0,0) = e$. Therefore, parts (a) and (b) follow. The proof of Part (c) will be deferred to other sections of the book. □

Example 3.12 Let $B = \begin{pmatrix} -1 & a \\ 0 & -1 \end{pmatrix}$ where $a \neq 0$. Then there is no matrix A such that $e^A = B$ for the following reason: $e^{\frac{1}{2}A} e^{\frac{1}{2}A} = e^A$, hence e^A is a square of a matrix. It is easy to check that B is not a square of any matrix. However, if $a = 0$, then $A = \begin{pmatrix} 0 & \pi \\ -\pi & 0 \end{pmatrix}$ is such that $e^A = B$.

3.2 Lie subgroups

Definition 3.13 A subgroup H of a Lie group G is a Lie subgroup of G if H is a submanifold of G and the inclusion map $i : H \to G$ is a group homomorphism.

Proposition 3.14 *Let $\mathcal{F}$ denote any family of left-invariant vector fields on a Lie group G. Then the orbit of $\mathcal{F}$ through the group identity is a Lie subgroup H of G. The orbit of $\mathcal{F}$ through any other point g is the left coset gH. The Lie algebra $\mathfrak{h}$ of H is equal to the evaluation $Lie_e(\mathcal{F})$.*

Proof This proposition is a paraphrase of the Orbit theorem for families of left-invariant vector fields. When $\mathcal{F}$ is a family of left-invariant vector fields, then $Lie(\mathcal{F})$ is a finite-dimensional algebra of left-invariant vector fields in G. Since the left-invariant vector fields are analytic, the tangent space at e of the orbit $\mathcal{O}_{\mathcal{F}}(e)$ is equal to $\mathfrak{h} = Lie_e(\mathcal{F})$. If $H = \mathcal{O}_{\mathcal{F}}(e)$, then the points of H are of the form

$$g = e^{A_m t_m} e^{A_{m-1} t_{m-1}} \ldots e^{A_1 t_1}$$

where $A_1, \ldots, A_m$ are in $\mathfrak{h}$ and where $t_1, \ldots, t_m$ are arbitrary real numbers. Evidently, H is a Lie subgroup of G and $\mathcal{O}_{\mathcal{F}}(g) = g\mathcal{O}_{\mathcal{F}}(e)$. □

Corollary 3.15 *Let G be a Lie group with its Lie algebra $\mathfrak{g}$. Each subalgebra $\mathfrak{h}$ of $\mathfrak{g}$ is the Lie algebra of a connected Lie subgroup H of G.*

Proof Let $\mathcal{F}$ denote the family of left-invariant vector fields with values $\mathfrak{h}$ at e. Then H is equal to the orbit of $\mathcal{F}$ through e. □

Proposition 3.16 *Any closed subgroup H of a Lie group G is a Lie subgroup of G with its topology equal to the relative topology inherited from G.*

Proof Let $\mathfrak{h} = \{A : \exp tA \in H\}$. For each element $A \in \mathfrak{h}$, let X_A denote the associated left-invariant vector field and let $\mathcal{F}$ denote the family vector fields X_A with $A \in \mathfrak{h}$. Then the orbit $\mathcal{O}_\mathcal{F}(e)$ of $\mathcal{F}$ is a Lie subgroup in G by Proposition 3.14. Let K denote the orbit $\mathcal{O}_\mathcal{F}(e)$. Evidently, $K \subseteq H$.

Let $\mathfrak{k}$ denote the Lie algebra of K. For each element $A \in \mathfrak{k}$, $\exp tA$ belongs to K hence, belongs to H for all t. Since each A in $\mathfrak{h}$ belongs to $\mathfrak{k}$, $Lie(\mathfrak{h})$ also belongs to $\mathfrak{k}$. Therefore, $Lie(\mathfrak{h}) = \mathfrak{h} = \mathfrak{k}$.

It remains to show that K is open in H when H is topologized by the relative topology inherited from G. Let H_0 denote the connected component of H through the group identity. Since K is connected, $K \subseteq H_0$. We will now show that K is open in H_0 from which it would follow that $K = H_0$.

Recall that the orbit topology is finer than the relative topology inherited from G. It suffices to show that for any neighborhood O of the identity, open in its orbit topology of K, there exists an open neighborhood U of the identity in G such that $U \cap H$ is contained in O. For then, every neighborhood of e that is open in the orbit topology would be also open in the relative topology, and the same would be true for any other open set in K since it is a group translate of an open neighborhood of the group identity.

Let $\mathfrak{p}$ denote any subspace of $\mathfrak{g}$ that is transversal to $\mathfrak{h}$, that is $\mathfrak{g} = \mathfrak{h} \oplus \mathfrak{p}$. If V in $\mathfrak{h}$ and W in $\mathfrak{p}$ are any neighborhoods of the origin then, according to Proposition 3.11, $(\exp V)(\exp W)$ contains an open neighborhood of the identity in G. Moreover, V can be chosen so that $\exp V \subseteq O$.

Suppose that $\exp V$ does not contain any sets of the form $H \cap U$ with U a neighborhood of e in G. Let V_n and W_n denote sequences of open sets in V and W that shrink to 0. Then there exist sequences of points h_n in V_n and p_n in W_n such that $p_n \neq 0$ and $\exp h_n \exp p_n$ belongs to H for all n. Since $\exp h_n$ belongs to H, $\exp p_n$ belongs to H as well.

To get the contradiction we will follow the argument used in [Ad] and assume that $\mathfrak{p}$ is equipped with a norm $||, ||$ (any vector space can be equipped with a Euclidean norm). The sequence $\left\{ \frac{p_n}{||p_n||} \right\}$ contains a convergent subsequence. There is no loss in generality if we assume that the sequence $\left\{ \frac{p_n}{||p_n||} \right\}$ itself converges to a point p with $||p|| = 1$.

The fact that $\{p_n\}$ converges to 0 implies that for any $t \neq 0$ there exist integers m_n such that

$$\frac{1}{m_n} < \frac{||p_n||}{t} < \frac{1}{m_n + 1}.$$

Evidently $m_n \to \infty$ and hence, $\lim_{n\to\infty} ||p_n|| m_n = t$. But then

$$(\exp p_n)^{m_n} = \exp m_n p_n = \exp m_n ||p_n|| \frac{p_n}{||p_n||} \to \exp tp.$$

This means that $\exp tp$ belongs to H, since $(\exp p_n)^{m_n}$ is in H, and H is closed. This contradicts the fact that p is in $\mathfrak{p}$.

Therefore, $\exp V$ contains a relatively open set $H \cap U$. Since $\exp V \subseteq O$, $H \cap U \subseteq O$, and hence, the orbit topology on K coincides with the relative topology inherited from G. This implies that K is both open and closed in H_0, and consequently it must be equal to H_0. But then the connected component through any other point $g \in H$ is equal to gK, and our proof is finished. $\square$

As a corollary of the preceeding proposition the following linear groups are Lie subgroups of the group $GL_n(R)$:

1. **The special linear group** $SL_n(R)$. Let $SL(\mathbb{R}^n)$ denote the group of all volume preserving linear automorphism of the Euclidean space $\mathbb{R}^n$ and let $Sl_n(\mathbb{R})$ be the subgroup of $Gl_n(\mathbb{R})$ of non-singular matrices T such that $Det(T) = 1$. Any oriented basis $e_1, \ldots, e_n$ sets up an isomorphism between $SL(\mathbb{R}^n)$ and $SL_n(R)$. Since $Det(e^{At}) = e^{Tr(A)t}$, it follows that the Lie algebra $sl_n(R)$ of $Sl_n(\mathbb{R})$ is equal to the vector space of all $n \times n$ matrices of zero trace.

2. **The orthogonal group** $O_n(R)$. Assume that $\mathbb{E}$ is a Euclidean vector space with $\langle x, y \rangle$ denoting its Euclidean scalar product. The orthogonal group denoted by $O(\mathbb{E})$ is a subgroup of all linear automorphisms T of $\mathbb{E}$ that satisfy $\langle Tx, Ty \rangle = \langle x, y \rangle$ for all x and y in $\mathbb{E}$. Any orthonormal basis in $\mathbb{E}$ sets up an isomorphism between $O(\mathbb{E})$ and the orthogonal group $O_n(\mathbb{R})$ consisting of $n \times n$ matrices T with real entries that satisfy $T^{-1} = T^*$, where T^* denotes the matrix transpose of T. Each matrix $T \in O_n(\mathbb{R})$ satisfies

$$1 = Det(TT^{-1}) = Det(TT^*) = Det(T)Det(T^*) = Det^2(T)$$

The subgroup of matrices with determinant equal to 1 is called the special group of rotations and is denoted by $SO_n(\mathbb{R})$. It is equal to the connected component of $O_n(\mathbb{R})$ through the identity matrix I. The Lie algebra of $O_n(\mathbb{R})$ is denoted by $so_n(\mathbb{R})$ and consists of the $n \times n$ skew-symmetric matrices. Therefore, the dimension of $O(\mathbb{E})$ is equal to $\frac{n(n-1)}{2}$.

3. **The Lorentzian group** $SO(p, q)$. The Lorentzian group, denoted by $SO(p, q)$ is the subgroup of $GL_n(R)$ that leaves the Lorentzian quadratic form $\langle x, y \rangle = \sum_i^p x_i p_i - \sum_{i=p+1}^n x_i p_i$ invariant. If $J_{p,q}$ is the matrix

$$\left(\begin{array}{cc} I_p & 0 \\ 0 & I_q \end{array} \right), q = n - p,$$

where I_p and I_q denote p- and q-dimensional identity matrices, then T belongs to $SO(p,q)$ if and only if $T^*JT = J$. The Lie algebra of $SO(p,q)$ is denoted by $so(p,q)$. Matrices A belong to $so(p,q)$ if and only if $A^*J + JA = 0$. An easy calculation shows that $so(p,q)$ consist of the matrices

$$A = \left(\begin{array}{cc} a & b \\ b^T & c \end{array} \right)$$

with a and c skew-symmetric $p \times p$ and $q \times q$ matrices and b an arbitrary $p \times q$ matrix. Here, b^T denotes the matrix transpose of b.

4. **The symplectic group** Sp_n. The quadratic form $[,]$ in $\mathbb{R}^{2n}$ defined by $[(x,p),(y,q)] = \sum_{i=1}^{n} q_i x_i - p_i y_i$ is called symplectic. The symplectic form can be related to the Euclidean quadratic form $(,)$ via the formula

$$[(x,p),(y,q)] = ((x,p),J(y,q)), J = \left(\begin{array}{cc} 0 & I \\ -I & 0 \end{array} \right).$$

The subgroup of $Gl_{2n}(\mathbb{R})$ that leaves the symplectic form invariant is called symplectic and is denoted by Sp_n. It follows that a matrix M is in Sp_n if and only if

$$M^T JM = J.$$

A matrix A belongs to the Lie algebra $sp_n(R)$ if and only if $e^{At} \in Sp_n$, or equivalently, if and only if $A^T J + JA = 0$. Therefore, matrices A in the Lie algebra sp_n have the following block form:

$$A = \left(\begin{array}{cc} a & b \\ c & -a^T \end{array} \right),$$

with a an arbitrary $n \times n$ matrix and b and c symmetric $n \times n$ matrices. Hence the dimension of Sp_n is equal to $n^2 + n(n + 1) = 2n^2 + n$.

5. **The unitary group** U_n. Let $\mathbb{E}$ denote a dimensional complex vector space with a Hermitian inner product $\langle z, w \rangle$. The unitary group $U(\mathbb{E})$ is the subgroup of complex linear automorphisms T that satisfy $\langle Tz, Tw \rangle = \langle z, w \rangle$ for all z, w in $\mathbb{E}$. Every orthonormal basis sets up an isomorphism between $U(\mathbb{E})$ and the space of $n \times n$ matrices T with complex entries that satisfy $T^{-1} = T^*$, where T^* denotes the Hermitian transpose of T. That is, the entries of T^{-1} are given by $\bar{T}_{ji}$, where T_{ij} denote the entries of T and where $\bar{a}$ stands for the complex conjugate of a. Then it follows that

$$1 = Det(TT^{-1}) = Det(TT^*) = Det(T)Det(T^*)$$
$$= Det(T)Det(\bar{T}) = |Det(T)|^2.$$

Therefore, the determinant of each member of the group lies on the unit circle, which accounts for its name.

The group of complex $n \times n$ matrices whose inverses are equal to their Hermitian transposes is denoted by U_n. The Lie algebra u_n consists of skew-Hermitian matrices A, that is, matrices A such that $A^* = -A$ with A^* denoting the Hermitian transpose of A. Any such matrix can be written as $A = B + iC$ with B equal to the real part of A and C equal to the imaginary part of A. It follows that B is skew-symmetric and C is symmetric. Note that u_n is a real Lie algebra and not a complex Lie algebra (iB is Hermitian, for any skew-Hermitian matrix B, hence does not belong to u_n). The dimension of u_n is equal to $\frac{n(n-1)}{2} + \frac{n(n+1)}{2} = n^2$.

Elements of U_n can be also represented by $2n \times 2n$ matrices with real entries. This representation, denoted by $U_{2n}(\mathbb{R})$ is obtained by treating $\mathbb{C}^n$ as $\mathbb{R}^{2n}$. In fact, $e_1, \ldots, e_n, ie_1, \ldots, ie_n$ is a basis for the real vector space obtained from $\mathbb{E}$ by restricting the scalars to real numbers. Then each point v of $\mathbb{E}$ is represented by $2n$ coordinates $(x_1, \ldots, x_n, p_1 \ldots, p_n)$ written more compactly as (x, p). Then the Hermitian product $\langle v, w \rangle$ of $v = (x, p)$ and $w = (y, q)$ is equal to

$$\langle v, w \rangle = (u, v) + i[v, w],$$

with $(\,,\,)$ the Euclidean quadratic form and $[\,,\,]$ the symplectic form. Each complex linear transformation T is real linear, and if T preserves the Hermitian product on $\mathbb{E}$ then its matrix $\hat{T}$ relative to the basis $e_1, \ldots, e_n, ie_1, \ldots, ie_n$ preserves both of the above quadratic forms. It then follows that

$$U_{2n}(\mathbb{R}) = O_{2n}(\mathbb{R}) \cap Sp_{2n}.$$

The subgroup of U_n of elements with determinant equal to 1 is denoted by SU_n and is called the special unitary group. Its Lie algebra is denoted by su_n. The case $n=2$ is particularly special since then SU_2 can be identified with the sphere S^3. The identification is simple: each matrix $\begin{pmatrix} z & w \\ -\bar{w} & \bar{z} \end{pmatrix}$ in SU_2 is identified with a point (z, w) on S^3 via the determinant $|z|^2 + |w|^2 = 1$.

6. **Semi-direct products: the Euclidean group of motions $SE_n(R)$.** Any Lie group K that acts linearly on a finite-dimensional topological vector space

V defines the semi-direct product $G = V \rtimes K$ consisting of points in $V \times K$ and the group operation

$$(v, S)(w, T) = (v + Sw, ST)$$

for all (v, S) and (w, T) in $V \times K$.

It follows that G is a group with the group identity e equal to $(0, I)$, where I is the identity in K, and the group inverse $(v, S)^{-1} = (-S^{-1}v, S^{-1})$. Topologized by the product topology of $V \times K$, G becomes a Lie group whose dimension is equal to $dim(V) + dim(K)$.

Each semi direct product $G = V \ltimes K$ acts on V by $(v, S)(x) = v + Sx$. The semi-direct product of a Euclidean space $\mathbb{E}^n$ with the group of rotations $O(\mathbb{E}^n)$ is called *the group of motions* of $\mathbb{E}^n$ because its action preserves the Euclidean distance. An orthonormal basis $e_1, \ldots, e_n$ in $\mathbb{E}^n$ identifies each point x of $\mathbb{E}^n$ with a column vector of coordinates $\begin{pmatrix} x_1 \\ \vdots \\ x_n \end{pmatrix}$ in $\mathbb{R}^n$, and identifies each $R \in O(\mathbb{E}^n)$ with the matrix (R_{ij}) in $O_n(\mathbb{R})$ via the formula $Re_i = \sum_{j=1}^{n} R_{ij}e_j$ for $i = 1, \ldots, n$.

The correspondence between elements (x, R) of $\mathbb{E}^n \ltimes O(\mathbb{E}^n)$ and the column vectors in $\mathbb{R}^n$ and matrices in $O_n(R)$ defines an isomorphism between $\mathbb{E}^n \rtimes O(\mathbb{E}^n)$ and the semi-direct product $\mathbb{R}^n \rtimes O_n(\mathbb{R})$. This isomorphism defines an embedding of $\mathbb{E}^n \ltimes O(\mathbb{E}^n)$ into $GL_{n+1}(\mathbb{R})$ by identifying (x, R) with a matrix $g = \begin{pmatrix} 1 & 0 \\ x & R \end{pmatrix}$ in $GL_{n+1}(\mathbb{R})$. The connected component of $\mathbb{E}^n \ltimes O(\mathbb{E}^n)$ that contains the group identity is equal to $\mathbb{E}^n \ltimes SO(\mathbb{E}^n)$ and is denoted by $SE_n(R)$.

3.3 Families of left-invariant vector fields and accessibility

Suppose now that $\mathcal{F}$ is a family of either right- or left-invariant vector fields in a Lie group G. The evaluation of $\mathcal{F}$ at the identity defines a set $\Gamma = \{X(e) : X \in \mathcal{F}\}$ in the Lie algebra $\mathfrak{g}$, which is called the trace of $\mathcal{F}$. We will use $Lie(\Gamma)$ to denote the Lie algebra generated by Γ. It follows that $Lie_e(\mathcal{F}) = Lie(\Gamma)$.

Then the orbit of $\mathcal{F}$ through the group identity e consists of elements g that can be written as

$$g = e^{t_p A_p} \cdots e^{t_2 A_2} e^{t_1 A_1}, \tag{3.2}$$

for some elements $A_1, A_2, \ldots, A_p$ in Γ and real numbers $t_1, t_2, \ldots, t_p$.

This orbit is a connected Lie subgroup K whose Lie algebra is $Lie(\Gamma)$. The orbit of $\mathcal{F}$ through any other point g_0 is either the right-translate Kg_0 or the left-translate g_0K depending whether $\mathcal{F}$ is right- or left-invariant. It follows that there is one orbit of $\mathcal{F}$ whenever $Lie(\Gamma) = \mathfrak{g}$, and G is connected.

For the purposes of controllabiity, however, it is the action of the semigroup $S(\mathcal{F})$ generated by $\{\exp tX : t \geq 0, X \in \mathcal{F}\}$ that is relevant. The reachable set $\mathcal{A}_\mathcal{F}(e)$ is the semigroup $S(\Gamma)$ consisting of points $g \in G$ that can be written in the form (3.2) but with $t_1 \geq 0$, $t_2 \geq 0, \ldots, t_p \geq 0$ and the reachable set $\mathcal{A}_\mathcal{F}(g_0)$ is either the right or the left translate of g_0 by $S(\Gamma)$ (depending whether $\mathcal{F}$ is right- or left-invariant).

It follows that $\mathcal{F}$ is controllable if and only if $S(\Gamma) = G$. This means that $Lie(\Gamma) = \mathfrak{g}$ is a necessary condition of controllability.

This question of sufficiency can be paraphrased in slightly more geometric terms by passing to the Lie saturate of $\mathcal{F}$, which means that Γ could be replaced by $LS(\Gamma) = LS_e(\mathcal{F})$. On the basis of Proposition 2.23 in Chapter 2 we could assume that $LS(\Gamma)$ enjoys the following properties:

1. $\Gamma + \Gamma \subseteq LS(\Gamma)$ and $R^+\Gamma \subseteq LS(\Gamma)$, and
2. If V is any vector subspace of $LS(\Gamma)$ then $e^{ad(V)}(\Gamma) \subseteq LS(\Gamma)$. This property implies that the largest vector subspace in $LS(\Gamma)$ is a Lie subalgebra of $\mathfrak{g}$.

Subsets of Lie algebras with these properties are called wedges in the literature on Lie semi-groups [Hg]. The largest vector subpace in a wedge is called an edge. The reachable sets e are the semigroups in G generated by $\{e^X : X \in \Gamma\}$. These semigroups will be denoted by $\mathcal{S}(\Gamma)$.

It follows that controllability questions of families of invariant vector fields reduce to finding conditions on Γ such that $\mathcal{S}(\Gamma) = G$. The simplest cases are given by our next theorem.

Proposition 3.17 *If G is either a compact and connected Lie group, or if G is a semi-direct product of a vector space V and a compact and connected Lie group K that admits no fixed non-zero points in V, then $\mathcal{S}(\Gamma) = G$ for any subset Γ such that $Lie(\Gamma) = \mathfrak{g}$.*

We will only prove the compact case. For the semi-direct products the reader is referred to [BJ].

Proof Assume first that G is compact and connected. The positive limit set $\Omega^+(A)$ of an element $A \in \mathfrak{g}$ is defined to be the closure of the set of all points X in G such that $e^{t_nA} \to X$ for some sequence $\{t_n\}$ such that $t_n \to \infty$. The positive limit set is non-empty when G is compact.

Let $h = \lim_{t_n \to \infty} e^{t_n A}$. There is no loss in generality if $\{t_n\}$ is chosen so that $t_{n+1} - t_n \to \infty$. Then,

$$\lim_{n \to \infty} e^{(t_{n+1} - t_n))A} = \lim_{n \to \infty} e^{t_{n+1} A} \lim_{n \to \infty} e^{t_n A} = h^{-1} h = e.$$

Hence, $e \in \Omega^+(A)$. It follows that $e^{-tA} = e^{-tA} e = e^{-tA} \lim_{t_n \to \infty} e^{t_n A} = \lim_{n \to \infty} e^{(t_n - t)A}$. Therefore, the closure of $\exp \Gamma$ contains $\exp \pm \Gamma$, and the latter is equal to G whenever $Lie(\Gamma) = \mathfrak{g}$. Hence, $S(\Gamma)$ is dense in G and therefore equal to G by Proposition 2.8 in Chapter 2. □

3.4 Homogeneous spaces

A Lie group G is said to act on a manifold M if there exists a smooth mapping $\phi : G \times M \to M$ such that:

1. $\phi(e, x) = x$ for all $x \in M$, where e denotes the group identity in G, and
2. $\phi(g_2 g_1, x) = \phi(g_2, \phi(g_1, x))$ for all g_1, g_2 in G and all $x \in M$.

Alternatively, group actions may be considered as the groups of transformations $\{\Phi_g, g \in G\}$ where $\Phi_g(x) = \phi(g, x)$, for $x \in M$, subject to $(\Phi_g)^{-1} = \Phi_{g^{-1}}$ and $\Phi_g \cdot \Phi_h = \Phi_{gh}$ for all g and h in G. The set $\{\Phi_g(x) : g \in G\}$ is called the orbit through x. The group G is said to act transitively on M if there is only one orbit, i.e., if for any pair of points x and y in M there is an element g in G such that $\Phi_g(x) = y$.

A manifold M together with a transitive action by a Lie group G is called *homogeneous*. Homogeneous spaces can be identified with the orbit through an arbitrary point x in which case M can be considered as the quotient spaces G/K where K is the isotropy group of the base point x, i.e., $K = \{g \in G : \Phi_g(x) = x\}$. The base space M is termed homogeneous because the isotropy groups corresponding to different base points are conjugate.

A Lie group G acts on any subgroup K by either the right action $(g, h) \to hg^{-1}$ or by the left action $(g, h) \to gh$. So homogenous spaces can be identified with the quotients G/K with K a closed subgroup of G.

3.4.1 Examples

Example 3.18 Positive-definite matrices The left action of $G = SL_n(R)$ on $K = SOn(R)$ identifies the space of left cosets gK with the quotient $SL_n(R)/SO_n(R)$. The space of left cosets can be identified with positive matrices $\mathcal{P}_n$ by the following argument. Every matrix $S \in G$ can be written in polar form as $S = PR$, where P is a positive matrix and R is a rotation. In fact, $P = \sqrt{SS^*}$ and $R = P^{-1}S$. Suppose now that $S_1 K = S_2 K$. Then $S_1 S_2^{-1}$ is a rotation. Hence, $(S_1 S_2^{-1})^{-1} = (S_1 S_2^{-1})^*$ which implies that $S_1 S_1^* = S_2 S_2^*$. Therefore the correspondence $SK \to \sqrt{SS^*}$ is one to one. Then $\mathcal{P}_n$ is topologized so that this correspondence is a homeomorphism.

The case $n = 2$ is special. Elements $\begin{pmatrix} a & b \\ c & d \end{pmatrix} \in SL_2(R)$ act on the points in the upper-half plane by the Moebius tranformations

$$\begin{pmatrix} a & b \\ c & d \end{pmatrix}(z) = \frac{az+b}{cz+d}.$$

It follows that $SO_2(R)$ is the isotropy subgroup associated with $z = i$. Since $x + iy = S(i)$ with $S = \begin{pmatrix} \sqrt{y} & \frac{x}{\sqrt{y}} \\ 0 & \frac{1}{\sqrt{y}} \end{pmatrix}$, the action is transitive. The above realizes the upper-half plane $\{z \in \mathbb{C} : Im(z) > 0\}$ as the quotient $SL_2(R)/SO_2(R)$.

Example 3.19 The generalized upper-half plane The generalized upper-half plane $\mathcal{H}_n^+$ consists of $n \times n$ complex matrices Z of the form $Z = X + iY$ with X and Y $n \times n$ matrices with real entries , X symmetric, and Y positive. It can be shown that the symplectic group Sp_n acts transitively on $\mathcal{H}_n^+$ by the fractional transformations

$$\begin{pmatrix} A & B \\ C & D \end{pmatrix}(Z) = (AZ+B)(CZ+D)^{-1}$$

and that the isotropy group of $Z_0 = iI$ is equal to SU_n. In this notation, $g \in Sp_n$ is written in the block form $g \begin{pmatrix} A & B \\ C & D \end{pmatrix}$. Therefore,

$$\mathcal{H}_n^+ = Sp_n/SU_n.$$

For $n = 2$, $Sp_2 = SL_2(R)$ and $SU_2 = SO_2(R)$. Hence, this formula agrees with the one obtained above.

Example 3.20 The spheres and the hyperboloids Any subgroup G of $GL_{n+1}(R)$ acts on column vectors in $\mathbb{R}^{n+1}$ by the matrix multiplications on the left. In the case that $G = SO_{n+1}(R)$ then this action preserves the Euclidean norm and therefore, the orbit through any non-zero vector x_0 is the sphere $||x|| = ||x_0||$. Most commonly, x_0 is taken to be e_1, in which case the isotropy group K consists of matrices of the form $\begin{pmatrix} 1 & 0 \\ 0^T & Q \end{pmatrix}$, where 0 denotes the zero row vector in $\mathbb{R}^n$, 0^T the column vector Q an arbitrary matrix in $SO_n(R)$. Evidently, the isotropy group is isomorphic to $SO_n(R)$, and thus

$$S^n = SO_{n+1}(R)/SO_n(R).$$

If the Euclidean inner product is replaced by the Lorentzian inner product $\langle x, y \rangle = x_1 y_1 - \sum_{i=2}^{n+1} x_i y_i$ then $||x||^2 = \langle x, x \rangle = 1$ is the hyperboloid $\mathbb{H}^n = \{x : x_1^2 - \sum_{i=2}^{n+1} = 1\}$. The group $SO(1, n)$ acts transitively on $\mathbb{H}^n$ and the isotropy subgroup of $x_0 = e_1$ is isomorphic to $SO_n(R)$, hence

$$\mathbb{H}^n = SO(1, n)(R)/SO_n(R).$$

Example 3.21 The Grassmannians The set of k-dimensional subspaces in an n-dimensional Euclidean space is denoted by $G(n, k)$. Each S in $G(n, k)$ can be identified with the orthogonal reflection P_S defined by $P_S(x) = x, , x \in S$, $P_S(x) = -x$, $x \in S^\perp$. Then $G(n, k)$ is topologized by the finest topology in which the correspondence $S \rightarrow P_S$ is a homeomorphism.

The group of rotations $O_n(R)$ acts on the reflections P_S by the conjugation, that is, $(T, P_S) \rightarrow TP_ST^*$. The action is transitive and the isotropy $\{T \in O_n(R) : TP_ST^* = P_S\}$ consists of the rotations that satisfy $T(S) = S$ and $T(S^\perp) = S^\perp$. This subgroup is isomorphic with $O_{n-k} \times O_k(R)$, hence

$$G(n, k) = O_n(R)/O_{n-k} \times O_k(R).$$

Example 3.22 The Steifel manifolds St_r The Stiefel manifold St_r consist of ordered orthonormal vectors $x_1, \ldots, x_r$ in a Euclidean vector space $\mathbb{E}^n$. The points of St_r can be represented by $r \times n$ matrices X with columns the coordinates of $x_1, \ldots, x_r$ relative to a fixed orthonormal basis in $\mathbb{R}^n$. Such matrices form an $nr - \frac{1}{2}r(r+1)$-dimensional manifold considered as points in $\mathbb{R}^{nr}$ subject to r^2 symmetric conditions $x_i x_j = \delta_{ij}$. If R is an element of $SO_n(R)$ and if X is an $n \times r$ matrix of orthonormal vectors, then RX is also a matrix of orthonormal vectors. This action is transitive, since any two $n \times r$ matrices X and Y can be completed to elements $\bar{X}$ and $\bar{Y}$ in $SO_n(R)$ in which case $R = \bar{X}(\bar{Y})^{-1}$ takes X onto Y. Therefore, St_r can be identified with the orbit of $SO_n(R)$ through the point $X_0 = (e_1, \ldots, e_r)$ modulo the isotropy group of X_0. Evidently, this isotropy group is isomorphic to $SO_{n-r}(R)$, hence

$$St_r = SO_n(R)/SO_{n-r}(R).$$

4

Symplectic manifolds: Hamiltonian vector fields

It is a common practice in applied mathematics to mix Hamiltonian vector fields with the underlying Riemannian structure and express Hamiltonian vector fields as the "skew-gradients" of functions. This tendency to identify tangent bundles with the cotangent bundles via the metric obscures the role of symplectic structure in Hamiltonian systems and also obscures the nature of symmetries present in Hamiltonian systems. For instance, on a given manifold there are many Riemannian structures, but there is only one canonical symplectic form on its cotangent bundle. So the passage from functions to Hamiltonian fields via the symplectic form is intrinsic, while the passage from functions to skew-gradients is not. For that reason we will be somewhat more formal in our introduction of Hamiltonian systems and will limit our discussion at first to vector spaces before going on to general manifolds.

4.1 Symplectic vector spaces

A finite-dimensional vector space V together with a bilinear form $\omega : V \times V \to \mathbb{R}$ is called *symplectic* if:

 (i) ω is skew-symmetric, i.e., $\omega(v, w) = \omega(w, v)$ for all v and w in V.
(ii) ω is non-degenerate, i.e., $\omega(v, w) = 0$ for all $w \in V$ can hold only for $v = 0$.

If $a_1, a_2, \ldots, a_n$ is a basis in V then $\omega(v, w) = \sum_{i=1}^{n} \sum_{j=1}^{n} w_i v_j \omega(a_i, a_j)$ where $w = \sum_{i=1}^{n} w_i a_i$ and $v = \sum_{i=1}^{n} v_i a_i$. The matrix A with entries $\omega(a_i, a_j)$ is skew-symmetric and hence its eigenvalues are imaginary. It follows that any symplectic space V must be even dimensional since A must be non-singular because of the non-degeneracy of ω.

The symplectic form provides means of identifying V with its dual. The correspondence

$$v \in V \to \omega(v, \cdot) \in V^*$$

is a linear isomorphism between V and V^*. A linear subspace S of a symplectic vector space V is called *isotropic* if ω vanishes on S, i.e., if $\omega(v, w) = 0$ for all v and w in S. Any one-dimensional subspace of V is isotropic. A linear subspace L of V is called *Lagrangian* if L is isotropic and not a proper linear subspace of any isotropic subspace of V.

A basis $a_1, \ldots, a_n, b_1, \ldots, b_n$ is called *symplectic* if

$$\omega(a_i, a_j) = \omega(b_i, b_j) = 0, \text{ and } \omega(a_i, b_j) = \delta_{ij}$$

for all i and j. We will use $\langle a_1, \ldots, a_k \rangle$ to designate the linear span of any vectors $a_1, \ldots, a_k$. It then follows that each of the spaces $L_1 = \langle a_1, \ldots, a_n \rangle$ and $L_2 = \langle b_1, \ldots, b_n \rangle$ is Lagrangian corresponding to any symplectic basis $a_1, \ldots, a_n, b_1, \ldots, b_n$.

A linear subspace S is said to be a *symplectic subspace of V* if the restriction of the symplectic form ω to S is non-degenerate. If S is any subset of V then $S^\perp = \{v : \omega(v, w) = 0, \text{ for all } w \in S\}$.

Proposition 4.1 *Let S be any linear subspace of V. Then* $\dim(S^\perp) + \dim(S) = \dim(V)$, *and*

$$(S^\perp)^\perp = S.$$

Proof The mapping $L : v \subset V \to \omega(v, \cdot)|_S \subset S^*$ is surjective, and its kernel is $S^\perp$. Since

$$\dim(ker(L)) + \dim(Range(L)) = \dim(V),$$

$$\dim(S^\perp) + \dim(S^*) = \dim(S^\perp) + \dim(S) = \dim(V) = \dim((S^\perp)^\perp) + \dim(S^\perp),$$

it follows that $\dim(S) = \dim(S^\perp)^\perp$. Since $S \subseteq (S^\perp)^\perp$, $S = (S^\perp)^\perp$. $\qquad \square$

Proposition 4.2 *Suppose that $a_1, \ldots, a_k, b_1, \ldots, b_k$ are any vectors in a symplectic space V that satisfy $\omega(a_i, a_j) = \omega(b_i, b_j) = 0, \omega(a_i, b_j) = \delta_{ij}$ for all $i \leq k$ and $j \leq k$. Let W denote the linear span of $a_1, \ldots, a_k, b_1, \ldots, b_k$. Then each of W and $W^\perp$ are symplectic subspaces of V and*

$$V = W \oplus W^\perp.$$

Proof If $w = \sum_{i=1}^{k} \alpha_i a_i + \beta_i b_i$ belongs to $W \cap W^\perp$ then

$$\alpha_i = \omega(w, b_i) = 0 \text{ and } \beta_i = \omega(a_i, w) = 0$$

for all i. Hence, $W \cap W^\perp = \{0\}$. Then it follows from Proposition 4.1 that $V = W \oplus W^\perp$. Evidently, the symplectic form must be non-degenerate on each factor and hence each of W and $W^\perp$ are symplectic.   □

Proposition 4.3   *Suppose that S is any isotropic subspace of V. Let $a_1, a_2, \ldots, a_k$ be any basis in S. Then there are vectors $b_1, b_2, \ldots, b_k$ in V such that*

$$\omega(a_i, a_j) = \omega(b_i, b_j) = 0, \text{ and } \omega(a_i, b_j) = \delta_{ij}$$

for all i and j.

Proof   Let S_1 denote the linear span of $a_2, \ldots, a_k$. Then a_1 cannot belong to $(S_1^\perp)^\perp$ by Proposition 4.1. Therefore there exists $b \in S_1^\perp$ such that $\omega(a_1, b) \neq 0$. Let $b_1 = \frac{1}{\omega(a_1, b)} b$. By Proposition 4.2, $V = W_1 \oplus W_1^\perp$ where $W_1 = \langle a_1, b_1 \rangle$. Since S_1 is contained in $W_1^\perp$ and $W_1^\perp$ is symplectic, the above procedure can be repeated with $S_2 = \langle a_3, \ldots, a_k \rangle$ to obtain b_2. The procedure stops at the k stage.   □

Corollary 4.4   *Each Lagrangian subspace of a 2n-dimensional symplectic space V is n dimensional.*

Corollary 4.5   *Every basis $a_1, \ldots, a_n$ of a Lagrangian space L corresponds to a symplectic basis $a_1, \ldots, a_n, b_1, \ldots, b_n$.*

Every symplectic basis $a_1, \ldots, a_n, b_1, \ldots, b_n$ gives rise to symplectic coordinates $(x_1, \ldots, x_n, p_1 \ldots, p_n)$ defined by $v = \sum_{i=1}^{n} x_i a_i + p_i b_i$ for any vector $v \in V$. If $(y_1, \ldots, y_n, q_1 \ldots, q_n)$ denote the symplectic coordinates of a vector w in V then

$$\omega(v, w) = \sum_{i=1}^{n} x_i q_i - y_i p_i. \tag{4.1}$$

We will use (x, p) and (y, q) to denote the symplectic coordinates $(x_1, \ldots, x_n, p_1, \ldots, p_n)$ and $(y_1, \ldots, y_n, q_1, \ldots, q_n)$ of any two points in V. If $((x, p), (y, q))$ designates the standard Euclidean product in $\mathbb{R}^{2n}$, then (4.1) can be written as $\omega(v, w) = ((x, p), J(y, q))$ where

$$J = \begin{pmatrix} 0 & I \\ -I & 0 \end{pmatrix},$$

with I equal to the $n \times n$ identity matrix.

The matrix J corresponds to the linear transformation defined by $J(a_i) = b_i$ and $J(b_i) = -a_i$ for each i. Then $J^2 = -I$ and hence J is the complex structure on V. Thus V together with J becomes an n-dimensional complex space with

$(\alpha+i\beta)v = \alpha v + \beta J(v)$ for any complex number $z = \alpha + i\beta$ and any vector $v \in V$. Then $a_1, \ldots, a_n$ is a basis for the complexification of V and $x_1 + ip_1, \ldots, x_n + ip_n$ are the complex coordinates of v relative to this basis.

The symplectic form then can be identified with the imaginary part of the Hermitian inner product on V given by

$$\langle x + ip, y + iq \rangle = \sum_{j=1}^{n} (x_j y_j + p_i q_j) + i(q_j x_j - p_j y_j). \tag{4.2}$$

Symplectic vector spaces arise naturally in the following setting. Let E^* designate the dual of a real n-dimensional vector space E. Then $V = E \times E^*$ has a natural symplectic structure given by

$$\omega((x,f),(y,g)) = g(x) - f(y) \tag{4.3}$$

for all (x,f) and (y,g) in V. Then both E and E^* can be embedded in V as $E \times \{0\}$ and $\{0\} \times E^*$, in which case they become Lagrangian subspaces, called horizontal and vertical Lagrangians. Any basis $a_1, \ldots, a_n$ in E gives rise to the dual basis $a_1^*, \ldots, a_n^*$ in E^*, and together they constitute a symplectic basis in V.

4.2 The cotangent bundle of a vector space

Any n-dimensional real vector space E is an n-dimensional manifold with $(x_1, \ldots, x_n)$ the global coordinates of a point x induced by a basis $a_1, \ldots, a_n$. Then it is natural to identify the tangent vectors at $x = \sum_{i=1}^{n} x_i a_i$ with $v = \sum_{i=1}^{n} \frac{dx_i}{dt} a_i$, for if $x(\epsilon)$ is a curve such that $x(0) = x_0$ then the induced tangent vector v at x_0 is identified with $v = \sum_{i=1}^{n} \frac{dx_i}{d\epsilon}(0)a_i$. In this context, the tangent space $T_x(E)$ is identified with the affine space $\{(x,v) : v \in E\}$. In this identification

$$T_x(E) = (x,0) + T_0(E),$$

with

$$\alpha(x,v) + \beta(x,w) = (x, \alpha v + \beta w).$$

Hence, the tangent space at the origin plays a special role: tangent vectors at x need to be first translated to the origin before they can be scaled or added.

Since $T_0(E)$ is equal to $(0) \times E$ it is then natural to identify the cotangent space $T_0^*(E)$ with $(0) \times E^*$ and the cotangent space at any point x with the affine space $\{(x,p) : p \in E^*\}$. If $a_1^*, \ldots, a_n^*$ denotes the dual basis corresponding to the basis $a, \ldots, a_n$ then every cotangent vector $(0, p)$ can be written as $p = \sum_{i=1}^{n} p_i a_i^*$, in which case the natural pairing between vectors and covectors is given by

$$\langle (x, p), (x, v) \rangle = \sum_{i-1}^{n} p_i v_i.$$

The above implies that the tangent bundle $T(E)$ and the cotangent bundle $T^*(E)$ are identified with $E \times E$ and $E \times E^*$. With this formalism at our disposal, then the tangent bundle $T(T^*(E))$ is identified with

$$(E \times E^*) \times (E \times E^*),$$

with the understanding that the first two factors designate the base point (x, p) in $T^*(E)$ and the second factors designate the tangent vectors $(\dot{x}, \dot{p})$ at this point.

There is a natural pairing between covectors p and tangent vectors $\dot{x}$ as well as between tangent vectors $\dot{p}$ and $\dot{x}$. This pairing gives rise to the following differential forms on T^*E.

Definition 4.6 Let $(\dot{x}_1, \dot{p}_1)$ and $(\dot{x}_2, \dot{p}_2)$ designate arbitrary tangent vectors at a point (x, p) in $T^*(E)$. Differential form θ defined by $\theta_{(x,p)}(\dot{x}, \dot{p}) = p(\dot{x})$ is called the Liouville form. If $d\theta$ denotes its exterior derivative, then $\omega = -d\theta$ is called the symplectic form on T^*E.

Coordinates $x_1, \ldots, x_n, p_1 \ldots, p_n$ relative to the bases $a_1, \ldots, a_n, a_1^*, \ldots, a_n^*$ are called symplectic. This choice of coordinates induces coordinates on tangent vectors $\dot{x}$ and $\dot{p}$ with $\dot{x} = \sum_{i=1}^{n} \dot{x}_i a_i$ and $\dot{p} = \sum_{i=1}^{n} \dot{p}_i a_i^*$. These relations could be expressed also by dual notation as $\dot{x} = \sum_{i=1}^{n} \dot{x}_i \frac{\partial}{\partial x_i}$ and $\dot{p} = \sum_{i=1}^{n} \dot{p}_i \frac{\partial}{\partial p_i}$. It follows that in these coordinates

$$\theta = \sum_{i=1}^{n} p_i dx_i \text{ and } \omega = \sum_{i=1}^{n} dx_i \wedge dp_i, \tag{4.4}$$

which further implies that

$$\omega_{(x,p)}(v, w) = \dot{q}(\dot{x}) - \dot{p}(\dot{y}). \tag{4.5}$$

for any tangent vectors $v = (\dot{x}, \dot{p})$ and $w = (\dot{y}, \dot{q})$ at a point (x, p) in $E \times E^*$. Expression (4.5) shows that ω is non-degenerate over each point $(x, p) \in E \times E^*$. Hence, ω defines a symplectic structure in the vector space of tangent vectors $E \times E^*$ over each base point (x, p). It is clear that the linear span of $(a_1, 0), \ldots, (a_n, 0)$ is a Lagrangian subspace of

$E \times E^*$ and so is the linear span of $(0, a_1^*), \ldots, (0, a_n^*)$. Combined vectors $(a_1, 0), \ldots, (a_n, 0), (0, a_1^*), \ldots, (0, a_n^*)$ form a symplectic basis for $E \times E^*$.

Definition 4.7 Vector field $\vec{H}$ is called Hamiltonian vector fields if there is a function H such that $dH(v) = \omega(\vec{H}, v)$ for all tangent vectors v.

Every function H induces a Hamiltonian vector field $\vec{H}$ given by

$$\vec{H} = \sum_{i=1}^{n} \frac{\partial H}{\partial p_i} \frac{\partial}{\partial p_i} - \frac{\partial H}{\partial x_i} \frac{\partial}{\partial x_i}$$

in each choice of symplectic coordinates.

The integral curves $(x(t), p(t))$ of $\vec{H}$ are conveniently written as

$$\frac{dx}{dt}(t) = \frac{\partial H}{\partial p}(x(t), p(t)), \frac{dp}{dt}(t) = -\frac{\partial H}{\partial x}(x(t), p(t)) \qquad (4.6)$$

with the understanding that it is a shorthand notation for the differential system

$$\frac{dx_i}{dt}(t) = \frac{\partial H}{\partial p_i}(x_1(t), \ldots, x_n(t), p_1(t), \ldots, p_n(t)) \text{ and}$$

$$\frac{dp_i}{dt}(t) = -\frac{\partial H}{\partial x_i}(x_1(t), \ldots, x_n(t), p_1(t), \ldots, p_n(t)) \qquad (4.7)$$

for all $i = 1, \ldots, n$.

4.3 Symplectic manifolds

We now extend the previous formalism to arbitrary cotangent bundles. The fundamental setting is the same as it was for cotangent bundles of vector spaces; the symplectic form is the exterior derivative of the differential form of Liouville. To explain in more detail, let π denote the natural projection from T^*M onto M carrying each covector ξ at x to its base point x. Then for each curve $\Sigma(t)$ in T^*M that originates at ξ at $t = 0$, the projected curve $\sigma(t) = \pi \circ \Sigma(t)$ originates at $x = \pi \circ \xi$.

The tangent map π_* is a linear mapping from $T_\xi(TM)$ onto T_xM that carries $\frac{d\Sigma}{dt}(0)$ onto $\frac{d\sigma}{dt}(0)$. The Liouville form θ is equal to the dual mapping π^* : $T^*M \to T(T^*M)$, defined by

$$\pi^*(\xi))\left(\frac{d\Sigma}{dt}(0)\right) = \xi \circ \pi_*\left(\frac{d\Sigma}{dt}(0)\right) = \xi \circ \frac{d\sigma}{dt}(0).$$

Thus θ maps a covector ξ at T_x^*M onto a covector θ_ξ at $T_\xi^*(T^*M)$ such that

$$\theta_\xi\left(\frac{d\Sigma}{dt}(0)\right) = \xi\left(\frac{d\sigma}{dt}(0)\right)$$

for each tangent vector $\frac{d\Sigma}{dt}(0)$ in $T_\xi(T^*M)$. This somewhat abstract definition of θ takes on the familiar form when expressed in local coordinates. Then, $\theta_\xi = \sum_{i=1}^n v_i p_i$, where $v_1, \ldots, v_n$ denote the coordinates of a tangent vector v relative to the basis $\frac{\partial}{\partial x_1}, \ldots, \frac{\partial}{\partial x_n}$ and $p_1, \ldots, p_n$ denote the coordinates of a covector at x relative to the dual basis $dx^1, \ldots, dx^n$. Therefore, $\theta = \sum_{i=1}^n p_i dx_i$ as in the previous section.

Definition 4.8 The symplectic form ω on T^*M is equal to the negative of the exterior derivative $d\theta$.

It then follows from (1.3) in Chapter 1 that

$$\omega_\xi(V(\xi), W(\xi)) = W(\theta(V) - V(\theta(W) - \theta([V, W]) \tag{4.8}$$

for any vector fields V and W on T^*M.

In symplectic coordinates, vector fields $\frac{\partial}{\partial x_1}, \ldots, \frac{\partial}{\partial x_n}, \frac{\partial}{\partial p_1}, \ldots, \frac{\partial}{\partial p_n}$ form a basis for tangent vectors on an open set on T^*M. Hence, it is sufficient to consider pairs of vector fields (V, W) of the form $V = \sum_{i=1}^n \dot{x}_i \frac{\partial}{\partial x_i} + \dot{p}_i \frac{\partial}{\partial p_i}$ and $W = \sum_{i=1}^n \dot{y}_i \frac{\partial}{\partial x_i} + \dot{q}_i \frac{\partial}{\partial p_i}$ for some points $(x_1, \ldots, x_n, p_1, \ldots, p_n)$ and $(y_1, \ldots, y_n, q_1, \ldots, q_n)$ in $\mathbb{R}^{2n}$. Then $[V, W] = 0$, and moreover, $\theta(V) = \sum_{i=1}^n p_i \dot{x}_i$ and $W(\theta(V)) = \sum_{i=1}^n \dot{q}_i \dot{x}_i$. Similarly, $V(\theta(W)) = \sum_{i=1}^n \dot{p}_i \dot{y}_i$. Hence, ω in (4.8) is given by

$$\omega = \sum_{i=1}^n \dot{q}_i \dot{x}_i - \dot{p}_i \dot{y}_i,$$

which agrees with expression (4.5) of the previous section.

Definition 4.9 If h is any function on M, then vector field $\vec{h}$, defined by

$$dh_\xi(X(\xi)) = \omega_\xi(\vec{h}(\xi), X(\xi))$$

for all tangent vectors $X(\xi) \in T_\xi M$, is the Hamiltonian vector field generated by h.

The above can be stated more succinctly in terms of the interior product $i_{\vec{h}}$ as $dh = i_{\vec{h}}$. The interior product denoted by i_X is a contraction of a differential form ω with a vector field X. It maps k-exterior forms $\Lambda^k(M)$ into $(k-1)$-forms in $\Lambda^{k-1}(M)$ by the formula

$$(i_X\omega)(X_1, \ldots, X_{k-1}) = \omega(X, X_1, \ldots, X_{k-1}), \tag{4.9}$$

with the understanding that the contraction of a zero form is zero.

It follows that Hamiltonian vector fields $\vec{h}$ are given by

$$\vec{h} = \sum_{i=1}^{n} \frac{\partial h}{\partial p^i} \frac{\partial}{\partial x^i} - \frac{\partial h}{\partial x^i} \frac{\partial}{\partial p^i} \tag{4.10}$$

in any choice of local coordinates, from which it follows that the integral curve of $\vec{h}$ are the solutions of

$$\frac{dx^i}{dt} = \frac{\partial h}{\partial p^i}, \frac{dp^i}{dt} = -\frac{\partial h}{\partial x^i}, i = 1, \ldots, n. \tag{4.11}$$

Cotangent bundles are a particular case of a more general class of manifolds called symplectic manifolds [Ar].

Definition 4.10 A manifold M together with a smooth, non-degenerate, closed 2-form ω on M is called symplectic. Any such 2-form is called symplectic. As in the case of the cotangent bundle, a vector field $\vec{h}$ on M is called Hamiltonian if there exists a function h on M such that $dh = i_{\vec{h}}$. The set of Hamiltonian vector fields on M will be denoted by $Ham(M)$.

Every symplectic manifold is even dimensional as a consequence of the non-degeneracy of the symplectic form.

Proposition 4.11 Darboux's theorem *Let (M, ω) denote a symplectic 2n-dimensional manifold. At each point of M there exists a coordinate neighborhood U with coordinates $(x_1, \ldots, x_n, p_1, \ldots, p_n)$ such that vectors*

$$\frac{\partial}{\partial x_1}, \ldots, \frac{\partial}{\partial x_n}, \frac{\partial}{\partial p_1}, \ldots, \frac{\partial}{\partial p_n} \tag{4.12}$$

form a symplectic basis on U. In this basis, the Hamiltonian vector fields are given by

$$\vec{f} = \sum_{i=1}^{n} \frac{\partial f}{\partial p_i} \frac{\partial}{\partial x_i} - \frac{\partial f}{\partial x_i} \frac{\partial}{\partial p_i}.$$

Proof We need to show that at any point q of M there exist a neighborhood U of q and a basis of vector fields $E_1, \ldots, E_n, E_{n+1}, \ldots, E_{2n}$ on U such that

$$\omega(E_i, E_j) = \omega(E_{n+i}, E_{n+j}) = 0, 1 \leq i, j \leq n, \omega(E_i, E_{n+j}) = \delta_{ij}.$$

Then, $x_1, \ldots, x_n, p_1, \ldots, p_n$ are the coordinates defined by the contractions

$$i_{E_i}\omega = dx_i, \ i_{E_{n+i}} = dp_i, i = 1, \ldots, n,$$

relative to which the vector fields are as in (4.12), and ω is given by

$$\omega = dx_1 \wedge dp_1 + \cdots + dx_n \wedge dp_n.$$

The proof is by induction on the dimension of M. Assume that the proposition is true for $dim(M) \leq 2(n-1)$ and let M be a $2n$-dimensional manifold with a symplectic form ω.

The proof is the same as in Proposition 4.3 in Section 4.1. Let E_1 be any vector field such that $E_1(q) \neq 0$. Then there exists a vector field V such that $\omega(E_i, V) \neq 0$, by the non-degeneracy condition. Let $E_{n+1} = \frac{1}{\omega(E_1, V)}V$. If $n = 1$, the proof is finished. Otherwise, pass to the skew-orthogonal complement $S^\perp = \{V : \omega(E_1, V) = \omega(V, E_{n+1}) = 0\}$. Since ω is non-degenerate on $S^\perp$, the induction hypothesis then implies that $S^\perp$ admits a basis with the above properties. □

It then follows that the Hamiltonian vector fields that correspond to the coordinate functions $f_i = p_i, f_{n+i} = -x_i, i = 1, \ldots, n$, span the tangent space at point of U. Therefore, each orbit of $Ham(M)$ is open and coincides with M whenever the latter is connected.

Our next propositions make use of the following generalization of the Lie derivative.

Definition 4.12 Let X denote a vector field on M and ω a k form in $\Lambda^k(M)$, with $k > 0$. Designate by Φ_ϵ the one-parameter group of diffeomorphisms induced by X. Then the Lie derivative $L_X\omega$ of ω along X is defined by the following formula:

$$(L_X\omega)_\xi(v_1, \ldots, v_k) = \frac{d}{d\epsilon}\omega_{\Phi_\epsilon(\xi)}((\Phi_\epsilon)_* v_1, \ldots, \Phi_{\epsilon*}v_k)|_{\epsilon=0}.$$

Evidently, the Lie derivative coincides with the directional derivative of a function along X. For general forms, the following remarkable formula holds.

Proposition 4.13 Cartan's formula   *The Lie derivative L_X is given by*

$$L_X = d \circ i_X + i_X \circ d,$$

where d denotes the exterior derivative and i_X is the contraction along X.

The proof of this proposition can be found in [AS].

Definition 4.14 If F is a diffeomorphism M and if ω is any form in $\Lambda^k(M)$, then $\hat{F}\omega$ is the form defined by

$$(\hat{F}\omega)_\xi(v_1, \ldots, v_k) = \omega_{F(\xi)}(F_*v_1, \ldots, F_*v_k).$$

A form ω is said to be invariant under a diffeomorphism F if $\hat{F}\omega = \omega$.

Definition 4.15 Let (M, ω) denote a symplectic manifold. A diffeomorphism that leaves the symplectic form ω invariant is called a symplectomorphism. The set of symplectomorphisms on M will be denoted by $Symp(M)$.

Evidently, $Symp(M)$ is a subgroup of the group of diffeomorphisms $Diff(M)$.

Proposition 4.16 *Let (M, ω) be a symplectic manifold. Then,*

(a) $\exp t\vec{h}$ is a symplectomorphism for each h in $C^\infty(M)$ and for each t in $\mathbb{R}$.
(b) $F \circ \vec{g} \circ F^{-1} = \vec{Fg}$ for each $F \in Symp(M)$ and each function g.
(c) $(\exp t\vec{h})\vec{g}(\exp -t\vec{h})$ belongs to $Ham(M)$ for each $t \in \mathbb{R}$ and all functions f and g.

Proof Part (a) follows from Cartan's formula $L_{\vec{h}}\omega = d \circ i_{\vec{h}}\omega + i_X \circ d\omega = d \circ dh = 0$. Part (b) follows from the calculation below:

$$\omega_\xi(X, \vec{Fg}F^{-1}) = \hat{F}\omega_\xi(X, \vec{Fg}F^{-1})$$
$$= \omega_{F(\xi)}(F^{-1} \circ X \circ F, \vec{g})$$
$$= F \circ (F^{-1} \circ X \circ F)g = X(Fg)$$
$$= \omega_\xi(X, (\vec{Fg}))$$

Therefore, $F \circ \vec{g} \circ F^{-1} = \vec{Fg}$. Part (c) is evident since $(\exp t\vec{h})$ belongs to $Symp(M)$. $\qquad\Box$

Definition 4.17 The Poisson bracket $\{f, g\}$ of any two functions f and g in $C^\infty(M)$ is defined by

$$\{f, g\}(\xi) = \frac{d}{dt}(\exp t\vec{g})f(\xi)|_{t=0} = (\vec{g}f)(\xi) = \omega_\xi(\vec{f}(\xi), \vec{g}(\xi)), \xi \in M.$$

Evidently, the Poisson bracket is skew-symmetric. Furthermore,

$$\{f_1f_2, g\}(\zeta) = \vec{g}(f_1f_2)(\xi) - f_1(\xi)\{f_2, g\}(\xi) + f_2(\xi)\{f_1, g\}(\xi).$$

Therefore, the Poisson bracket acts on functions as a derivation.

Proposition 4.18 *If $F \in Symp(M)$ then $F \circ \{f, g\} = \{Ff, Fg\}$ for any functions f and g.*

Proof

$$F \circ \{f, g\} = F \circ \omega(\vec{f}, \vec{g}) = \hat{F}\omega(F \circ \vec{f} \circ F^{-1}, F \circ \vec{g} \circ F^{-1})$$
$$= \omega((F \circ \vec{f} \circ F^{-1}, F \circ \vec{g} \circ F^{-1})$$
$$= \omega(\vec{Ff}, \vec{Fg}) = \{Ff, Fg\}$$

$\qquad\Box$

Then $\frac{d}{dt}(\exp t\vec{h})\{f, g\}|_{t=0} = \frac{d}{dt}\{(\exp t\vec{h})f, (\exp t\vec{h})g\}|_{t=0}$ reduces to the Jacobi identity:

$$\{f, \{g, h\}\} + \{h, \{f, g\}\} + \{g, \{h, f\}\} = 0.$$

Corollary 4.19 $C^\infty(M)$ *is a Lie algebra under the Poisson bracket, that is, the Poisson bracket is bilinear, skew-symmetric, and satisfies the Jacobi's identity.*

Proposition 4.20 *Ham(M) is a Lie subalgebra of Vec(M), and the correspondence*

$$f \to \vec{f}$$

is a Lie algebra homomorphism from $C^\infty(M)$ onto Ham(M). The kernel of this homomorphism consists of constant functions, whenever M is connected.

Proof Let f and g be arbitrary functions on M. Then,

$$[\vec{f}, \vec{g}]h = \vec{g} \circ \vec{f}(h) - \vec{f} \circ \vec{g}(h) = \{\{h,f\}, g\} - \{\{h,g\},f\} = \{\{f,g\}, h\} = \vec{\{f,g\}}h.$$

Therefore, $[\vec{f}, \vec{g}] = \vec{\{f,g\}}$. This shows that $Ham(M)$ is a Lie algebra, and the correspondence $f \to \vec{f}$ a Lie algebra homomorphism.

In any system of symplectic coordinates, $\vec{f}$ is given by

$$\vec{f} = \sum_{i=1}^{n} \frac{\partial f}{\partial p_i} \frac{\partial}{\partial x_i} - \frac{\partial f}{\partial x_i} \frac{\partial}{\partial p_i}.$$

If $\vec{f} = 0$, then $\frac{\partial f}{\partial p_i} = 0$ and $\frac{\partial f}{\partial x_i} = 0$ and therefore f is constant in each such coordinate neighborhood. But then f is constant when M is connected. $\square$

5

Poisson manifolds, Lie algebras, and coadjoint orbits

The correspondence between functions and vector fields is articulated more naturally by the Poisson bracket rather than the symplectic form. This subtle distinction leads to the notion of a Poisson manifold [Wn]. The shift from symplectic to Poisson structure leads to remarkable discoveries fundamental for the theory of integrable systems and the geometry of Lie groups. Let us begin with the basic concepts.

5.1 Poisson manifolds and Poisson vector fields

Definition 5.1 A Poisson bracket on a manifold M is a mapping

$$\{,\} : C^\infty(M) \times C^{\infty(M)} \to C^\infty(M)$$

that is bilinear and skew-symmetric, and satisfies

$$\{fg, h\} = f\{g, h\} + g\{f, h\},$$
$$\{f, \{g, h\}\} + \{\{h, \{f, g\}\} + \{g, \{h, f\}\} = 0,$$

for all functions f, g, h in $C^\infty(M)$. A manifold M together with a Poisson bracket $\{,\}$ is called a Poisson manifold.

Definition 5.2 If $(M, \{,\})$ is a Poisson manifold then vector field $\vec{f}$ defined by $\vec{f}(g) = \{g, f\}$ will be called a Poisson vector field. The set of all Poisson vector fields will be denoted by $Poiss(M)$.

It is an easy consequence of the Jacobi identity that $[\vec{f}, \vec{g}] = \{\vec{f}, g\}$ for all functions f and g, which implies that the Poisson vector fields form a Lie subalgebra of $Vec(M)$. Moreover, the mapping

$$f \to \vec{f}$$

is a homomorphism from the Poisson algebra C^∞ onto the Lie subalgebra $Poiss(M)$ in $V^\infty(M)$.

Proposition 5.3   *Let* $(M, \{,\})$ *be a symplectic manifold. and let* $\vec{h}$ *be a Poisson vector field on M. Then,*

$$\exp t\vec{h}\{f, g\} = \{(\exp t\vec{h})f, (\exp t\vec{h})g\},$$

for all functions f and g.

Proof   Let $\phi(t) = \exp -t\vec{h}(\{(\exp t\vec{h})f, (\exp t\vec{h})g\}$. Then, $\frac{d\phi}{dt}$ is equal to

$$\exp -t\vec{h}(-\vec{h}\{(\exp t\vec{h})f + (\exp t\vec{h})g\} + \frac{d}{dt}\{(\exp t\vec{h})f + (\exp t\vec{h})g\})$$

$$= \exp -t\vec{h}(-\vec{h}\{(\exp t\vec{h})f + (\exp t\vec{h})g\} + \vec{h}\{(\exp t\vec{h})f + (\exp t\vec{h})g\}) = 0$$

because of the Jacobi identity. Therefore, $\phi(t) = \{f, g\}$ for all t.   □

Proposition 5.4   *Let f and g be any functions on a Poisson manifold M. Then* $(\exp t\vec{f})\vec{g}(\exp -t\vec{f})$ *is a Poisson vector field that corresponds to* $(\exp t\vec{f})g$.

Proof   Let h denote a function on M. Then,

$$(\exp t\vec{f})\vec{g}(\exp -t\vec{f})h = \exp t\vec{f}\{(\exp -t\vec{f})h, g\} = \{h, (\exp t\vec{f})g\}.$$

□

Proposition 5.5   *Let* $(M, \{,\})$ *be a Poisson manifold and let* $\mathcal{F}$ *denote the family of all Poisson vector fields on M. Then each orbit of* $\mathcal{F}$ *is a symplectic submanifold of M.*

Proof   Let N denote an orbit of $\mathcal{F}$. It follows from (1.15) in Chapter 1 that the tangent space at each point z of N is a linear combination of vectors

$$\hat{z} \circ (\exp -t_1\vec{h}_1 \circ \cdots \exp -t_m\vec{h}_m)\vec{g}(\circ \exp t_m\vec{h}_m \circ \exp t_{m-1}\vec{h}_{m-1} \circ \cdots \exp t_1\vec{h}_1)$$

for some functions $g, h_1, \ldots, h_m$ on M. According to Proposition 5.4,

$$(\exp -t_1\vec{h}_1 \circ \cdots \exp -t_m\vec{h}_m)\vec{g}(\circ \exp t_m\vec{h}_m \circ \exp t_{m-1}\vec{h}_{m-1} \circ \cdots \exp t_1\vec{h}_1) = \vec{h},$$

where $h = (\exp -t_1\vec{h}_1 \circ \cdots \exp -t_m\vec{h}_m)g$. Hence each tangent space T_zN of N is equal to the evaluation $\mathcal{F}_z$ of $\mathcal{F}$ at z.

To show that N is symplectic, let $\{,\}_N$ denote the restriction of the Poisson bracket to N. Recall that all smooth functions on N are the restrictions of smooth function on M. It follows $(N, \{,\}_N)$ is a Poisson submanifold of $(M, \{,\})$. Define

$$\omega_\xi(\vec{f}(\xi), \vec{g}(\xi)) = \{f, g\}_N(\xi) \tag{5.1}$$

for all points ξ in N.

To verify that ω is non-degenerate, suppose that $\omega_\xi(\vec{f}(\xi), \vec{g}(\xi)) = 0$ for some function g, and all functions f on N. Since $\omega_\xi(\vec{f}(\xi), \vec{g}(\xi)) = \{f, g\}_N(\xi) = \vec{g}(f)(\xi)$ it follows that $\vec{g} f = 0$ for any function on N. But this means that $\vec{g} = 0$ and consequently, ω is non-degenerate. Since the Jacobi identity holds on N, ω is closed. $\qquad\square$

The above proposition may be restated in the language of distributions as follows.

Proposition 5.6 *The distribution $\mathcal{D}$ defined by $\mathcal{D}(\xi) = \{\vec{f}(\xi) : f \in C^\infty(M)\}$ is integrable and each maximal integral manifold of $\mathcal{D}$ is symplectic.*

We will refer to the maximal integral manifolds of $\mathcal{D}$ as the symplectic leaves of M.

Corollary 5.7 *A connected Poisson manifold M is symplectic if and only if the family of all Poisson vector fields has only one orbit in M.*

5.2 The cotangent bundle of a Lie group: coadjoint orbits

Let us now incorporate certain group symmetries into the symplectic structure of the cotangent bundle of a Lie group G. Each Lie group G admits a global frame of either left- or right-invariant vector fields, which implies that both the tangent bundle TG and the cotangent bundle T^*G can be written as the products $G \times \mathfrak{g}$ and $G \times \mathfrak{g}^*$, where $\mathfrak{g}^*$ denotes the dual of the Lie algebra $\mathfrak{g}$ of G.

In fact, every basis $A_1, \ldots, A_n$ in the Lie algebra $\mathfrak{g}$ of a Lie group G defines a global frame of either left- or right-invariant vector fields $X_1, \ldots, X_n$ such that $X_i(e) = A_i$ for each $i = 1, \ldots, n$. We will work with left-invariant vector fields and identify the tangent bundle TG with $G \times \mathfrak{g}$. This identification is done as follows: points $(g, A) \in G \times \mathfrak{g}$ are identified with vectors $V(g) \in T_g(G)$ via the formula

$$V(g) = (L_g)_*(A).$$

Similarly, points (g, ℓ) of $G \times \mathfrak{g}^*$ are identified with points $\xi \in T_g^*(G)$ via the formula $L_g^* \xi = \ell$, that is, ξ and ℓ are related by $\xi(V(g)) = \ell(L_{g^{-1}} V(g))$ for every tangent vector $V(g)$ at g. In particular, $\xi(V(g)) = \ell(A)$ for left-invariant vector fields $V(g) = (L_g)_*(A)$, and $\xi(V(g)) = \ell((L_{g_*}^{-1} R_{g_*}(A))$ for any right-invariant vector field $V(g) = R_{g_*}(A)$.

Each product $G \times \mathfrak{g}$ and $G \times \mathfrak{g}^*$ is a product of a Lie group with a vector space, and hence is a Lie group in its own right. On $G \times \mathfrak{g}^*$ the group operation is given by

$$(g_1, \ell_1)((g_2, \ell_2) = (g_1 g_2, \ell_1 + \ell_2).$$

Having identified T^*G with the product $G \times \mathfrak{g}^*$, the tangent bundle of T^*G will be identified with the product $(G \times \mathfrak{g}^*) \times (\mathfrak{g} \times \mathfrak{g}^*)$, with the understanding that the second product denotes the tangent vectors at the base points described by the first product. Relative to this decomposition, vector fields on T^*G will be written as pairs $(X(g, \ell), Y(g, \ell))$ with $X(g, \ell) \in \mathfrak{g}$, $Y(g, \ell) \in \mathfrak{g}^*$, and (g, ℓ) the base point in $G \times \mathfrak{g}^*$. In particular, the left-invariant vector fields V on $G \times \mathfrak{g}^*$ are then represented by the pairs $(A, \ell) = V(e, 0)$. It follows that

$$\Phi_t(g_0, \ell_0) = \{(g_0 e^{tA}, \ell_0 + t\ell) : (g_0, \ell_0) \in G \times \mathfrak{g}^*\}$$

is the one-parameter group generated by a left-invariant vector field V on $G \times \mathfrak{g}^*$ whose value at the identity is equal to (A, ℓ).

Let us now evaluate the canonical symplectic form on T^*G in the representation $G \times \mathfrak{g}^*$. In this representation the Liouville's differential form θ is given by

$$\theta_{(g,\ell)}((X(g, \ell), Y(g, \ell)) = \ell(X(g, \ell)).$$

When V is left-invariant, then $\theta_{(g,\ell)}(V) = \ell(A)$, where A denotes the projection of $V(e, 0)$ on $\mathfrak{g}$.

If V_1 and V_2 are any left-invariant fields such that $V_1(e, 0) = (A_1, \ell_1)$ and $V_2(e, 0) = (A_2, \ell_2)$ then $\exp V_i t(g_0, \ell_0) = (g_0 \exp tA_i, \ell_0 + t\ell_i), i = 1, 2$. Therefore,

$$V_2(\theta(V_1))(g, \ell) = \frac{d}{dt}(\theta_{\exp tV_2(g,\ell)}(V_1)|_{t=0} = \frac{d}{dt}(\ell + t\ell_2)(A_1)|_{t=0} = \ell_2(A_1),$$

and similarly, $V_1 \theta(V_2)(g, \ell) = \ell_1(A_2)$. Moreover, $\theta_{(g,\ell)}([V_1, V_2]) = \ell([A_1, A_2])$.

Let us now return to the symplectic form $\omega = -d\theta$. We have

$$\omega_{(g,\ell)}(V_1, V_2) = -d\theta_{(g,\ell)}(V_1, V_2) = V_2(\theta(V_1))(g, \ell)$$
$$- V_1(\theta(V_2))(g, \ell) - \theta_{(g,\ell)}([V_1, V_2])$$
$$= \ell_2(A_1) - \ell_1(A_2) - \ell([A_1, A_2])$$

We will refer to

$$\omega_{(g,\ell)}((A_1, l_1), (A_2, l_2)) = l_2(A_1) - l_1(A_2) - \ell([A_1, A_2]) \qquad (5.2)$$

as the left-invariant representation of ω.

The left-invariant representation is different from the one obtained through the canonical coordinates, hence the usual formulas $\frac{dg}{dt} = \frac{\partial H}{\partial \ell}, \frac{d\ell}{dt} = -\frac{\partial H}{\partial g}$ are no longer valid. The correct expressions are obtained by identifying a function

H on $G \times \mathfrak{g}^*$ with its Hamiltonian vector field $\vec{H}((g, \ell)) = (A(g, \ell), a(g, \ell))$ through the formula

$$\omega_{(g,\ell)}\left(A(g, \ell), a(g, \ell)\right), (B, b)) = dH_{(g,\ell)}(B, b). \tag{5.3}$$

Then,

$$dH_{g,\ell}(B, 0) = \frac{d}{dt}H(g \exp(Bt), \ell)|_{t=0} = \partial H_g \circ L_{g_*}(B)$$

and

$$dH_{g,\ell}(0, b) = \frac{d}{dt}H(g, \ell + tb))|_{t=0} = \partial H_l(b).$$

Since $\partial H_l(g, \ell)$ is a linear function on $\mathfrak{g}^*$, it is naturally identified with an element of $\mathfrak{g}$. It follows that

$$A(g, \ell) = \partial H_l(g, \ell) \text{ and } l(g, \ell) = -\partial H_g(g, \ell) \circ L_{g_*} - ad^* \partial H_l(g, \ell)(\ell),$$

and therefore the integral curves $(g(t), \ell(t))$ of $\vec{H}$ are the solution curves of

$$\frac{dg}{dt}(t) = g(t)\partial H_l(g(t), \ell(t)),$$

$$\frac{d\ell}{dt}(t) = -\partial H_g(g(t), \ell(t)) \circ L_{g_*} - ad^* \partial H_l(g(t), \ell(t))(\ell(t)) \tag{5.4}$$

where $ad^*A : \mathfrak{g}^* \to \mathfrak{g}^*$ is defined by $(ad^*A(\ell))(B) = \ell[A, B]$ for all $B \in \mathfrak{g}$.

Definition 5.8 Functions F on $G \times \mathfrak{g}^*$ are said to be left-invariant if $F(L_g(h), \ell) = F(h, \ell)$ for all g and h in G and all ℓ in $\mathfrak{g}^*$.

It follows that the left-invariant functions on $G \times \mathfrak{g}^*$ are in exact correspondence with the functions on $\mathfrak{g}^*$. The integral curves of the Hamiltonian vector fields generated by the left invariant functions are the solutions of the following differential equations

$$\frac{dg}{dt}(t) = g(t)dH_{\ell(t)}, \frac{d\ell}{dt}(t) = -ad^* dH_{(\ell(t)}(\ell(t)). \tag{5.5}$$

Proposition 5.9 *[Ki] The dual $\mathfrak{g}^*$ of a Lie algebra $\mathfrak{g}$ is a Poisson manifold with the Poisson bracket*

$$\{f, h\}(\ell) = -\ell([df, dh])$$

for any functions f and h on $\mathfrak{g}^*$.

Proof Functions on $\mathfrak{g}^*$ coincide with the left-invariant functions on $G \times \mathfrak{g}^*$. Hence,

$$\omega_{(g,\ell)}(\vec{f}, \vec{h}) = \omega_{(g,\ell)}((df, 0), (dh, 0)) = -ad^*([df, dg])\ell = -\ell([df, dh]).$$

It follows that $\{\,,\,\}$ is the restriction of the canonical Poisson bracket on $G \times \mathfrak{g}^*$ to the left-invariant functions. Hence, it automatically satisfies the properties of a Poisson manifold.   □

Remark 5.10 In the literature on integrable systems, the Poisson bracket $\{f, h\}(\ell) = \ell([df, dh])$ is often referred as the Lie–Poisson bracket [Pr]. We have taken its negative so that the Poisson vector fields agree with the projections to $\mathfrak{g}^*$ of the Hamiltonian vector fields generated by the left-invariant functions. This choice of sign is also dictated by the choice of the sign for our Lie bracket [Jc].

It follows that each function H on $\mathfrak{g}^*$ defines a Poisson vector field $\vec{H}$ through the formula $\{f, H\} = \vec{H}(f)$, and it also defines a Hamiltonian vector field on $G \times \mathfrak{g}^*$. Each integral curve $(g(t), \ell(t))$ of the Hamiltonian field associated with H is a solution of (5.5) and the projection $\ell(t)$ is an integral curve of $\vec{H}$, and conversely, each solution of $\vec{H}$ is the projection of a solution of (5.5). Therefore, integral curves of $\vec{H}$ are the solutions of

$$\frac{d\ell}{dt}(t) = -ad^* dH_{(\ell(t))}(\ell(t)). \qquad (5.6)$$

Solutions of equation (5.6) are intimately linked with the coadjoint orbits of G. To explain in more detail, let us introduce the relevant concepts first. The diffeomorphism $\Phi_g(x) = L_g \circ R_g^{-1}(x) = gxg^{-1}$ satisfies $\Phi(e) = e$, and $\Phi_{gh} = \Phi_g \circ \Phi_h$ for all g and h. The tangent map of Φ_g at the group identity is denoted by Ad_g. It follows that Ad_g is a linear automorphism of $\mathfrak{g}$.

Moreover, $Ad_{gh} = Ad_g \circ Ad_h$, and therefore $g \to Ad_g$ is a representation of G into $Gl(\mathfrak{g})$. It is called *the adjoint representation* of G. The dual representation Ad^*, called *the coadjoint representation*, is a representation of G into $Gl(\mathfrak{g}^*)$. It is defined by

$$Ad_g^*(\ell) = \ell \circ Ad_{g^{-1}}, \ell \in \mathfrak{g}^*. \qquad (5.7)$$

Definition 5.11 The set $\{Ad_g^*(\ell) : g \in G\}$ is called the coadjoint orbit of G through ℓ.

The following proposition is of central importance in the theory of integrable systems.

Proposition 5.12 *Let $\mathcal{F}$ denote the family of Poisson vector fields on $\mathfrak{g}^*$ and let M be any coadjoint orbit of G. Then the orbit of $\mathcal{F}$ through each point ℓ in M is an open submanifold of M.*

Proof Let h be a function on $\mathfrak{g}^*$. The integral curves $\ell(t)$ of the associated Poisson field $\vec{h}$ are the solutions of equation (5.6) and $\ell(t)$ is the projection of a solution $(g(t), \ell(t))$ of (5.5), that is, $g(t)$ is a solution of $\frac{dg}{dt} = L_{g(t)}{}_*(dh(\ell(t)))$.

Let $Ad_{g(t)}$ denote the curve in $Gl(\mathfrak{g})$ defined by the solution $g(t)$ of $\frac{dg}{dt} = L_{g(t)}{}_*(dh(\ell(t)))$ with $g(0) = e$. Then according to (1.11) of Chapter 1,

$$\frac{d}{dt}(\ell_0 \circ Ad_{g(t)}(X)) = -(\ell_0 \circ Ad_{g(t)}[dh, X]$$

for any $X \in \mathfrak{g}$ and any $\ell_0 \in \mathfrak{g}^*$. This means that $Ad^*_{g^{-1}(t)}(\ell_0)$ satisfies equation (5.6). Hence, $\ell(t) = Ad^*_{g^{-1}(t)}(\ell_0)$, which implies that $\ell(t)$ evolves on the coadjoint orbit through ℓ_0.

It follows that the orbit of $\mathcal{F}$ through a point ℓ in M is contained in M. It remains to show that the orbit of $\mathcal{F}$ through ℓ contains an open neghborhood of ℓ in M.

Let $\mathcal{F}_0$ denote the subfamily of $\mathcal{F}$ consisting of linear functions $h_A(\ell) = \ell(A)$ defined by an element A in $\mathfrak{g}$. An integral curve $\ell(t)$ of $\vec{h}_A$ is of the form $\ell(t) = Ad^*_{(e^{At})}(\ell_0)$ for some ℓ_0. If $h_1, \ldots, h_m$ is any subset of linear functions defined by $A_1, \ldots, A_m$ in $\mathfrak{g}$, then

$$\ell \circ \exp t_m \vec{h}_m \circ \cdots \exp t_1 \vec{h}_1 = Ad^*_{e^{A_m t_m}} \circ \cdots \circ Ad^*_{e^{A_1 t_1}}(\ell) = Ad^*_{(e^{A_m t_m} \ldots e^{A_1 t_1})}(\ell).$$

Our proof is now finished since the transformation $(t_1, \ldots, t_m) \to e^{A_m t_m} \cdots e^{A_1 t_1}$ covers an open neighborhood of the identity for any basis $A_1, \ldots, A_m$ of $\mathfrak{g}$. □

Corollary 5.13 *Each coadjoint orbit of G is a symplectic submanifold of $\mathfrak{g}^*$. The tangent space at a point ℓ on a coadjoint orbit M consists of vectors $\ell \circ adX, X \in \mathfrak{g}$ and*

$$\omega_\ell(\ell \circ adX, \ell \circ adY) = \ell[X, Y] \tag{5.8}$$

is the symplectic form on M.

This corollary is a direct consequence of Proposition 5.5.

Corollary 5.14 *Each coadjoint orbit of G is even dimensional.*

Proof Symplectic manifolds are even-dimensional. □

In contrast to the coadjoint orbits, adjoint orbits can be odd dimensional, as the following example shows.

Example 5.15 Let G be the group of the upper-diagonal matrices $\begin{pmatrix} a & b \\ 0 & c \end{pmatrix}$ with $ac \neq 0$. The Lie algebra $\mathfrak{g}$ of G consists of 2×2 matrices $X = \begin{pmatrix} x_1 & x_2 \\ 0 & x_3 \end{pmatrix}$. Let A_1^*, A_2^*, A_3^* be the dual basis in $\mathfrak{g}^*$ relative to the basis

$$A_1 = \begin{pmatrix} 1 & 0 \\ 0 & 0 \end{pmatrix}, A_2 = \begin{pmatrix} 0 & 1 \\ 0 & 0 \end{pmatrix}, A_3 = \begin{pmatrix} 0 & 0 \\ 0 & 1 \end{pmatrix}.$$

Then,

$$Ad_G(X) = \left\{ x_1 A_1 - \left(\frac{b^2}{ac} x_1 + \frac{b}{a} x_2 + \frac{b}{a} x_3 \right) A_2 + x_3 A_3 : ac \neq 0 \right\}$$

and

$$Ad_G^*(l_1 A_1^* + l_2 A_2^* + l_3 A_3^*) = \left\{ \left(l_1 - \frac{b^2}{ac} l_2 \right) A_1^* + cl_2 A_2^* + l_3 A_3^* : ac \neq 0 \right\}.$$

Evidently, $Ad_G(X)$ is one dimensional for all matrices X with $x_2 x_3 \neq 0$, but $Ad_G(A^*)$ is either zero dimensional when $l_2 = 0$, or two dimensional in other cases.

Definition 5.16 A non-degenerate quadratic form $\langle , \rangle$ on $\mathfrak{g}$ is said to be invariant if

$$\langle [A, B], C \rangle = \langle A, [B, C] \rangle$$

for all elements A, B, C in $\mathfrak{g}$.

Proposition 5.17 *Adjoint orbits are symplectic on any Lie algebra $\mathfrak{g}$ that admits an invariant form $\langle , \rangle$. The symplectic form ω is given by $\omega_L(adA(L), adB(L)) = \langle L, [A, B] \rangle$ for all tangent vectors $adA(L)$ and $adB(L)$ at L.*

Proof Since $\langle , \rangle$ is non-degenerate it can be used to identify $\mathfrak{g}^*$ with $\mathfrak{g}$ via the formula $\ell \in \mathfrak{g}^* \Leftrightarrow L \in \mathfrak{g}$ whenever $\ell(X) = \langle L, X \rangle$ for all $X \in \mathfrak{g}$. Then $Ad_G^*(\ell)$ is identified with $Ad_G(L)$ and tangent vectors ad^*A at $Ad_\mathfrak{g}(\ell)$ are identified with tangent vectors $ad(A)$ at $Ad_\mathfrak{g}(L)$. It follows that the symplectic form $\omega_\ell(ad^*A(\ell), ad^*B(\ell)) = \ell([A, B])$ is taken to $\omega_L(ad(A)(L), ad(B)(L)) = \langle L, [A, B] \rangle$. $\qquad\square$

Definition 5.18 Let $\mathfrak{g}$ be any Lie algebra. The bilinear form $Kl(A, B) = Tr(ad(A) \circ ad(B))$, where $Tr(X)$ denotes the trace of a matrix X, is called the Killing form [Hl].

The Killing form is a symmetric quadratic form on $\mathfrak{g}$ that enjoys several noteworthy properties. To begin, it is invariant relative to any automorphism

ϕ on $\mathfrak{g}$, that is $Kl(A, B) = Kl(\phi(A), \phi B))$ for any A, B in $\mathfrak{g}$. The argument is simple:

$$Kl(\phi(A), \phi(B)) = Tr(ad(\phi(A)) \circ ad(\phi(B)))$$
$$= Tr(\phi \circ ad(A) \circ \phi^{-1} \circ \phi \circ ad(B) \circ \phi^{-1})$$
$$= Kl(A, B).$$

In particular, Kl is Ad_G invariant. Then,

$$0 = \frac{d}{dt} Kl(Ad_{\exp tA}(B), Ad_{\exp tA}(C))|_{t=0} = Kl([A, B], C + Kl(B, [A, C]),$$

and therefore, Kl is invariant in the sense of Definition 5.5.

Definition 5.19 A Lie algebra is called semi-simple if its Killing form is non-degenerate.

It follows that in semi-simple Lie groups every adjoint orbit is symplectic. Most studies of coadjoint orbits have been carried out in a semi-simple setting (apart from three-dimensional Lie groups, where the coadjoint orbits can be computed explicitly). One of the more outstanding results in this theory is that every coadjoint (adjoint) orbit is Kähler on simple compact Lie groups [Bo].

It was the work of A. Kirillov [Ki] that initiated the interest in coadjoint orbits. This interest was further amplified with the discovery of solitons in the Korteveg de Vries equation and their relation to Toda lattices [Ks; Sy].

Remarkably, many coadjoint orbits are symplectomorphic [Al] to the cotangent bundles of other manifolds, and that observation sheds new light on the theory of integrable systems. This observation with its implications deserves special attention and will be deferred to Chapter 10. In the meantime, we will first make another detour to control theory and its Maximum Principle as a bridge between Hamiltonian systems and optimal control.

6

Hamiltonians and optimality: the Maximum Principle

Let us now return to control systems and their reachable sets with a more detailed interest in their boundary properties. Recall the basic setting initiated in Chapter 2: the control system

$$\frac{dx}{dt} = F(x(t), u(t)), \ u(t) \in U \tag{6.1}$$

on a manifold M and the associated family of vector fields $\mathcal{F}$ defined by the constant controls. We will assume that the control system (6.1) conforms to the same assumptions as in Chapter 2, so that the Cauchy problem $x(\tau) = x_0$ associated with the time-varying vector field $X_u(t) = F(x, u(t))$ admits a unique solution $x(t)$ that satisfies $\frac{dx}{dt}(t) = X_u(t)(x(t))$ for almost all t for each admissible control $u(t)$, and varies smoothly relative to the initial point x_0. We will use $\mathcal{U}$ to denote the space of bounded and measurable control functions on compact intervals $[t_0, t_1]$ that take values in the control set U in $\mathbb{R}^m$. Such controls will be referred to as admissible controls. As before, $\mathcal{A}(x, T)$, $\mathcal{A}(x, \leq T)$ and $\mathcal{A}(x)$ will denote the reachable sets (at exactly T units of time, up to T units of time, and reachable at any time) by the trajectories generated by the control functions in $\mathcal{U}$. Analogously, $\mathcal{A}_{\mathcal{F}}(x, T)$, $\mathcal{A}_{\mathcal{F}}(x, \leq T)$, and $\mathcal{A}_{\mathcal{F}}(x)$ will denote the reachable sets by the elements in $\mathcal{F}$, i.e., points reachable by piecewise constant controls in $\mathcal{U}$. Since each control function in $\mathcal{U}$ is the pointwise limit of a sequence of piecewise constant controls, the reachable sets defined by the trajectories generated by the elements of $\mathcal{U}$ have the same closure as the reachable sets generated by the family $\mathcal{F}$.

In addition, we will use $\bar{A}$ to denote the closure of set A, ∂A to denote the boundary of A, and A^0 or $\text{int}(A)$ to denote the interior of A.

6.1 Extremal trajectories

In this chapter we will be interested in the boundary points of the reachable sets. To avoid the situation in which every point is a boundary point, we will assume that the interiors of the reachable sets are not empty. Of course, if the interior of $\mathcal{A}(x_0, T)$ is not empty, then the interior of $\mathcal{A}(x_0)$ is not empty, but the converse may not be true, as the following simple example shows.

Example 6.1 The reachable set at exactly T units of time of the control system $\frac{dx}{dt} = u(t)e_1 + (1 - u(t))e_2$ in $\mathbb{R}^2$ is equal to the line $x_2(T) - x_2(0) + x_1(T) - x_1(0) = T$, while $\mathcal{A}(x_0)$ is the half space $x_2(T) - x_2(0) + x_1(T) - x_1(0) \geq 0$.

We have already seen that the interior of $\mathcal{A}(x_0)$ is not empty whenever $Lie(\mathcal{F})$ is of full rank at x_0, in which case the set of interior points of $\mathcal{A}_{\mathcal{F}}(x_0)$ is dense in its closure. In this regard, an even stronger statement is given by the following:

Proposition 6.2 *If* $dim(Lie_x(\mathcal{F})) = dim(M)$ *then*

$$int(\mathcal{A}_{\mathcal{F}}(x)) = int(\bar{\mathcal{A}}_{\mathcal{F}}(x)).$$

For the proof see [Jc, p. 68]. To state the conditions under which the interior of $\mathcal{A}(x, T)$ is not empty, we need to introduce another geometric object, called the zero-time ideal $\mathcal{I}(\mathcal{F})$.

Definition 6.3 The vector space spanned by the iterated Lie brackets of $\mathcal{F}$, called the derived algebra of $\mathcal{F}$, will be denoted by $\mathcal{D}(\mathcal{F})$. The vector space spanned by $\mathcal{D}(\mathcal{F})$ and all differences $X - Y$ of elements in $\mathcal{F}$ is called the zero-time ideal of $\mathcal{F}$ and is denoted by $\mathcal{I}(\mathcal{F})$.

It is easy to see that $Lie(\mathcal{F}) = RX + \mathcal{I}(\mathcal{F})$, where X is any fixed vector field in $\mathcal{F}$ and RX is the line through X, because any linear combination $\alpha_1 X_1 + \cdots + \alpha_k X_k$ of elements in $\mathcal{F}$ can be written as $\left(\sum_{i=1}^{k} \alpha_i\right) X + \alpha_1(X_1 - X) + \cdots + \alpha_k(X_k - X)$, and $Lie(\mathcal{F}) = \langle \mathcal{F} \rangle + \mathcal{D}(\mathcal{F})$ where $\langle \mathcal{F} \rangle$ denotes the linear span of $\mathcal{F}$.

The above implies that either $dim(Lie_x(\mathcal{F})) - 1 = dim(\mathcal{I}_x(\mathcal{F}))$, or $dim (Lie_x(\mathcal{F}) = dim(\mathcal{I}_x(\mathcal{F}))$. The second case occurs whenever $X(x)$ belongs to $\mathcal{I}_x(\mathcal{F})$ for some $X \in \mathcal{F}$.

Definition 6.4 The zero-time orbit of $\mathcal{F}$ through x_0 will be denoted by $N_0(x_0)$. It consists of points x which can be written as

$$\hat{x} = \hat{x}_0 \circ \exp t_k X_k \circ \cdots \exp t_2 X_2 \circ \exp t_1 X_1,$$

for some elements $X_1, \ldots, X_k$ in $\mathcal{F}$ and $(t_1, \ldots, t_k) \in \mathbb{R}^k$ that satisfy $t_1 + t_2 + \cdots + t_k = 0$.

Zero-time orbits are associated with the group of diffeomorhisms of the form $\exp t_k X_k \circ \cdots \exp t_2 X_2 \circ \exp t_1 X_1$ for some elements $X_1, \ldots, X_k$ in $\mathcal{F}$ and $(t_1, \ldots, t_k) \in \mathbb{R}^k$ that satisfy $t_1 + t_2 + \cdots + t_k = 0$. This group constitutes a normal subgroup of the group $G(\mathcal{F})$ generated by $\{\exp tX : X \in \mathcal{F}, t \in \mathbb{R}\}$.

Suppose now that X is a complete vector field in a family $\mathcal{F}$ and suppose that $N_0(x)$ is the zero-time orbit of $\mathcal{F}$ through a point x. Then, each time translate $N_t = \exp tX(N_0(x))$ is the zero-time orbit through $\hat{y} = \hat{x} \circ \exp tX(x)$. The relevant properties of the zero-time orbits are assembled in the proposition below.

Proposition 6.5 *Suppose that $\mathcal{F}$ is a family of vector fields such that the zero-time ideal $\mathcal{I}(\mathcal{F})$ is Lie determined, and suppose that $\mathcal{F}$ contains at least one complete vector field X. Then,*

(a) *the orbit of $\mathcal{I}(\mathcal{F})$ through a point x coincides with the zero-time orbit through x.*

(b) *If N is an orbit of $\mathcal{F}$ then the dimension of $\mathcal{I}_x(\mathcal{F})$ is constant over N. This dimension is either equal to the dimension of $Lie_x(\mathcal{F})$, in which case the zero-time orbit through any point in N coincides with N, or it is equal to $dim(Lie_x(\mathcal{F})) - 1$, in which case, each zero-time manifold is s submaniold of N of codimension 1 and N is foliated by the time translates of $N_0(x)$.*

(c) *Each reachable set $\mathcal{A}(x, T)$ is contained in N_T and has a non-empty interior in the topology of N_T. Moreover, the set of interior points in $\mathcal{A}(x, T)$ is dense in the closure of $\mathcal{A}(x, T)$.*

We refer the reader to [Jc] for the proof of this theorem.

Corollary 6.6 $\mathcal{A}(x, T)$ *has a non-empty interior in M if and only if $\mathcal{I}_x(\mathcal{F}) = T_x M$.*

Even though the interior of $\mathcal{A}(x, T)$ is dense in $\bar{\mathcal{A}}(x, T)$, it is not true in general that the interior of $\bar{\mathcal{A}}(x, T)$ is equal to the interior of $\mathcal{A}(x, T)$ as the following example shows.

Example 6.7 Consider

$$\frac{dx}{dt} = \frac{v(t)}{\sqrt{x^2 + y^2}} x - u(t)y, \quad \frac{dy}{dt} = \frac{v(t)}{\sqrt{x^2 + y^2}} y + u(t)x,$$

in the punctured plane $M = \mathbb{R}^2 - \{0\}$, where $|v(t)| \leq \frac{1}{2\pi}$ and $|u(t)| < 1$. In polar coordinates, $r = \sqrt{x^2 + y^2}$, $\phi = \tan^{-1} \frac{y}{x}$, the above system takes on a paticularly simple form:

$$\frac{dr}{dt} = v(t), \quad \frac{d\phi}{dt} = u(t).$$

The reachable set from $r = 1$, $\phi = 0$ at time $T = \pi$ is given by $\{(r, \phi) : \frac{1}{2} \le r \le \frac{3}{2}, -\pi < \phi < \pi\}$. The interior of the closure of this set includes the open segment $-\frac{3}{2} < x < -\frac{1}{2}$ which is not reachable at time $T = \pi$.

With these preliminary considerations behind us, let us now come to the main issues of this chapter, the extremal trajectories, the Maximum Principle, and the relations to optimal control and the calculus of variations. To begin with:

Definition 6.8 A trajectory $x(t)$ that emanates from x_0 at $t = 0$ and belongs to the boundary of $\mathcal{A}(x_0)$ at some time $T > 0$ is called extremal. The corresponding control is also called extremal.

Let us now introduce typical optimal control problems and then relate them to the extremal properties of the trajectories. Optimal control problems are defined by the control system (6.1) and a cost functional $f : M \times U \to \mathbb{R}$. There are two optimal problems defined by this data. The first problem is defined over the trajectories of indefinite time duration; it consists of finding a control $\bar{u}(t) \in \mathcal{U}$ that generates a trajectory $\bar{x}(t)$ on some interval $[0, T]$ that satisfies $\bar{x}(0) = x_0$ $\bar{x}(T) = x_1$ and attains the minimum value for the integral $\int_0^S f(x(t), u(t)) \, dt$ relative to any other trajectory $(x(t), u(t))$ on $[0, S]$ that satisfies the same boundary conditions $x(0) = x_0$ and $x(S) = x_1$. That is,

$$\int_0^T f(\bar{x}(t), \bar{u}(t)) \, dt \le \int_0^S f(x(t), u(t)) \, dt.$$

The prototype of these problems are time optimal problems in which $f = 1$ and T is the least positive time that x_1 can be reached from x_0 by a trajectory of the system.

The second optimal problem is defined over the trajectories of fixed time duration. Its formulation is similar to the one above, except that all the competing trajectories are defined over the same time interval $[0, T]$. Conceptually, these two problems are the same, since the optimal problem with fixed time duration can be turned into optimal problems over indefinite time duration by adding an extra equation $\frac{d\tau}{dt} = 1$ to the dynamics $\frac{dx}{dt} = F(x(t), u(t))$. In the extended space $M \times R$, the point (x_1, T) can be reached from $(x_0, 0)$ only at time T.

A control $\bar{u}(t)$ that generates a trajectory $\bar{x}(t)$ that attains the minimum value for $\int_0^S f(x(t), u(t)) \, dt$ is called optimal and so is the corresponding trajectory $\bar{x}(t)$. Sometimes it is convenient to refer to $(\bar{x}(t), \bar{u}(t))$ as an optimal pair. Thus an optimal control problem is a two point boundary value problem of finding the minimum of the cost functional $\int_0^T f(u(t), x(t)) \, dt$ over the trajectories $x(t)$ that satisfy the given boundary conditions. Suppose now that the dynamics F, the control set U and the cost functional f are given. The pair (f, F) defines another control system $\tilde{\mathcal{F}}$ on $\tilde{M} = \mathbb{R} \times M$ with the trajectories $\tilde{x}(t) = (y(t), x(t))$ the solutions of

$$\frac{dy}{dt}(t) = f(x(t), u(t)), \quad \frac{dx}{dt}(t) = F(x(t), u(t)), u(t) \in U. \qquad (6.2)$$

System (6.2) will be referred to as the cost-extended control system. The passage to the cost-extended system turns the optimal problem into an extremal problem, in the sense that every optimal pair $(\bar{x}(t), \bar{u}(t))$ generates a cost-extended trajectory $(\bar{y}(t), \bar{x}(t))$ which at the terminal time T is on the boundary of the cost-extended reachable set, because the line segment $\{(\alpha, \bar{x}(T)) : 0 < \alpha < y(T)\}$ does not intersect the cost-extended reachable set $\tilde{\mathcal{A}}(\tilde{x}(0))$.

The Maximum Principle then provides a necessary condition that a trajectory be extremal. This necessary condition states that an extremal trajectory is the projection of an integral curve of the associated Hamiltonian lift in the cotangent bundle T^*M. We will come back to this statement with complete details, but in the meantime let us digress briefly into the classical theory of the calculus of variations and make some parallels between its basic concerns and the problems of optimal control.

6.2 Optimal control and the calculus of variations

Let us go back to the central question of the calculus of variations of finding the conditions under which a given curve $\bar{x}(t), t \in [t_0, t_1]$ in $\mathbb{R}^n$ yields a minimum for the integral $\int_{t_0}^{t_1} f(t, x(t), \frac{dx}{dt}) \, dt$ in some neighborhood of $\bar{x}(t)$, subject to fixed boundary conditions [Cr] and [GF]. In this context, $f(t, x, u)$ is a given real-valued function on $\mathbb{R} \times \mathbb{R}^n \times \mathbb{R}^n$.

Let us first consider the most common case in which the minimum is taken in the weak sense, that is, over continuously differentiable curves $x(t)$ on the interval $[t_0, t_1]$ that satisfy the boundary conditions $\bar{x}(t_0) = x(t_0)$ and $\bar{x}(t_1) = x(t_1)$ and conform to

$$\sup \left\{ \|x(t) - \bar{x}(t)\| + \left\| \frac{dx}{dt}(t) - \frac{d\bar{x}}{dt}(t) \right\| : t \in [t_0, t_1] \right\} < \epsilon$$

for some $\epsilon > 0$. Then the most common necessary condition of optimality, based on the premise that a local minimum of a given functional occurs at the points where the derivative is zero, leads to the Euler–Lagrange equation

$$\frac{d}{dt}\frac{\partial f}{\partial u}(t, x(t), u(t)) + \frac{\partial f}{\partial x}(t, x(t), u(t)) = 0, u(t) = \frac{dx}{dt}(t). \quad (6.3)$$

In the situations where the strong Legendre condition $\sum_{ij}^{n} \frac{\partial f^2}{\partial u_i \partial u_j}(t, x, u)$ $u_i u_j > 0$ holds in some neighborhood of the solution curve $\left(\bar{x}(t), \frac{d\bar{x}}{dt}\right)$, the equation

$$p = \frac{\partial f}{\partial u}(t, x, u)$$

can be solved to yield $u = \phi(t, x, p)$. In that case, $(\bar{x}(t), \bar{p}(t))$ is a solution of the Hamiltonian equation

$$\frac{dx}{dt} = \frac{\partial H}{\partial p}, \frac{dp}{dt} = -\frac{\partial H}{\partial x}, \quad (6.4)$$

associated with $H = -f(t, x, \phi(t, x, p)) + \sum_{j=1}^{n} p_j \phi_j(t, x, p)$, where $\bar{p}(t) = \frac{\partial f}{\partial u}\left(t, \bar{x}(t), \frac{d\bar{x}}{dt}\right)$.

Equations (6.3) and (6.4) are very different from each other even though at a first glance they may seem completely equivalent. On manifolds, the Euler–Lagrange equation is an equation on the tangent bundle, and as a second-order differential equation it requires some notion of a connection or a covariant derivative for its formulation. In contrast, the Hamiltonian equations seem more natural, in the sense that, once the Hamiltonian is obtained, its equations are written intrinsically on the cotangent bundle thanks to the canonical symplectic form on the cotangent bundle.

On the other hand, the Euler–Lagrange equation has a conceptual advantage, in the sense that it corresponds to the critical point of the functional. In the case of the Hamiltonian equations, the connection to the original variational problem is less clear. This conceptual discrepancy left its imprint on the existing literature. For the physicist, the Legendre transform was good enough since it led to the Hamiltonian that represents the total energy of the system. For that reason, the Hamiltonian formulation was always preferable in the physical world. That is not so much the case for a geometer, who preferred to see his geodesics through the Euler–Lagrange equation [DC; Gr].

Leaving such philosophical matters aside, there is another issue that sharply divides the calculus of variations – the contributions of C. Weierstrass and his theory of strong extrema [Cr; Yg]. Let us now go briefly into this theory and to the "excess function" of Weierstrass, which, according to L. C. Young, "revolutionized the calculus of variations" [Yg].

In the theory of strong extrema, parametrized curves defined on an interval $I = [t_0.t_1]$ are called addmisible if they are continuous and differentiable almost everywhere on I with the derivative belonging to $L^\infty([t_0, t_1])$. It then follows that admissible curves satisfy the Lipschitz condition $\|x(t) - x(s)\| \le K|t - s|$ for all s and t in I, and therefore are absolutely continuous, that is, $x(t) - x(s) = \int_s^t \frac{dx}{d\tau}(\tau) d\tau$ [Ro].

On the other hand, every absolutely continuous curve $x(t)$ on I is differentiable almost everywhere on I and is admissible when $\|\frac{dx}{dt}\|$ is bounded on I. Hence, admissible curves coincide with absolutely continuous curves with bounded derivatives, i.e., they coincide with the Lipschitzian curves.

A strong neighborhood of an admissible curve $x(t)$, $t \in [t_0, t_1]$, consists of admissible curves $y(t)$ all defined on the interval $[t_0, t_1]$ such that $\sum_{i=1}^n |x_i(t) - y_i(t)| < \epsilon$ for some $\epsilon > 0$ and all $t \in [t_0, t_1]$. Every strong neighborhood of a curve $x(t)$ defines an open neighborhood $\mathcal{O}_\epsilon(x)$ of $\mathbb{R}^n$ defined by $\mathcal{O}_\epsilon(x) = \{y \in \mathbb{R}^n : \sup_{t \in [0,T]} \sum_{i=1}^n |y_i - x_i(t)| < \epsilon\}$.

Any smooth function L on $\mathbb{R} \times \mathbb{R}^n \times \mathbb{R}^n$ defines a functional $\int_{t_0}^{t_1} L(t, x(t), \frac{dx}{dt}(t)) dt$ over the admissible curves $x(t)$ in $\mathbb{R}^n$. The basic problem of the calculus of variations (in the strong sense) is to find conditions under which a curve $\bar{x}(t)$ on an interval $I = [t_0, t_1]$ attains a minimum of $\int_{t_0}^{t_1} L(t, x(t), \frac{dx}{dt}(t)) dt$ over all admissible curves $x(t)$ that satisfy the same boundary conditions as $\bar{x}(t)$ in some strong neighborhood of $\bar{x}(t)$.

This problem admits a natural reformulation in control theoretic terms: addmissible curves in $\mathcal{O}_\epsilon(x)$ are the solutions in $\mathcal{O}_\epsilon(x)$ of the "control" system

$$\frac{dx}{dt} = \sum_{i=1}^n u_i(t) e_i, \tag{6.5}$$

where $e_1, \dots, e_n$ denote the standard basis in $\mathbb{R}^n$, and where the controls functions $u_1(t), \dots, u_n(t)$ are measurable curves on the interval $[t_0, t_1]$. Every admissible curve in $\mathcal{O}_\epsilon(\bar{x})$ is a trajectory of (6.5) generated by the control $u(t) = \frac{dx}{dt}(t)$, but the converse may not be true, since a solution of (6.5) that originates in $\mathcal{O}_\epsilon(\bar{x})$ at t_0 may not remain in $\mathcal{O}_\epsilon(\bar{x})$ for all $t \in [t_0, t_1]$.

The problem of finding conditions under which $\bar{x}(t)$ yields a minimum for $\int_{t_0}^{t_1} L\left(t, x(t), \frac{dx}{dt}\right) dt$ over all admissible curves in some strong neighborhood $\mathcal{O}_\epsilon(\bar{x})$ of $\bar{x}(t)$ subject to the boundary conditions $x(t_0) = \bar{x}(t_0)$ and $x(t_1) = \bar{x}(t_1)$, is the same as the problem of finding conditions under which $\bar{u}(t) = \frac{d\bar{x}}{dt}$ is an optimal control for the control system (6.5) on $\mathcal{O}_\epsilon(\bar{x})$ relative to the boundary conditions $x(t_0) = \bar{x}(t_0)$, $x(t_1) = \bar{x}(t_1)$ and the cost functional $\int_{t_0}^{t_1} L(t, x(t), u(t))) dt$.

Weierstrass' revolutionary discovery was that along an optimal trajectory $\bar{x}(t)$ the following inequality must hold:

$$L\left(t, \bar{x}(t), \frac{d\bar{x}}{dt}\right) - L(t, \bar{x}(t), u(t)) - \left(\frac{d\bar{x}}{dt} - u(t)\right) \cdot \frac{\partial L}{\partial u}\left(t, \bar{x}(t), \frac{d\bar{x}}{dt}\right) \ge 0, \tag{6.6}$$

for any curve $u(t) \in \mathbb{R}^n$. The function

$$E(t, x, y, u) = L(t, x, y) - L(t, x, u) - (y - u)\frac{\partial L}{\partial u}(t, x, u)|_{u=y}$$

became known as the excess function of Weierstrass [Cr].

So, in contrast to the Euler–Lagrange equation, which appears as a necessary condition for optimality in the weak sense, the inequality of the excess function of Weierstrass is a necessary condition of optimality in the strong sense. As far as I can tell, the two conditions have never been properly integrated in the classical literature on this subject. [Cr; Yg]. We will presently show that the inequality of Weierstrass may be considered as a harbinger of the Maximum Principle.

In regard to weak versus strong optimality, it is clear that an optimal curve in the strong sense that is continuously differentiable, is also optimal in the weak sense, but the converse may not be true. For instance, $x(t) = 0$ is weakly optimal for the functional defined by $L = ||x(t)||^2 \left(1 - ||\frac{dx}{dt}||^2\right)$, because $L \geq 0$ in a sufficiently small weak neighborhood of $x(t) = 0$. That is no longer the case in the strong sense, no matter how small the neighborhhood.

6.3 The Maximum Principle

The Maximum Principle states that each extremal trajectory of a control system is the projection of an integral curve of a certain Hamiltonian system on the cotangent bundle of M. In order to be more explicit, we will restore the notations and the basic material introduced in the previous chapter concerning the symplectic structure of the cotangent bundle. Then θ will denote the Liouville form, and $\omega = -d\theta$ will denote the symplectic form on T^*M. Points of T^*M will be generally denoted by ξ and their projections on M by x. The starting point is the observation that each control system can be lifted to a Hamiltonian system on the cotangent bundle. This lifting is based on the following:

Definition 6.9 The Hamiltonian lift of a time-varying vector field X_t on a manifold M is the Hamiltonian vector field $\vec{h}_t$ on $T^*(M)$ that corresponds to the function $h_t(\xi) = \theta_\xi(X_t(x)) = \xi \circ X_t(x)$ for $\xi \in T_x^*M$.

Remark 6.10 Time-varying vector fields make a smoother transition to control systems.

Let us first examine this definition in terms of the canonical coordinates. If $X_t = \sum_{i=1}^{n} X^i(t, x_1, \ldots, x_n)\frac{\partial}{\partial x_i}$ then $h_t(\xi) = \xi \circ X_t(x)$ is given by

$$h(t, x_1, \ldots, x_n, p_1, \ldots, p_n) = \sum_{i=1}^{n} p_i X^i(t, x_1, \ldots, x_n).$$

It follows that the integral curves of $\vec{h}_t$ are the solutions of the following system of equations:

$$\frac{dx_i}{dt}(t) = X^i(t, x_1(t) \ldots, x_n(t)), \quad \frac{dp_i}{dt}(t)$$

$$= -\sum_{j=1}^{n} p_j(t)\frac{\partial}{\partial x_i}X^j(t, x_1(t) \ldots, x_n(t)), i = 1, \ldots, n \qquad (6.7)$$

This system is the adjoint companion of the variational equation

$$\frac{dx_i}{dt}(t) = X_t^i, (x_1(t) \ldots, x_n(t)), \quad \frac{dv_i}{dt}(t) = \sum_{j=1}^{n} v_j(t)\frac{\partial}{\partial x_j}X^i(t, x_1(t) \ldots, x_n(t)).$$

$$(6.8)$$

We now highlight some facts which will be relevant for the general situation. We first note that the projection of the variational equation on the second factor is a linear time-varying system $\frac{dv}{dt} = A(t)v(t)$, with $A(t)$ the matrix with entries $A_{ij}(t) = \left(\frac{\partial}{\partial x_j}X^i(t, x(t))\right)$. It then follows from the theory of differential equations that $v(t)$ is defined over the same time interval as the solution curve $x(t)$ [CL]. Moreover, the companion curve $p(t)$ in (6.7), being the solution of the adjoint system $\frac{dp}{dt}(t) = -A^*(t)p(t)$, is also defined over the same interval as the underlying curve $x(t)$. Hence the integral curves of $\vec{h}_t$ exist over the same intervals as the integral curves of X_t.

Secondly, the solutions $v(t)$ of the variational equation and the solutions $p(t)$ of the Hamiltonian equation are related through a simple relation

$$\sum_{i=1}^{n} p^i(t)v^i(t) = \text{constant}. \qquad (6.9)$$

Condition (6.9) gives rise to the "forward-backward" principle: *The solutions of the variational equations propagate forward in time, while the solutions of the adjoint equation propagate backward in time, and their scalar product remains constant for all times.*

Let us now elevate these observations to the general setting independent of the particular choice of coordinates. Let $\Phi_{t,\tau}$ denote the time-varying flow of X_t defined by $x(t) = \Phi_{t,\tau}(x_0)$, where $x(t)$ denotes the solution of $\frac{dx}{dt} = X_t(x(t))$

with $x(\tau) = x_0$. The uniqueness theorem of solutions with given initial data implies that the flow satisfies the following composition rule:

$$\Phi_{t,\tau} = \Phi_{t,s} \circ \Phi_{s,\tau}, \tag{6.10}$$

whenever this composition is well defined. Conversely, every time-varying flow that satisfies (6.10) is generated by a time-varying vector field X_t.

The tangent map $(\Phi_{\tau,t})_*$ maps $T_{x(\tau)}M$ onto $T_{x(t)}M$, and the dual mapping $(\Phi_{\tau,t})^*$ maps $T_{x(t)}^*M$ onto $T_{x(\tau)}^*M$. Let $\Psi_{\tau,t} = \Phi_{t,\tau}^*$. Since the order of the variables is reversed, $\Psi_{\tau,t}$ maps $T_{x(\tau)}^*M$ onto $T_{x(t)}^*M$. Moreover,

$$\Psi_{\tau,s} \circ \Psi_{s,t}(\xi) = \Psi_{\tau,t}(\xi \circ (\Phi_{t,s})_* = \xi \circ (\Phi_{t,s})_*(\Phi_{s,\tau})_*$$
$$= \xi \circ (\Phi_{t,\tau})_* = \Psi_{\tau,t}(\xi).$$

It follows that $\Psi_{\tau,t}$ is a time-varying flow on T^*M, and hence is generated by a time-varying vector field on T^*M.

Proposition 6.11 $\Psi_{\tau,t}$ is the flow of the Hamiltonian lift $\vec{h}_t$ of X_t.

Proof Let V_t denote the infinitesimal generator of the flow $\Psi_{\tau,t}$, i.e., $V_t = \frac{\partial}{\partial t}\Psi_{\tau,t}|_{\tau=t}$. Then $\pi(\Psi_{\tau,t}(\xi)) = x(t)$ for every $\xi \in T_{x(\tau)}^*$, where π denotes the natural projection from T^*M onto M. Therefore,

$$\pi_*(V_t(\xi)) = X_t(\pi(\xi)), \xi \in T_x^*M.$$

If $w(s)$ is any curve in T^*M such that $w(s) \in T_{y(s)}^*(M)$ then $\Psi_{\tau,t}(w(s))$ belongs to the cotangent space of M at $\Phi_{\tau,t}(y(s))$. Hence the projection $\pi(\Psi_{\tau,t}w(s))$ is equal to $\Phi_{\tau,t}(y(s))$.

It follows that

$$\pi_*((\Psi_{\tau,t})_*W) = (\Phi_{\tau,t})_*v,$$

where $W = \frac{dw}{ds}(0)$ and $\frac{dy}{ds}(0) = v$. Let now $\xi(t)$ denote the curve $t \to \Psi_{\tau,t}(w(0))$. Then,

$$\theta_{\xi(t)}((\Psi_{\tau,t})_*W) = \xi(t) \circ \pi_*((\Psi_{\tau,t}))_*W) = w(0) \circ \Phi_{t,\tau_*} \circ \{\Phi_{\tau,t}\}_*v = w(0)(v)$$

and, hence, $\frac{d}{dt}(\theta_{\xi(t)}((\Psi_{\tau,\epsilon})_*W)|_{t=\tau} = 0$. Since $\frac{d}{dt}(\theta_{\xi(t)}((\Psi_{\tau,\epsilon})_*W)|_{t=\tau}$ is the same as the the the Lie derivative $L_{V_t}(\theta)(W)$, we can use Cartan's formula to write the preceding equality as

$$0 = L_{V_t}(\theta) = d \circ i_{V_t}\theta + i_{V_t} \circ d\theta.$$

But $i_{V_t}\theta_\xi = \xi(X_t(x)$ for all $\xi \in T_x^*(M)$ because $\pi_*(V_t) = X_t$. Hence

$$dh_t = i_{V_t} \circ \omega.$$

□

Let us now consider the Hamiltonian lift of the control system (6.11). Each control functions $u(t)$ defines a time-varying Hamiltonian $h_{u(t)}$, with $h_{u(t)}(\xi) = \xi \circ F(x, u(t))$, ξ in $T_x^* M$, which, in turn, induces the Hamiltonian vector field $\vec{h}_{u(t)}$. The family of time-varying Hamiltonian vector fields $\vec{\mathcal{H}} = \{\vec{h}_{u(t)} : u \in \mathcal{U}\}$ is the Hamiltonian lift of the control system (6.1). Every flow in $\vec{\mathcal{H}}$ conforms to the same regularity conditions as the underlying system (6.1), and hence,

$$\frac{d\xi}{dt}(t) = \vec{h}_{u(t)}(\xi(t))$$

admits a unique solutions $\xi(t)$ through each initial point ξ_0, with the understanding that the differential equation is satisfied only at the points where $\xi(t)$ is differentiable.

Theorem 6.12 Maximum Principle–version 1   *Suppose that $\bar{x}(t)$ is a trajectory generated by a control $\bar{u}(t)$ such that $\bar{x}(T)$ belongs to $\partial \mathcal{A}(x_0)$. Then $\bar{x}(t)$ is the projection of a curve $\bar{\xi}(t)$ in $T^* M$ defined on $[0, T]$ such that:*

1. $\frac{d\bar{\xi}}{dt}(t) = \vec{h}_{\bar{u}(t)}(\bar{\xi}(t))$ *for almost all $t \in [0, T]$.*
2. $\bar{\xi}(t) \neq 0$ *for any $t \in [0, T]$.*
3. $h_{\bar{u}(t)}(\bar{\xi}(t)) = 0$ *for almost all $t \in [0, T]$.*
4. $h_u(\bar{\xi}(t)) \leq 0$ *for any u in U for almost all $t \in [0, T]$.*

Curves $\xi(t)$ in $T^* M$ that satisfy conditions (1), (2), (3), and (4) of the Maximum Principle are called *extremals*. The Maximum Principle then says that each exremal trajectory $x(t)$ in the base space M is the projection of an extremal curve in $\xi(t)$ in $T^* M$.

Proof   The proof of the Maximum Principle can be divided into three essentially independent parts. The first part deals with the infinitesimal variations along an extremal curve and the cone of attainability at the terminal point $\bar{x}(T)$. The second part consists in showing that the cone of attainability cannot be the entire tangent space $T_{\bar{x}(T)}M$ when $\bar{x}(T)$ belongs to the boundary of the attainable set. The third, and final part, consists of choosing an integral curve $\bar{\xi}$ of $\vec{h}_{\bar{u}(t)}$ that satisfies the Maximum Principle. We break up the proof accordingly, into three separate parts.

Throughout the proof we will assume that $\bar{x}(t)$ is a trajectory generated by $\bar{u}(t)$ that belongs to the boundary of $\mathcal{A}(x_0)$ at some time $T > 0$. Then $\Psi_{t,s}(\bar{x}(s)) = \bar{x}(t)$ will denote the local flow of the time-varying vector field $X_{\bar{u}(t)} = F(\bar{x}(t), \bar{u}(t))$. We will make use of the following auxiliary facts:

(a) There exists a tubular neighborhood $\cup \{O_s : 0 \leq s \leq T\}$ with each O_s an open neighborhood of $\bar{x}(s)$ such that $x(t) = \Psi_{t,s} y$ is defined for all $t \in [0, T]$, for each $y \in O_s$ (this fact follows from a theorem on the continuity of solutions of differential equations with respect to the initial data [CL]).

(b) The next statement concerns the nature of regular points along the curve $\bar{x}(t)$. A point τ is a regular point of $\bar{x}(t)$ if $\bar{x}(t)$ is differentiable at $t = \tau$ and

$$\frac{d}{ds}(\bar{x}(\tau + \lambda(s)))|_{s=0} = \frac{d\lambda}{ds}(0)X_{\bar{u}(\tau)}(\bar{x}(\tau))$$

for any differentiable curve $\lambda(s) \in \mathbb{R}$ defined in an open interval around zero such that $\lambda(0) = \tau$. Every Lebesgue point of $\bar{u}(t)$ is a regular point of $\bar{x}(t)$. Since Lebesgue points of any integrable function have full measure on any compact interval, it follows that almost every point $t \in [0, T]$ is a regular point of $\bar{x}(t)$. In particular, regular points are dense in $[0, T]$.

Recall that a point τ is called a Lebesgue point of an integrable curve $u(t)$ in $\mathbb{R}^m$ if $\lim_{t \to \tau} \frac{1}{|t-\tau|} \int_t^\tau |u_i(s) - u_i(\tau)|\, ds = 0, i = 1, \ldots, m$. Then, $\frac{d}{d\tau} \int_a^\tau u(s)\, ds = u(\tau)$ at every Lebesgue point of $u(t)$. But then it can be shown that

$$\lim_{\epsilon \to 0} \frac{1}{\epsilon} \int_{\tau - \epsilon}^\tau X_{\bar{u}(s)}(\bar{x}(s))\, ds = \lim_{\epsilon \to 0} \frac{1}{\epsilon} \int_{\tau - \epsilon}^\tau F(\bar{x}(s), \bar{u}(s))\, ds = X_{\bar{u}(\tau)}(\bar{x}(\tau)).$$

(The basic theory of Lebesgue points can be found in *Theory of Functions of a Real Variable* by I. P. Natanson [Na]. The relevant facts are also discussed in the original publication on the Maximum Principle [Pt].) ☐

Part 1: The cone of attainability

We shall use the family $\mathcal{F} = \{X_u : X_u(x) = F(x, u), x \in M, u \in U\}$ to generate the perturbations of of the flow along $\bar{x}(t)$. Vector fields X_u in $\mathcal{F}$ are time invariant, hence, their flows $\{\Phi_s^u : s \in \mathbb{R}\}$ are defined by a single parameter s, in contrast to the flow defined by a time-varying control $u(t)$.

Each X_u in $\mathcal{F}$ defines a perturbed trajectory $t \to \sigma(t, \tau, u, \lambda_1, \lambda, s)$ with

$$\sigma(t, \tau, u, \lambda_1, \lambda, s) = \Psi_{t - s\lambda, \tau} \circ \Phi_{\lambda_1 s}^u \circ \bar{x}(\tau - s\lambda_1), \tag{6.11}$$

for s sufficiently small and positive, and arbitrary numbers λ and λ_1 with λ_1 positive. The perturbed trajectory follows $\bar{x}(t)$ up to $\tau - s\lambda$ units of time, then switches to X_u, which it follows for $s\lambda_1$ units of time, and then it switches back to $X_{\bar{u}(t)}$, which it follows for the rest of the time. At each time $t > \tau$, this perturbed trajectory is in the reachable set from x_0 and, moreover, $\sigma(t, \tau, u, \lambda, s) \to \bar{x}(t)$ uniformly as s tends to zero.

At regular points τ and t, the curve $s \to \sigma(t, \tau, u, \lambda, s)$ is differentiable at $s = 0$ and defines a tangent vector at $T_{\bar{x}(t)}M$ given by

$$v(t, \tau, u) = \lambda X_{\bar{u}(t)}(\bar{x}(t)) + \lambda_1 \Psi_{t, \tau *}(X_u - X_{\bar{u}(\tau)})(\bar{x}(\tau)). \tag{6.12}$$

This vector is called an elementary perturbation vector of $\bar{x}(t)$. These vectors are a subset of a more general class of tangent vectors obtained by differentiating more complex perturbations. These perturbed trajectories are of the form $x_s(\tau) = \Psi_{\tau - s\lambda, \tau_m} \circ \Phi(s)$, where $\Phi(s)$ is a local diffeomorphism generated by the composition of elementary perturbations

$$\Phi(s) = \Phi^{u_m}_{s\lambda_m} \circ \Psi_{\tau_m - s\lambda_m, \tau_{m-1}} \circ \cdots \circ \Psi_{\tau_3 - s\lambda_3, \tau_2}$$
$$\circ \, \Phi^{u_2}_{s\lambda_2} \circ \Psi_{\tau_2 - s\lambda_2, \tau_1} \circ \Phi^{u_1}_{s\lambda_1} \circ \bar{x}(\tau_1 - s\lambda_1), \tag{6.13}$$

at regular points $0 < \tau_1 \leq \tau_2 \cdots \leq \tau_m$, $\tau_m \leq \tau$. The perturbed trajectory $x_s(t)$ follows $\bar{x}(t)$ for $\tau_1 - s\lambda_1$ units of time, then switches to X_{u_1} for $s\lambda_1$ units of time. The terminal point of this piece of trajectory is then moved along $X_{\bar{u}(t)}$ until the time $t = \tau_2 - s\lambda_2$, after which the switch is made to X_{u_2} for $s\lambda_2$ units of time. Then the flow reverts back to $X_{\bar{u}(t)}$ for $t = \tau_2 - s\lambda_3$ units of time, and then the trajectory switches to X_{u_3} for $s\lambda_3$ units of time. This process is repeated in time succession until the last switch to X_{u_m} is completed, upon which the trajectory follows $X_{\bar{u}(t)}$ until the terminal time $\tau - s\lambda$.

The terminal point $x_s(\tau)$ is in $\mathcal{A}(x_0)$ for any positive numbers $\lambda_1, \ldots, \lambda_m$ and any number λ, provided that s is a sufficiently small positive number. The tangent vector v at $\bar{x}(\tau)$, obtained by differentiating $x_s(\tau)$ with respect to s at $s = 0$, is of the form

$$v = \lambda X_{\bar{u}(\tau)}(\bar{x}(\tau)) + \sum_{i=1}^{m} \lambda_i \Psi_{\tau, \tau_i *}(X_{u_i} - X_{\bar{u}(\tau_i)})(\bar{x}(\tau_i)). \tag{6.14}$$

as can be easily verified by induction on the number of regular points.

The set of vectors v given by (6.14) may be regarded as the positive convex cone generated by the elementary perturbation vectors (for the same terminal point τ). We shall use $C(\tau)$ to denote its topological closure. Since $X_{\bar{u}(\tau_2)}(\bar{x}(\tau_2)) = \Psi_{\tau_2, \tau_1 *} X_{\bar{u}(\tau_1)}(\bar{x}(\tau_1))$,

$$\Psi_{\tau_2, \tau_1 *} C(\tau_1) \subseteq C(\tau_2), \tag{6.15}$$

for any regular points τ_1 and τ_2 with $\tau_1 < \tau_2$.

Definition 6.13 The cone $K(T) = \bigcup \{C(t), t \leq T, t \text{ regular}\}$ is called the cone of attainability at $\bar{x}(T)$.

Lemma 6.14 $K(T)$ is a closed convex cone.

Proof   If $v \in K(T)$, then $v \in C(t)$ for some regular time t. Since $C(t)$ is a cone, $\alpha v \in C(t)$ for all $\alpha \geq 0$. Hence, $\alpha v \in K(T)$. If $v_1 = \Psi_{T, \tau_1 *} c_1$ for some

$c_1 \in C(t_1)$ and if $v_2 = \Psi_{T,\tau_2 *} c_2$ for some $c_2 \in C(t_2)$ with $\tau_1 < \tau_2$, then $v_1 = \Psi_{T,\tau_2 *} \Psi_{\tau_2,\tau_1 *} c_1$. Now both vectors c_2 and $\Psi_{\tau_2,\tau_1 *} c_1$ belong to $C(\tau_2)$ hence, their sum belongs to $C(\tau_2)$. This implies that $v_1 + v_2$ belongs to $K(T)$. $\qquad\square$

Part 2

Lemma 6.15 *The cone of attainability $K(T)$ is not equal to $T_{\bar{x}(T)}M$ whenever $\bar{x}(T) \in \partial \mathcal{A}(x_0)$.*

Proof Suppose that $K(T) = T_{\bar{x}(T)}M$. Let $w_1, \ldots, v_n$ be any points in $K(T)$ such that the convex cone generated by these points is $T_{\bar{x}(T)}M$. Then there exist regular points $\tau_1, \ldots, \tau_n$ and vectors $c_1, \ldots, c_n$, such that each $c_i \in C(\tau_i)$ and $w_i = (\Psi_{T,\tau_i})_* c_i = w_i$.

There is no loss in generality in assuming that $\tau_1 \leq \tau_2 \leq \cdots \leq \tau_n$. Then vectors $v_i = (\Psi_{\tau_n,\tau_i})_* c_i$, $i = 1, \ldots, n$ all belong to $C(\tau_n)$ and the positive convex cone generated by these vectors is equal to $T_{\bar{x}(\tau_n)}M$.

The preceding argument shows that $K(T) = T_{\bar{x}(T)}M$ implies that $C(\tau) = T_{\bar{x}(\tau)}M$ for some regular point $\tau \leq T$. Since $C(\tau)$ is the closure of the convex cone spanned by the elementary perturbation vectors, there exist elementary vectors $v_1, \ldots, v_m$ in $C(\tau)$ such that the convex cone generated by $v_1, \ldots, v_m$ is equal to $T_{\bar{x}(\tau)}M$. Let $\tau_1, \ldots, \tau_m$ denote regular points in the interval $[0, \tau]$ such that

$$v_i = \lambda X_{\bar{u}(\tau)}(\bar{x}(\tau)) + \lambda_i (\Psi_{\tau,\tau_i})_* (X_{u_i} - X_{\bar{u}(\tau_i)})(\bar{x}(\tau_i)), \; i = 1, \ldots, m.$$

There is no loss in generality in assuming that all these regular points are distinct, and renumbered, so that $0 = \tau_0 < \tau_1, \tau_2 < \cdots < \tau_m$. Let $F_i(s_i) = \Phi_{s_i}^{u_i} \Psi_{\tau_i - s_i, \tau_{i-1}}, i = 1, \ldots, m$, and then let

$$F(s_1, \ldots, s_m) = \Psi_{\tau + s_1 + \cdots + s_m, \tau_m} \circ F_m(s_m) \circ F_{m-1}(s_{m-1}) \circ \cdots \circ F_1(s_1)(x_0).$$

$$(6.16)$$

It follows that $F(0) = \bar{x}(T)$ and that there is neighborhood $\mathcal{O}$ of 0 in $\mathbb{R}^m$ such that $F(\mathcal{O} \cap \mathbb{R}_+^m) \subseteq \mathcal{A}(x_0)$, where $\mathbb{R}_+^m$ denotes the orthant $\{s \in \mathbb{R}^m : s_i \geq 0, i = 1, \ldots, m\}$.

Since $\frac{\partial F}{\partial s_i}(0) = v_i$, $i = 1, \ldots, m$, and since the convex cone spanned by $v_1, \ldots, v_m$ is equal to $T_{\bar{x}(\tau)}M$, $dF_0(\mathbb{R}_+^m) = T_{\bar{x}(\tau)}M$, where dF_0 denotes the tangent map of F at $s = 0$. But then, according to the Generalized Implicit Function theorem [[AS], Lemma 12.4, p. 172], $\bar{x}(T) = F(0) \in int F(\mathcal{O} \cap \mathbb{R}_+^m)$, which contradicts our assumption that $\bar{x}(T)$ is on the boundary of $cA(x_0)$. $\qquad\square$

Part 3: The extremal curve

We will now revert to the notations established earlier, with $\Psi_{t,\tau}$ the flow induced by $X_{\bar{u}(t)}$, $(\Psi_{t,\tau})_*$ its tangent flow along by $\bar{x}(t)$, defined by $(\Psi_{t,\tau})_*(v) = \frac{d}{ds}(\Psi_{t,\tau})(\sigma(s)|_{s=0}$, where $\sigma(s)$ is a curve that satisfies $s(0) = \bar{x}(\tau)$ and $\frac{d\sigma}{ds}(0) = v$, and $\Psi_{t,\tau}^*$ the dual of $\Psi_{t,\tau}$.

Since $K(T)$ is a proper closed convex cone there exists a covector $\xi_1 \in T_{\bar{x}(T)}^* M$ such that $\xi_1(v) \leq 0$ for each $v \in K(T)$. Moreover, $\xi_1(X_{\bar{u}(T)}(\bar{x}(T)) = 0$, since $\alpha X_{\bar{u}(T)}(\bar{x}(T))$ is in $K(T)$ for any real number α.

Let $\bar{\xi}(t) = \Phi_{T,t}^*(\xi_1)$. It then follows from Proposition 6.11 that $\bar{\xi}(t)$ is an integral curve of $\vec{h}_{\bar{u}(t)}$ whose projection on M is equal to $\bar{x}(t)$. It remains to show that $\bar{\xi}(t)$ satisfies the conditions of the Maximum Principle. If t is a regular point and if $v \in C(t)$, then $v_T = \Phi_{T,t_*}(v)$ belongs to $K(T)$ and hence, $\bar{\xi}(t)(v) = \bar{\xi}(T)(v_T) \leq 0$ by the "forward-backward principle." This shows that $\bar{\xi}(t)(X_u - X_{\bar{u}(t)}(\bar{x}(t)) \leq 0$ and that $\bar{\xi}(t)X_{\bar{u}(t)}(\bar{x}(t)) = 0$ for any regular point t. Since almost all points are regular,

$$0 = h_{\bar{u}(t)}(\bar{\xi}(t)) \geq h_u(t)(\bar{\xi})(t), \; u \in U$$

almost everywhere in $[0, T]$.   □

There is another version of the Maximum Principle pertaining to the trajectories $x(t)$ whose terminal point $x(T)$ belongs to the boundary of $\mathcal{A}(x_0, T)$. This version is given by our next theorem.

Theorem 6.16 The Maximum Principle–version 2 *Suppose that $x(t)$ is a trajectory generated by a control $\bar{u}(t)$ such that the terminal point $x(T)$ belongs to $\partial \mathcal{A}(x_0, T)$. Then $x(t)$ is the projection of a curve $\xi(t)$ in T^*M defined on $[0, T]$ such that:*

1. *$\frac{d\xi}{dt}(t) = \vec{h}_{\bar{u}(t)}(\xi(t))$ for almost all $t \in [0, T]$.*
2. *$\xi(t) \neq o$ for any $t \in [0, T]$.*
3. *$h_{\bar{u}(t)}(\xi(t))$ is constant on $[0, T]$.*
4. *$h_{\bar{u}(t)}(\xi(t)) \geq h_u(\xi(t))$ for any u in U for almost all $t \in [0, T]$.*

Proof Let (x, y) denote the points of $\tilde{M} = M \times \mathbb{R}$, and then consider an extended control system on $\tilde{M}$ given by

$$\frac{dx}{dt} = F(x(t), u(t)), \frac{dy}{dt} = 1, \tag{6.17}$$

over the bounded and measurable control functions that take values in a given set U in $\mathbb{R}^m$. Let $\tilde{\mathcal{A}}(\tilde{x}_0)$ denote the set of points in $\tilde{M}$ that are reachable from $\tilde{x}_0 = (x_0, 0)$ by the trajectories of (6.17). A point $\tilde{x}_1 = (x_1, T)$ is reachable by a trajectory $\tilde{x}(t) = (x(t), y(t))$ from $\tilde{x}_0$ if and only if $x_1 = x(T)$. Morever, $\tilde{x}(T) \in \partial \tilde{\mathcal{A}}(\tilde{x}_0)$ whenever $x(T) \in \partial \mathcal{A}(x_0, T)$.

Therefore, Theorem 6.12 is applicable to the trajectory $\tilde{x}(t) = (x(t), y(t))$ generated by the control $\bar{u}(t)$. Let $\tilde{\xi}$ denote the points of $T^*\tilde{M}$. Since $T^*\tilde{M} = T^*M \times T^*\mathbb{R}$, $\tilde{\xi} = (\xi, \lambda)$ with $\xi \in T^*M$ and $\lambda \in T^*\mathbb{R}$. If $T^*\mathbb{R}$ is represented by $\mathbb{R}^* \times \mathbb{R}$, then each point $\lambda \in T_y^*\mathbb{R}$ can be written as $\lambda = (p, y)$. Then $\tilde{h}_{\bar{u}(t)}(\tilde{\xi}) = h_{\bar{u}(t)}(\xi) + p$. Let $\tilde{\xi}(t) = (\xi(t), \lambda(t))$ denote an integral curve of $\tilde{h}_{\bar{u}(t)}$ that satisfies the Maximum Principle in Theorem 6.12. Then,

$$\frac{d\xi}{dt} = h_{\bar{u}(t)}(\xi(t)), \frac{dp}{dt} = -\frac{\partial \tilde{h}_{\bar{u}(t)}}{\partial y} = 0.$$

Thus, p is constant. But then $h_{\bar{u}(t)}(\xi(t))$ is constant, since $\tilde{h}_{\bar{u}(t)}(\tilde{\xi}(t)) = 0$. Finally, $\tilde{h}_{\bar{u}(t)}(\tilde{\xi}(t)) \geq \tilde{h}_u(\tilde{\xi}(t))$ implies that $h_{\bar{u}(t)}(\xi(t)) \geq h_u(\xi(t))$ almost everywhere. □

Alternatively, one could have proved this version of the Maximum Principle by modifying the perturbations in the proof of Theorem 6.12 so that the endpoint of the perturbed trajectory traces a curve in $\mathcal{A}(x_0, T)$, as was done in [AS]. Simply restrict to $\lambda = 0$ in (6.11), and (6.13). Then, the proof in Theorem 6.12 leads to the inequality

$$h_{\bar{u}(t)}(\xi(t)) \geq h_u(\xi(t))$$

for any u in U for almost all $t \in [0, T]$. The remaining part of the proof follows from the following proposition.

Proposition 6.17 *Let $\xi(t)$ be an integral curve of the Hamiltonian $\vec{h}_{u(t)}$ defined in the interval $[0, T]$ by a bounded and measurable control $u(t)$. If $\xi(t)$ satisfies $h_{u(t)}(\xi(t)) \geq h_v(\xi(t)), v \in U$, at all points t of $[0, T]$ where $h_{u(t)}(\xi(t))$ is differentiable, then $h_{u(t)}(\xi(t))$ is constant in $[0, T]$.*

Proof Let K equal to the closure of $\{u(t) : t \in [0, T]\}$. Since $u(t)$ is bounded, K is compact. By the usual estimates on vector fields over compact domains one can show that function $\phi(\xi) = Max_{u \in K}\{h_u(\xi) : u \in K\}$ is Lipschitzian as a function on T^*M. Therefore, $h_{u(t)}(\xi(t))$ is a Lipschitzian curve and hence, absolutely continuous.

We will next show that $\frac{d}{dt}\phi(\xi(t)) = 0$ a.e. in $[0, T]$. Since $\phi(\xi(t)) = h_{u(t)}(\xi(t))$ almost everywhere, this would imply that $\frac{d}{dt}h_{u(t)}(\xi(t))$ is equal to 0 almost everywhere, which in turn would imply that $h_{u(t)}(\xi(t))$ is constant.

Let τ denote a Lebesgue point for $u(t)$. Then $\phi(\xi(t))$ is differentiable at $t = \tau$ and $\phi(\xi(\tau)) = h_{u(\tau)}(\xi(\tau))$. For points $t > \tau$,

$$\frac{\phi(\xi(t)) - \phi(\xi(\tau))}{t - \tau} \geq \frac{h_{u(\tau)}(\xi(t)) - h_{u(\tau)}(\xi(\tau))}{t - \tau}.$$

After taking the limit as $t \to \tau$, the above yields $\frac{d}{dt}\phi(\xi(t))|_{t=\tau} \geq \frac{d}{dt}\phi_{u(\tau)}$ $(\xi(t))|_{t=\tau} = \{h_{u(\tau)}, h_{u(\tau)}\}(\xi(t)) = 0$. The same argument with $t < \tau$ shows that $\frac{d}{dt}\phi(\xi(t))|_{t=\tau} \leq 0$. Therefore, $\frac{d}{dt}\phi(\xi(t))|_{t=\tau} = 0$.   □

We will now adapt the Maximum Principle to problems of optimal control. The Hamiltonian lift of the cost-extended system

$$\frac{dy}{dt}(t) = f(x(t), u(t)), \quad \frac{dx}{dt}(t) = F(x(t), u(t))$$

takes place in $T^*\tilde{M}$, where $\tilde{M} = \mathbb{R} \times M$. Then $T^*\tilde{M} = T^*\mathbb{R} \times T^*M$. Points $\tilde{\xi}$ in $T^*\tilde{M}$ will be written as $\tilde{\xi} = (\xi_0, \xi)$. In these notations, the Hamiltonian lift is given by

$$h_{\bar{u}(t)}(\tilde{\xi}) = \xi_0 f_{u(t)}(x) + \xi F_{u(t)}(x), \tilde{\xi} \in T^*_{(y,x)}\tilde{M}.$$

Since this Hamiltonian does not depend explicitly on the variable y, $\xi_0(t)$ is constant along each integral curve $(\xi_0(t), \xi(t))$ of $\vec{h}_{u(t)}$. According to the Maximum Principle, $\tilde{x}(t)$ is the projection of an extremal curve $\tilde{\xi}(t)$ defined by the Hamiltonian lift

$$h_{\bar{u}(t)}(\tilde{\xi}) = \xi_0 f_{u(t)} + \xi F_{u(t)}$$

for $\tilde{\xi} \in T^*_{(y,x)}\tilde{M}$. For optimal control problems in which the cost functional is minimized $\xi_0 \leq 0$. Since the Hamiltonian is a homogeneous function, this constant can be reduced to -1 whenever it is not equal to zero.

The Hamiltonian $h_{u(t)(\tilde{\xi})} = \xi_0 f_{u(t)} + \xi F_{u(t)}$ is usually identified with its projection on T^*M in which ξ_0 is regarded as a parameter. The values of ξ_0 give rise to two kinds of extremal curves – normal and abnormal. Extremal curves which are generated by the Hamiltonian vector field that correspond to h_u with $\xi_0 = -1$ are called *normal* and the extremals generated by $h_u(\xi) = \xi \circ F_u$ are called *abnormal*. In this context the Maximum Principle lends itself to the following formulation.

Proposition 6.18 The Maximum Principle of optimality   *The optimal trajectories $(x(t), u(t))$ are either the projections of normal extremal curves $\xi(t)$ generated by the Hamiltonian $h_u(\xi) = -f(x, u) + \xi \circ F_u(x)$, or they are the projections of abnormal extremal curves generated by the Hamiltonian $h_u(\xi) = \xi \circ F_u(x)$. The abnormal extremal curves cannot be equal to zero at any time $t \in [0, T]$. In either case, the extremal curves conform to the maximality condition*

$$h_{u(t)}(\xi(t)) = \xi_0 f(x(t), u(t)) + \xi(t) \circ F_u(t)(x(t)) \geq \xi_0 f(x(t), v) + \xi(t) \circ F_v(x(t))$$
$$(6.18)$$

for any $v \in U$ and almost all $t \in [0, T]$. For optimal problems with a variable time interval, $h_{u(t)}(\xi(t)) = 0$, a.e. on $[0, T]$.

The Maximum Principle also covers the case where a two-point boundary value optimal problem is replaced by an optimal control problem of finding a trajectory $(x(t), u(t))$ that originates on a submanifold S_0 of M at $t = 0$ and terminates on a submanifold S_1 of M at $t = T$ and minimizes the integral $\int_0^T f(x(t), u(t)) \, dt$ among all such trajectories. For such problems, the Maximum Principle imposes additional conditions known as the transversality conditions:

If $(x(t), u(t))$ is the projection of an extremal curve $\xi(t)$ then $\xi(0)(v) = 0$ for all tangent vectors v in the tangent space of S_0 at $x(0)$ and $\xi(0)(v) = 0$ for all tangent vectors v in the tangent space of S_1 at $x(T)$.

6.4 The Maximum Principle in the presence of symmetries

For Lie determined systems, the perturbations could be extended to vector fields in the Lie saturate to obtain the following version of the Maximum Principle.

Theorem 6.19 The saturated Maximum Principle *Suppose that $\mathcal{F}$ consisting of vector fields $X_u(x) = F(x, u)$, $u \in U$ is Lie determined, and suppose that $\bar{x}(t)$ is a solution of $\frac{d\bar{x}}{dt} = F(\bar{x}(t), \bar{u}(t))$ such that $x(T) \in \partial \mathcal{A}(x_0)$. Then $\bar{x}(t)$ is the projection of a curve $\bar{\xi}(t)$ in T^*M defined on $[0, T]$ such that:*

1. *$\frac{d\bar{\xi}}{dt}(t) = \vec{h}_{\bar{u}(t)}(\bar{\xi}(t))$ for almost all $t \in [0, T]$.*
2. *$\bar{\xi}(t) \neq o$ for any $t \in [0, T]$.*
3. *$h_{\bar{u}(t)}(\bar{\xi}(t)) = 0$ for all $t \in [0, T]$.*
4. *$h_X(\bar{\xi}(t)) \leq 0$ for any X in $LS(\mathcal{F})$ for almost all $t \in [0, T]$ where h_X denotes the Hamiltonian lift of X.*

Proof If the cone of attainability $K(T)$ corresponding to the perturbations in the Lie saturate were equal to $T_{\bar{x}(t)}M$, then, as in the proof of the Maximum Principle, that would imply that the terminal point $\bar{x}(T)$ is in the interior of the closure of $\mathcal{A}(x_0)$, which then would imply that $\bar{x}(T)$ is in the interior of $\mathcal{A}(x_0)$, as a consequence of Proposition 6.2. This implication violates the original assumption that $\bar{x}(T)$ belongs to the boundary of $\mathcal{A}(x_0)$. $\qquad\square$

In general, it is not true that $cl(int(\mathcal{A}(x, T))) = int(\mathcal{A}(x, T))$, and therefore, $cl(\mathcal{A}(x, T))$ is not a useful object to generate an analogous extension of the Maximum Principle–version 2. However, in the presence of symmetries, condition (4) of the Maximum Principle–version 2 could be extended to a larger family to give additional information about the extremal trajectory.

To elaborate, let us first define the basic terms.

Definition 6.20 A diffeomorphism Φ is a symmetry for control system (6.1) if $\Phi^{-1}(\mathcal{A}(\Phi(x), T)) = \mathcal{A}(x, T)$, $x \in M$.

If Φ is a symmetry, then $\Phi^{-1}\mathcal{A}(x, T) = \mathcal{A}(\Phi^{-1}x, T)$, which in turn implies that Φ^{-1} is a symmetry as well. We will use $Sym(\mathcal{F})$ to denote the group of symmetries associated with control system (6.1). Evidently, any diffeomorphism Φ which is a symmetry also satisfies a weaker condition $\Phi^{-1}(\mathcal{A}(\Phi(x))) = \mathcal{A}(x)$, $x \in M$; hence, $Sym(\mathcal{F})$ is contained in the group of normalizers for $\mathcal{F}$. Therefore, $\{\Phi_\sharp(\mathcal{F}); \Phi \in Sym(\mathcal{F}), X \in \mathcal{F}\} \subseteq LS(\mathcal{F})$.

However, a normalizer need not be a symmetry, as the following example shows.

Example 6.21 Let M be the group of motions of the plane. Then each matrix g in M can be represented as $g = \begin{pmatrix} 1 & 0 \\ x & R \end{pmatrix}$, $x \in \mathbb{R}^2$, $R \in SO_2(R)$. Consider the Serret–Frenet system

$$\frac{dg}{dt} = g(t) \begin{pmatrix} 0 & 0 & 0 \\ 1 & 0 & -u(t) \\ 0 & u(t) & 0 \end{pmatrix},$$

with $u(t)$ playing the role of control. For any solution $g(t) = \begin{pmatrix} 1 & 0 \\ x(t) & R(t) \end{pmatrix}$, T is the length of $x(t)$ in the interval $[0, T]$ and $u(t)$ is the signed curvature of $x(t)$.

It follows that $\mathcal{A}(g_0, T)$ is not equal to M for any $T > 0$. Since $\mathcal{A}(g_0) = M$ for any $g_0 \in M$, any diffeomorphism is a weak symmetry for the system, but only the left-translations are the symmetries of the system.

Proposition 6.22 *Suppose that the control system admits a group of symmetries* $Sym(\mathcal{F})$. *Then each extremal curve* $\xi(t)$ *in Theorem 6.16 satisfies the strengthened condition;*

$$h_{\bar{u}(t)}(\xi(t)) \geq h_X(\xi(t)), \tag{6.19}$$

for any $X \in \{\Phi_\sharp(\mathcal{F}); \Phi \in Sym(\mathcal{F}), X \in \mathcal{F}\}$, *and for almost all* $t \in [0, T]$.

The proof is analogous to the proof of the saturated Maximum Principle and will be omitted.

Definition 6.23 A vector field X is said to be a symmetry for system (6.1) if its one-parameter group of diffeomorphisms $\{\Phi_t : t \in R\}$ is a subgroup of $Sym(\mathcal{F})$).

Proposition 6.24 Noether's theorem *Suppose that a vector field X is a symmetry for system (6.1). Then the Hamiltonian h_X associated with the Hamiltonian lift of X is constant along each extremal curve $\xi(t)$ defined by Theorem 6.16.*

Proof Let $\xi(t)$ denote an extremal curve generated by a control function $\bar{u}(t)$ that projects onto the trajectory $x(t)$ and let $\{\Phi_s : s \in \mathbb{R}\}$ denote the group of diffeomorphisms generated by X. As before, $(\Phi_s)_\sharp Y = (\Phi_s)_* Y\Phi_{-s}$. The strengthened maximality condition implies that

$$h_{\bar{u}(t)}(\xi(t)) \geq h_{(\Phi_s)_\sharp Y}(\xi(t)), \ a.e. \in [0, T], \tag{6.20}$$

for any $Y \in \mathcal{F}$ and any s. This inequality is preserved if Y is replaced by any time-varying vector field $Y_{u(t)}(x) = F(x, u(t))$, and in particular it is preserved when $u(t) = \bar{u}(t)$. Let $X_t(x) = F(x, \bar{u}(t))$.

The resulting inequality shows that $h_{\Phi_s \sharp X_t}(\xi(t))$ attains its maximum for $s = 0$ and therefore, $\frac{d}{ds} h_{\Phi_s \sharp X_t}(\xi(t))|_{s=0} = 0$. This means that

$$0 = \frac{d}{ds} h_{\Phi_s \sharp X_t}(\xi(t))|_{s=0} = \xi(t)([X, X_t](x(t)) = \{h_X, h_{\bar{u}(t)}\}(\xi(t).$$

Since $\frac{d}{dt} h_X(\xi(t)) = \{h_X, h_{\bar{u}(t)}\}(\xi(t)) = 0$, $h_X(\xi(t))$ is constant. □

To recognize Noether's theorem in its more familiar form we need to return to optimal control problem and the cost-extended system (6.2)

$$\frac{dy}{dt}(t) = f(x(t), u(t)), \ \frac{dx}{dt}(t) = F(x(t), u(t))$$

associated with the optimal control problem of minimizing $\int_0^T f(x(t), u(t)) \, dt$ over the trajectories of $\frac{dx}{dt} = F(x(t), u(t))$, $u(t) \in U$ that satisfy the given boundary conditions on a fixed time interval $[0, T]$. The following proposition is a corollary of the preceding proposition.

Proposition 6.25 *Suppose that a vector field X on M with its flow $\{\Phi_s\}$ satisfies $(\Phi_s)_* F_u \circ \Phi_{-s} = F_{u_s}$ for some $u_s \in U$ and all $s \in \mathbb{R}$, and $f(\Phi_s(x), u_s) = f(x, u)$ for all $s \in \mathbb{R}$ and all $u \in U$. Then the vector field $\tilde{X} = (0, X)$ is a symmetry for the extended system (6.2). Consequently, the Hamiltonian $h_X(\xi_0, \xi) = \xi \circ X(x)$ is constant along each extremal curve $(\xi_0, \xi(t))$.*

Proof Let $\{\tilde{\Phi}_s\}$ denote the flow for the extended vector field $\tilde{X}$. If $z(t) = (y(t), x(t))$ is any trajectory of (6.2), let $\tilde{w}(t) = (w_0(t), w(t)) = \tilde{\Phi}_s(z(t))$. Then

$\frac{dw_0}{dt} = f(x(t), u(t)) = f(\Phi_s(x(t), u_s(t)))$ and $\frac{dw}{dt} = \Phi_{s*}F(\Phi_{-s}(x(t), u(t))) = F(w(t), u_s(t))$.

Therefore, $\tilde{w}(t)$ is a solution of (6.2) with $\tilde{w}(0) = \tilde{\Phi}_s(z(0))$. It follows that $\tilde{\Phi}_s\mathcal{A}(\tilde{x}_0, T) \subseteq \mathcal{A}(\tilde{\Phi}_s\tilde{x}_0, T)$. Hence, $\tilde{X}$ is a symmetry for (6.2).

Since $h_{\tilde{x}}(\xi_0, \xi) = \xi(X(x) = h_X(\xi), h_X(\xi(t)))$ is constant along any extremal curve $\tilde{\xi}(t) = (\xi_0, \xi(t))$ by the previous proposition. □

Suppose now that G is a Lie group which acts on a manifold M. If $\phi : G \times M \to M$ denotes this action, let Φ_g denote the induced diffeomorphism $\Phi_g(x) = \phi(g, x)$. The mapping $g \to \Phi_g$ is a homomorphism from G into the group of diffeomorphisms on M, that is, $\Phi_h \circ \Phi_g = \Phi_{hg}$ for any h, g in G.

If $\mathfrak{g}$ denotes the Lie algebra of G, the above implies that $\{\Phi_{e^{tA}} : t \in \mathbb{R}\}$ is a one-parameter group of diffeomorphisms on M for each exponential e^{tA} in G defined by $A \in \mathfrak{g}$. Let X_A to denote its infinitesimal generator. It follows that X_A is a complete vector field on M. In the terminology of [Jc], X_A is a vector field subordinated to G. It is easy to show that the correspondence $A \to X_A$ satisfies

$$X_{\alpha A + \beta B} = \alpha X_A + \beta X_B \text{ and } X_{[A,B]} = [X_A, X_B].$$

Let h_A denote the Hamiltonian lift of X_A, that is, $h_A(\xi) = \xi(X_A(\pi(\xi)))$, $\xi \in T^*M$. For a fixed ξ, the mapping $A \to h_A(\xi)$ is linear, hence can be identified with an element $\ell(\xi)$ in the dual $\mathfrak{g}^*$ of $\mathfrak{g}$.

Definition 6.26 The mapping $\ell : T^*M \to \mathfrak{g}^*$ is called the moment map [AM].

Proposition 6.27 *Suppose that G is a symmetry group for a control system $\frac{dx}{dt} = F(x(t), u(t))$, $u(t) \in U$. Then the moment map is constant along each extremal curve $\xi(t)$ defined by Theorem 6.16.*

Proof A paraphrase of the proof in Proposition 6.24 shows that $h_A(\xi(t))$ is constant along each extremal curve $\xi(t)$ for any $A \in \mathfrak{g}$. This means that $\ell(\xi(t))$ is constant. □

The action of G on M can be extended to an action on T^*M by the formula $\Phi_g^*(\xi) = \xi \circ \Phi_{g^{-1}*}$ for each $\xi \in T_x^*M$, where Φ_{g*} denotes the tangent map of Φ_g. It follows that $\Phi_h^* \circ \Phi_g^* = \Phi_{hg}^*$, and hence the mapping $(g, \xi) \in G \times T_xM \to (\Phi_g^*(\xi) \in T_x^*M$ is an action under G. The moment map carries the following equivariance property:

$$\ell(\Phi_g^*(\xi)) = Ad_g^*(\ell)(\xi),$$

where Ad_g^* denotes the coadjoint action of G on $\mathfrak{g}^*$.

The above implies that G is a symmetry group under the right action for any left-invariant control system F. Then G itself is a symmetry group for F under the right action. The fact that F is left-invariant means that $F(x, u) = L_{x*}F(e, u)$ for each $x \in G$ and each $u \in U$. In this notation, L_x denotes the left translation by x, L_{x*} is its tangent map, and e is the group identity in G. This implies that $\mathcal{A}(x, T) = L_x \mathcal{A}(e, T)$.

Hence, the group of left translations is a symmetry group for any left-invariant control system. If $\Phi_g = L_g$ then $\Phi_{e^{tA}}x = xe^{tA}$. Hence, X_A is the right-invariant vector field with $X_A(e) = A$.

To preserve the left-invariant symmetries, T^*G will be realized as $G \times \mathfrak{g}^*$ as explained in the previous chapter. In this representation, points ξ in T^*G are represented by pairs (g, ℓ); the Hamiltonians of left-invariant vector fields $X_A = L_{g*}(A)$ are given by $h_A = \ell(A)$ and $h_A(g, \ell) = \ell(L_{g^{-1}*} \circ R_{g*}(A))$ are the Hamiltonians of right-invariant fields $X_A(g) = R_{g*}(A)$.

It then follows from Proposition 6.24 that $\ell(t)(L_{g^{-1}(t)*} \circ R_{g}(t)_*(A))$ is constant along each extremal curve $\xi(t) = (g(t), \ell(t))$. But this means that $Ad^*_{g(t)}(\ell(t)) = Ad^*_{g(0)}(\ell(0))$, where Ad^* denotes the coadjoint action of G, or, equivalently,

$$\ell(t) = Ad^*_{g^{-1}(t)g(0)}(\ell(0)).$$

The above can be assembled into the following proposition.

Proposition 6.28 *The projection $\ell(t)$ of each extremal curve $\xi(t) = (g(t), \ell(t))$ generated by a left-invariant control system evolves on the coadjoint orbit of G through $\ell(0)$.*

6.5 Abnormal extremals

For optimal control problems in which the dimension of the control space is less than the dimension of the state space it may happen that an optimal trajectory is not a projection of a normal extremal curve. The presence of "abnormal curves" for the problems of the calculus of variations was noticed by A. G. Bliss in the early 1920s [Bl]. Carathéodory rediscovered these curves in the late 1930s in his treatment of Zermelo's navigational problem, which he called anomalous [Cr]. Much later, L. C. Young drew attention to the "rigid curves," curves that do not admit variations and hence can not be detectable by the classical methods of the calculus of variations [Yg]. In some instances, particularly in differential geometry literature, abnormal extremals are ignored which resulted in a number of false claims [Rn; Sz].

It was not until the late 1990s that it became clear that there was nothing "abnormal" about abnormal extremals and that they may occur in many instances for the following reasons. Each optimal trajectory $x(t)$ that minimizes the cost $\int_0^T f(x(t), u(t))\, dt$ defines an extended trajectory $\tilde{x} = (y(t), x(t))$ which terminates on the boundary of extended reachable set $\tilde{A}(\tilde{x}_0, T)$ at time T because $y(T) = \int_0^T f(x(t), u(t))\, dt$ is minimal. Therefore, the cone of attainability along an optimal trajectory cannot be the entire tangent space. In the simplest of all situations, the cone of attainability has a non-empty interior in the extended tangent space and its projection on $T_x M$ is equal to $T_x M$. Then there are no abnormal extremals.

However, it may happen that the projection of the cone of attainability on $T_x M$ is a proper closed convex cone, possibly of lower dimension, even though the reachable set $A(x_0, T)$ is open for each $x_0 \in M$ and any $T > 0$. Since the projection of the cone of attainability on $T_* M$ corresponds to the zero multiplier in front of the cost, the projection is independent of the cost and depends only on the control system. This is the situation that results in abnormal extremals. In some cases the projection of these extremals is optimal for a given cost functional, and not optimal for some other cost functionals. Determining the optimality status of the projected trajectory is in general a difficult problem [Ss1]. Below we will illustrate the general situation with some problems of sub-Riemannian geometry initiated by a famous paper of R. Montgomery [AB; BC; Mt].

Let $\mathcal{D}$ denote the kernel of $\Theta = dz - A(y)dx$, where x, y, z denote the coordinates of a point $q \in \mathbb{R}^3$ and where A is a function with a single non-degenerate critical point at $y = 0$, that is, A is a function with $\frac{dA}{dy}(0) = 0$ and $\frac{d^2A}{dy^2}(0) \neq 0$. Then $\mathcal{D}$ is a two-dimensional distribution spanned by vector fields $X_1 = \frac{\partial}{\partial x} + A(y)\frac{\partial}{\partial z}$ and $X_2 = \frac{\partial}{\partial y}$ The trajectories of $\mathcal{D}$ are the solutions of

$$\frac{dq}{dt} = u(t)X_1(q) + v(t)X_2(q), \tag{6.21}$$

where $u(t)$ and $v(t)$ are bounded and measurable control functions. In coordinates,

$$\frac{dx}{dt} = u(t), \frac{dy}{dt} = v(t), \frac{dz}{dt} = A(y)u(t). \tag{6.22}$$

The paper of Agrachev *et al.* deals with the Martinet form $\Theta = dz - \frac{y^2}{2}dx$ [AB]. In the paper of Montgomery [Mt], θ is given by $\Theta = dz - A(r)d\theta$, where r, θ, z are the cylindrical coordinates $x = r\cos\theta$, $y = r\sin\theta$, $r = \sqrt{x^2 + y^2}$ and A is a function of r that has a single non-degenerate critical point at $r = 1$. If we introduce new coordinates $x = \theta$, $y = r + 1$, and redefine $A(y)$ as $A(y + 1)$, then the Martinet distribution becomes a special case of the Montgomery distribution.

It is easy to verify that $[X_1, X_2] = A'(y))\frac{\partial}{\partial z}$ and that $[X_2, [X_1, X_2]] = -A''\frac{\partial}{\partial z}$. Hence, $Lie_q\{X_1, X_2\} = T_q\mathbb{R}^3$ at all points q, and therefore, any pair of points in $\mathbb{R}^3$ can be connected by a trajectory of (6.21).

A two-dimensional distribution in a three-dimensional ambient space is called *contact* if for any point q in the ambient space M and any choice of vector fields X_1 and X_2 that span $\mathcal{D}$ in a neighborhood of a point q, $X_1(q), X_2(q), [X_1, X_2](q)$ are linearly independent. The distribution defined by Θ is not contact since $[X_1, X_2](q) = 0$ on the plane $y = 0$. This plane is called the Martinet plane.

Since $dz = A(y)dx$ on $\mathcal{D}$, any metric on $\mathcal{D}$ can be written in the form $a(q)dx^2 + 2b(q)dxdy + c(q)dy^2$ for some functions a, b, c. Over the solutions $q(t)$ of (6.21) such a metric can be written as $\left\langle \frac{dq}{dt}, \frac{dq}{dt} \right\rangle = a(q(t))u^2(t) + 2b(q(t))u(t)v(t) + c(q(t))v^2(t)$. In terms of this notation, the length of $q(t)$ on an interval $[0, T]$ is given by $\int_0^T \sqrt{\left\langle \frac{dq}{dt}, \frac{dq}{dt} \right\rangle}\, dt$, while its energy is given by $\frac{1}{2}\int_0^T \left\langle \frac{dq}{dt}, \frac{dq}{dt} \right\rangle dt$.

Let us now consider the extremals associated with the sub-Riemannian problem of finding a trajectory of shortest length that connects two given points in $\mathbb{R}^3$. This problem can be phrased either as a time optimal problem over the solutions of (6.21) subject to the constraint $\left\langle \frac{dq}{dt}, \frac{dq}{dt} \right\rangle \leq 1$, or as the problem of minimizing the energy over a fixed time interval $[0, T]$. The latter results in the optimal control problem of minimizing $\int_0^T (a(q(t)u^2(t) + 2b(q(t)u(t)v(t) + c(q(t)v^2(t))\, dt$ over the trajectories of (6.21) that satisfy $q(0) = q_0$ and $q(T) = q_1$.

For simplicity of exposition we will assume that $b = 0$, and we will also assume that a and c depend only on x and y, and are of the form $a(x, y) = 1 + yf(x, y)$ and $c(x, y) = 1 + g(x, y)$ for some smooth functions f and g. The case where a, b, c do not depend on z is called *isoperimetric* in [AB]. The same paper shows that in the case $A(y) = \frac{y^2}{2}$ any sub-Riemannian isoperimetric problem can be reduced to the above normal form. The metric in this normal form is said to be *flat* if $a = c = 1$. In the paper of Montgomery [Mt], the metric $dr^2 + r^2d\theta^2$ corresponds to $a = 1 + y(2 + y)$ and $c = 1$.

Let us now consider the abnormal extremals associated with (6.21). As we already remarked, those extremals depend on the distribution only, and not on the metric. Let p_x, p_y, p_z denote the dual variables corresponding to the coordinates x, y, z. Then the Hamiltonian lift h of (6.22) is given by

$$h = u(t)h_1 + v(t)h_2,$$

where $h_1 = p_x + A(y)p_z$ and $h_2 = p_y$. An abnormal extremal is a solution of the following constrained Hamiltonian system:

$$\frac{dx}{dt} = u(t), \frac{dy}{dt} = v(t), \frac{dz}{dt} = A(y)v(t),$$

$$\frac{dp_x}{dt} = 0, \frac{dp_y}{dt} = -\frac{dA}{dy}p_z, \frac{dp_z}{dt} = 0, p_y = 0, p_x + A(y)p_z = 0. \tag{6.23}$$

It is easy to see that $x(t) = x_0 + t$, $y(t) = 0, z(t) = z_0 + A(0)t, p_x = -A(0)p_z, p_y = 0$, $p_z = c$, with c an arbitrary constant, are the only solutions of the above system. The projections of these abnormal curves are straight lines $x(t) = x_0 + t$, $z(t) = z_0 + A(0)t$ in the Martinet plane $y = 0$; they are generated by $u(t) = 1$ and $v(t) = 0$ (in the cylindrical coordinates of Montgomery, these lines are helices $r(t) = 1, \theta(t) = \theta_0 = t, z(t) = z_0 + A(1)t$).

The normal extremals are generated by the Hamiltonan

$$H = \frac{1}{2}\left(\frac{1}{a}(p_x + A(y)p_z)^2 + \frac{1}{c}p_y^2\right),$$

defined by the extremal controls $u = \frac{1}{a}(p_x + A(y)p_z)$ and $v = \frac{1}{c}p_y$. It follows that they are the solutions of

$$\frac{dx}{dt} = \frac{1}{a}(p_x + A(y)p_z), \frac{dy}{dt} = \frac{p_y}{c}, \frac{dz}{dt} = A(y)\frac{1}{a}(p_x + A(y)p_z),$$

$$\frac{dp_x}{dt} = \frac{a_x}{a^2}(p_x + A(y)p_z)^2 + \frac{c_x}{c^2}p_y^2, \tag{6.24}$$

$$\frac{dp_y}{dt} = \frac{a_y}{a^2}(p_x + A(y)p_z)^2 - \frac{A'(y)p_z}{a}(p_x + A(y)p_z) + \frac{c_y}{c^2}p_y^2, \frac{dp_z}{dt} = 0.$$

Definition 6.29 The projections of the extremals on $\mathbb{R}^3$ are called geodesics. A geodesic is called abnormal (respectively normal) if it is a projection of an abnormal (respectively normal) extremal curve. It is called strictly abnormal if it is a projection of an abnormal extremal curve but not the projection of a normal one.

Proposition 6.30 *Lines $x(t) = x_0 + t$, $z(t) = z_0 + A(0)t$ in the Martinet plane are strictly abnormal if and only if $f(x, 0) \neq 0$.*

Proof The above lines in the Martinet plane are the projections of normal extremal curves if and only if $p_x + A(0)p_z = 1$ and $a_y(x, 0) = 0$. Since $a(x, y) = 1 + yf(x, y)$, $a(x, 0) = 1$ and $a_y(x, 0) = f(x, 0)$). It follows that $a_y(x, 0) = 0$ if and only if $f(x, 0) = 0$. □

It follows that the abnormal geodesics are not strictly abnormal in the flat case, but that they are strictly abnormal in Montgomery's example. The abnormal geodesics are optimal in the flat case, and remarkably, they are also optimal (at least for small time intervals) in Montgomery's example. In both cases, point $q_1 = (x_0 + c, 0, A(0)c)$ with $c > 0$ can be reached from $q_0 = (x_0, 0, z_0)$ in time $t = c$ with controls that satisfy $u^2 + v^2 \leq 1$ only with $u(t) = 1$ and $v(t) = 0$. Hence, this trajectory is optimal for any optimal problem and not just for the sub-Riemannian one. This observation is evident in the flat case, but is not so obvious in the example of Montgomery. The proposition below provides the necessary arguments.

Proposition 6.31 *There exists an interval $[0, T]$ with $T < 1$ such that the helix $q(t) = (1, t, A(1)t)$ is optimal on this interval.*

Proof It will be convenient to work with the cylindrical coordinates. Then $X_1(r, \theta) = \frac{1}{r} \left(\frac{\partial}{\partial \theta} + A(r) \frac{\partial}{\partial z} \right)$ and $X_2 = \frac{\partial}{\partial r}$ are othonormal vector fields for the metric $dr^2 + r^2 d\theta^2$. Consider now the reachable set $\mathcal{A}(q_0, T)$ associated with

$$\frac{dr}{dt} = v(t), \quad \frac{d\theta}{dt} = \frac{1}{r} u(t), \quad \frac{dz}{dt} = \frac{1}{r} A(r) u(t), \quad u^2(t) + v^2(t) \leq 1 \qquad (6.25)$$

with $q_0 = (1, 0, 0)$. Since the control set U is compact and convex, each velocity set $\{ u X_1(x) + b X_2(x), u^2 + v^2 \leq 1 \}$ is compact and convex. Moreover, each trajectory that originates at q_0 at $t = 0$ exists for all t in $[0, T]$ with $T < 1$ (for $T = 1$ the trajectory with $u = -1$ escapes M at time 1 since $r(1) = 0$). Therefore, $\mathcal{A}(x_0, T)$ is closed, as a consequence of Fillipov's theorem [Jc, p. 120; AS, p. 141]. It is also compact, but that is not relevant for the computation below.

We will now consider an auxiliary problem of finding a trajectory $q(t)$ that originates at q_0, terminates at $r(T) = 1, \theta(T) = T$ and minimizes $z(T)$. Such a trajectory exists since the reachable set is closed. Moreover, the terminal point $q(T)$ is on the boundary of $\mathcal{A}(q_0, T)$.

Our claim is that the helix is the only solution for this auxiliary problem. The following observation is crucial: if $q(t) = (r(t), \theta(t), z(t))$ is any trajectory that solves the auxiliary problem such that $r(t)$ is not equal to 1 for all t, then $\frac{d\theta}{dt}$ must be negative somewhere in the interval $[0, T]$. The argument is simple: since $A(1)$ is a global mimimum for $A(r)$, $A(r) = A(1) + B(r)$ for some positive function $B(r)$. Then

$$z(T) = \int_0^T A(r(t)) \frac{d\theta}{dt} dt = A(1)T + \int_0^T B(r(t)) \frac{d\theta}{dt} dt.$$

Hence, $z(T) > A(1)T$ unless $\frac{d\theta}{dt}$ is negative on a set of positive measure.

According to the Maximum Principle, $x(t)$ is the projection of an extremal curve $\xi(t) = (r(t), \theta(t), p_r(t), p_\theta(t))$ associated with the Hamiltonian

$$h_{(u(t), v(t))} = -\frac{\lambda}{r} A(r) u(t) + p_r v(t) + \frac{1}{r} p_\theta u(t), \quad \lambda = 0, 1.$$

In the abnormal case $\lambda = 0$, $p_r^2 + p_\theta^2 > 0$ and hence the extremals are generated by the Hamiltonian $h = \sqrt{p_r^2 + \frac{1}{r^2}p_\theta^2}$. But then $\frac{d\theta}{dt} = \frac{1}{r^2 h}p_\theta$ and p_θ is constant. The projections of these curves cannot meet the boundary conditions in view of the above remark, and hence are ruled out.

We now pass to the normal case $\lambda = 1$. There are two possibilities: either

$$p_r^2(t) + (p_\theta - A(r))^2 > 0, \qquad (6.26)$$

and the extremal curves are generated by the Hamiltonian $h = \sqrt{p_r^2 + \frac{1}{r^2}(p_\theta - A(r))^2}$ corresponding to the extremal controls $u(t) = \frac{1}{hr}(p_\theta - A(r))$ and $v(t) = \frac{p_r}{h}$, or

$$p_r(t) = 0 \text{ and } p_\theta - A(r) = 0, a.e. \text{ on } [0, T]. \qquad (6.27)$$

In the latter case the extremal solutions are given by

$$\frac{dr}{dt} = v(t), \quad \frac{d\theta}{dt} = \frac{1}{r}u(t), \quad \frac{dp_r}{dt} = \left(-\frac{1}{r^2}A(r) + \frac{1}{r}A'(r)\right) + \frac{1}{r^2}p_\theta u(t), \quad \frac{dp_\theta}{dt} = 0.$$
$$(6.28)$$

Since p_θ is constant, $A(r(t)) = A(1)$ and $p_\theta = A(1)$. It follows that $r(t) = 1$, and hence, $v(t) = 0$. But then $u(t) = 1$, and the resulting solution is the helix.

We will show that the helix is the only solution by showing that the projections of the integral curves of $h = \sqrt{p_r^2 + \frac{1}{r^2}(p_\theta - A(r))^2}$ given by

$$\frac{dr}{dt} = \frac{p_r}{h}, \quad \frac{d\theta}{dt} = \frac{p_\theta - A(r)}{r^2 h}, \quad \frac{dp_r}{dt} = \frac{p_\theta - A}{hr^2}\left(\frac{p_\theta - A}{r} + A'(r)\right), \quad \frac{dp_\theta}{dt} = 0$$
$$(6.29)$$

cannot meet the given boundary conditions.

System (6.29) is integrable. In fact, p_θ is constant and

$$\frac{dr}{dt} = \pm\sqrt{h^2 - \frac{(p_\theta - A(r))^2}{r^2}}. \qquad (6.30)$$

For $\frac{d\theta}{dt}$ to change sign, $p_\theta - A(1)$ must be positive and there must be an interval $[t_0, t_1]$ such that $p_\theta - A(r(t)) < 0$ on (t_0, t_1) with $p_\theta - A(r(t_0)) = p_\theta - A(r(t_1)) = 0$. On this interval, $\frac{d\theta}{dr} = \pm\frac{p_\theta - A(r)}{hr^2}\frac{h}{\sqrt{h^2 - \frac{(p_\theta - A(r))^2}{r^2}}}$, or

$$\Delta\theta = \int \frac{(p - A(r)) \, dr}{r^2\sqrt{h^2 - \frac{(p_\theta - A(r))^2}{r^2}}}.$$

Let $\frac{p_\theta - A(r)}{r} = -h\cos\phi$. Then, $dr = \frac{\sin\phi}{h\cos\phi - A'(r)}d\phi$, and the above integral becomes

$$\Delta\theta = -\int \frac{\cos\phi \, d\phi}{r(h\cos\phi - A'(r))}. \qquad (6.31)$$

On the interval $[t_0, t_1]$, $-\frac{\pi}{2} \leq \phi \leq \frac{\pi}{2}$ and $\Delta\theta$ is given by

$$-\int_{-\frac{\pi}{2}}^{\frac{\pi}{2}} \frac{\cos\phi \, d\phi}{r(h\cos\phi - A'(r))}.$$

For small T, r is close to 1 and $A'(r)$ is close to 0, hence the above integral is close to π. Hence $\theta(T)$ cannot be equal to T. □

As we have already remarked, the above distributions are not contact, because $[X_1, X_2]$ is linearly dependent on X_1 and X_2 on the Martinet plane. The fact that abnormal geodesics are confined to the Martinet plane is a particular case of a more general situation described by the following proposition.

Proposition 6.32 *Suppose that a distribution $\mathcal{D}$ is spanned by two vector fields X_1, X_2 in an open set M. Suppose further that $[X_1, X_2](x)$ is linearly independent from $X_1(x)$ and $X_2(x)$ at each $x \in M$. Then no integral curve $x(t)$ of $\mathcal{D}$ in M is the projection of an abnormal extremal curve.*

Proof If $x(t)$ is an integral curve of $\mathcal{D}$ then $\frac{dx}{dt} = u(t)X_1(x(t)) + v(t)X_2(x(t))$ for some functions $u(t)$ and $v(t)$. Every abnormal extremal curve $\xi(t)$ in T^*M is an integral curve of the Hamiltonian vector field $\vec{h}$ associated with

$$h(\xi) = u(t)h_1(\xi) + v(t)h_2(\xi),$$

such that $\xi(t) \neq 0$ and is subject to the constraints $h_1(\xi(t)) = h_2(\xi(t)) = 0$. In this notation h_1 and h_2 denote the Hamiltonian lifts of X_1 and X_2, that is, $h_1(\xi) = \xi(X_1(x))$, $h_2(\xi) = \xi(X_2(x))$ for all $\xi \in T_x^*M$.

But then $0 = \frac{d}{dt}h_1(\xi(t)) = \{h_1, h\}(\xi(t)) = h_{[X_1, u(t)X_1 + v(t)X_2]}(\xi(t)) = v(t)h_{[X_1, X_2]}(\xi(t))$. Similarly, $0 = \frac{d}{dt}h_2(\xi(t)) = u(t)h_{[X_2, X_1]}(\xi(t))$. Since $x(t)$ is not a stationary curve, either $u(t) \neq 0$ or $v(t) \neq 0$. Therefore, $h_{[X_1, X_2]}(\xi(t)) = 0$. But that contradicts $\xi(t) \neq 0$ since $X_1(x), X_2(x), [X_1, X_2](x)$ span T_xM. □

6.6 The Maximum Principle and Weierstrass' excess function

Having the Maximum Principle at our disposal, let us now return to the necessary conditions for the strong extrema in the problem of the calculus of variations. To make an easy transition to the control problem, let M denote a strong neighborhood of $\bar{x}(t)$. Secondly, pass to the extended space $\tilde{M} = \mathbb{R} \times M$, by treating time as another spacial variable x_0, in order to make L dependent on the space variables only. Additionally, use $\tilde{x}(t) = (t + t_0, x(t + t_0))$ to shift the boundary conditions to $\tilde{x}(0) = (t_0, \bar{x}(t_0))$ and $\tilde{x}(t_1 - t_0) = (t_1, \bar{x}(t_1))$.

To convert to optimal control problem, let $X_1, \ldots, X_n$ be any linearly independent set of vector fields that span $T_x M$ at each point $x \in M$. Then every admissible curve $x(t)$ is a solution of $\frac{dx}{dt} = \sum_{i=1}^{n} u_i(t) X(x(t))$, and every extended curve $\tilde{x}(t) = (x_0(t), x(t))$ is a solution of the "affine control system"

$$\frac{dx_0}{dt} = 1, \frac{dx}{dt} = \sum_{i=1}^{n} u_i(t) X_i(x(t), \tag{6.32}$$

with controls $u(t)$ taking values in $U = \mathbb{R}^n$. If a trajectory $\tilde{x}(t)$ satisfies the boundary conditions $\tilde{x}(0) = (t_0, \bar{x}(t_0))$ and $\tilde{x}(s) = (t_1, \bar{x}(t_1))$ for some $s > 0$ then $s = t_1 - t_0$. This implies that if $\bar{x}(t)$ is to attain the minimum of $\int_{t_0}^{t_1} L\left(t, x(t), \frac{dx}{dt}\right) dt$ subject to the above boundary conditions, then the extended trajectory $\tilde{x}(t) = (t + t_0, \bar{x}(t + t_0))$ is a solution for the optimal problem defined by $\int_0^T L(\tilde{x}(t), u(t)) dt$ with variable time interval $[0, T]$.

Let $\bar{u}(t)$ denote the control that generates $\bar{x}(t)$. Then the optimal trajectory $(\tilde{x}(t), \bar{u}(t))$ is the projection of an extremal curve $(\lambda, (\xi_0(t), \xi(t))$, where (ξ_0, ξ) is a point in $T^* \tilde{M}$, realized as the product $T^* \mathbb{R} \times T^* M$, and $\lambda = 0, -1$, depending on whether $\xi(t)$ is normal or abnormal. This extremal curve is an integral curve of $\vec{h}_{\bar{u}(t)}$ associated with

$$h_{\bar{u}(t)}(\xi_0(t), \xi(t)) = \lambda L(\tilde{x}, \bar{u}(t)) + \xi_0 + \sum_{i=1}^{n} \bar{u}_i(t) X_i(x), (\xi_0, \xi) \in T_{x_0}^* \mathbb{R} \times T_x^* M.$$

$$\tag{6.33}$$

Moreover, $h_{\bar{u}(t)}(\xi_0(t), \xi(t)) = 0$, for almost all t along the extremal curve.

There are two immediate consequences of the Maximum Principle. First, there are no abnormal extremals. For, if $\lambda = 0$, then $h_u(\xi_0, \xi(t)) \leq 0$ for any $u \in \mathbb{R}^n$ would imply that $\xi(t) = 0$. But then $\xi_0 = 0$ since $h_{\bar{u}(t)}((\xi_0, \xi(t)) = \xi_0$. However, $(\xi_0, \xi(t)) = 0$ violates the non-degeneracy condition of the Maximum Principle. Second, $\xi_0(t)$ is constant, because $x_0(t)$ is a cyclic coordinate, that is, the vector fields in (6.32) do not depend explicitly on x_0.

Consequently, we may pass to the reduced Hamiltonian

$$H_{\bar{u}(t)}(\xi) = -L(t, x(t), \bar{u}(t)) + \sum_{i=1}^{n} \bar{u}_i(t) h_i(\xi), \tag{6.34}$$

where h_i denotes the Hamiltonian lift of X_i for $i = 1, \ldots, n$. This Hamiltonian is the projection of the original Hamiltonian (6.33). As such, it is constant almost everywhere along each extremal curve $\xi(t)$, and is maximal, in the sense that $H_{\bar{u}(t)}(\xi(t)) = Max_{u \in \mathbb{R}^n} H_u(\xi(t))$ a.e. in $[t_0, t_1]$.

It is important to note that the Hamiltonian in (6.34) is intrinsic, that is, does not depend on the choice of the frame $X_1, \ldots, X_n$. For if $Y_1, \ldots, Y_n$ is another frame on M, then every point $y \in T_x M$ is represented by the sum $\sum_{i=1}^n v_i Y_i(x)$ for some numbers $v_1, \ldots, v_n$. It then follows that $Y_i(x) = \sum_{j=1}^n U_{ij}(x)X_j(x)$ for some transformation $U(x) \in GL_n(\mathbb{R})$ and that $u_i = \sum_{j=1}^n U_{ji} v_j$. Along $\bar{x}(t)$, $\bar{u}_i(t) = \sum_{j=1}^n U_{ji}(\bar{x}(t))\bar{v}_j(t)$ where $\frac{d\bar{x}}{dt} = \sum_{i=1}^n \bar{v}_i(t)Y_i(\bar{x}(t))$. Furthermore,

$$H_{\bar{u}(t)}(\xi) = -L\left(t, x, \sum_{i=1}^n \bar{u}_i(t)X_i(x)\right) + \sum_{i=1}^n \bar{u}_i(t)\xi(X_i(x))$$

$$= -L\left(t, x, \sum_{i=1}^n \sum_{ij} U_{ji}(x)\bar{v}_j(t)X_i(x)\right) + \sum_{ij} U_{ji}\bar{v}_j(t)\xi(X_i(x))$$

$$= L\left(t, x, \sum_{i=1}^n \bar{v}_i(t)Y_i(x)\right) + \sum_{i=1}^n \bar{v}_i(t)\xi(Y_i(x)) = H_{\bar{v}(t)}(\xi).$$

Hence $\vec{H}_{\bar{u}(t)}$ and $\vec{H}_{\bar{v}(t)}$ have the same integral curves.

Since $H_{\bar{u}(t)}$ is maximal along an extremal curve $\xi(t)$

$$0 = -\frac{\partial L}{\partial u_i}(t, x(t), u) + h_i(\xi(t)), \quad i = 1, \ldots, n, \tag{6.35}$$

and $\frac{\partial^2 L}{\partial u_i \partial u_j}(t, \bar{x}(t), \bar{u}(t)) \geq 0$. This condition is known as the Legendre condition in the literature on the calculus of variations. As we have already remarked earlier, when $\frac{\partial^2 L}{\partial u_i \partial u_j}(t, \bar{x}(t), \bar{u}(t)) 0$ is strictly positive-definite in some neighborhood of $\bar{x}(t), t \in [t_0, t_1]$, then equation (6.35) is solvable, in the sense that there is a function ϕ defined in some neighborhood of the graph $\{(t, \xi(t)) : t \in [t_0, t_1]\}$ such that

$$\bar{u}(t) = \phi(t, \xi(t)). \tag{6.36}$$

Under these conditions, each extremal curve $\xi(t)$ is an integral curve of a single Hamiltonian vector field defined by

$$H(t, \xi) = -L(t, x, \phi(t, \xi)) + \sum_{i=1}^n \phi_i(t, \xi)\frac{\partial L}{\partial u_i}(t, x, \phi(t, \xi)). \tag{6.37}$$

Suppose now that the Hamiltonian H given by (6.37) is defined on the entire cotangent bundle T^*M. That is, suppose that the equation

$$0 = -\frac{\partial L}{\partial u_i}(t, x, u) + h_i(\xi), \ i = 1, \ldots, n$$

admits a global smooth solution $u = \phi(t, \xi), \xi \in T_x^*M$. This condition is equivalent to saying that L induces a transformation $\Phi : \mathbb{R} \times T^*M \to \mathbb{R} \times TM$ such that $\Phi(t, \xi) = \sum_{i=1}^n \phi_i(t, \xi) X_i(x)$ for each $(t, \xi) \in \mathbb{R} \times T_x^*M$. It follows that

$$H(t, \xi) = Max_{u \in \mathbb{R}^n}\{H_u(t, \xi), \xi \in T^*M\}. \tag{6.38}$$

The famous condition of Weierstrass,

$E(t, \bar{x}(t)), \bar{u}(t), u)$

$$= L(t, \bar{x}(t), \bar{u}(t)) - L(t, \bar{x}(t), u) + \sum_{i=1}^n (\bar{u}_i(t) - u_i)\frac{\partial L}{\partial u}(t, \bar{x}(t), \bar{u}(t)) \geq 0,$$

is nothing more than a paraphrase of the maximality condition $H_{\bar{u}(t)}(\xi(t)) \geq H_u(\xi(t)), u \in \mathbb{R}^n$.

In the literature on the calculus of variations, the excess function of Weierstrass is usually associated with the sufficient conditions of optimality under some additional assumptions [Cr; Yg]. The first assumption is that the Hilbert–Cartan form $-Hdt + \Theta$, with Θ the Liouville form $\xi \circ \pi_*$, is exact in some neighborhood of an extremal curve $\bar{\xi}(t), t \in [t_0, t_1]$. The Hilbert–Cartan form is usually written in the canonical coordinates as $-H(t, x, p)dt + pdx$ [C2]. To say that this form is exact, means that there exists a function $S(t, x)$ such that the differential dS is equal to $-Hdt + pdx$. But then, $\frac{\partial S}{\partial x_i} = p_i$ and $\frac{\partial S}{\partial t} = -H$. That is, $S(t, x)$ is a solution of the Hamilton–Jacobi equation

$$\frac{\partial S}{\partial t} + H\left(t, x, \frac{\partial S}{\partial x}\right) = 0. \tag{6.39}$$

The second assumption is that some neighborhood of $\bar{\xi}(t), t \in [t_0, t_1]$ projects diffeomorphically onto a neighborhood of the projected curve $\bar{x}(t) = \pi(\bar{\xi}(t))$. This assumption implies that every curve $x(t)$ in this neighborhood is the projection of of a unique curve $\xi(t)$ in T^*M. Moreover, $\xi(t)$ and $\bar{\xi}(t)$ have the same initial and terminal points whenever $x(t)$ and $\bar{x}(t)$ have the same terminal points. Under these assumptions, $\bar{x}(t)$ is optimal relative to the curves $x(t)$ in this neighborhood that satisfy the same boundary conditions $x(t_0) = \bar{x}(t_o), x(t_1) = \bar{x}(t_1)$.

The proof is simple and goes as follows. Let $u(t) = \frac{dx}{dt}$ and $\bar{u}(t) = \frac{d\bar{x}}{dt}$. Then,

$$\int_{t_0}^{t_1} (L(t, x(t), u(t)) - L(t, \bar{x}(t), \bar{u}(t)))\, dt$$

$$= \int_{t_0}^{t_1} (H_{\bar{u}(t)}(t, \bar{x}(t), \bar{p}(t)) - \bar{p}(t)\bar{u}(t) - H_{u(t)}(t, x(t), p(t)) + p(t)u(t))\, dt$$

$$\geq \int_{t_0}^{t_1} (H_{\bar{u}(t)}(t, \bar{x}(t), \bar{p}(t)) - \bar{p}(t)\bar{u}(t) - H_{\bar{u}(t)}(t, x(t), p(t)) + p(t)u(t))\, dt$$

$$= \int_{t_0}^{t_1} \left(H(t, \bar{x}(t), \bar{p}(t)) - \bar{p}(t)\frac{d\bar{x}}{dt} - H(t, x(t), p(t)) + p(t)\frac{dx}{dt} \right) dt$$

$$= \int_{\bar{\gamma}} (H\,dt - p\,dx) - \int_{\gamma} (H\,dt - p\,dx) = \int_{\bar{\gamma}} dS - \int_{\gamma} dS = 0,$$

where $\bar{\gamma}(t) = (t, (\bar{x}(t), \bar{p}(t)))$ and $\gamma(t) = (t, (x(t), p(t)))$.

7

Hamiltonian view of classic geometry

We have now arrived at a transition point at which the preceding theory begins to be directed to problems of geometry and applied mathematics. We will begin with the applications to the classic geometry, as a segue to further investigations into the geometry of symmetric spaces and the problems of mechanics. We will proceed in the spirit of Felix Klein's Erlangen Program of 1782, and start with a Lie group G, which acts on a manifold M with the aim of arriving at the natural geometry on M induced by the structure of G.

7.1 Hyperbolic geometry

Consider first the group $G = SU(1,1) = \left\{ \begin{pmatrix} a & b \\ \bar{b} & \bar{a} \end{pmatrix} : a, b \text{ in } \mathbb{C}, |a|^2 - |b|^2 = 1 \right\}$, and its action on the unit disk $\mathbb{D} = \{z \in \mathbb{C} : |z| = 1\}$ by the Moebius transformations $gz = \frac{az+b}{\bar{b}z+\bar{a}}$. Our aim is to show that the classic hyperbolic metric $\frac{4|dz|^2}{1-|z|^2}$ on $\mathbb{D}$ is an immediate consequence of the canonical sub-Riemannian problem on G.

Since the above action is transitive, $\mathbb{D}$ can be regarded as the homogeneous space G/K where K is the isotropy group of a fixed point z_0. It is convenient to take $z_0 = 0$, in which case, $K = \left\{ \begin{pmatrix} a & 0 \\ 0 & \bar{a} \end{pmatrix}, |a| = 1 \right\}$; the isotropy group through any other point z_0 is the conjugate group gKg^{-1}, where $g \in G$ such that $g(0) = z_0$.

96

The Lie algebra $\mathfrak{g}$ of G consists of the matrices $\left\{ \begin{pmatrix} i\alpha & \beta \\ \bar{\beta} & -i\alpha \end{pmatrix}, \alpha \in \mathbb{R} \right.$

and $\beta \in \mathbb{C} \left. \right\}$. Then $\mathfrak{g}$ can be regarded as the sum $\mathfrak{g} = \mathfrak{p} \oplus \mathfrak{k}$, where $\mathfrak{k} =$

$\left\{ \alpha \begin{pmatrix} i & 0 \\ 0 & -i \end{pmatrix} : \alpha \in \mathbb{R} \right\}$ and $\mathfrak{p} = \left\{ \begin{pmatrix} 0 & \beta \\ \bar{\beta} & 0 \end{pmatrix} : \beta \in \mathbb{C} \right\}$. It is easy to check

that $[A_1, A_2] \in \mathfrak{k}$ for any matrices A_1 and A_2 in $\mathfrak{p}$.

Consider now the bilinear form $\langle A, B \rangle = 2Tr(AB)$, where Tr denotes the trace of a matrix. It follows that $\langle A_1, A_2 \rangle = -\alpha_1\alpha_2 + \beta_1\bar{\beta}_2 + \bar{\beta}_1\beta_2$ for any matrices

$A_1 = \begin{pmatrix} i\alpha_1 & \beta_1 \\ \bar{\beta}_1 & -i\alpha_1 \end{pmatrix}$ and $A_2 = \begin{pmatrix} i\alpha_1 & \beta_2 \\ \bar{\beta}_2 & -i\alpha_2 \end{pmatrix}$. Hence $\langle , \rangle$ is positive-

definite on $\mathfrak{p}$ and can be used to define a left-invariant metric on the space of curves $g(t)$ in G that satisfy $g^{-1}(t)\frac{dg}{dt}(t) = A(t)$, with $A(t) \in \mathfrak{p}$. We can put this observation in the sub-Riemannian context by introducing the matrices

$$A_1 = \frac{1}{2}\begin{pmatrix} 0 & i \\ -i & 0 \end{pmatrix}, A_2 = \frac{1}{2}\begin{pmatrix} 0 & 1 \\ 1 & 0 \end{pmatrix}, A_3 = \frac{1}{2}\begin{pmatrix} -i & 0 \\ 0 & i \end{pmatrix}. \quad (7.1)$$

Matrices A_1 and A_2 form an orthonormal basis for $\mathfrak{p}$, hence the length of any curve $g(t)$ in G that satisfies $g^{-1}(t)\frac{dg}{dt}(t) \in \mathfrak{p}$ is given by $\int_0^T \sqrt{u_1^2(t) + u_2^2(t)}\, dt$, where $u_1(t)$ and $u_2(t)$ are defined by $g^{-1}(t)\frac{dg}{dt} = u_1(t)A_1 + u_2(t)A_2$. Since $[A_1, A_2] = A_3$ any two points g_0 and g_1 in G can be connected by a horizontal curve $g(t)$, that is, a curve that satisfies $g^{-1}(t)\frac{dg}{dt} \in \mathfrak{p}$.

So there is a natural time optimal control problem on G that consists of finding the trajectories $g(t)$ of

$$\frac{dg}{dt} = g(t)(u_1(t)A_1 + u_2(t)A_2), u_1^2(t) + u_2^2(t) \le 1$$

that connect g_0 to g_1 in the least possible time T.

To see the connection with the hyperbolic metric on M, consider the lifting of curves in $\mathbb{D}$ to the horizontal curves in G. Any curve $z(t) = x(t) + iy(t)$ lifts to a curve $g(t) = \begin{pmatrix} a(t) & b(t) \\ \bar{b}(t) & \bar{a}(t) \end{pmatrix}$ in G via the formula

$$z(t) = \frac{a(t)(0) + b(t)}{\bar{b}(t)(0) + \bar{a}(t)} = \frac{b(t)}{\bar{a}(t)}.$$

An easy calculation shows that $g(t) = \frac{1}{\sqrt{1-|z(t)|^2}}\begin{pmatrix} 1 & z(t) \\ \bar{z}(t) & 1 \end{pmatrix}\begin{pmatrix} e^{-i\theta(t)} & 0 \\ 0 & e^{i\theta(t)} \end{pmatrix}$,

where $\theta(t)$ is an arbitrary curve.

Proposition 7.1   *a.* $g^{-1}(t)\frac{dg}{dt}(t) \in \mathfrak{p}$ *for all t if and only if*

$$\frac{d\theta}{dt} = \frac{1}{1 - |z(t)|^2}\left(\frac{dx}{dt}y(t) - \frac{dy}{dt}x(t)\right).$$

If $\theta(t)$ is any curve that satisfies this condition (they differ by an initial condition) then,

$$g^{-1}(t)\frac{dg}{dt} = \frac{2}{1 - |z(t)|^2}\left(\left(\frac{dx}{dt}\cos 2\theta - \frac{dy}{dt}\sin 2\theta\right)A_1\right.$$
$$\left. + \left(\frac{dx}{dt}\sin 2\theta + \frac{dy}{dt}\cos 2\theta\right)A_2\right). \tag{7.2}$$

The energy functional associated with $g(t)$ is given by

$$\frac{1}{2}\left\langle g^{-1}(t)\frac{dg}{dt}, g^{-1}(t)\frac{dg}{dt}\right\rangle = \frac{2}{(1 - |z(t)|^2)^2}\left|\frac{dz}{dt}\right|^2. \tag{7.3}$$

Proof Let $\alpha = \sqrt{1 - |z|^2}$, $Z = \begin{pmatrix} 1 & z \\ \bar{z} & 1 \end{pmatrix}$, and $\Theta = \begin{pmatrix} e^{-i\theta} & 0 \\ 0 & e^{i\theta} \end{pmatrix}$, so that $g(t) = \frac{1}{\alpha(t)}Z(t)\Theta(t)$. A short calculation shows that

$$g^{-1}(t)\frac{dg}{dt} = -\frac{\dot{\alpha}}{\alpha}I + \frac{1}{\alpha^2}\begin{pmatrix} -\dot{z}\bar{z} & \dot{z}e^{2i\theta} \\ \dot{\bar{z}}e^{-2i\theta} & -\bar{z}\dot{z} \end{pmatrix} + \dot{\theta}\begin{pmatrix} -i & 0 \\ 0 & i \end{pmatrix}.$$

Hence, $g^{-1}\frac{dg}{dt} = \frac{1}{\alpha^2}\begin{pmatrix} 0 & \dot{z}e^{2i\theta} \\ \dot{\bar{z}}e^{-2i\theta} & 0 \end{pmatrix}$ whenever $\dot{\theta} = \frac{1}{\alpha^2}(\dot{x}y - x\dot{y})$. Upon separating the real and the imaginary parts of $\dot{z}e^{2i\theta}$ we get the formula above.   □

The above shows that the sub-Riemannian metric on G induces the hyperbolic metric $ds = 2\frac{|dz|}{1-|z|^2}$ on the unit disk. The lifting to horizontal curves in G uncovers the differential form $d\theta = \frac{1}{1-|z|^2}(ydx - xdy)$ relevant for evaluating the hyperbolic area enclosed by closed curves in M.

There are two other models in the classical hyperbolic geometry: the upper half plane $\mathcal{P} = \{z \in \mathbb{C} : \Im z \geq 0\}$ and the hyperboloid $\mathbb{H}^3 = \{x \in \mathbb{R}^3 : -x_0^2 + x_1^2 + x_2^2 = 1, x_0 > 0\}$. All of these models can be seen in a unified way through the actions of Lie groups. In the first case, $\mathcal{P}$ can be realized as the quotient $SL_2(R)/SO_2(R)$, while in the second case $\mathbb{H}^3$ can be realized as the quotient $SO(1,2)/SO_2(R)$.

In the first case, $SL_2(R)$ acts on $\mathcal{P}$ via the Moebius transformations. If $g(z) = \frac{az+b}{cz+d}$ then $g(i) = i$ if and only if $g \in SO_2(R)$. Hence, points $z \in \mathcal{P}$ can be identified with matrices g in $SL_2(R)$ via the formula $g(i) = z$ up to the matrices in $SO_2(R)$. Since any matrix g in $SL_2(R)$ can be written as the product

$\begin{pmatrix} a & b \\ 0 & c \end{pmatrix} \begin{pmatrix} \alpha & \beta \\ -\beta & \alpha \end{pmatrix}$ with $\alpha^2 + \beta^2 = 1$, the above identification can be reduced to the upper triangular matrices in $SL_2(R)$. It then follows that any curve $z(t)$ in $\mathcal{P}$ is identified with

$$g(t) = \begin{pmatrix} \frac{1}{\sqrt{y(t)}}y(t) & \frac{1}{\sqrt{y(t)}}x(t) \\ 0 & \frac{1}{\sqrt{y(t)}} \end{pmatrix} \begin{pmatrix} \cos\theta(t) & \sin\theta(t) \\ -\sin\theta(t) & \cos\theta(t) \end{pmatrix},$$

where $\theta(t)$ is an arbitrary curve. In this case,

$$A_1 = \frac{1}{2}\begin{pmatrix} 1 & 0 \\ 0 & -1 \end{pmatrix}, A_2 = \frac{1}{2}\begin{pmatrix} 0 & 1 \\ 1 & 0 \end{pmatrix}, A_3 = \frac{1}{2}\begin{pmatrix} 0 & -1 \\ 1 & 0 \end{pmatrix}, \quad (7.4)$$

is the basis that is isomorphic to the matrices in (7.1). For then, A_1, A_2, A_3 conform to the same Lie bracket table as in the case of $su(1,1)$, and the trace form $\langle A, B \rangle = 2Tr(AB)$ is positive-definite on the linear span of A_1 and A_2, with A_1 and A_2 orthonormal relative to $\langle\,,\,\rangle$.

As in the first case, the sub-Riemannian metric on the two-dimensional left-invariant distribution spanned by matrices A_1 and A_2 induces the hyperbolic metric on $\mathcal{P}$. It is easy to check that $z(t)$ lifts to a horizontal curve $g(t)$ if and only if $\frac{d\theta}{dt}(t) = \frac{1}{y(t)}\frac{dy}{dt}$. The sub-Riemannian metric on the space of horizontal curves coincides with the hyperbolic metric on $\mathcal{P}$ in the sense that

$$\sqrt{\left\langle g^{-1}(t)\frac{dg}{dt}, g^{-1}(t)\frac{dg}{dt} \right\rangle} = \frac{\sqrt{\dot{x}^2 + \dot{y}^2}}{y}. \quad (7.5)$$

These two hyperbolic models are isometric as can be easily demonstrated through the Cayley transform $C = \frac{1}{\sqrt{2}}\begin{pmatrix} i & 1 \\ -1 & -i \end{pmatrix}$. The Moebius transformation defined by C takes $\mathbb{D}$ onto $\mathcal{P}$. Hence, $C^{-1}gC$ acts on the unit disk $\mathbb{D}$ for any $g \in SL_2(R)$. In fact if $g = \begin{pmatrix} a & b \\ c & d \end{pmatrix}$ then $C^{-1}gC = \begin{pmatrix} p & q \\ \bar{q} & \bar{p} \end{pmatrix}$, where

$$p = \frac{1}{2}(a+d+i(b-c)), q = \frac{1}{2}(-b-c+i(d-a)).$$

Since $C \in SL_2(\mathbb{C})$, $|p|^2 - |q|^2 = 1$ and hence $\begin{pmatrix} p & q \\ \bar{q} & \bar{p} \end{pmatrix}$ belongs to $SU(1,1)$. This conjugation provides the desired isomorphism between $SU(1,1)$ and $SL_2(R)$ and the corresponding Lie algebras $sl_2(R)$ and $su(1,1)$.

One can easily check that if $\tilde{A}_i = CA_iC^{-1}$, where A_1, A_2, A_3 are the matrices in (7.4), then $\tilde{A}_1 = -A_1, \tilde{A}_2 = -A_2, \tilde{A}_3 = A_3$, where now A_1, A_2, A_3 are the matrices in (7.1). Since $\tilde{A}_1$ and $\tilde{A}_2$ remain orthonormal relative to the trace metric, the above shows that the sub-Riemannian problems on $SL_2(R)$ and $SU(1, 1)$ are isometric.

To complete this hyperbolic landscape, let us shift to the hyperboloid $\mathbb{H}^2 = \{x \in \mathbb{R}^3 : x_1^2 - x_2^2 - x_3^2 = 1, x_1 > 0\}$. Since $G = SO(1, 2)$ is the isometry group for the Lorentzian metric $(x, y) = x_1 y_1 - x_2 y_2 - x_3 y_3$, G acts on $\mathbb{H}^2$ by the matrix multiplication $(g, x) \rightarrow gx$. Then $\mathbb{H}^2$ can be represented as the quotient G/K where K is the isotropy group $Ke_1 = e_1$. The matrices in K are

of the form $\begin{pmatrix} 1 & 0 & 0 \\ 0 & a & b \\ 0 & -b & a \end{pmatrix}$ with $a^2 + b^2 = 1$. Hence, K is isomorphic with

$SO_2(R)$.

The Lie algebra $\mathfrak{g}$ of G consist of the matrices $\begin{pmatrix} 0 & \alpha & \beta \\ \alpha & 0 & -\gamma \\ \beta & \gamma & 0 \end{pmatrix}$. Hence,

$$A_1 = \begin{pmatrix} 0 & 1 & 0 \\ 1 & 0 & 0 \\ 0 & 0 & 0 \end{pmatrix}, A_2 = \begin{pmatrix} 0 & 0 & 1 \\ 0 & 0 & 0 \\ 1 & 0 & 0 \end{pmatrix}, A_3 = \begin{pmatrix} 0 & 0 & 0 \\ 0 & 0 & -1 \\ 0 & 1 & 0 \end{pmatrix} \quad (7.6)$$

form a basis for $\mathfrak{g}$. The reader can readily check that the trace form $\langle A, B \rangle = \frac{1}{2} Tr(AB)$ is positive-definite on the linear span of A_1 and A_2.

Analogous to the previous cases, A_1 and A_2 define a two-dimensional, left-invariant distribution $\mathcal{D}$ whose integral curves $g(t)$, again called horizontal, are the solutions of $\frac{dg}{dt} = g(t)(u_1(t)A_1 + u_2(t)A_2)$. Any curve $x(t)$ in $\mathbb{H}^2$ can be lifted to a horizontal curve $g(t)$ via the formula $g(t)e_1 = x(t)$. Then, $\frac{dx}{dt} = \frac{dg}{dt}e_1 = g(t)(u_1 A_1 + u_2 A_2)e_1$ and consequently,

$$-\dot{x}_1^2 + \dot{x}_2^2 + \dot{x}_3^2 = -(g(t)(u_1 A_1 + u_2 A_2)e_1, g(t)(u_1 A_1 + u_2 A_2)e_1) = u_1^2 + u_2^2.$$

So the hyperbolic metric coincides with the sub-Riemannian metric defined by the trace form. All three of these Lie algebras are isomorphic and conform to the following Lie bracket table:

$$[A_1, A_2] = A_3, \ [A_1, A_3] = A_2, \ [A_2, A_3] = -A_1,$$

but the corresponding groups are not all isomorphic: $SL_2(R)$ is a double cover of $SO(1, 2)$.

In fact, any matrix any matrix $X \in sl_2(R)$ can be written as the sum of a symmetric matrix $\begin{pmatrix} -x_3 & x_2 \\ x_2 & x_3 \end{pmatrix}$ and a skew-symmetric matrix $\begin{pmatrix} 0 & -x_1 \\ x_1 & 0 \end{pmatrix}$, in which case,

$$\langle X, X \rangle = -\frac{1}{2} Tr(X^2) = x_1^2 - x_2^2 - x_3^2.$$

Hence, $sl_2(R)$, as a vector space, can be regarded as a three-dimensional Lorentzian space. It follows that $Ad_g(X) = gXg^{-1}$ belongs to $SO(1, 2)$ for any $g \in SL_2(R)$. The correspondence $g \Rightarrow Ad_g$, called the adjoint representation of $SL_2(R)$, shows that $SL_2(R)$ is a double cover of $SO(1, 2)$ since $Ad_{\perp I} = I$. It also shows that $sl_2(R)$ and $so(1, 2)$ are isomorphic, with $-ad(a_1A_1 + a_2A_2 + a_3A_3)$

corresponding to the matrix $\begin{pmatrix} 0 & a_1 & a_2 \\ a_1 & 0 & -a_3 \\ a_2 & a_3 & 0 \end{pmatrix}$.

7.2 Elliptic geometry

There are two ways to proceed. One can either begin with the action of $SO_3(R)$ on the sphere S^2, and then identify the sphere with the quotient $SO_3(R)/SO_2(R)$, or alternatively, one may start with the action of SU_2 on the projective complex line $\mathbb{C}P^1$. These methods, similar in principle, but different in appearance, are in fact isometric, as can be seen through the stereographic projection of the sphere S^2 onto the extended complex plane $\mathbb{C}P^1$.

In the first case, $SO_3(R)$ acts on the sphere S^2 by the matrix multiplication $(g, x) \rightarrow gx$. Then $SO_2(R)$ is the isotropy group of the point $x_0 = e_1$. As in the hyperbolic situations, $x(t)$ on the sphere can be lifted to horizontal curves $g(t)$ in $SO_3(R)$ via the formula $g(t)e_1 = x(t)$. The horizontal curves are defined relative to the left-invariant distribution spanned by the

matrices $A_1 = \begin{pmatrix} 0 & 0 & 1 \\ 0 & 0 & 0 \\ -1 & 0 & 0 \end{pmatrix}$ and $A_2 = \begin{pmatrix} 0 & -1 & 0 \\ 1 & 0 & 0 \\ 0 & 0 & 0 \end{pmatrix}$. It follows that

$[A_1, A_2] = -A_3$, where $A_3 = \begin{pmatrix} 0 & 0 & 0 \\ 0 & 0 & -1 \\ 0 & 1 & 0 \end{pmatrix}$.

Matrices A_1, A_2, A_3 are orthonormal relative to the bilinear form $\langle A, B \rangle = -\frac{1}{2}Tr(AB)$. If $g(t)$ is a solution of $\frac{dg}{dt} = g(t)(u_1(t)A_1 + u_2(t)A_2)$, then

$$\dot{x}_1^2 + \dot{x}_2^2 + \dot{x}_3^2 = u_2^2(t) + u_2^2(t).$$

The reader may verify this formula by a calculation similar to the one used on the hyperboloid. Hence, the sub-Riemannian metric on the horizontal curves coincides with the spherical metric on the sphere inherited from the Euclidean metric in the ambient space $\mathbb{R}^3$.

Let us now turn to SU_2 and its action $\left(\begin{pmatrix} a & b \\ -\bar{b} & \bar{a} \end{pmatrix}, z \right) \rightarrow \frac{az+b}{\bar{b}z+\bar{a}}$ on $\mathbb{C}P^1$. It is easy to verify that this action is transitive and that $K =$

$\left\{ \begin{pmatrix} a & 0 \\ 0 & \bar{a} \end{pmatrix} : |a| = 1 \right\}$ is the isotropy group of $z = 0$ (K is also the isotropy group of $z = \infty$). Evidently, K is isomorphic to S^1 and $SO_2(R)$.

The Lie algebra $\mathfrak{g}$ of SU_2 consists of matrices $\begin{pmatrix} i\alpha & \beta \\ -\bar{\beta} & -i\alpha \end{pmatrix}$ with $\alpha \in R$ and $\beta \in C$. Let now $\langle A, B \rangle = -2Tr(AB)$. Relative to this form, the Pauli matrices

$$A_1 = \frac{1}{2} \begin{pmatrix} 0 & 1 \\ -1 & 0 \end{pmatrix}, A_2 = \frac{1}{2} \begin{pmatrix} 0 & i \\ i & 0 \end{pmatrix}, A_3 = \frac{1}{2} \begin{pmatrix} -i & 0 \\ 0 & i \end{pmatrix} \quad (7.7)$$

form an orthonormal basis in su_2. Since SU_2 is a double cover of $SO_3(R)$, su_2 and $so_3(R)$ are isomorphic. In fact, matrices A_1, A_2, A_3 on $so_3(R)$ are in exact correspondence with the Pauli matrices on su_2.

Analogous to the situation on the disk, $g(t) = \frac{1}{\sqrt{1+|z(t)|^2}} \begin{pmatrix} 1 & z \\ \bar{z} & 1 \end{pmatrix}$

$\begin{pmatrix} e^{i\theta(t)} & 0 \\ 0 & e^{-i\theta(t)} \end{pmatrix}$ is a lift of a curve $z(t)$ in CP^1. The proposition below describes the horizontal lifts.

Proposition 7.2 *Let $\mathfrak{p}$ denotes the real linear span of A_1 and A_2. Then $g^{-1}(t)\frac{dg}{dt}(t) \in \mathfrak{p}$ for all t if and only if*

$$\frac{d\theta}{dt} = \frac{1}{1+|z(t)|^2} \left(\frac{dx}{dt} y(t) - \frac{dy}{dt} x(t) \right).$$

If $\theta(t)$ is any curve that satisfies this condition (they differ by an initial condition), then

$$g^{-1}(t)\frac{dg}{dt} = \frac{2}{1+z(t)^2} \left(\left(\frac{dx}{dt} \cos 2\theta - \frac{dy}{dt} \sin 2\theta \right) A_1 \right.$$

$$\left. + \left(\frac{dx}{dt} \sin 2\theta + \frac{dy}{dt} \cos 2\theta \right) A_2 \right). \quad (7.8)$$

The energy functional associated with $g(t)$ is given by

$$\frac{1}{2} \left\langle g^{-1}(t)\frac{dg}{dt}, g^{-1}(t)\frac{dg}{dt} \right\rangle = \frac{2}{1+|z(t)|^2} \left| \frac{dz}{dt} \right|^2. \quad (7.9)$$

Traditionally, these two realizations of the elliptic metric are reconciled through the stereographic projection of S^2 onto CP^1, in which points $x = (x_1, x_2, x_3)$ on S^2 are projected onto points z in CP^1 via the formula

$$x_1 = \frac{2}{1+|z|^2}x, \ y = \frac{2}{1+|z|^2}y, \ x_3 = \frac{|z|^2-1}{1+|z|^2}.$$

Then an easy calculation shows that

$$\dot{x}_1^2 + \dot{x}_2^2 + \dot{x}_3^2 = \frac{4|dz|^2}{1 + |z|^2}.$$

Let us now extract the essential properties of the sub-Riemannian problems uncovered by these geometric models. We have shown that each non-Euclidean space M is a homogenous space G/K. In the hyperbolic case, G is either $SU(1, 1)$, $SL_2(R)$, or $SO_0(1, 2)$, and in the elliptic case G is either $SO_3(R)$ or SU_2. In the first case the groups are non-compact, while in the second case they are compact. In each case, a scalar multiple of the trace form $Tr(AB)$ is positive-definite on the two-dimensional vector space $\mathfrak{p}$ in the Lie algebra $\mathfrak{g}$ of G and defines a metric $\langle\,,\,\rangle$ on the corresponding left-invariant distribution $\mathcal{D}$. Then matrices A_1, A_2, A_3 denote a basis in $\mathfrak{g}$ such that A_1, A_2 form an orthonormal basis in $\mathfrak{p}$ and $[A_1, A_2] = -\epsilon A_3$ with $\epsilon = 1$ in the elliptic case and $\epsilon = -1$ in the hyperbolic case.

Curves $g(t)$ in G are called horizontal if $g(t)^{-1}\frac{dg}{dt} \in \mathfrak{p}$ for all t. Horizontal curves are the solutions of a "control system"

$$\frac{dg}{dt} = g(t)(u_1(t)A_1 + u_2(t)A_2) \tag{7.10}$$

where $u_1(t)$ and $u_2(t)$ are arbitrary bounded and measurable functions playing the role of controls.

Vertical curves are the curves that satisfy $g^{-1}\frac{dg}{dt} = u(t)A_3$ for some function $u(t)$. Vertical curve that satisfy $g(0) = I$ define a group K, which i is isomorphic to $SO_2(R)$ in all cases. We have shown that each curve in $M = G/K$ is a projection of a horizontal curve and that any two horizontal curves g_1 and g_2 that project onto the same curve $x(t)$ in M satisfy $g_2(t) = g_1(t)h$ for some $h \in K$.

Any two points of G can be connected by a horizontal curve, since $[A_1, A_2] = \pm A_3$, and any horizontal curve has length $\int_0^T \sqrt{u_1^2(t) + u_2^2(t)}\, dt$ and energy $\frac{1}{2}\int_0^T (u_1^2(t) + u_2^2(t))\, dt$. The fundamental sub-Riemannian problem consists of finding a horizontal curve of shortest length that connects two given points in G.

We have shown that the projection of the sub-Riemannian metric in G coincides with the Riemannian metric in M. We have also shown that both the hyperbolic and the elliptic models are isometric, so it suffices to pick a representative in each group. We will pick the simply connected models, $G = SU(1, 1)$ in the hyperbolic case, and $G = SU_2$ in the elliptic case. These groups are double covers of $SO_0(1, 2)$ and $SO_3(R)$.

We will use G_ϵ to denote these two groups with $G_\epsilon = SU(1, 1)$ for $\epsilon = -1$ and $G_\epsilon = SU_2$ for $\epsilon = 1$. On occasions we will refer to these groups as the non-compact case ($\epsilon = -1$) and compact case ($\epsilon = 1$). Then $\mathfrak{g}_\epsilon$ will denote the Lie algebra of G_ϵ. If A_1, A_2, A_3 denote the bases in $\mathfrak{g}_\epsilon$ given by (7.1) and (7.7) then these matrices conform to the Lie bracket Table 7.1.

Table 7.1

[,]	A_1	A_2	A_3
A_1	0	$-\epsilon A_3$	A_2
A_2	ϵA_3	0	$-A_1$
A_3	$-A_2$	A_1	0

Since $Tr(AB) = Tr(BA)$, and since the sub-Riemannian metric is a scalar multiple of the trace form,

$$\langle A, [B, C] \rangle_\epsilon = \langle [A, B], C \rangle_\epsilon, A, B, C \text{ in } \mathfrak{g}_\epsilon.$$

The fact that $[\mathfrak{p}_\epsilon, \mathfrak{k}_\epsilon] = \mathfrak{p}_\epsilon$ implies that $gAg^{-1} \in \mathfrak{p}_\epsilon$ for $A \in \mathfrak{p}_\epsilon$ and any $g \in K$. This implies that the sub-Riemannian metric is invariant under K, that is,

$$\langle gAg^{-1}, gBg^{-1} \rangle_\epsilon = \langle A, B \rangle_\epsilon, g \in K.$$

The precise relation between the sub-Riemannian problem on G_ϵ and the induced Riemannian problem on M_ϵ is described by the following proposition.

Proposition 7.3 *Suppose that $g(t)$ is the shortest horizontal curve that connects a coset $g_0 K$ to the coset $g_1 K$. Then the projected curve $x(t)$ in M_ϵ is the shortest curve that connects $x_0 = g_0 K$ to $x_1 = g_1 K$.*

Conversely, if $x(t)$ be the shortest curve in M_ϵ that connects x_0 and x_1, then $x(t)$ is the projection of a horizontal curve $g(t)$ of minimal sub-Riemannian length that connects the coset $x_0 = g_0 K$ to the coset $x_1 = g_1 K$.

Proof The proof is obvious.   □

It follows that the geodesic problem on M_ϵ is equivalent to finding a horizontal curve of minimal length that connects the initial manifold $S_0 = g_0 K$ to the terminal manifold $S_1 = g_1 K$.

Next to the geodesic problem we will also consider another natural geometric problem on M_ϵ, the problem of minimizing the integral $\frac{1}{2} \int_0^T \kappa^2(t) \, dt$, where $\kappa(t)$ denotes the geodesic curvature of a curve $x(t)$ in M_ϵ that satisfies the given tangential conditions at $t = 0$ and $t = T$. We will refer to this problem as the *elastic problem* for the following historical reasons.

This problem originated with a study of Daniel Bernoulli, who in 1742 suggested to L. Euler that the differential equation describing the equilibrium shape of a thin inextensible beam subject to bending torques at its ends can be found by making the integral of the square of the curvature along the beam a minimum. Euler, acting on this suggestion, obtained the differential equation for this problem in 1744 and was able to describe its solutions, known since then as the *elasticae* [E; Lv]. Its modern treatment was initiated by P. Griffiths, who obtained the Euler–Lagrange equation on simply connected spaces of constant curvature [Gr]. This study inspired several other papers [BG; LS; Je], each of which tackled the problem through its own, but somewhat disparate methods.

The elastic problem can be naturally formulated as a two point boundary value problem in the isometry group G_ϵ as an optimal control problem over the Serret–Frenet system

$$\frac{dg}{dt} = g(t)(A_1 + u(t)A_3) \tag{7.11}$$

in G_ϵ that seeks the minimum of the integral $\frac{1}{2}\int_0^T u^2(t)\,dt$ among the solutions that satisfy the boundary conditions $g(0) = g_0$ and $g(T) = g_1$.

This formulation draws attention to a more general class of variational problems, called affine-quadratic, that plays an important role in the theory of integrable systems and also lends itself to quick and elegant solutions.

The focus on Lie groups, rather than on the spaces on which the groups act, reveals a curious twist in our understanding of the classical geometry. Our literature on geometry, dominated by its historical origins, tends to favor the Euclidean geometry over its non-Euclidean analogues. In this mindset, the leap to non-Euclidean geometries is drastic, and requires a great deal of intellectual adjustment. However, had the history started with non-Euclidean geometry first, the passage to Euclidean geometry would require only a minor algebraic shift.

To be more explicit, note that K acts linearly and irreducibly on $\mathfrak{p}_\epsilon$ by conjugation. Let G_0 denote the semi-direct product $\mathfrak{p}_\epsilon \ltimes K$. Evidently, G_0 is isomorphic to $SE_2(R)$, the group of motions of the plane. The restriction of $\langle\,,\,\rangle_\epsilon$ to $\mathfrak{p}_\epsilon$ is K-invariant and hence defines a Euclidean metric on $\mathfrak{p}_\epsilon$. Thus $\mathfrak{p}_\epsilon$ together with $\langle\,,\,\rangle_\epsilon$ is a two-dimensional Euclidean space $\mathbb{E}^2$. Then $\mathfrak{p}_\epsilon$ can be realized as the quotient space G_0/K through the action $(x, R)(y) = x+Ry$ of G_0 on $\mathfrak{p}_\epsilon$: $\mathfrak{p}_\epsilon$ is the orbit of G_0 through the origin in $\mathfrak{p}_\epsilon$, and K is the isotropy group.

If $\mathfrak{g}_0$ denotes the Lie algebra of G_0, then its points are pairs (A, B), $A \in \mathfrak{p}_\epsilon$ and $B \in \mathfrak{k}$ and $[(A_1B_1), (A_2, B_2)] = (adB_1(A_2) - adB_2(A_1), [B_1, B_2])$ is the Lie bracket in $\mathfrak{g}_0$. Then (A, B) in $\mathfrak{g}_0$ can be i identified with $A + B$ in $\mathfrak{g}_\epsilon$, in which case the semi-direct Lie bracket is given by

$$(adB_1(A_2) - adB_2(A_1), [B_1, B_2]) = [B_1, A_2] - [B_2, A_1] + [B_1, B_2]. \tag{7.12}$$

Of course, in this case $\mathfrak{k}$ is a one-dimensional Lie algebra and $[B_1, B_2] = 0$. With this identification $\mathfrak{g}_\epsilon$, as a vector space, carries a double Lie algebra $\mathfrak{g}_\epsilon$ and $\mathfrak{g}_0$. However, relative to the semi-direct Lie bracket, any two elements in $\mathfrak{p}_\epsilon$ commute. The reader may easily check that A_1, A_2, A_3 in the semi-direct case conform to the Lie bracket table (Table 7.1) with $\epsilon = 0$.

As in the non-Euclidean setting, horizontal curves in G_0 are the curves that satisfy $g(t)^{-1}\frac{dg}{dt} \in \mathfrak{p}_\epsilon$, or $\frac{dg}{dt} = g(t)(u_1(t)A_1 + u_2(t)A_2)$ for some control functions $u_1(t)$ and $u_2(t)$. If we denote $g(t)$ by the pair $(x(t), R(t)) \in \mathfrak{p}_\epsilon \ltimes K$, then

$$\frac{dx}{dt} = u_1(t)Ad_R(A_1) + u_2(t)Ad_R(A_2), \quad \frac{dR}{dt} = 0,$$

and $||\frac{dx}{dt}||^2 = u_1^2 + u_2^2$.

An analogous argument applies to the Euclidean elastic problem, and therefore we may conclude that for any left-invariant optimal problem on G_ϵ, there is a corresponding Euclidean problem on $G_0 = \mathfrak{p}_\epsilon \ltimes K$.

7.3 Sub-Riemannian view

We will now show that the above classical problems are easily solvable by the theoretic ingredients provided by the previous chapters. Let h_1, h_2, h_3 denote the Hamiltonian lifts of the left-invariant vector fields $X_i(g) = gA_i$, $i = 1, 2, 3$. In the left-invariant realization of T^*G as the product $G_\epsilon \times \mathfrak{g}_\epsilon^*$, these functions are given by $h_i(\ell) = \ell(A_i)$, $i = 1, 2, 3$, where $\ell \in \mathfrak{g}_\epsilon^*$. Functions h_1, h_2, h_3 may be regarded as the coordinates of a point $\ell \in \mathfrak{g}_\epsilon^*$ relative to the dual basis A_1^*, A_2^*, A_3^*. The sub-Riemannian problem of finding the horizontal curves of minimal length leads to the energy Hamiltonian

$$H = \frac{1}{2}(h_1^2 + h_2^2). \tag{7.13}$$

Recall now that any function H on $\mathfrak{g}_\epsilon^*$ may be considered as a left-invariant Hamiltonian on G_ϵ in which case its Hamiltonian equations can be written

$$\frac{dg}{dt} = g(t)dH, \quad \frac{d\ell}{dt} = -ad^*dH(\ell(t))(\ell(t) \tag{7.14}$$

with $(g, \ell) \in G \times \mathfrak{g}_\epsilon$, where $dH = \frac{\partial H}{\partial h_1}A_1 + \frac{\partial H}{\partial h_2}A_2 + \frac{\partial H}{\partial h_3}A_3$. The following proposition is basic.

Proposition 7.4 $C = h_1^2 + h_2^2 + \epsilon h_3^2$ *is an integral of motion for any left-invariant Hamiltonian H.*

Proof It suffices to prove that the Poisson bracket $\{I, H\} = 0$:

$$\{C, H\}(\ell) = \ell([dI, dH])$$

$$= \ell \left[h_1 A_1 + h_2 A_2 + \epsilon h_3 A_3, \frac{\partial H}{\partial h_1} A_1 + \frac{\partial H}{\partial h_2} A_2 + \frac{\partial H}{\partial h_3} A_3 \right]$$

$$= \ell \left(-\epsilon h_1 \frac{\partial H}{\partial h_2} A_3 + h_1 \frac{\partial H}{\partial h_3} A_2 + \epsilon h_2 \frac{\partial H}{\partial h_1} A_3 - h_2 \frac{\partial H}{\partial h_3} A_1 \right.$$

$$\left. + \epsilon h_3 \left(-\frac{\partial H}{\partial h_1} A_2 + \frac{\partial H}{\partial h_2} \right) \right)$$

$$= -\epsilon h_1 h_3 \frac{\partial H}{\partial h_2} + h_1 h_2 \frac{\partial H}{\partial h_3} + \epsilon h_2 h_3 \frac{\partial H}{\partial h_1} - h_1 h_2 \frac{\partial H}{\partial h_3}$$

$$+ \epsilon h_3 \left(-h_2 \frac{\partial H}{\partial h_1} + h_1 \frac{\partial H}{\partial h_2} \right) = 0.$$

$\square$

Function C is called a Casimir or an invariant function. It follows that the Hamiltonian $H = \frac{1}{2}(h_1^2 + h_2^2)$ is a scalar multiple of the Casimir function in the Euclidean case, and hence, Poisson commutes with any function on $\mathfrak{g}_0^*$. But this means that each of h_1, h_2, h_3 are constant along the flow of $\vec{H}$.

Since the extremal controls satisfy $u_1 = h_1$ and $u_2 = h_2$, the solutions $g(t) = (x(t), R(t)) \in \mathfrak{p}_\epsilon \ltimes K$ are the solutions of

$$\frac{dx}{dt} = Ad_{R(t)}(h_1 A_1 + h_2 A_2), \quad \frac{dR}{dt} = 0.$$

Hence, $R(t)$ is constant and $x(t) = x_0 + t h_1 Ad_R(A_1) + t h_2 Ad_R(A_1)$ are straight lines in $\mathfrak{p}_\epsilon$.

In the non-Euclidean cases the solutions reside on the intersection of the energy cylinder $H = \frac{1}{2}(h_1^2 + h_2^2)$ and the hyperboloid $C = h_1^2 + h_2^2 - h_3^2$, in the hyperbolic case, and the sphere $C = h_1^2 + h_2^2 + h_3^2$, in the elliptic case. This implies that h_3 is constant along the flow of $\vec{H}$ (which we already knew from the symmetry considerations, since the metric is invariant under K). The extremals which project onto the geodesics in M_ϵ satisfy the transversality conditions $\ell(0)(\mathfrak{k}_\epsilon) = 0$ and $\ell(T)(\mathfrak{k}_\epsilon) = 0$. This means that $h_3 = 0$.

Equations (7.14) imply that

$$\frac{dh_1}{dt} = \epsilon h_3 h_2, \quad \frac{dh_2}{dt} = -\epsilon h_3 h_1. \tag{7.15}$$

If we now write $h = h_1 + i h_2$ then

$$\frac{dh}{dt} = -i\epsilon h_3 h. \tag{7.16}$$

It follows that

$$h(t) = ae^{-\epsilon h_3 t}, \text{ with } a = h(0).$$

Then,

$$\frac{dg}{dt} = g(t)(h_1(t)A_1 + h_2(t)A_2) = g(t)\frac{1}{2}\begin{pmatrix} 0 & h(t) \\ -\epsilon \bar{h}(t) & 0 \end{pmatrix}$$

$$= g(t)\begin{pmatrix} e^{\frac{i}{2}\epsilon h_3} & 0 \\ 0 & e^{-\frac{i}{2}\epsilon h_3} \end{pmatrix}\frac{1}{2}\begin{pmatrix} 0 & a \\ -\epsilon\bar{a} & 0 \end{pmatrix}\begin{pmatrix} e^{-\frac{i}{2}\epsilon h_3} & 0 \\ 0 & e^{\frac{i}{2}\epsilon h_3} \end{pmatrix} \quad (7.17)$$

The preceding equality implies that

$$g(t) = g_0 e^{(P_\epsilon + \epsilon Q)t} e^{-\epsilon Q t}, \quad (7.18)$$

with $P_\epsilon = \frac{1}{2}\begin{pmatrix} 0 & a \\ -\epsilon\bar{a} & 0 \end{pmatrix}$ and $Q = \epsilon h_3 A_3$, are the solutions to the sub-Riemannian problem. The sub-Riemannian geodesics satisfy an additional condition $H = \frac{1}{2}$, which implies that $|a|^2 = a_1^2 + a_2^2 = 1$

The above formula can be simplified by calculating the exponentials of matrices: let $L = \frac{1}{2}\begin{pmatrix} -i\epsilon a_3 & a \\ -\epsilon\bar{a} & -i\epsilon a_3 \end{pmatrix}$, where h_3 is replaced by a_3 for the uniformity of notation. Then, $L^2 = -\epsilon\frac{|a|^2 + \epsilon a_3^2}{4}I = -\epsilon\frac{\alpha^2}{4}I$, where $\alpha^2 = |a|^2 + \epsilon a_3^2 = C$. Therefore,

$$e^{Lt} = I + Lt + \frac{t^2}{2!}L^2 + \frac{t^3}{3!}L^3 + \cdots$$

$$= I\left(1 + (-\epsilon)\frac{1}{2!}\left(\frac{t}{2}\alpha\right)^2 + (-\epsilon^2)\frac{1}{4!}\left(\frac{t}{2}\alpha\right)^4 + (-\epsilon)^3\frac{1}{6!}\left(\frac{t}{2}\alpha\right)^6 \cdots\right)$$

$$+ \frac{2}{\alpha}L\left(\left(\frac{t}{2}\alpha\right) - \epsilon\frac{1}{3!}\left(\frac{t}{2}\alpha\right)^3 + (-\epsilon)^2\frac{1}{5!}\left(\frac{t}{2}\alpha\right)^5 + \cdots\right).$$

In the elliptic case,

$$e^{Lt} = \cos\left(\frac{t}{2}\alpha\right)I + \frac{\sin\left(\frac{t}{2}\alpha\right)}{\alpha}2L, \quad (7.19)$$

while, in the hyperbolic case, there are three distinct possibilities depending on the sign of C. If $C > 0$ then $\alpha > 0$ and

$$e^{Lt} = \cosh\left(\frac{t}{2}\alpha\right)I + \frac{\sinh\left(\frac{t}{2}\alpha\right)}{\alpha}2L.$$

In the case that $C < 0$ then $\alpha = i\sqrt{-C}$ and $\cosh(t\alpha) = \cos(t\sqrt{-C})$. The remaining case, $C = 0$, yields $e^{Lt} = I + Lt$. When $|a|^2 = 1$, then

$$e^{Lt} = \cosh\frac{t}{2}\sqrt{1-a_3^2}I + \frac{1}{\sqrt{1-a_3^2}}\sinh\frac{t}{2}\sqrt{1-a_3^2}L, \ 1-a_3^2 > 0,$$

$$e^{Lt} = I + Lt, \ 1 - a_3^2 = 0,$$

$$e^{Lt} = \cos\frac{t}{2}\sqrt{a_3^2-1}I + \frac{1}{\sqrt{a_3^2-1}}\sin\frac{t}{2}\sqrt{a_3^2-1}L, \ 1-a_3^2 < 0.$$

If we write $e^{Lt} = \begin{pmatrix} u(t) & v(t) \\ v(t) & u(t) \end{pmatrix}$ for some complex numbers $u(t), v(t)$, then,

$$u(t) = e^{\frac{it}{2}a_3}\left(\cosh\left(\frac{t}{2}\alpha - \frac{ia_3}{\alpha}\sinh\frac{t}{2}\alpha\right)\right), \ v(t) = \frac{a}{\alpha}e^{\frac{it}{2}a_3}\sinh\frac{t}{2}\alpha, \ \alpha^2 = 1 - a_3^2,$$

$$u(t) = e^{\frac{it}{2}}\left(1 - \frac{it}{2}a_3\right), \ v(t) = \frac{at}{2}e^{\frac{it}{2}}, \ a_3^2 = 1, \qquad (7.20)$$

$$u(t) = e^{-\frac{it}{2}a_3}\left(\cos\left(\frac{t}{2}\alpha - \frac{ia_3}{\alpha}\sin\frac{t}{2}\alpha\right)\right), \ v(t) = \frac{a}{\alpha}e^{\frac{it}{2}a_3}\sin\frac{t}{2}\alpha, \ \alpha^2 = a_3^2 - 1.$$

The projected curves $z(t)$ on the unit disk $\mathbb{D}$ are given by

$$z(t) = \begin{cases} \dfrac{a\sinh\frac{t}{2}\sqrt{1-a_3^2}}{\sqrt{1-a_3^2}\cosh\frac{t}{2}\sqrt{1-a_3^2}+ia_3\sinh t\sqrt{1-a_3^2}}, & 1-a_3^2 > 0, \\[4mm] \dfrac{at}{1+it}, & a_3^2 = 1, \qquad (7.21) \\[4mm] \dfrac{a\sin\frac{t}{2}\sqrt{a_3^2-1}}{\sqrt{a_3^2-1}\cos\frac{t}{2}\sqrt{a_3^2-1}+ia_3\sin t\sqrt{a_3^2-1}}, & 1-a_3^2 < 0. \end{cases}$$

This family of curves in the unit disk move along the generalized circles (circles and lines) through the origin with centers at $c = \frac{-ia}{2a_3}$. For each curve $z(t)$ in the above family, a_3 is the geodesic curvature of $z(t)$. Curves with $a_3^2 < 1$ satisfy

$$|z(t)|^2 = \frac{\sinh^2\frac{t}{2}\sqrt{1-a_3^2}}{\sinh^2\frac{t}{2}\sqrt{1-a_3^2}+1-a_3^2}.$$

Since the center c of these circles is greater than $\frac{1}{2}$, $z(t)$ moves along the circular segment from $z(\infty) = \frac{-a}{\alpha-ia_3}$ to $z(\infty) = \frac{a}{\alpha+ia_3}$, with $\alpha^2 = 1 - a_3^2$. The limiting case $a_3 = 0$ results in the line

$$z(t) = a\tanh\frac{t}{2}, \qquad (7.22)$$

which is the hyperbolic geodesic through the origin.

The nilpotent case, $a_3^2 = 1$ corresponds to the circle with center at $\frac{ia}{2}$. This special class of circles, which are tangential to the boundary $|z| = 1$ are called the class of *horocycles*. The point on the boundary on such a circle is identified with $z(\pm\infty)$. The remaining circles, $a_3^2 > 1$ are contained entirely in $\mathcal{D}$. Along such circles, $z(t)$ undergoes circular periodic motion with period $T = \frac{(2n+1)\pi}{\sqrt{a_3^2-1}}$.

In the elliptic case, $e^{Lt} = \begin{pmatrix} u(t) & v(t) \\ -\bar{v}(t) & \bar{u}(t) \end{pmatrix}$ and the solutions are given by

$$u(t) = e^{\frac{it}{2}a_3} \left(\cos\left(\frac{t}{2}\alpha - \frac{ia_3}{\alpha} \sin\frac{t}{2}\alpha \right)\right), \; v(t) = \frac{a}{\alpha}e^{\frac{it}{2}a_3} \sin\frac{t}{2}\alpha, \; \alpha^2 = 1 + a_3^2.$$

(7.23)

The projected curves

$$z(t) = \frac{B}{A} = \frac{a \sin\frac{t}{2}\alpha}{\alpha \cos\frac{t}{2}\alpha + ia_3 \sin\frac{t}{2}\alpha}$$

are circles through the origin, having centers at $\frac{ia}{2a_3}$, with the exception of the geodesic, which is given by the line

$$z(t) = a \tan\frac{t}{2},$$

(7.24)

when $a_3 = 0$.

Let us now comment on some phenomena in sub-Riemannian geometry that are not present in Riemannian geometry. First, some definitions which are common to both of these geometries. The set of all points $g \in G_\epsilon$ whose distance from the origin is T is called the sub-Riemannian sphere of radius T centered at I and will be denoted by $S_T(I)$. The sub-Riemannian ball of radius T, denoted by $B_T(I)$, is the set of all points whose distance from I is less or equal than T. Similar to these notions is the notion of the exponential map, or the wave front, which assigns to each sub-Riemannian geodesic its endpoint. In our situation, the exponential map at T is the mapping from the cylinder $|a|^2 = 1, a_3 \in \mathbb{R}$ onto the point $g = g_0 e^{(P_\epsilon + \epsilon Q)t} e^{-\epsilon Qt}$ where P_ϵ and Q_ϵ are the matrices defined by (7.18). It will be denoted by $W_I(T)$.

In the problems of Riemannian geometry, the exponential map coincides with the sphere of radius T for sufficiently small T. In the problems above, the Riemannian geodesics are given by (7.22) in the hyperbolic case and by (7.23) in the elliptic case. In the hyperbolic case, T is the shortest distance from I and $z(T) = a \tanh\frac{t}{2}$; hence the wave front coincides with the Riemannian sphere for any $T > 0$. In the elliptic case, the Riemannian sphere of radius T coincides with the wave front up to $T = \pi$, beyond which the geodesic stops being optimal. This is evident on the Riemann sphere: the geodesic in (7.24) traces the great circle through the south pole of the Riemann sphere and reaches the north pole at $T = \pi$. Every point beyond the north pole is reachable by the geodesic traced in the opposite direction in time less than π.

In problems of sub-Riemannian geometry the wave front $W_I(T)$ is never equal to $S_I(T)$, no matter how small T. The argument is essentially the same in both the elliptic and the hyperbolic case. In the hyperbolic case, the sub-Riemannian geodesics generated by (a, a_3) with $a_3^2 > 1$ are of the same form as the geodesics in the elliptic case. For any $T > 0$, $v = 0$ for $\frac{T}{2}\alpha = n\pi$, $n = 1, 2, \ldots$. For such values of α, the endpoint of the geodesic $g(T)$ generated by any a, $|a| = 1$, reaches the isotropy group $K = \left\{ \begin{pmatrix} A & 0 \\ 0 & \bar{A} \end{pmatrix} : |A| = 1 \right\}$. In fact, if α_n is defined by

$$\frac{T}{2}\sqrt{\alpha_n^2 + \epsilon} = n\pi, \ \epsilon = \pm 1,$$

then

$$A_n = (-1)^n e^{-i\alpha_n \frac{T}{2}}.$$

This means that all the geodesics generated by $\{(a, \alpha_n) : |a| = 1\}$ arrive at the same point $g_n(T) = \begin{pmatrix} A_n & 0 \\ 0 & \bar{A}_n \end{pmatrix}$ in K. Therefore $g_n(T)$ is the intersection of several geodesics, and hence $g_n(t)$ cannot be optimal for $t > T$. This implies that for any $T > 0$, there exist geodesics $g(t)$ that emanate from I at $t = 0$ and arrive at K in any time less than T. Hence $g(T)$ is in $W_I(T)$ but not in $S_I(T)$.

Furthermore, $W_I(T)$ is not closed for any sufficiently small T. To see this, note that $\lim_{n \to \infty} (-1)^n e^{-i\alpha_n \frac{T}{2}} = 1$ whenever $\frac{T}{2}\sqrt{\alpha_n^2 + \epsilon} = n\pi$, $\epsilon = \pm 1$. That means that the sequence of geodesics $g_n(t)$ defined by any (a, α_n), subject to $|a| = 1$, converges to the identity at time T. But I is not reachable by a geodesic for sufficiently small time T.

Let us end this discussion with the hyperbolic geodesics in $\mathbb{D}$. An easy calculation shows that if $z(t) = a \tanh \frac{t}{2}$ then

$$t = \log \frac{1 + |z(t)|}{1 - |z(t)|}.$$

To find the hyperbolic distance between any two points z_1 and z_2 use the isometry $w = \frac{z - z_1}{1 - \bar{z}_1 z}$ to transform z_1 to the origin. Hence, the distance between z_1 and z_2 is the same as the distance from 0 to $\frac{z_2 - z_1}{1 - \bar{z}_1 z_2}$. It follows that this distance is given by

$$\log \frac{1 + \left| \frac{z_2 - z_1}{1 - \bar{z}_1 z_2} \right|}{1 - \left| \frac{z_2 - z_1}{1 - \bar{z}_1 z_2} \right|}.$$

Hyperbolic geometry plays an important and profound role in the theory of analytic functions on the unit disk largely due to the following proposition [Os].

Proposition 7.5 The Schwarz–Pick lemma   *Let $d(z_1, z_2)$ denote the hyperbolic distance between points z_1 and z_2 in $\mathbb{D}$. Then*

$$d(f(z_1), f(z_2)) \le d(z_1, z_2) \qquad (7.25)$$

for any analytic function on $\mathcal{D}$. The equality occurs only for the Moebius transformations defined by the action of $SU(1, 1)$.

The following lemma is crucial.

Lemma 7.6 The Schwarz lemma   *Let F be an analytic function in $\mathbb{D}$ such that $F(0) = 0$. The $|F(z)| \le |z|$ for all $z \in \mathbb{D}$. If the equality occurs for some $z_0 \in \mathbb{D}$ then $F(z) = \alpha z$ with $|\alpha| = 1$.*

The proof can be found in any book on complex function theory. We will also need the following observation: function $f(x) = \log \frac{1+x}{1-x}$ is monotone increasing in the interval $(0, 1)$. We now turn to the proof of the proposition.

Proof   Suppose that $S(z) = \frac{az+b}{bz+\bar{a}}$ with $|a|^2 - |b|^2 = 1$ satisfies $S(z_1) = 0$, then $S(z) = \frac{z - z_1}{1 - \bar{z}_1 z} \alpha$ for some complex number α with $|\alpha| = 1$. The inverse $S^{-1}(z) = \frac{z + z_1}{1 + \bar{z}_1 z} \bar{\alpha}$ takes 0 to z_1. The composition $F(z) = S_1 \circ f \circ S^{-1}(z)$, where $S_1(z) = \frac{z - f(z_1)}{1 - \overline{f(z_1)} z}$, is an analytic function in $\mathbb{D}$ that satisfies $F(0) = 0$. By Schwarz's lemma, $|F(z)| \le |z|$, which means that $|S_1 \circ f(z)| \le |S(z)|$. Therefore,

$$\left| \frac{f(z_2) - f(z_1)}{1 - \bar{f}(z_1) f(z_2)} \right| \le \left| \frac{z_2 - z_1}{1 - \bar{z}_1 z_2} \right|.$$

But then

$$d(f(z_1), f(z_2)) = \log \frac{1 + \left| \frac{f(z_2) - f(z_1)}{1 - \bar{f}(z_1) f(z_2)} \right|}{1 - \left| \frac{f(z_2) - f(z_1)}{1 - \bar{f} z_1 f(z_2)} \right|} \le \log \frac{1 + \left| \frac{z_2 - z_1}{1 - \bar{z}_1 z_2} \right|}{1 - \left| \frac{z_2 - z_1}{1 - \bar{z}_1 z_2} \right|} = d(z_1, z_2).$$

In the case $F(z) = \alpha z$ with $|\alpha| = 1$, $f(z) = \alpha S_1^{-1} \circ S(z)$. $\square$

7.4 Elastic curves

The Hamiltonian $H = \frac{1}{2}h_3^2 + h_1$ associated with the elastic problem is also completely integrable, because H, the Casimir $C = h_1^2 + h_2^2 + \epsilon h_3^2$, and the Hamiltonian lift of any right-invariant vector field form an involutive family on T^*G. In fact, the same can be said for any left-invariant Hamiltonian on T^*G. Recall that $H_X(g,\ell) = \ell(g^{-1}Ag)$ is the Hamiltonian lift of a right-invariant vector field $X(g) = Ag$ in the left-invariant trivialization of T^*G. In this setting, integrability of left-invariant Hamiltonians on T^*G coincides with the integrability on coadjoint orbits.

Each coadjoint orbit in $\mathfrak{g}_\epsilon^*$ is identified with the hypersurface $h_1^2 + h_2^2 + \epsilon h_3^2 = c$ and the integral curves of $\vec{H}$ reside in the intersection of the surfaces $H = c_1$ and $C = c_2$ for some constants c_1 and c_2. $H = c_1$ is a cylindrical paraboloid in the coordinates h_1, h_2, h_3 and $C = c_2$ is a cylinder $h_1^2 + h_2^2 = c_2, h_3 \in \mathbb{R}$ in the Euclidean case, the sphere $h_1^2 + h_2^2 + h_3^2 = c_2$ in the elliptic case and the hyperboloid $h_1^2 + h_2^2 - h_3^2 = c_2$ in the hyperbolic case.

The Hamiltonian equation $\frac{d\ell}{dt} = - \, ad^*dH(\ell(t))(\ell(t))$ means that $\frac{d\ell}{dt}(X) = - \ell[dH, X]$ for any $X \in \mathfrak{g}_\epsilon$, where $dH = A_1 + h_3 A_3$. It follows that

$$\frac{dh_1}{dt} = \frac{d\ell}{dt}(A_1) = -\ell([A_1 + h_3 A_3, A_1]) = h_3 h_2,$$

$$\frac{dh_2}{dt} = \frac{d\ell}{dt}(A_2) = -\ell[A_1 + h_3 A_3, A_2]) = \epsilon h_3 - h_1 h_3,$$

$$\frac{dh_3}{dt} = \frac{d\ell}{dt}(A_3) = -\ell[A_1 + h_3 A_3, A_3]) = -h_2.$$

Together with $\frac{dg}{dt} = g(t)(A_1 + h_3(t)A_3)$ these equations constitute the extremal curves of the elastic problem. The projections $x(t) = g(t)K$ on the quotient $M_\epsilon = G_\epsilon/K$ are called elastic. It follows from the previous section that $h_3(t)$ is the signed geodesic curvature of $x(t)$.

Proposition 7.7 $\xi(t) = h_3^2(t)$ is a solution of the following equation:

$$\left(\frac{d\xi}{dt}\right)^2 = -\xi^3(t) + 4\xi^2(t)(H - \epsilon) - 4\xi(t)(H^2 - C). \tag{7.26}$$

Proof

$$\left(\frac{d\xi}{dt}\right)^2 = 4h_2^2 h_3^2 = 4(C - h_1^2 - \epsilon h_3^2)h_3^2$$

$$= 4\left(C - \left(H - \frac{1}{2}h_3^2\right)^2 - \epsilon h_3^2\right)h_3^2 = 4\left(C - \left(H - \frac{1}{2}\xi\right)^2 - \epsilon\xi\right)\xi$$

$$= -\xi^3 + 4\xi^2(H - \epsilon) - 4\xi(H^2 - C)$$

$\square$

Change of variable $\xi = -4p - \frac{4}{3}(H - \epsilon)$ transforms equation (7.26) into

$$\left(\frac{dp}{dt}\right)^2 = 4p^3 - g_2 p - g_3, \tag{7.27}$$

where

$$g_2 = \frac{4}{3}(H - \epsilon)^2 + (H^2 - C) \text{ and } g_3 = \frac{8}{27}(H - \epsilon)^3 + \frac{1}{3}(H - \epsilon)(H^2 - C).$$

The solutions of this equation are well known in the literature on elliptic functions. They are of the form $p(t) = \wp(t - c)$ where $\wp$ is the function of Weierstrass [Ap] (meromorphic and doubly periodic in the complex plane having a double pole at $z = 0$).

There is a natural angle θ that links the above equation to the equation of a mathematical pendulum. The pendulum equation is obtained by introducing an angle θ defined by

$$\epsilon - h_1(t) = J \cos\theta(t), \quad h_2(t) = J \sin\theta(t)$$

along each extremal curve $h_1(t), h_2(t), h_3(t)$. Here J is another constant obtained as follows

$$(\epsilon - h_1)^2 + h_2^2 = 1 - 2\epsilon h_1 + h_1^2 + h_2^2 = 1 - \epsilon(2H - h_3^2) + C - \epsilon h_3^2 = J^2,$$

that is, $J^2 = 1 - 2\epsilon H + C$. Then,

$$-J\sin\theta h_3 = -h_3 h_2 = -\frac{dh_1}{dt} = -J\sin\theta\frac{d\theta}{dt},$$

hence, $\frac{d\theta}{dt} = h_3(t)$. This equation then can be written as

$$\frac{d\theta}{dt} = \pm\sqrt{2(H - \epsilon) + 2J\cos\theta}, \tag{7.28}$$

since $\frac{d\theta}{dt}^2 = h_3^2$, and $H = \frac{1}{2}\left(\frac{d\theta}{dt}\right)^2 + \epsilon - J\cos\theta$.

Equation (7.28) is the equation is the equation of the mathematical pendulum with energy E equal to $H - \epsilon$. According to A. E. Love, G. Kirchhoff was the first to notice the connection between the elastic problem and the pendulum (see [Lv] for further details). Kirchhoff referred to the pendulum as the "kinetic analogue" of the elastic problem.

The connection with the pendulum reveals two distinct cases, which Love calls inflectional and non-inflectional. The inflectional case corresponds to $E < J$. In this case there is a cut-off angle θ_c and the pendulum oscillates between $-\theta_c$ and θ_c. The curvature changes sign at the cut-off angle (which accounts for the name). The remaining case $E \geq J$ corresponds to the non-inflectional case, where the curvature does not change sign. The case $E = H$ is the critical case, where the pendulum has just enough energy to reach the top (at $t = \infty$). Otherwise, the pendulum has sufficient energy to go around the top (see [Je] for further details).

The passage from equation (7.27) to the pendulum equation (7.28) transforms the solutions expressed in terms of the elliptic function of Weierstrass to the solutions expressed in terms of Legendre's function $sn(u, k^2)$. This transformation is achieved via the following change of variables: if $2\phi = \theta$, then (7.28) becomes

$$2\dot{\phi} = \sqrt{2(E + J)}\sqrt{1 - k^2 \sin^2 \phi}, \tag{7.29}$$

where $k^2 = \frac{2J}{E+J}$. Under the substitution $x = \sin \phi$, equation (7.29) becomes

$$\dot{x} = \sqrt{\frac{E + J}{2}}\sqrt{(1 - x^2)(1 - k^2 x^2)},$$

or

$$\int \frac{dx}{\sqrt{(1 - x^2)(1 - k^2 x^2)}} = \sqrt{\frac{E + J}{2}}t.$$

The function $z = f(u)$ defined by $u = \int_0^z \frac{dz}{\sqrt{(1-z^2)(1-k^2 z^2)}}$ is called Legendre's function and is denoted by $z = sn(u, k^2)$. It follows that $x = sn\left(\sqrt{\frac{E+J}{2}}t, k^2\right)$, and therefore the extremal curvature $\kappa(t)$ is given by

$$\kappa(t) = \sqrt{2(E + J)}\sqrt{1 - k^2 sn^2\left(\sqrt{\frac{E + J}{2}}t, k^2\right)}.$$

7.5 Complex overview and integration

We will now show that the remaining equations are solvable by quadrature in terms of $\kappa(t)$. The integration procedure is essentially the same in both the elliptic and the hyperbolic case. This fact is best demonstrated by passing to complex Lie group $G = SL_2(\mathbb{C})$ and to the real forms of the complex Lie algebra $\mathfrak{g} = sl_2(\mathbb{C})$.

The Lie algebra $\mathfrak{g}$ of $SL_2(\mathbb{C})$ consists of matrices $M = \begin{pmatrix} a & b \\ c & -a \end{pmatrix}$ with complex entries a, b, c. Any such matrix can be written as $M = a_1 A_1 + a_2 A_2 + a_3 A_3$ where A_1, A_2, A_3 are the Pauli matrices defined by (7.7). In fact, $a = -\frac{i}{2}a_3, \frac{b+c}{2} = ia_2, \frac{b-c}{2} = a_1$. A real Lie algebra $\mathfrak{h}$ is said to be a real form for a complex Lie algebra $\mathfrak{g}$ if $\mathfrak{g} = \mathfrak{h} + i\mathfrak{h}$. Both $\{\mathfrak{g}_\epsilon, \epsilon = \pm 1\}$ are real forms of $sl_2(\mathbb{C})$ because su_2 is the linear span of matrices with real coefficients a_1, a_2, a_3 and $su(1,1)$ it the linear span of matrices with coefficients ia_1, ia_2, a_3, as a_1, a_2, a_3 range over all real numbers.

The real Poisson structure also extends to complex Poisson structure on $\mathfrak{g}^*$ with $\{f, h\}(\ell) = \ell([df, dh]$ for any complex functions f and h on $\mathfrak{g}^*$ and any point ℓ in $\mathfrak{g}^*$. Then any complex function H on $\mathfrak{g}^*$ defines a Hamiltonian vector field $\vec{H}$ on $G \times \mathfrak{g}^*$ with integral curves $(g(t), \ell(t))$ the solutions of

$$\frac{dg}{dt} = g(t)dH(\ell(t)), \quad \frac{d\ell}{dt} = -ad^* dH(\ell(t))(\ell(t)). \tag{7.30}$$

In particular, $H = \frac{1}{2}h_3^2 + h_1$, where h_1, h_2, h_3 now denote the dual (complex) coordinates of a point $\ell \in \mathfrak{g}^*$ relative to the dual basis A_1^*, A_2^*, A_3^* in $\mathfrak{g}^*$, may be considered as the complexification of the Hamiltonian generated by the elastic problem. It follows that the Hamiltonian equations of H are given by

$$\frac{d}{dt} = g(A_1 + h_3A_3), \quad \frac{dh_1}{dt} = h_3h_2, \quad \frac{dh_2}{dt} = h_3 - h_1h_3, \quad \frac{dh_3}{dt} = -h_2. \tag{7.31}$$

In this setting angle θ is defined in the same manner as in the real case with

$$1 - h_1 = J\cos\theta, \quad h_2 = J\sin\theta,$$

with $J^2 = 1 - 2H + C^2$ and $C^2 = h_1^2 + h_2^2 + h_3^2$. The corresponding pendulum equation $\frac{d\theta}{dt} = \sqrt{2}\sqrt{E + J\cos\theta}$, with $E = H - 1$, is now complex valued.

We will make use of the fact that $\ell \in \mathfrak{g}^*$ can be identified with $L \in \mathfrak{g}$ via the formula $\ell(X) = \langle L, X\rangle$ for all $X \in \mathfrak{g}$, where $\langle A, B\rangle = -\frac{1}{2}Tr(AB)$. Since the Pauli matrices remain orthonormal relative to this quadratic form, $\langle A, B\rangle = a_1b_1 + a_2b_2 + a_3b_3$, for any matrices A and B in $\mathfrak{g}$. Moreover, the invariance property $\langle [A, B], C\rangle = \langle A, [B, C]\rangle$ extends to the complex setting in $\mathfrak{g}$. Then $\ell = h_1A_1^* + h_2A^* + h_3A^*$ corresponds to $L = h_1A_1 + h_2A_2 + h_3A_3$. Under this identification equations (7.30) are identified with

$$\frac{dg}{dt} = g(t)(A_1 + h_3A_3), \quad \frac{dL}{dt} = [dH(\ell(t)), L(t)] = [A_1 + h_3A_3, L(t)], \tag{7.32}$$

and each coadjoint orbit is identified with an adjoint orbit $g(t)L(t)g^{-1}(t) = \Lambda$ for some matrix $\Lambda \in \mathfrak{g}$.

It follows that $C^2 = \langle L(t), L(t)\rangle$ is the Casimir on $\mathfrak{g}$. On each adjoint orbit, $\langle L(t), L(t)\rangle = \langle \Lambda, \Lambda\rangle$. Since $SL_2(\mathbb{C}$ acts transitively by conjugation on the complex variety $\{X \in \mathfrak{g} : \langle X, X\rangle = \langle \Lambda, \Lambda\rangle\}$, there is no loss in generality in assuming that $\Lambda = CA_3$. Corresponding to this choice of normalization there is an adapted choice of coordinates $\theta_1, \theta_2, \theta_3$ on G defined by

$$g = e^{\theta_1 A_3}e^{\theta_2 A_1}e^{\theta_3 A_3}. \tag{7.33}$$

Then $gLg^{-1} = CA_3$ implies that

$$e^{\theta_3 A_3} L(t) e^{-\theta_3 A_3} = Ce^{-\theta_2 A_1} A_3 e^{\theta_2 A_1}.$$

Since

$$e^{\theta_3 A_3} = \begin{pmatrix} e^{-\frac{i}{2}\theta_3} & 0 \\ 0 & e^{\frac{i}{2}\theta_3} \end{pmatrix}, \ e^{\theta_2 A_1} = \begin{pmatrix} \cos\frac{1}{2}\theta_2 & \sin\frac{1}{2}\theta_2 \\ -\sin\frac{1}{2}\theta_2 & \cos\frac{1}{2}\theta_2 \end{pmatrix},$$

the above implies that $h_3 = C\cos\theta_2$ and

$$h_1 + ih_2 = -iC(\cos\theta_3 + i\sin\theta_3)\sin\theta_2, \ -h_1 + ih_2$$
$$= -iC(\cos\theta_3 - i\sin\theta_3)\sin\theta_2.$$

Therefore,

$$h_1(t) = C\sin\theta_3\sin\theta_2, \ h_2(t) = -C\cos\theta_3\sin\theta_2, \ h_3 = C\cos\theta_2. \quad (7.34)$$

The remaining angle θ_1 will be now calculated from the equation $g^{-1}\frac{dg}{dt} = A_1 + h_3 A_3$ and the representation (7.33). Differentiation of (7.33) along a curve $g(t)$ results in

$$g^{-1}\frac{dg}{dt} = \frac{d\theta_1}{dt} g^{-1} A_3 g + \frac{d\theta_2}{dt} e^{\theta_3 A_3} A_1 e^{\theta_3 A_3} + \frac{d\theta_3}{dt} A_3. \quad (7.35)$$

An easy calculation shows that $e^{-\theta_3 A_3} A_1 e^{\theta_3 A_3} = \cos\theta_3 A_1 + \sin\theta_3 A_2$. If $g(t)$ is a solution of $g^{-1}\frac{dg}{dt} = A_1 + h_3 A_3$, then (7.33) yields

$$\frac{h_1}{C}\frac{d\theta_1}{dt} + \cos\theta_3\frac{d\theta_2}{dt} = 1, \ \frac{h_2}{C}\frac{d\theta_1}{dt} + \sin\theta_3\frac{d\theta_2}{dt} = 0, \ \frac{h_3}{C}\frac{d\theta_1}{dt} + \frac{d\theta_3}{dt} = h_3.$$
$$(7.36)$$

Therefore,

$$\frac{d\theta_1}{dt} = \frac{C\sin\theta_3}{h_1\sin\theta_3 - h_2\cos\theta_3} = \frac{\sin\theta_3}{\sin\theta_2} = \frac{Ch_1}{\sin^2\theta_2}. \quad (7.37)$$

Since $h_3 = C\cos\theta_2$, $\sin^2\theta_2 = C^2 - h_3^2$, and hence,

$$\theta_1(t) = C\int\frac{H - \frac{1}{2}h_3^2}{C^2 - h_3^2}\,dt = C\int\frac{H - \frac{1}{2}\kappa^2(t)}{C^2 - \kappa^2(t)}\,dt.$$

Let us finally note that $z(t) = g(t)(0) = e^{\theta_1 A_3} e^{\theta_2 A_1}(0) = e^{-i\theta_1}\tan\theta_2$ is the projection of an extremal curve all the way down to $SL_2(\mathbb{C})/SO_2(\mathbb{C})$. This projection is called complex elastica in [Jm]. Then elliptic elasticae are the projections of the extremal curves with real valued h_1, h_2, h_3, while hyperbolic elasticae are the projections of the extremal curves in which h_1 and h_2 are imaginary and h_3 is real.

8

Symmetric spaces and sub-Riemannian problems

Remarkably, the group theoretic treatment of the classical geometry described in the previous chapter extends to a much wider class of Riemannian spaces, known as symmetric spaces in the literature on differential geometry [C1; Eb; Hl; Wf]. Our exposition of this subject matter is somewhat original in the sense that it explains much of the basic theory through symplectic formalism associated with canonical sub-Riemannian problems on Lie algebras $\mathfrak{g}$ that admit a Cartan decomposition $\mathfrak{g} = \mathfrak{p} \oplus \mathfrak{k}$.

The presentation is motivated by two main goals. Firstly, we want to demonstrate the relevance of the mathematical formalism described earlier in this text for this beautiful area of mathematics. Secondly, we want to present the subject in a self-contained way that lends itself naturally to our study of integrable systems on Lie algebras and the problems of applied mathematics.

8.1 Lie groups with an involutive automorphism

We will begin with Lie groups with an involutive automorphism, rather than the symmetric spaces themselves as is often done in the existing literature on this subject [Eb; Hl]. An automorphism σ on a Lie group G that is not equal to the identity which satisfies $\sigma^2 = I$ is called *involutive*. More explicitly, an involutive automorphism σ satisfies:

1. $\sigma : G \to G$ is analytic.
2. $\sigma(gh) = \sigma(g)\sigma(h)$ for all g and h in G.
3. $\sigma(g) = e$ only for $g = e$.
4. $\sigma^2(g) = g$ for all g in G.

The group of fixed points $H_0 = \{g : \sigma(g) = g\}$ is a closed subgroup of G, and hence a Lie subgroup of G. We will use H to denote the connected component of H_0 that contains the group identity.

118

The tangent map σ_* at the identity is an automorphism of the Lie algebra $\mathfrak{g}$, that is, satisfies $\sigma_*([A, B]) = [\sigma_*(A), \sigma_*(B)]$ for any A and B in $\mathfrak{g}$. It follows that

$$(\sigma_* + I)(\sigma_* - I) = 0,$$

because $\sigma_*^2 = I$. Therefore, $\sigma_*(M) = M$, or $\sigma_*(M) = -M$ for any M in $\mathfrak{g}$. Let

$$\mathfrak{p} = \{M : \sigma_*(M) = -M\} \text{ and } \mathfrak{h} = \{M : \sigma_*(M) = M\}.$$

Since $\sigma_*([A, B]) = [\sigma_*(A), \sigma_*(B)]$, the following Lie bracket conditions hold:

$$[\mathfrak{p}, \mathfrak{p}] \subseteq \mathfrak{h}, [\mathfrak{h}, \mathfrak{p}] \subseteq \mathfrak{p}, [\mathfrak{h}, \mathfrak{h}] = \mathfrak{h} \tag{8.1}$$

We will refer to the above Lie algebra decomposition as the Cartan decomposition induced by σ, or simply the Cartan decomposition. The Lie algebraic conditions (8.1) imply that $\mathfrak{h}$ is a Lie subalgebra of $\mathfrak{g}$.

Proposition 8.1 $\mathfrak{h}$ *is the Lie algebra of H.*

Proof For any M in the Lie algebra of H, $\exp tM$ is in H, and therefore $\sigma(\exp tM) = \exp tM$ for all t. It follows that $\sigma_*(M) = M$, and hence M belongs to $\mathfrak{h}$. We have shown that $\mathfrak{h}$ contains the Lie algebra of H.

Suppose now that M is any point in $\mathfrak{h}$. Then

$$\sigma(\exp tM) = \exp t \sigma(M) = \exp tM.$$

The above implies that $\{\exp tM : t \in \mathbb{R}\}$ is a curve in H. But then M belongs to the Lie algebra of H. □

Proposition 8.2 $Ad_h(\mathfrak{p})$ *is contained in* $\mathfrak{p}$ *for any h in H.*

Proof Let B denote an arbitrary element of $\mathfrak{h}$. Then for any A in $\mathfrak{p}$, $Ad_{\exp tB}(A) = \exp t(adA(B))$. Since $adA(B)$ belongs to $\mathfrak{p}$ it follows that $Ad_{\exp tB}(A) \in \mathfrak{p}$ for all t. But then any h in H can be written as the product

$$h = \exp t_m B_m \circ \exp t_{m-1} B_{m-1} \circ \cdots \exp t_1 B_1.$$

Then $Ad_h(A) = Ad_{\exp t_m B_m} \circ \cdots Ad_{\exp t_1 B_1}(A)$, and therefore $Ad_h(A)$ belongs to $\mathfrak{p}$ for any A in $\mathfrak{p}$. □

Definition 8.3 We shall write $Ad_h^{\mathfrak{p}}$ for the restriction of Ad_h to $\mathfrak{p}$, and $Ad_H^{\mathfrak{p}}$ will denote the subgroup of $Gl(\mathfrak{p})$ generated by $\{Ad_h^{\mathfrak{p}} : h \in H\}$.

Definition 8.4 The pair (G, H) will be called a symmetric pair if G is a Lie group with an involutive automorphism and H is a closed and connected subgroup of fixed points of the automorphism. If in addition $Ad_H^{\mathfrak{p}}$ is a compact subgroup of $Gl(\mathfrak{p})$ then (G, H) will be called a Riemannian symmetric pair.

8.2 Symmetric Riemannian pairs

The following proposition marks a point of departure for a kind of variational problems that characterize symmetric spaces.

Proposition 8.5 *Let (G, H) be a symmetric pair. There exists a positive-definite quadratic form $\langle , \rangle$ on $\mathfrak{p}$ that is invariant under $Ad_H^{\mathfrak{p}}$ if and only if $Ad_H^{\mathfrak{p}}$ is compact in $Gl(\mathfrak{p})$.*

Proof Any positive-definite form $\langle , \rangle$ on $\mathfrak{p}$ turns $\mathfrak{p}$ into a Euclidean vector space. Then the orthogonal group $O(\mathfrak{p})$ is the largest group that leaves $\langle , \rangle$ invariant. If a positive-definite form $\langle , \rangle$ is invariant under $Ad_H^{\mathfrak{p}}$ then $Ad_H^{\mathfrak{p}}$ is contained in $O(\mathfrak{p})$. Since $O(\mathfrak{p})$ is compact and $Ad_H^{\mathfrak{p}}$ is closed, $Ad_H^{\mathfrak{p}}$ is compact.

Suppose now that $Ad_H^{\mathfrak{p}}$ is a compact subgroup of $Gl(\mathfrak{p})$. Let $E(\mathfrak{p})$ denote the vector space of all symmetric quadratic forms on $\mathfrak{p}$. Then $Ad_H^{\mathfrak{p}}$ acts on $E(\mathfrak{p})$ by

$$T(Q)(x, y) = \langle T^{-1}x, T^{-1}y \rangle$$

for each quadratic form $Q = \langle , \rangle$ in $E(\mathfrak{p})$, and each T in $Ad_H^{\mathfrak{p}}$.

Since $Ad_H^{\mathfrak{p}}$ is compact, continuous functions on $Ad_H^{\mathfrak{p}}$ with values in $E(\mathfrak{p})$ can be integrated so as to obtain values in $E(\mathfrak{p})$. In this situation, integration commutes with linear maps [Ad]. It then follows that the integral

$$\int_{Ad_H^{\mathfrak{p}}} g(Q) \, dg$$

defines a positive-definite, $Ad_H^{\mathfrak{p}}$-invariant quadratic form for any positive-definite quadratic form Q on $E(\mathfrak{p})$.   □

We will now suppose that (G, H) is a Riemannian symmetric pair with a positive-definite, Ad_H-invariant quadratic form $\langle , \rangle$ in $\mathfrak{p}$, where invariance means

$$\langle Ad_H(A), Ad_H(B) \rangle = \langle A, B \rangle$$

for any A and B in $\mathfrak{p}$. It then follows, by differentiating $\langle Ad_{\exp tC}(A), Ad_{\exp tC}(B) \rangle = \langle A, B \rangle$ at $t = 0$, that $\langle [C, A], B \rangle + \langle A, [C, B] \rangle = 0$ for any element C in $\mathfrak{h}$, or that

$$\langle [A, C], B \rangle = \langle A, [C, B] \rangle \tag{8.2}$$

for A and B in $\mathfrak{p}$ and C in $\mathfrak{h}$.

Recall now the Killing form $Kl(A, B) = Tr(ad(A) \circ ad(B))$ defined in Chapter 5 and its basic invariance property relative to the automorphisms

ϕ on $\mathfrak{g}$. This invariance property implies that Kl is Ad_H-invariant, and secondly it implies that $\mathfrak{p}$ and $\mathfrak{h}$ are orthogonal, because

$$Kl(A, B) = Kl(\sigma_*(A), \sigma_*(B)) = Kl(-A, B) = -Kl(A, B).$$

The following proposition is a variant of a well-known theorem about the simultaneous diagonalization of two quadratic forms, one of which is positive-definite.

Proposition 8.6 *Let* $\mathfrak{p}_0 = \{A \in \mathfrak{p} : Kl(A, X) = 0, X \in \mathfrak{p}\}$. *Then* $\mathfrak{p}_0$ *is* $ad(\mathfrak{h})$-*invariant. Moreover, there exist non-zero real numbers* $\lambda_1, \lambda_2 \ldots, \lambda_m$ *and linear subspaces* $\mathfrak{p}_1, \mathfrak{p}_2, \ldots, \mathfrak{p}_m$ *such that:*

(i) $\mathfrak{p} = \mathfrak{p}_0 \oplus \mathfrak{p}_1 \oplus \cdots \oplus \mathfrak{p}_m$.
(ii) $\mathfrak{p}_0, \ldots, \mathfrak{p}_m$ *are pairwise orthogonal relative to* Kl.
(iii) *Each* $\mathfrak{p}_i$, $i \geq 1$ *is* $ad(\mathfrak{h})$-*invariant and contains no proper* $ad(\mathfrak{h})$-*invariant subspaces.*
(iv) *For each* $i = 1, \ldots, m$

$$Kl(A, B) = \lambda_i \langle A, B \rangle$$

for all A and B in $\mathfrak{p}_i$.

Proof If $\mathfrak{p}_0 = \{A \in \mathfrak{p} : Kl(A, X) = 0, X \in \mathfrak{p}\}$ then, $Kl(ad(\mathfrak{h})(\mathfrak{p}_0), X) = Kl(\mathfrak{p}_0, ad(\mathfrak{h})(X)) = 0$; hence, $ad(\mathfrak{h})(\mathfrak{p}_0) \subseteq \mathfrak{p}_0$. Let $\mathfrak{p}_0^\perp$ denote the orthogonal complement in $\mathfrak{p}$ relative to $\langle\,,\,\rangle$. It follows that $\mathfrak{p}_0^\perp$ is also invariant under $ad(\mathfrak{h})$. Since $\langle\,,\,\rangle$ is positive-definite, $\mathfrak{p}_0 \cap \mathfrak{p}_0^\perp = 0$. Therefore, Kl is non-degenerate on $\mathfrak{p}_0^\perp$.

Let $\mathfrak{p}_1$ denote the set theoretic intersection of all $ad(\mathfrak{h})$-invariant subspaces of $\mathfrak{p}_0^\perp$. Then $\mathfrak{p}_1$ is an $ad(\mathfrak{h})$-invariant subspace of $\mathfrak{p}_0^\perp$ that admits no proper subspace that is invariant under $ad(\mathfrak{h})$. Now proceed inductively: if $\mathfrak{p}_0 \oplus \mathfrak{p}_1$ is not equal to $\mathfrak{p}$ relative to $\langle\,,\,\rangle$ then let $\mathfrak{p}_2$ be equal to the intersection of all $ad(\mathfrak{h})$-invariant subspaces in $\mathfrak{p}_1^\perp$. Continue until the m-th stage at which $\mathfrak{p}_m^\perp = 0$. Then, $\mathfrak{p} = \mathfrak{p}_0 \oplus \mathfrak{p}_1 \oplus \cdots \oplus \mathfrak{p}_m$. It is clear from the construction that each $\mathfrak{p}_i, i = 1, \ldots, m$ does not admit any proper $ad(\mathfrak{h})$-invariant subspaces. Therefore, up to a reordering, this decomposition is unique.

If we now repeated this process with the orthogonal complements taken relative to Kl, the resulting decomposition would be the same as the one above, up to a possible renumbering of the spaces. This shows that $Kl(\mathfrak{p}_i, \mathfrak{p}_j) = 0$, $i \neq j$.

It remains to show part (iv). Let A denote the point in $\mathfrak{p}_i$ that yields the maximum of $Kl(x, x)$ over the points in $\mathfrak{p}_i$ that satisfy $\langle x, x \rangle = 1$. According to the Lagrange multiplier rule there exists a number λ_i such that A is a critical point of $F(x) = Kl(x, x) + \lambda_i(1 - \langle x, x \rangle)$, that is, $Kl(A, X) = \lambda_i \langle A, X \rangle$, for all $X\mathfrak{p}_i$. The multiplier λ_i cannot be equal to zero because $\mathfrak{p}_i$ is orthogonal to $\mathfrak{p}_0$.

Then $Kl(ad(\mathfrak{h})(A), X) = \lambda_i \langle ad(\mathfrak{h})(A), X \rangle$. Since $\mathfrak{p}_i$ does not admit any proper $ad(\mathfrak{h})$-invariant subspaces, the action $h \to ad(h)(A)$ is irreducible. This implies that $ad(h)A$ contains an open set in $\mathfrak{p}_i$. Therefore, $Kl(Y, X) = \lambda_i \langle Y, X \rangle$ for all X and Y in $\mathfrak{p}_i$.   □

Proposition 8.7 *Let H be a Lie group such that Ad_H is a compact subgroup of $Gl(\mathfrak{g})$. Then the Killing form is negative-definite on the Lie algebra $\mathfrak{h}$ provided that $\mathfrak{h}$ has zero center.*

Proof Let $\langle , \rangle$ denote any positive-definite Ad_H invariant quadratic form on $\mathfrak{h}$, and let $A_1, \ldots, A_n$ denote any orthonormal basis relative to this form. Then for any B in $\mathfrak{h}$ the matrix of $ad(B)$ relative to this basis is skew-symmetric. If $M = (m_{ij})$ and $N = (n_{ij})$ are any skew-symmetric matrices. then the trace of MN is equal to $-2(\sum_{i=1}^{n} \sum_{j=1}^{n} m_{ij} n_{ij}$. Therefore,

$$Kl(B, B) = Tr(ad(B) \circ ad(B)) = -2 \left(\sum_{i=1}^{n} \sum_{j=1}^{n} m_{ij}^2 \right) \leq 0.$$

If $Kl(B, B) = 0$ then $adB = 0$, and hence B belongs to the center of $\mathfrak{h}$. But then $B = 0$.   □

Corollary 8.8 *If (G, H) is a Riemannian symmetric pair and if the Lie algebra $\mathfrak{h}$ of H has zero center, then Kl is negative-definite on $\mathfrak{h}$.*

Proposition 8.9 *Let $\mathfrak{p} = \mathfrak{p}_0 \oplus \cdots \oplus \mathfrak{p}_m$ denote the decomposition described by Proposition 8.6 associated with a Riemannian symmetric pair (G, H) where the Lie algebra $\mathfrak{h}$ of H has zero center. Let $\mathfrak{g}_i = \mathfrak{p}_i + [\mathfrak{p}_i, \mathfrak{p}_i]$ for each $i = 1, \ldots, m$. Then,*

(i) *Each $\mathfrak{g}_i$ is a an ideal of $\mathfrak{g}$ that is also invariant under the involutive automorphism σ_*.*

(ii) *$[\mathfrak{g}_i, \mathfrak{g}_j] = 0$ and $Kl(\mathfrak{g}_i, \mathfrak{g}_j) = 0$ for $i \neq j$.*

Proof Let $A_i \in \mathfrak{p}_i$, $A_j \in \mathfrak{p}_j$ with $i \neq j$. Since $ad(\mathfrak{h})(\mathfrak{p}_i) \subseteq \mathfrak{p}_i$ and $\mathfrak{p}_i$ and $\mathfrak{p}_j$ are orthogonal relative to Kl, $Kl(ad(\mathfrak{h})(A_i), A_j) = Kl(\mathfrak{h}, [A_i, A_j]) = 0$. But $Kl[A_i, A_j], \mathfrak{h}) = 0$ implies that $[A_i, A_j] = 0$ since Kl is negative-definite on $\mathfrak{h}$. Then $[\mathfrak{p}_i, [\mathfrak{p}_j, \mathfrak{p}_j]] = 0$ by the Jacobi's identity, and the same goes for $[[\mathfrak{p}_i, \mathfrak{p}_i], [\mathfrak{p}_j, \mathfrak{p}_j]]$. Hence $[\mathfrak{g}_i, \mathfrak{g}_j] = 0$. Evidently, $Kl(\mathfrak{g}_i, \mathfrak{g}_j) = 0$, $i \neq j$.

To prove that each $\mathfrak{g}_i$ is an ideal of $\mathfrak{g}$, let $A + B$ be an arbitrary element of $\mathfrak{g}$ with $A \in \mathfrak{p}$ and $B \in \mathfrak{h}$. Since each $\mathfrak{p}_i$ is $ad(\mathfrak{h})$-invariant, it follows that $adB(\mathfrak{p}_i) \subseteq \mathfrak{p}_i$. This fact, together with the Jacobi's identity implies that $adB([\mathfrak{p}_i, \mathfrak{p}_i]) \subseteq [\mathfrak{p}_i, \mathfrak{p}_i]$. Therefore, $\mathfrak{g}_i$ is $ad(\mathfrak{h})$-invariant.

Let $A = A_1 + \cdots + A_m$ with $A_i \in \mathfrak{p}_i$ for each $i = 1, \ldots, m$. Then $adA(\mathfrak{g}_i) = adA_i(\mathfrak{g}_i)$ because $[A_j, \mathfrak{g}_i] = 0$. Relations (8.1) together with the

Jacobi's identity imply that $[A_i, \mathfrak{g}_i] \subseteq \mathfrak{g}_i$. Since it is evident that each $\mathfrak{g}_i$ is σ_* invariant, our proof is finished. $\qquad\square$

Proposition 8.10 The fundamental decomposition *Let $\mathfrak{g}_1, \ldots, \mathfrak{g}_m$ be as in the previous proposition. Define $\mathfrak{p}_0 = \{A \in \mathfrak{p} : Kl(A, X) = 0, X \in \mathfrak{p}\}$ and $\mathfrak{g}_0 = \mathfrak{p}_0 \oplus \{B \in \mathfrak{h} : adB(\mathfrak{p}) \subseteq \mathfrak{p}_0\}$. Then,*

(i) $[\mathfrak{p}, \mathfrak{p}_0] = 0$.
(ii) $\mathfrak{g}_0$ *is an ideal in $\mathfrak{g}$.*
(iii) $\mathfrak{g} = \mathfrak{g}_0 \oplus \mathfrak{g}_+ \oplus \mathfrak{g}_-$ *where* $\mathfrak{g}_+ = \cup\{\mathfrak{g}_i : Kl|_{\mathfrak{p}_i} = \lambda_i \langle\,,\,\rangle|_{\mathfrak{p}_i}, \lambda_i > 0\}$, *and*
$\mathfrak{g}_- = \cup\{\mathfrak{g}_i : Kl|_{\mathfrak{p}_i} = \lambda_i \langle\,,\,\rangle|_{\mathfrak{p}_i}, \lambda_i < 0\}$.

Proof Let $A \in \mathfrak{p}, A_0 \in \mathfrak{p}_0$ and $B \in \mathfrak{h}$. Then

$$Kl([A, A_0], B) = Kl(A_0, adB(A)) = 0.$$

Therefore, $[A, A_0] = 0$ since Kl is negative-definite on $\mathfrak{h}$. To prove (ii) take first $A \in \mathfrak{p}$. Because of (i), it suffices to show that $ad[A, B](\mathfrak{p}) \subseteq \mathfrak{p}_0$ for any $B \in \mathfrak{g}_0 \cap \mathfrak{h}$. But $Kl([A, B], \mathfrak{p}) = Kl(A, [B, \mathfrak{p}]) = 0$, since $[B, \mathfrak{p}] \subseteq \mathfrak{p}_0$. So $[A, B] \in \mathfrak{g}_0$.

Now take $A \in \mathfrak{h}$ and an element $A_0 + B_0$ in $\mathfrak{g}_0$ with $A_0 \in \mathfrak{p}_0$ and $B_0 \in \mathfrak{h}$. Since $Kl([A, A_0], \mathfrak{p}) = Kl(A, [A_0, \mathfrak{p}]) = 0, [A, A_0] \in \mathfrak{p}_0$. To show that $[A, B_0] \in \mathfrak{g}_0$, we need to show that $Kl([[A, B_0], X], Y) = 0$ for any X and Y in $\mathfrak{p}$. But that follows immediately from Jacobi's identity applied to $[[A, B_0], X]$.

To show (iii), let

$$\mathfrak{h}_1 = \cup\{[\mathfrak{p}_i, \mathfrak{p}_i] : Kl|_{\mathfrak{p}_i} = \lambda_i \langle\,,\,\rangle|_{\mathfrak{p}_i}, \lambda_i \neq 0\}.$$

Note that $\mathfrak{p}_0$ coincides with the space of eigenvectors corresponding to zero eigenvalue of Kl. It is easy to see that $\mathfrak{h}_1$ is an ideal in $\mathfrak{h}$. Let $\mathfrak{h}_0$ denote its orthogonal complement relative to the Killing form. Claim that $\mathfrak{h}_0 \subseteq \{B : adB(\mathfrak{p}) \subseteq \mathfrak{p}_0\}$. Indeed, if $A \in \mathfrak{p}$ and $X \in \mathfrak{p}$ are written as $A = \sum A_i$ and $X = \sum X_i$ with A_i and X_i in $\mathfrak{p}_i$ then for any $B \in \mathfrak{h}_0$,

$$Kl(adB(A), X) = Kl(B, [A, X]) = Kl\left(B, \sum[A_i, X_i]\right) = 0.$$

It follows that $\mathfrak{h} \subseteq \mathfrak{g}_0 \oplus \mathfrak{g}_+ \oplus \mathfrak{g}_-$, and therefore (iii) is true. $\qquad\square$

Recall now that a Lie algebra is said to be semi-simple if the Killing form is non-degenerate. A Lie algebra is said to be simple if it is semi-simple and contains no proper ideals.

Proposition 8.11 *Every semi-simple Lie algebra $\mathfrak{g}$ is a direct sum of simple ideals.*

Proof Suppose that $\mathfrak{g}_0$ is a simple ideal in $\mathfrak{g}$ not equal to $\mathfrak{g}$. Let $\mathfrak{g}_0^\perp$ denote the orthogonal complement of $\mathfrak{g}_0$ relative to Kl. Then, $\mathfrak{g} = \mathfrak{g}_0 \oplus \mathfrak{g}_0^\perp$ because Kl

is non-degenerate on $\mathfrak{g}$. Moreover, $\mathfrak{g}_0^\perp$ is an ideal in $\mathfrak{g}$. The proof is simple: if $X \in \mathfrak{g}$, $A \in \mathfrak{g}_0^\perp$ and $B \in \mathfrak{g}_0$, then

$$Kl([X,A],B) = Kl(X,[A,B]) = 0.$$

Hence, $[X,A] \in \mathfrak{g}_0^\perp$. Since Kl is non-degenerate on $\mathfrak{g}_0^\perp$, the argument can be repeated whenever $\mathfrak{g}_0^\perp$ is not a simple ideal. □

Corollary 8.12   *Every semi-simple Lie algebra has zero center.*

Proof   If C denotes the center of $\mathfrak{g}$ then C is an ideal in $\mathfrak{g}$ on which Kl is zero. By the argument above, $\mathfrak{g} = C \oplus C^\perp$. But then $C \neq 0$ would violate the non-degeneracy of Kl. □

Proposition 8.13   *Suppose that (G,H) is a Riemannian symmetric pair where the Lie algebra $\mathfrak{h}$ of H has a zero center. Let $\mathfrak{g}_i = \mathfrak{p}_i + [\mathfrak{p}_i, \mathfrak{p}_i]$ be the ideals in $\mathfrak{g}$ such that Kl restricted to $\mathfrak{p}_i$ is given by $\lambda_i \langle \,, \rangle$ with $\lambda_i \neq 0$. Then each $\mathfrak{g}_i$ is a simple ideal, i.e., it contains no proper non-zero ideals.*

Proof   Any proper ideal of $\mathfrak{g}_i$ would split $\mathfrak{p}_i$ into a direct sum of two orthogonal $ad(\mathfrak{h})$-invariant spaces. Since $ad(\mathfrak{h})$ acts irreducibly on $\mathfrak{p}_i$ one of these subspaces must be (0). □

Corollary 8.14   *If $\mathfrak{g}_0$ in Proposition 7.10 is equal to zero, then $\mathfrak{g}$ is semi-simple.*

Remark 8.15   It is not necessary for $\mathfrak{g}_0$ to be equal to zero for $\mathfrak{g}$ to be semi-simple. It may happen that $\mathfrak{g}_0$ is a subalgebra of $\mathfrak{h}$, in which case $\mathfrak{g}$ is semi-simple, even though $\mathfrak{g}_0 \neq 0$.

Definition 8.16   Let $\mathfrak{g} = \mathfrak{g}_0 \oplus \mathfrak{g}_+ \oplus \mathfrak{g}_-$ denote the fundamental decomposition of $\mathfrak{g}$. Then $M = G/H$ is said to be of Euclidean type if $\mathfrak{g}_0 = \mathfrak{g}$, it is said to be of compact type if $\mathfrak{g}_0 = \mathfrak{g}_+ = 0$, and is said to be of non-compact type if $\mathfrak{g}_0 = \mathfrak{g}_- = 0$.

The simplest type is the Euclidean type. It is characterized by the decomposition $\mathfrak{g} = \mathfrak{p} \oplus \mathfrak{h}$ where $\mathfrak{p}$ is a commutative algebra. In such a case, G is isomorphic to the semi-direct product $\mathfrak{p} \rtimes Ad_H$ and $G/H \simeq \mathfrak{p}$.

Let us now consider some particular cases covered by the above general framework.

Example 8.17 Euclidean spaces   Let K denote a Lie group that acts linearly on a finite-dimensional vector space V and let G denote the semi-direct product $V \rtimes K$. Then G admits an involutive automorphism $\sigma(x,k) = (-x,k)$ for each

(x, k) in G. It follows that $H = \{0\} \ltimes K$ is the group of fixed points of σ, and that the corresponding Lie algebra decomposition is given by

$$\mathfrak{p} = V \ltimes \{0\} \text{ and } \mathfrak{h} = \{0\} \ltimes \mathfrak{k},$$

where $\mathfrak{k}$ denotes the Lie algebra of K. For (G, K) to be a symmetric Riemannian pair it is necessary and sufficient that $Ad_H|^{\mathfrak{p}}$ be a compact subgroup of $Gl(\mathfrak{p})$. It is easy to check that $Ad_k(A, 0) = (k(A), 0)$ for each $(A, 0)$ in $\mathfrak{p}$ and each $k \in K$. It follows that $K = Ad_H|^{\mathfrak{p}}$. Hence, for (G, H) to be a symmetric Riemannian pair K needs to be a compact subgroup of $Gl(V)$.

Any positive-definite Ad_H-invariant quadratic form on $\mathfrak{p}$ turns V into a Euclidean space $\mathbb{E}^n$ and identifies K with a closed subgroup of the rotation group $O(\mathbb{E}^n)$. Evidently,

$$G/H = \mathbb{E}^n.$$

Example 8.18 $(SL_n(R), SO_n(R))$ **and positive-definite matrices** Let σ be the automorphism on $G = SL_n(R)$ defined by $\sigma(g) = (g^T)^{-1}, g \in G$, where g^T denotes the matrix transpose of g. Then $\sigma(g) = g$ if and only if $g^T = g^{-1}$, that is, whenever g belongs to $SO_n(R)$. Thus, the isotropy group H is equal to $SO_n(R)$.

The Lie algebra $\mathfrak{g}$ of G consists of $n \times n$ matrices with real entries having zero trace. Then σ_* decomposes $\mathfrak{g}$ into the sum $\mathfrak{p} \oplus \mathfrak{h}$ with $\mathfrak{p}$ the space of symmetric matrices in $\mathfrak{g}$, and $\mathfrak{h}$ the algebra of skew-symmetric matrices in $\mathfrak{g}$. This Cartan decomposition recovers the well-known facts that every matrix in $\mathfrak{g}$ is the sum of a symmetric and a skew-symmetric matrix while Cartan relations reaffirm that the Lie bracket of two symmetric matrices is skew-symmetric.

The Killing form on $sl_n(\mathbb{R})$ is given by $Kl(A, B) = 2nTr(AB)$. The form $\langle A, B \rangle = \frac{1}{2}Tr(AB) = \frac{1}{4n}Kl(A, B)$ is more convenient, since then $\langle A, B \rangle = \frac{1}{2}\sum_{i,j}^n a_{ij}b_{ji}$ for any matrices A and B in $sl_n(R)$. In particular, $\langle A, A \rangle = \sum_{i \geq j}^n a_{ij}^2$ for any symmetric matrix A, and $\langle A, A \rangle = -\sum_{i>j}^n a_{ij}^2$ for any skew-symmetric matrix A. It follows that $\langle , \rangle$ is a positive-definite, $SO_n(R)$-invariant quadratic form on the space of symmetric matrices. This inner product turns $(SL_n(R), SO_n(R))$ into a symmetric Riemannian pair.

It is easy to verify that the ideal $\mathfrak{g}_0$ in Proposition 8.10 is equal to zero, which then implies that $[\mathfrak{p}, \mathfrak{p}] = \mathfrak{k}$. Since Kl is a positive multiple of $\langle , \rangle$, $(SL_n(R), SO_n(R))$ is a symmetric Riemannian pair of non-compact type.

The quotient space G/H can be identified with the space of positive-definite $n \times n$ matrices $\mathcal{P}(n, R)$ via the action $(g, P) \to gPg^T, P \in \mathcal{P}(n, R^n), g \in SL_n(R)$. For then, $SO_n(R)$ is the isotropy group of the orbit through the identity matrix, and therefore, $\mathcal{P}(n, R^n)$ can be identified with the quotient $SL_n(R)/SO_n(R)$.

Example 8.19 Self-adjoint subgroups of $SL_n(R)$ A subgroup G of $SL_n(R)$ is said to be *self-adjoint* if g^T is in G for each g in G. It can be easily verified that $G = SO(p,q)$ and Sp_n are self adjoint subgroups of $SL_n(R)$. Recall that $SO(p,q)$ is the group that leaves the Lorentzian form $\langle x, y \rangle_{p,q} = -\sum_{i=1}^{p} x_i y_i + \sum_{i=1}^{n} x_{i+n} y_i$ invariant, while Sp_n is the group that leaves the symplectic form $\langle x, y \rangle = \sum_{i=1}^{n} x_i y_{n+i} - \sum_{i=1}^{n} x_{n+i} y_i$ invariant.

For self-adjoint subgroups G, the restriction of $\sigma(g) = (g^T)^{-1}$ to G can be taken as an involutive automorphism on G, in which case $H = G \cap SO_n(\mathbb{R})$ becomes the group of fixed points. Moreover, $\mathfrak{p}$, the subspace of symmetric matrices in the Lie algebra $\mathfrak{g}$ of G, and $\mathfrak{h}$, the subalgebra of skew-symmetric matrices in $\mathfrak{g}$, become the Cartan factors induced by this automorphism. Then the restriction of the trace metric $\langle A, B \rangle = \frac{1}{2} Tr(AB)$ to $\mathfrak{p}$ defines an $ad(\mathfrak{h})$-invariant metric on $\mathfrak{p}$. With this metric (G, H) becomes a symmetric Riemannian pair.

When $G = SO(p,q)$, the Lie algebra $\mathfrak{g}$ consists of all $n \times n, n = p + q$ matrices M having the block form $M = \begin{pmatrix} A & B \\ B^T & C \end{pmatrix}$ with A and C skew-symmetric $p \times p$ and $q \times q$ matrices, and B an arbitrary $p \times q$ matrix. The Cartan space $\mathfrak{p}$ consists of matrices $M = \begin{pmatrix} 0 & B \\ B^T & 0 \end{pmatrix}$, and $\mathfrak{h}$ is the complementary space $M = \begin{pmatrix} A & 0 \\ 0 & C \end{pmatrix}$. The isotropy group H is the group of matrices

$$\begin{pmatrix} g_1 & 0 \\ 0 & g_2 \end{pmatrix}$$

with $g_1 \in O_p(\mathbb{R})$, $g_2 \in O_q(\mathbb{R})$ such that $\mathrm{Det}\,(g_1)\mathrm{Det}(g_2) = 1$. We will denote this group by $S(O_p(\mathbb{R}) \times O_q(\mathbb{R}))$. The symmetric pair (G, H) together with the quadratic form $\langle A, B \rangle = \frac{1}{2} Tr(AB), A, B \in \mathfrak{p}$ becomes a Riemannian symmetric pair (of non-compact type).

The homogeneous space $M = SO(p,q)/S(O_p(\mathbb{R}) \times O_q(\mathbb{R}))$ can be identified with the open subset of the Grassmannians consisting of all q-dimensional subspaces in $\mathbb{R}^{p+q}$ on which the quadratic form $\langle x, x \rangle_{p,q} = -\sum_{i=1}^{p} x_i^2 + \sum_{i=p+1}^{p+q} x_i^2$ is positive-definite.

Example 8.20 $(SO(1,n), 1 \times SO_n(R))$ **and the hyperboloid** $\mathbb{H}^n$ When $p = 1$, $q = n$, the isotropy group $S(O_1 \times O_n)$ is equal to $\{1\} \times SO_n(\mathbb{R})$ and the quotient $SO(1,n)/\{1\} \times SO_n(R)$ can be identified with the hyperboloid $\mathbb{H}^n$ via the following realization.

Let $x_0, x_1, \dots, x_n$ denote the coordinates of a point $x \in \mathbb{R}^{n+1}$ relative to the standard basis of column vectors $e_0, e_1, \dots, e_n$ in $\mathbb{R}^{n+1}$. The group $SO(1,n)$

acts on the points of the hyperboloid $\mathbb{H}^n = \{x \in \mathbb{R}^{n+1} : x_0^2 - (x_1^2 + \cdots + x_n^2) = 1, x_0 > 0\}$ by the matrix multiplication. The action is transitive, and $\{1\} \times SO)n(R)$ is the isotropy group of the orbit through e_0. Hence, $\mathbb{H}^n$ can be identified with the orbit through the point e_0.

In this identification, curves $x(t)$ on $\mathbb{H}^n$ are identified with curves $g(t)$ in $SO(1, n)$ that satisfy $x(t) = g(t)e_0$ for all t. Then,

$$\frac{dx}{dt}(t) = \frac{dg}{dt}(t)e_0 = g(t)\begin{pmatrix} 0 & B \\ B^T & C \end{pmatrix}e_0,$$

where $B = (b_1, b_2 \ldots, b_n)$ and C is an $n \times n$ skew-symmetric matrix. It follows that

$$\frac{dx}{dt}(t) = \sum_{i=1}^{n} b_i g(t)e_i,$$

and therefore,

$$\left\langle \frac{dx}{dt}, \frac{dx}{dt} \right\rangle_{1,n} = -\frac{dx_0}{dt}^2 + \sum_{i=1}^{n} \frac{dx_i}{dt}^2 = \sum_{i=1}^{n} b_i^2(t) = \frac{1}{2}TrM^2,$$

where M denotes the matrix $\begin{pmatrix} 0 & B \\ B^T & 0 \end{pmatrix}$ in $\mathfrak{p}$.

The above development shows that the canonical hyperbolic metric on $\mathbb{H}^n$ can be extracted from the trace metric on the Cartan space $\mathfrak{p}$ in $\mathfrak{g}$.

Example 8.21 (Sp_n, SU_n) and the generalized Poincaré plane Matrices g in Sp_n are defined by $g^T Jg = J$, where $J = \begin{pmatrix} 0 & -I \\ I & 0 \end{pmatrix}$. Since $J^T = -J$, Sp_n is a self-adjoint subgroup of $SL_{2n}(\mathbb{R})$. The isotropy group H is equal to $SO_{2n}(\mathbb{R}) \cap Sp_n$, which is equal to SU_n. Therefore, (Sp_n, SU_n) is a symmetric Riemannian pair.

We will presently show that the quotient space Sp_n/SU_n can be identified with $\mathcal{P}_n$, the set of $n \times n$ complex matrices Z of the form $Z = X + iY$ with X, and Y symmetric matrices with real entries and Y positive-definite. This space is called the generalized Poincaré plane.

Let us first note that each g in Sp_n can be written in block form $\begin{pmatrix} A & B \\ C & D \end{pmatrix}$ satisfying

$$AB^T = BA^T, CD^T = DC^T, AD^T - BC^T = I.$$

For $n = 2$, matrices A, B, C, D are numbers, and the preceeding conditions reduce to $AD - BC = 1$, from which it follows that $SL_2(\mathbb{R}) = Sp_1$, and

$SU_1 = SO_2(\mathbb{R})$. Therefore, the generalized Poincaré plane coincides with the hyperbolic upper half plane discussed in the preceding chapter.

For a general n, matrices

$$g = \begin{pmatrix} A & 0 \\ 0 & (A^T)^{-1} \end{pmatrix}, A \in GL_n(\mathbb{R}), \ g = \begin{pmatrix} I & B \\ 0 & I \end{pmatrix}, B = B^T, \ g = \begin{pmatrix} 0 & I \\ -I & 0 \end{pmatrix}$$

$$(8.3)$$

generate Sp_n. To see this, first note that $\begin{pmatrix} 0 & -I \\ I & 0 \end{pmatrix}\begin{pmatrix} I & C \\ 0 & I \end{pmatrix}\begin{pmatrix} 0 & I \\ -I & 0 \end{pmatrix} =$

$\begin{pmatrix} I & 0 \\ C & I \end{pmatrix}$, therefore, every matrix $\begin{pmatrix} I & 0 \\ C & I \end{pmatrix}$ with C an arbitrary symmetric matrix is generated by the matrices in (8.3). But, then, the exponentials

$$\begin{pmatrix} e^{tA} & 0 \\ 0 & e^{-tA^T} \end{pmatrix}, \begin{pmatrix} I & tB \\ 0 & I \end{pmatrix}, \begin{pmatrix} I & 0 \\ Ct & I \end{pmatrix}$$

are the generators of Sp_n, by the Orbit theorem (Proposition 1.6).

We shall presently show that the fractional transformations

$$W = (AZ + B)(CZ + D)^{-1}$$

defined by the matrices $g = \begin{pmatrix} A & B \\ C & D \end{pmatrix}$ in Sp_n define an action on $\mathcal{P}_n$. It suffices to restrict our proof to matrices in (8.3) since they are the generators of Sp_n. Let $Z = X + iY$ with X symmetric and Y positive-definite. Then,

$$\begin{pmatrix} A & 0 \\ 0 & (A^T)^{-1} \end{pmatrix}(Z) = AZA^T = AXA^T + iAYA^T,$$

$$\begin{pmatrix} I & B \\ 0 & I \end{pmatrix}(Z) = Z + B = (X + B) + iY.$$

Evidently, both $AXA^T + iAYA^T$, and $(X + B) + iY$ belong to $\mathcal{P}_n$.

When $g = \begin{pmatrix} 0 & I \\ -I & 0 \end{pmatrix}$ then $G(Z) = -Z^{-1}$, so it remains to show that $-Z^{-1}$ belongs to $\mathcal{P}_n$ for any Z in $\mathcal{P}_n$.

Let V be a matrix in $GL_n(\mathbb{R})$ such that $VYV^T = I$. Then

$$VZV^T = VXV^T + iI = S + iI.$$

It follows that

$$-Z^{-1} = -(X + iY)^{-1} = -((V^{-1}(S + iI)(V^T)^{-1})^{-1} = -V^T(S + iI)^{-1}V.$$

Since $S^2 + I = (S + iI)(S - iI)$,

$$-Z^{-1} = -V^T(S + iI)^{-1}V$$
$$= V^T((S^2 + I)(S - iI)^{-1})^{-1})V$$
$$= -V^T(S - iI)(S^2 + I)^{-1}V = P + iQ,$$

with $P = -V^T S(S^2 + I)^{-1}V$ and $Q = V^T(S^2 + I)^{-1}V$. Evidently $S^2 + I$ is positive-definite. Since the inverse of a positive-definite matrix is positive-definite, Q is a positive-definite matrix. Matrix S is symmetric and commutes with $S^2 + I$, and therefore S commutes with $(S^2 + I)^{-1}$. Hence P is symmetric. This argument shows that $-Z^{-1}$ is in $\mathcal{P}_n$.

We have now shown that Sp_n acts on $\mathcal{P}_n$. Any point on the orbit through iI is of the form $g(iI) = (iA + B)(iC + D)^{-1}$. If $Z = X + iY$, then

$$\begin{pmatrix} A & 0 \\ 0 & A^{-1^T} \end{pmatrix}\begin{pmatrix} I & B \\ o & I \end{pmatrix}(iI) = \begin{pmatrix} A & AB \\ 0 & (A^T)^{-1} \end{pmatrix}(iI)$$
$$= iAA^T + ABA^T = X + iY$$

if A and B conform to $Y = AA^T$ and $X - ABA^T$. Hence, Sp_n acts transitively on $\mathcal{P}_n$.

The reader can easily verify that the isotropy group consists of matrices $g = \begin{pmatrix} A & B \\ -B & A \end{pmatrix}$. But then $gg^T = \begin{pmatrix} I & 0 \\ 0 & I \end{pmatrix}$, and therefore g belongs to $SO_{2n}(\mathbb{R}) \cap Sp_n$. It follows that

$$Sp_n/SU_n = \{X + iY \ X^T = X, Y > 0\}.$$

Example 8.22 ($SO(p,q), S(O_p(R) \times SO_q(R))$ **and oriented Grassmann manifolds** Let p and q be positive integers such that $p + q = n$ and let D denote the diagonal matrix with its diagonal entries equal to -1 in the first p rows, and equal to 1 in the remaining q rows. Then $D^{-1} = D$.

Let $\sigma : SO_n(\mathbb{R}) \to SO_n(\mathbb{R})$ denote the mapping $\sigma(g) = DgD^{-1} = DgD$. It is easy to verify that σ is an involutive automorphism on G. The set of matrices g in G that are fixed by σ are of the form $g = \begin{pmatrix} A & 0 \\ 0 & D \end{pmatrix}$ where A is a $p \times p$ matrix and D a $q \times q$ matrix. Denote this group by $S(O_p(\mathbb{R}) \times O_q(\mathbb{R}))$. It consists of pairs of matrices (A, D) with $A \in O_p(\mathbb{R})$ and $D \in O_q(\mathbb{R})$ such that $Det(A)Det(D) = 1$.

Then σ_* splits the Lie algebra $\mathfrak{g} = so_n(\mathbb{R})$ into the Cartan factors $\mathfrak{p} \oplus \mathfrak{h}$, with $\mathfrak{p}$ equal to the vector space of matrices P of the form $P = \begin{pmatrix} 0 & B \\ -B^T & 0 \end{pmatrix}$, and $\mathfrak{h}$

the Lie algebra of matrices $Q = \begin{pmatrix} A & 0 \\ 0 & D \end{pmatrix}$, with both A and D antisymmetric; $\mathfrak{h}$ is the Lie algebra of the isotropy group of $H = S(O_p(\mathbb{R}) \times O_q(\mathbb{R}))$.

The pair (G, H) is a symmetric Riemannian pair with the metric on $\mathfrak{p}$ defined by

$$\langle P_1, P_2 \rangle = -\frac{1}{2} Tr(P_1 P_2).$$

We leave it to the reader to show that the Killing form on $SO_n(\mathbb{R})$ is given by

$$Kl(A, B) = (n - 2)Tr(AB).$$

Thus the Riemannian metric on $\mathfrak{p}$ is a scalar multiple of the restriction of the Killing form. Since the Killing form is negative-definite on $\mathfrak{g}$, the pair (G, H) is of compact type.

The homogeneous space $Gr_p = SO_n/SO_p(\mathbb{R}) \times SO_q(\mathbb{R})$ is the space of all oriented p- dimensional linear subspaces in $\mathbb{R}^{p+q}$. So the symmetric space $SO_n/S(O_p(\mathbb{R}) \times SO_q(\mathbb{R}))$ is a double cover of Gr_p.

Example 8.23 $(SO_{1+n}(R), \{1\} \times SO_n(R))$ **and the sphere** S^n When $p = 1$ and $q = n$, then $SO_p(\mathbb{R}) = 1$ and the isotropy group H is equal to $\{1\} \times SO_n(\mathbb{R})$. The set of oriented lines in $\mathbb{R}^{n+1}$ is identified with the n-dimensional sphere S^n. More explicitly, the sphere is identified with the orbit of SO_{n+1} through the point $e_0 = (1, 0, \ldots, 0)$ of $\mathbb{R}^{n+1}$ under the action of matrices in SO_{n+1} on the column vectors of $\mathbb{R}^{n+1}$.

The Riemannian metric induced by the Killing form on $\mathfrak{p}$ in $\mathfrak{g}$ coincides with the standard metric on S^n inherited from the Euclidean metric on $\mathbb{R}^{n+1}$. Indeed, any curve $x(t)$ on S^n is identified with a curve $g(t)$ in $SO_{n+1}(\mathbb{R})$ via the formula $x(t) = g(t)e_0$. Let $\frac{dg}{dt}(t) = g(t)(P(t) + Q(t))$ with

$$P(t) = \begin{pmatrix} 0 & -p_1 & \cdots & & -p_n \\ p_1 & 0 & 0 \ldots & & 0 \\ p_2 & 0 & 0 \ldots & & 0 \\ \vdots & \vdots & & \cdots & \vdots \\ p_n & 0 & 0 \ldots & & 0 \end{pmatrix}$$

in $\mathfrak{p}$ and $Q(t) \in \mathfrak{h}$. Then, $\frac{dx}{dt}(t) = \frac{dg}{dt}(t)e_0 = g(t)(P + Q)e_0 = \sum_{i=1}^{n} p_i(t)g(t)e_i$, and therefore,

$$\sum_{0=1}^{n} \frac{dx_i}{dt}^2 (t) = \sum_{i=1}^{n} p_i(t)^2 = \langle P(t), P(t) \rangle.$$

8.3 The sub-Riemannian problem

Let us now return the symmetric pairs (G, H) and the associated Cartan decomposition $\mathfrak{g} = \mathfrak{p} \oplus \mathfrak{h}$.

Definition 8.24 A curve $g(t)$ in G defined over an open interval (a, b) will be called horizontal if $g^{-1}(t)\frac{dg}{dt}(t)$ belongs to $\mathfrak{p}$ for all t in (a, b).

Horizontal curves can also be considered as the integral curves of a left-invariant distribution $\mathcal{P}$ defined by $\mathcal{P}(g) = \{gA : A \in \mathfrak{p}\}$, $g \in G$, and as such they conform to the accessibility theory of families of analytic vector fields discussed in the previous chapters. In particular, the orbit of $\mathcal{P}$ through the group identity is a connected Lie subgroup K of G, and the orbit through any other point $g \in G$ is the left coset gK. Let $\mathfrak{k}$ denote the Lie algebra generated by $\mathfrak{p}$. The Herman–Nagano theorem implies that $\mathfrak{k}$ is the Lie algebra of K. It follows that the group identity can be connected to any point g_1 in K by a horizontal curve. More precisely, for any $T > 0$ there exists a horizontal curve $g(t)$ defined on the interval $[0, T]$ such that $g(0) = e$ and $g(T) = g_1$.

In the language of bundles, G is a principal H bundle over G/H, with H acting on G by the right action $(h, g) \rightarrow gh$. The distribution $\mathcal{P}$ is called a *connection*. We will use π to denote the natural projection from G onto G/H.

We will now show that for any curve $x(t)$ in M and any $g_0 \in \pi^{-1}(x(0))$ there is a unique horizontal curve $g(t)$ such that $g(0) = g_0$ and $\pi(g(t) = x(t)$ for all t. We shall call such a curve *the horizontal lift to g_0 of a base curve $x(t)$*.

Let $x(t) = y(t)H$ for some representative curve $y(t) \in G$. Then $\pi^{-1}(x(0)) = y(0)H$ and $g_0 = y(0)h_0$ for some $h_0 \in H$. Every curve $y(t) \in G$ is a solution of

$$\frac{dy}{dt}(t) = y(t)(A(t) + B(t))$$

for some elements $A(t) \in \mathfrak{p}$ and $B(t) \in \mathfrak{h}$. In our case take $g(t) = y(t)h(t)$, where $h(t)$ is the solution of $\frac{dh}{dt}(t) = -B(t)h(t)$ with $h(0) = h_0$. Evidently $\pi(g(t)) = x(t)$ and $g(0) = y(0)h_0 = g_0$. To show that $g(t)$ is a horizontal curve we need to differentiate:

$$\frac{dg}{dt}(t) = y(t)(A(t) + B(t))h(t) - y(t)B(t)h(t)$$

$$= g(t)(h^{-1}(t)A(t)h(t) + h^{-1}(t)B(t)h(t)) - g(t)(h^{-1}(t)B(t)h(t))$$

$$= g(t)(h^{-1}(t)A(t)h(t)).$$

Since $h^{-1}(t)A(t)h(t)$ belongs to $\mathfrak{p}$, $g(t)$ is horizontal. It is clear from the construction that such a lift is unique.

If $g_1(t)$ and $g_2(t)$ are two horizontal lifts of the same base curve $x(t)$, then $g_2(0) = g_1(0)h$ for some $h \in H$. Then $g(t) = g_1(t)h$ is a horizontal curve that projects onto $x(t)$ and satisfies $g(0) = g_2(0)$. Therefore, $g_2(t) = g_1(t)h$, by the uniqueness of lifts. Horizontal curves that differ by a right multiple in H are called *parallel*. It follows that the horizontal lifts of a curve $x(t)$ are parallel translates of a horizontal lift to any point $g_0 \in \pi^{-1}(x(0))$.

Definition 8.25 $Hol(\mathcal{P}) = K \cap H$ is called the holonomy group of $\mathcal{P}$.

Proposition 8.26 *The holonomy group is isomorphic to the loop group at any point x in the base space G/H.*

Proof Let $x_0 = g_0 H$ be any base point, and let $x(t)$ be any curve such that $x(0) = x_0 = x(T)$. Let $g(t)$ denote the horizontal lift of $x(t)$ to a point $g_0 h_0$ in the fiber over x_0. Since $x(T) = x_0$, the terminal point $g(T) = g_0 h_0 h$ for some $h \in H$.

The correspondence $x(t) \rightarrow h$ is one to one. In this correspondence, h^{-1} corresponds to the lift $g(T - t)h^{-1}$ of the reversed curve $x(T - t)$. Moreover, if $x_1(t), t \in [0, T_1]$ and $x_2(t), t \in [0, T_2]$ are any two loops, their concatenation $x(t) = x_1(t), t \in [0, T_1], x((t) = x_2(t - T_1), t \in [T_1, T_1 + T_2]$ is a loop at x_0. If $g_1(t)$ and $g_2(t)$ denote the horizontal lifts of $x_1(t)$ and $x_2(t)$ then $g_1(T_1) = g_0 h_0 h_1$ and $g_2(T_2) = g_0 h_0 h_2$, for some h_1 and h_2 in H. It follows that $g(t) = g_1(t), t \in [0, T_1], g(t) = g_2(t - T_1)(g_0 h_0)^{-1} g_1(T_1)$ is the horizontal lift of $x(t)$ to $g_0 h_0$ that satisfies $g(T_2 + T_1) = g_0 h_0 h_1 (g_0 h_0)^{-1}(g_0 h_o)h_1 = g_0 h_0 h_2 h_1$.

If $g(t)$ is a horizontal curve that originates at $g_0 h_0$, then $g(t) = g_0 h_o g_0(t)$, where $g_0(t)$ is a horizontal curve that originates at e for $t = 0$, since $\mathcal{P}$ is left-invariant. Then $g(T) = g_0 h_o g_0(T) = g_0 h_0 h$ reduces to $h = g_0(T)$. Hence, $h \in K \cap H$. Conversely, any point $h \in K \cap H$ corresponds to the terminal point $g_0(T)$ of some horizontal curve that satisfies $g_0(0) = e$. Then $g(t) = g_0 h_o g_0(t)$ is the horizontal curve at $g_0 h_0$ whose projected loop corresponds to h. $\qquad \square$

The previous discussion shows that there is no loss in generality in assuming that a symmetric Riemannian pair (G, H) satisfies the condition that any pair of points in G can be connected by a horizontal curve. In such a situation H is equal to the holonomy group of the connection $\mathcal{P}$.

Definition 8.27 A symmetric Riemannian pair (G, H) satisfies the controllability condition if any two points in G can be connected by a horizontal curve.

Proposition 8.28 *Suppose that (G, H) is a Riemannian symmetric pair such that the Lie algebra $\mathfrak{h}$ of H has zero center. Then any two points of G can be connected by a horizontal curve if and only if $[\mathfrak{p}, \mathfrak{p}] = \mathfrak{h}$.*

Proof Any two points of G can be connected by a horizontal curve if and only if $Lie(\mathfrak{p}) = \mathfrak{g}$ (the Orbit theorem). According to the fundamental decomposition of $\mathfrak{g}$ (Proposition 7), $Lie(\mathfrak{p}) = \mathfrak{g}$ if and only if $\{B \in \mathfrak{h} : adB(\mathfrak{p}) \subseteq \mathfrak{p}_0\} = 0$. But $\{B \in \mathfrak{h} : adB(\mathfrak{p}) \subseteq \mathfrak{p}_0\} = 0$ is a necessary and sufficient condition that $\mathfrak{h} = \cup\{[\mathfrak{p}_i, \mathfrak{p}_i] : Kl|_{\mathfrak{p}_i} = \lambda_i\langle , \rangle|_{\mathfrak{p}_i}, \lambda_i \neq 0\}$. $\qquad\square$

Corollary 8.29 *Suppose that (G, H) is controllable such that $\mathfrak{h}$ has zero center. If $\mathfrak{g}$ has also zero center, then $\mathfrak{g}$ is semi-simple.*

Proof We will continue with the notations in Proposition 8.10. If (G, H) is controllable, then $\mathfrak{g}_0 = \mathfrak{p}_0$. Since $\mathfrak{g}_0$ is an ideal in $\mathfrak{g}$, $adh(\mathfrak{p}_0) \subseteq \mathfrak{p}_0$. But, $[\mathfrak{p}, \mathfrak{p}] = \mathfrak{h}$ and $[\mathfrak{p}, \mathfrak{p}_0] = 0$. Therefore, $ad\mathfrak{h}(\mathfrak{p}_0) = 0$ by Jacobi's identity. Hence, $\mathfrak{p}_0$ is in the center of $\mathfrak{g}$ and by our assumption must be zero. But then the Killing form is non-degenerate on $\mathfrak{g}$. $\qquad\square$

When the pair (G, H) is Riemannian and controllable, then there is a canonical sub-Riemannian problem on G induced by a positive-definite Ad_H-invariant quadratic form $\langle , \rangle$ on $\mathfrak{p}$. Any horizontal curve $g(t)$ in G can be assigned the length in an interval $[0, T]$ given by

$$L_T(g(t)) = \int_0^T (\langle U(t), U(t)\rangle)^{\frac{1}{2}} \, dt, \qquad (8.4)$$

where $U(t) = g^{-1}(t)\frac{dg}{dt}(t)$. Assuming that $Lie(\mathfrak{p}) = \mathfrak{g}$, any two points can be connected by a horizontal curve. Therefore the problem of finding a horizontal curve of minimal length that connects two given points in G is well defined. This minimal length is called the sub-Riemannian distance. The sub-Riemannian problem is left-invariant, so the solutions through any point will be the left-translates of the solution through the group identity e.

The sub-Riemannian metric on G induces a Riemannian metric on the base curves in G/H with the Riemannian length of a curve $x(t)$ in M equal to the sub-Riemannian length of a horizontal curve $g(t)$ that projects onto $x(t)$. The length of $x(t)$ is well defined because the sub-Riennian lengths of parallel horizontal curves are all equal, since $\langle , \rangle$ is Ad_H-invariant. That is,

$$\left\langle \frac{dx}{dt}, \frac{dx}{dt} \right\rangle = \langle A(t), A(t)\rangle,$$

where $A(t) = g^{-1}(t)\frac{dg}{dt}$ defined by a horizontal curve that projects onto $x(t)$. As we have already stated, it suffices to consider curves of minimal length from a fixed initial point x_0, since the curves of minimal length from any other point are the translates by an element of G under the left action. In what follows we will consider the horizontal curves $g(t)$ that satisfy $g(0) = e$. Then their projections $x(t) = g(t)H$ satisfy $x(0) = H$.

A horizontal lift of a curve $x(t)$ that is of minimal length in an interval $[0, T]$ is also of minimal sub-Riemannian length on $[0, T]$, but the converse may not be true. To get the converse, we need to consider the totality of horizontal curves that connect H to a given left coset g_1H. Then the projection $x(t)$ of the curve of minimal sub-Riemannian length in this class is of minimal length in M.

We shall now recast each of the above variational problems as optimal control problems in G and obtain the extremal curves through the Maximum Principle. For simplicity of exposition we will assume that $\mathfrak{g}$ has zero center, which in turn implies that $\mathfrak{g}$ is semi-simple. It will be advantageous to adapt the problem to the decomposition $\mathfrak{g} = \mathfrak{g}_1 \oplus \cdots \oplus \mathfrak{g}_m$ with $\mathfrak{g}_i = \mathfrak{p}_i + [\mathfrak{p}_i, \mathfrak{p}_i]$ explained in Propositions 8.9 and 8.10. Then let

$$A_1^{(1)}, \ldots, A_{m_1}^{(1)}, A_1^{(2)}, \ldots, A_{m_2}^{(2)}, \ldots, A_1^{(m)}, \ldots, A_{m_m}^{(m)}, \; m_1 + m_2 + \cdots + m_m = n,$$

be an orthonormal basis in $\mathfrak{p}$ such that $A_1^{(k)}, \ldots, A_{m_k}^{(k)}$ is an orthonormal basis for $\mathfrak{p}_k$, $k = 1, \ldots, m$. Then horizontal curves $g(t)$ are the solutions of the control problem

$$\frac{dg}{dt}(t) = g(t) \sum_{k=1}^{m} \sum_{i=1}^{m_k} u_i^{(k)}(t) A_i^{(k)} \tag{8.5}$$

with control functions $u(t) = \left(u_1^{(1)}(t), \ldots, u_{m_m}^{(m)}(t) \right)$ measurable and bounded on compact intervals $[0, T]$. It will be convenient to work with the energy functional

$$E(g(t)) = \frac{1}{2} \int_0^T \langle u(t), u(t) \rangle \, dt = \frac{1}{2} \int_0^T \left(\sum_{k=1}^{m} \sum_{i=1}^{m_k} \left(u_i^{(k)} \right)^2(t) \right) dt, \tag{8.6}$$

instead of the length functional. Our variational problems are reformulated as follows:

1. *The sub-Riemannian problem.* Let $T > 0$ and g_1 in G be fixed. Find a trajectory $(\hat{g}(t), \hat{u}(t))$ of the control system (8.5) in the interval $[0, T]$ that satisfies $\hat{g}(0) = e, \hat{g}(T) = g_1$ such that

$$E(\hat{g}(t)) \leq E(g(t))$$

   for any trajectory $(g(t), u(t))$ in $[0, T]$ that satisfies the same boundary conditions $g(0) = e$ and $g(T) = g_1$.
2. *The Riemannian problem.* Let $T > 0$ and g_1 in G be fixed. Find a trajectory $(\hat{g}(t), \hat{u}(t))$ of the control system (8.5) in the interval $[0, T]$ that satisfies $\hat{g}(0) \in H, \hat{g}(T) \in g_1H$ such that

$$E(\hat{g}(t)) \leq E(g(t))$$

for any trajectory $(g(t), u(t))$ in $[0, T]$ that satisfies $g(0) \in H$ and $g(T) \in g_1 H$.

The only difference between the two problems is that the terminal points e, g_1 in the sub-Riemannian problem are replaced by the terminal manifolds H and $g_1 H$ in the Riemannian problem.

8.4 Sub-Riemannian and Riemannian geodesics

To take advantage of the left-invariant symmetries, the cotangent bundle T^*G will be realized as the product $G \times \mathfrak{g}^*$. Recall that a linear function ξ at the tangent space $T_g G$ is identified with (g, ℓ) in $G \times \mathfrak{g}^*$ via the formula $\ell(A) = \xi(L_g)_*(A)$ for all $A \in \mathfrak{g}$. In this representation the Hamiltonians of the left-invariant vector fields $X(g) = gA$ are linear functions $\ell(A)$ on $\mathfrak{g}^*$.

Let $h_1^{(k)}, \ldots, h_{m_k}^{(k)}$ denote the Hamiltonians of the left-invariant vector fields $X_1^{(k)}(g) = gA_1^{(k)}, \ldots, X_{m_k}^{(k)}(g) = gA_{m_k}^{(k)}, k = 1, \ldots, m$. According to the Maximum Principle, optimal trajectories $(g(t), u(t))$ of the preceeding variational problems are the projections of the extremal curves $(g(t), \ell(t))$ in T^*G, which can be of two kinds, normal and abnormal.

The fact that our distribution is of contact type, namely that $\mathfrak{p} + [\mathfrak{p}, \mathfrak{p}] = \mathfrak{g}$, implies that the abnormal extremals can be ignored, because every optimal trajectory is the projection of a normal extremal curve (the Goh condition [AS, p. 319]).

Normal extremals are the integral curves of a single Hamiltonian

$$H(\ell) = \frac{1}{2} \sum_{k=1}^{m} \sum_{i=1}^{m_k} \left(h_i^{(k)} \right)^2 (\ell). \tag{8.7}$$

Its integral curves are the solutions of the following differential system:

$$\frac{dg}{dt}(t) = g(t)dH(\ell(t)), \frac{d\ell}{dt}(t) = -(ad^* dH(\ell(t))(\ell(t)), dH = \sum_{k=1}^{m} \sum_{i=1}^{m_k} h_i^{(k)} A_i^{(k)}. \tag{8.8}$$

Sub-Riemannian geodesics are the projections on G of the extremal curves on energy level $H = \frac{1}{2}$. The projections to G/H of the sub-Riemannian geodesics that satisfy the transversality condition $\ell(T)(\mathfrak{h}) = 0$ are called Riemannian geodesics.

Proposition 8.30 *Each sub-Riemannian geodesic $g(t)$, $g(0) = e$, is of the form*

$$g(t) = e^{(P-Q)t}e^{Qt}, \tag{8.9}$$

for some matrices $P \in \mathfrak{p}$ and $Q \in \mathfrak{h}$ with $\langle P, P \rangle = 1$. Each Riemannian geodesic that originates at $\pi(e)$ is the projection of a curve $g(t) = e^{Pt}$ for some matrix $P \in \mathfrak{p}$ with $\langle P, P \rangle = 1$.

Proof Let $\mathfrak{g}^* = \mathfrak{g}_1^* \oplus \cdots \oplus \mathfrak{g}_m^*$, where each $\mathfrak{g}_k^*$ is identified with the annihilator of the complementary space $\mathfrak{g}_k^\perp$, and then let $\ell = \ell_1 \oplus \ell_2 \cdots \oplus \ell_m$ denote the corresponding decomposition of $\ell \in \mathfrak{g}^*$ into the factors $\mathfrak{g}_i^*$. Also, let $dH = \sum_{k=1}^m dH_k$, where $dH_k = \sum_{i=1}^{m_k} h_i^{(k)} A_i^{(k)}, k = 1, \ldots, m$. Then for any $X = X_1 + X_2 \cdots + X_m$ in $\mathfrak{g}$,

$$\frac{d\ell}{dt}(X) = -\ell \left[\sum_{k=1}^m dH_k, X \right] = -\ell \sum_{k=1}^m [dH_k, X_k] = -\sum_{k=1}^m \ell_k [dH_k, X_k],$$

because $[\mathfrak{g}_i, \mathfrak{g}_j] = 0$ for $i \neq j$. Hence, equation (8.8) decomposes into a system of equation

$$\frac{d\ell_k}{dt} = -ad^* dH_k(\ell)(\ell_k(t)), k = 1, \ldots, m_m. \tag{8.10}$$

Since $\mathfrak{g}$ is semi-simple, $\ell \in \mathfrak{g}^*$ can be identified with $L \in \mathfrak{g}$ via the formula $\ell(X) = Kl(L, X)$ for all $X \in \mathfrak{g}$. If $\ell(t)$ is a solution of (8.8), then $L(t)$ is a solution of

$$\frac{dL}{dt} = [dH, L(t)], \tag{8.11}$$

as can be verified easily:

$$Kl\left(\frac{dL}{dt}, X\right) = \frac{d\ell}{dt}(X) = -\ell[dH, X] = Kl([X, dH], L) = Kl(X, [dH, L]).$$

Since X is arbitrary, equation (8.11) follows.

Let $L_1, \ldots, L_m$ denote the projections of $L \in \mathfrak{g}$ onto the factors $\mathfrak{g}, \ldots, \mathfrak{g}_m$. Since the factors $\mathfrak{g}_1, \mathfrak{g}_2, \ldots, \mathfrak{g}_m$ are orthogonal relative to the Killing form, each element $\ell_k \in \mathfrak{g}_k^*$ corresponds to L_k, in the sense that

$$\ell_k(X_k) = Kl(L_k, X_k), X_k \in \mathfrak{g}_k. \tag{8.12}$$

Therefore, equation (8.11) breaks up into the invariant factors

$$\frac{dL_k}{dt} = [dH_k, L_k], k = 1, \ldots, m. \tag{8.13}$$

Let now $L_k = P_k + Q_k$ with $P_k \in \mathfrak{p}_k$ and Q_k in $\mathfrak{h}_k = [\mathfrak{p}_k, \mathfrak{p}_k]$. Recall that $\mathfrak{p}_k$ and $\mathfrak{h}_k$ are orthogonal relative to the Killing form. Then,

$$\frac{dP_k}{dt} = [dH_k, Q_k], \quad \frac{dQ_k}{dt} = [dH_k, P_k]. \tag{8.14}$$

On $\mathfrak{p}_k$ the Killing form is a constant multiple of $\langle \, , \, \rangle$, that is, $Kl(A, B) = \lambda_k \langle A, B \rangle$. If we now write $P_k = \sum_{i=1}^{m_k} l_i^{(k)} A_i^{(k)}$, then

$$l_i^{(k)} = \left\langle P_k, A_i^{(k)} \right\rangle = \frac{1}{\lambda_k} Kl\left(P_k, A_i^{(k)}\right) = \frac{1}{\lambda_k} l_k\left(A_i^{(k)}\right) = \frac{1}{\lambda_k} h_i^{(k)}(\ell).$$

Hence, $P_k = \frac{1}{\lambda_k} dH_k$, $k = 1, \ldots, m$, and therefore $[dH_k, P_k] = 0$. It follows that the solutions of (8.14) are given by

$$\frac{dP_k}{dt} = \lambda_k [P_k, Q_k], \quad \frac{dQ_k}{dt} = 0. \tag{8.15}$$

Equation (8.15) is readily solvable: Q_k is a constant matrix, and $P_k(t) = e^{-\lambda_k Q_k t} P_k(0) e^{\lambda_k Q_k t}$. Then $dH_k = e^{-\lambda_k Q_k t} \lambda_k P_k(0) e^{\lambda_k Q_k t}$, and

$$dH = e^{-Qt} P e^{Qt}, \text{ where } Q = \lambda_1 Q_1 \oplus \cdots \oplus \lambda_m Q_m,$$

$$P = \lambda_1 P_i(0) \oplus \cdots \oplus \lambda_m P_m(0).$$

The sub-Riemannian geodesics are the solutions of

$$\frac{dg}{dt} = g(t)(dH(t)) = g(t)\left(e^{-Qt} P e^{Qt}\right). \tag{8.16}$$

Let $h(t) = e^{-Qt}$ and $g_0 = gh$. Then, $\frac{dg_0}{dt} = g\left(e^{-Qt} P e^{Qt}\right) h - g_0 h Q = g_0(P - Q)$. Hence,

$$g(t) = e^{(P-Q)t} e^{Qt}.$$

If $\ell(t)$ is to satisfy the transversality condition $\ell(T)(\mathfrak{h}) = 0)$, then $Q_k(T) = 0$ for each k. Since $Q_k(t)$ is constant, $Q_k = 0$ and $g(t) = e^{Pt}$. $\qquad\Box$

The sub-Riemannian sphere of radius r centered at the identity consists of all points in G which are a distance r away from e. The set $W_T(e) = \{g : g = e^{(P-Q)T} e^{QT}, P \in \mathfrak{p}, Q \in \mathfrak{h}\}$, $\langle P, P \rangle = 1$ is called the wave front at T or the exponential mapping at T.

These concepts are natural extensions of their counterparts in Riemannian geometry. In particular, in this situation, the exponential mapping at $\pi(e)$ is given by $P \to e^{PT} \pmod{H}$, $\langle P, P \rangle = 1$, and the exponential mapping at any other point $x = g \pmod{H}$ is equal to the left-translate by g of the exponential mapping at $x = I \pmod{H}$. It follows from the above proposition that each Riemannian symmetric space $M = G/H$ is complete, in the sense that the exponential map through any point x is defined for all times T. For small T, the exponential mapping is surjective and coincides with the Riemannian sphere of radius T. That is not the case for sub-Riemannian problems: each wave front $W_T(e)$ contains points whose distance from e is greater than T, as we have already seen in the preceding chapter.

Definition 8.31 A diffeomorphism Φ on a Riemannian manifold M with its metric $\langle , \rangle$ is called an isometry if $\langle \Phi_* v_1, \Phi_* v_2 \rangle_{\Phi(x)} = \langle v_1, v_2 \rangle_x$ for all x in M and all tangent vectors v_1 and v_2 in $T_x M$.

Each left translation by an element in G and each right translation by an element in H is an isometry for the symmetric Riemannian space G/H, and so is the restriction of the involutive automorphism σ.

Therefore, S_p defined by $S_p(gH) = (g_0(\sigma(g_0)))^{-1}\sigma(g)H$ is an isometry for each point $p = g_0 H$. It follows that $S_p(p) = g_0 H = p$. Let $\gamma(t) = g_0(\exp tA)H$ be any geodesic from p. An easy calculation shows that $\sigma(\exp tA) = \exp t\sigma_*(A) = \exp -tA$, and therefore,

$$S_p(\gamma(t)) = g_0(\sigma(g_0))^{-1}\sigma(g_0 \exp tA)H) = g_0\sigma(\exp -tA)\sigma(H) = S_p(\gamma(-t)).$$
(8.17)

The symmetry S_p is called *the geodesic symmetry at p*. Every Riemannian manifold that admit a geodesic symmetry at each of its points is called a *symmetric space* [Eb; Hl].

The literature on symmetric spaces begins with the geodesic symmetries as the point of departure and then arrives at the symmetric Riemannian pairs (G, H) via the isometry arguments ([Eb] or [Hl]). We have shown that one can take the opposite path, and start with the symmetric pair (G, H) and finish with the geodesic symmetry at the end.

8.5 Jacobi curves and the curvature

Consider now the second-order conditions associated with the above geodesic problems. Any parametrized curve $x(t)$ in G/H is the projection $\pi(g)$ of a horizontal curve $g(t)$ such that $\frac{dx}{dt} = \pi_*(g(t)A(t))$, where $A(t) = g^{-1}(t)\frac{dg}{dt}$. If $\left\| \frac{dx}{dt} \right\|$ denotes the Riemannian norm of $\frac{dx}{dt}$, then $\left\| \frac{dx}{dt} \right\|^2 = \langle A(t), A(t) \rangle$. To every curve of tangent vectors $v(t)$ defined along a curve $x(t)$ there corresponds a curve $B(t)$ in $\mathfrak{p}$ such that $X(t) = \pi_*(g(t)B(t))$.

Definition 8.32 The covariant derivative $D_x(v(t))$ of a curve of tangent vectors $v(t)$ along $x(t)$ is given by $\pi_*(g(t)\frac{dB}{dt}(t))$.

The geodesic curvature $\kappa(t)$ along along a curve $x(t)$ parametrized by arc length is equal to $\kappa(t) = \left\| D_{x(t)} \left(\frac{dx}{dt} \right) \right\|$. If $g(t)$ is a horizontal curve that projects onto a curve $x(t)$ parametrized by arc length then $\|A(t)\| = 1$, where $A(t)$ is a curve in $\mathfrak{p}$ defined by $g^{-1}(t) \frac{dg}{dt} = A(t)$. It then follows that $\kappa(t) = \left\| \frac{dA}{dt} \right\|$.

Proposition 8.33 *The projections of sub-Riemannian geodesics on G/H have constant geodesic curvature. Riemannian geodesics are the projections of sub-Riemannian geodesics having zero curvature.*

Proof The sub-Riemannian geodesics are the horizontal curves that are the solutions of $\frac{dg}{dt} = g(t)A(t)$, where $A(t) = e^{-Qt}Pe^{Qt}$ for some constant matrices $P \in \mathfrak{p}$ with $\|P\| = 1$ and $Q \in \mathfrak{h}$. Then the geodesic curvature of the projected curve in G/H is given by $\kappa = \left\| \frac{dA}{dt} \right\| = \|[Q, P]\|$. The Riemannian geodesics are the projections of sub-Riemannian geodesics with $Q = 0$. $\square$

With each geodesic curve $g(t) = e^{t\hat{A}}$ consider now the variational curve

$$v_{\hat{A}}(t)(A) = \pi_* \left(\frac{d}{d\epsilon} e^{-t\hat{A}} e^{t(\hat{A}+\epsilon A)}|_{\epsilon=0} \right), v_{\hat{A}}(0)(A) = 0, \qquad (8.18)$$

generated by a fixed direction A in $\mathfrak{p}$. Let $g(t, \epsilon) = e^{-t\hat{A}} e^{t(\hat{A}+\epsilon A)}$ and $x(t, \epsilon) = \pi(g(t, \epsilon))$. Since $g(t, 0) = e$, $\frac{\partial g}{\partial \epsilon}|_{\epsilon=0}$ is a curve in $\mathfrak{g}$. The following proposition is fundamental.

Proposition 8.34 *Let $J_{\hat{A}}(t)(A)$ denote the projection of $\frac{\partial g}{\partial \epsilon}|_{\epsilon=0}$ on $\mathfrak{p}$. Then,*

$$J_{\hat{A}}(t)(A) = \sum_{k=0}^{\infty} \frac{t^{2k+1}}{(2k+1)!} ad^{2k}\hat{A}(A).$$

Proof

$$\frac{\partial g}{\partial t}(t, \epsilon) = -e^{-\hat{A}t} \hat{A} e^{(\hat{A}+\epsilon A)t} + e^{-\hat{A}t}(\hat{A} + \epsilon A)e^{(\hat{A}+\epsilon A)}$$

$$= g(t, \epsilon) \left(e^{-t(\hat{A}+\epsilon A)} \epsilon A e^{t(\hat{A}+\epsilon A)} \right).$$

Then, $e^{-t(\hat{A}+\epsilon A)} \epsilon A e^{t(\hat{A}+\epsilon A)}) = \epsilon(P(t, \epsilon)(A) + Q(t, \epsilon)(A))$, where

$$P(t, \epsilon) = \sum_{k=0}^{\infty} \frac{t^{2k}}{(2k)!} ad^{2k}(\hat{A} + \epsilon A), Q(t, \epsilon) = \sum_{k=0}^{\infty} \frac{t^{2k+1}}{(2k+1)!} ad^{2k+1}(\hat{A} + \epsilon A).$$

Since $ad^{2k}(\hat{A} + \epsilon A)(A) \in \mathfrak{p}$ and $ad^{2k+1}(\hat{A} + \epsilon A)(A) \in \mathfrak{h}$, it follows that $P(t, \epsilon)(A)$ belongs to $\mathfrak{p}$ and $Q(t, \epsilon)(A)$ belongs to $\mathfrak{h}$.

Let $\hat{g}(t, \epsilon) = g(t, \epsilon)\hat{h}(t, \epsilon)$, where $\hat{h}(t, \epsilon)$ denotes the field of solution curves in H defined by $\frac{\partial \hat{h}}{\partial t}(t, \epsilon) = -\hat{h}(t, \epsilon)Q(t, \epsilon), \hat{h}(t, 0) = I$. It then follows that

$$\frac{\partial}{\partial t}\hat{g}(t, \epsilon) = \hat{g}(t, \epsilon)(\epsilon\hat{h}^{-1}(t, \epsilon)P(t, \epsilon)(A)\hat{h}(t, \epsilon)).$$

Since $\pi(g(t, \epsilon)) = \pi(\hat{g}(t, \epsilon))$, $\hat{g}$ is a horizontal field of curves that projects onto $x(t, \epsilon)$. Hence, $\frac{\partial x}{\partial t} = \pi_*(\hat{g}(t, \epsilon)(\epsilon\hat{h}^{-1}(t, \epsilon)P(t, \epsilon)\hat{h}(t, \epsilon))$ and

$$\frac{D}{\partial \epsilon}\left(\frac{\partial x}{\partial t}\right)(t, \epsilon) = \pi_*\left(\hat{g}(t, \epsilon)\frac{\partial}{\partial \epsilon}\epsilon\left(\hat{h}^{-1}P(t, \epsilon)(A)\hat{h}\right)\right).$$

Let now $\frac{\partial}{\partial \epsilon}g(t, \epsilon) = g(t, \epsilon)((V(t, \epsilon) + W(t, \epsilon))$ with $V(t, \epsilon) \in \mathfrak{p}$ and $W(t, \epsilon) \in \mathfrak{h}$. If $\bar{g}(t, \epsilon) = g(t, \epsilon)\bar{h}(t, \epsilon)$ with $\bar{h}(t, \epsilon)$ the solution of $\frac{\partial \bar{h}}{\partial \epsilon}(t, \epsilon) = -\bar{h}(t, \epsilon)W(t, \epsilon), \bar{h}(t, 0) = I$ then,

$$\frac{\partial \bar{g}}{\partial \epsilon} = \bar{g}(t, \epsilon)(\bar{h}^{-1}V\bar{h}).$$

It follows that $\bar{g}(t, \epsilon)$ is a horizontal field of curves over the field of curves $x(t, \epsilon)$ in G/H, and therefore, $\frac{\partial x}{\partial \epsilon}(t, \epsilon) = \pi_*(\bar{g}(t, \epsilon)(\bar{h}^{-1}V\bar{h}))$. Then,

$$\frac{D}{\partial t}\left(\frac{\partial x}{\partial \epsilon}\right)(t, \epsilon) = \pi_*\left(\bar{g}(t, \epsilon)\frac{\partial}{\partial t}(\bar{h}^{-1}(t, \epsilon)V(t, \epsilon)\hat{h})(t, \epsilon)\right).$$

Since $\frac{D}{\partial t}\left(\frac{\partial x}{\partial \epsilon}\right)(t, \epsilon) = \frac{D}{\partial \epsilon}\left(\frac{\partial x}{\partial t}\right)(t, \epsilon)$,

$$\pi_*\left(\bar{g}(t, \epsilon)\frac{\partial}{\partial t}\left(\bar{h}^{-1}(t, \epsilon)V(t, \epsilon)\hat{h}(t, \epsilon)\right)\right) = \pi_*\left(\hat{g}(t, \epsilon)\frac{\partial}{\partial \epsilon}\epsilon\left(\hat{h}^{-1}P(t, \epsilon)(A)\hat{h}\right)\right).$$

At $\epsilon = 0$, the left-hand side of this equation is equal to $\pi_*\left(\frac{dJ_{\hat{A}}}{dt}(t)(A)\right)$, while the right-hand side is equal to $\pi_*(P(t, 0)(A))$. But then,

$$\frac{dJ_{\hat{A}}}{dt}(t)(A) = P(t, 0)(A) = \sum_{k=0}^{\infty}\frac{t^{2k}}{(2k)!}ad^{2k}\hat{A}(A),$$

hence, $J_{\hat{A}}(t) = \sum_{k=0}^{\infty}\frac{t^{2k+1}}{(2k+1)!}ad^{2k}\hat{A}$.　　　□

Corollary 8.35 $J_{\hat{A}}(t)$ *is the solution of the equation*

$$\frac{d^2 J_{\hat{A}}}{dt^2}(t) = ad^2\hat{A}\left(J_{\hat{A}}(t)\right), J_{\hat{A}}(0) = 0. \tag{8.19}$$

Equation (8.19) is called *Jacobi's equation*, and its solutions are called *Jacobi's curves*.

Corollary 8.36

$$\|J_{\hat{A}}(t)(A)\|^2 = t^2 \|A\|^2 + \frac{1}{3}t^4 \langle A, ad^2\hat{A}(A) \rangle + R(t), \qquad (8.20)$$

where the remainder term $R(t)$ satisfies $\lim_{t \to 0} \frac{R(t)}{t^4} = 0$.

The second term $\langle A, ad^2\hat{A}(A) \rangle$ in the Taylor series expansion of $J_{\hat{A}}(t)(A)$ is the negative of the Riemannian curvature of the underlying manifold G/H, a fact true on any Riemannian manifold M [DC].

Let us now consider the curvature in its own right and investigate the implications on the structure of the symmetric space.

Definition 8.37 The bilinear form $\kappa : \mathfrak{p} \times \mathfrak{p} \to \mathbb{R}$ defined by

$$\kappa(A, B) = \langle [[A, B], A], B \rangle$$

for each A and B in $\mathfrak{p}$ is called the Riemannian curvature of the Riemannian space G/H.

It follows that $\kappa(A, B) = -\langle ad^2 A(B), B \rangle$ and that $\kappa(A, A) = 0$. The fact that $\langle adA(\mathfrak{h}), B \rangle = -\langle A, adB(\mathfrak{h}) \rangle$ easily implies that κ is symmetric, in the sense that $\kappa(A, B) = \kappa(B, A)$.

Definition 8.38 The sectional curvature of a two-dimensional linear subspace P in $\mathfrak{p}$ is equal to $\kappa(A, B)$, where A and B are any orthonormal vectors in P.

We leave it to the reader to show that the sectional curvature is well defined, in the sense that it is independent of the choice of a basis for P.

Definition 8.39 A symmetric Riemannian space G/H is said to be of constant curvature if the sectional curvatures of any pair of two-dimensional linear subspaces of $\mathfrak{p}$ are equal.

Proposition 8.40 *Let G/H be a symmetric Riemannian space and let K_A denote the restriction of ad^2A to $\mathfrak{p}$ for $A \in \mathfrak{p}$, $\|A\| = 1$. Then,*

1. *G/H is of zero constant curvature if and only if $K_A = 0$ for each A. This happens if and only if $\mathfrak{p}$ is a commutative Lie subalgebra of $\mathfrak{g}$.*
2. *G/H is a space of non-zero constant curvature if and only if the restriction of K_A to the orthogonal complement of A is a scalar multiple of the identity independent of A.*

Proof We will first show that K_A is a symmetric linear operator on $\mathfrak{p}$. Note that for any X, Y, Z, W in $\mathfrak{p}$

$$\langle [[X, Y], Z], W \rangle = -\langle [Z, [X, Y]], W \rangle = -\langle Z, [[X, Y], W] \rangle = -\langle [[X, Y], W], Z \rangle,$$

from which it follows that

$$\langle [[X, Y], Z], Z \rangle = 0. \tag{8.21}$$

Then for arbitrary elements X, Y of $\mathfrak{p}$,

$$
\begin{aligned}
\langle K_A(X), Y \rangle &= \langle [A, [A, X]], Y \rangle = \langle A, [[A, X], Y] \rangle \\
&= \langle A, [[A, Y], X] \rangle + \langle A, [[Y, X], A] \rangle \\
&= \langle [A, [A, Y]], X \rangle + \langle [[X, Y]], A], A \rangle \\
&= \langle ad^2 A(Y), X \rangle = \langle K_A(Y), X \rangle.
\end{aligned}
$$

Let now $E_1 = A$, and lef $E_2, \dots, E_n$ be an orthonormal basis in the orthogonal complement of A in $\mathfrak{p}$ with respect to which K_A is diagonal, that is,

$$K_A(E_1) = 0, \text{ and } K_A(E_j) = \lambda_j E_j, j = 2, \dots, n.$$

Then the sectional curvature of the plane spanned by E_1, E_j is given by λ_j. The space G/H is of constant curvature if and only if $\lambda_2 = \cdots = \lambda_n = \lambda$, that is, if and only if $K_A = \lambda I$ on the orthogonal complement to A.

The curvature is zero if and only if $\lambda = 0$ for each A. In such a case $\mathfrak{p}$ must be a commutative algebra because the Killing form is negative-definite on $\mathfrak{h}$. In fact,

$$Kl([A, B], [A, B]) = Kl([[A, B], A], B) = -Kl\left(ad^2 A(B), B\right) = 0$$

for any A, B in $\mathfrak{p}$, Therefore, $[A, B] = 0$. □

8.6 Spaces of constant curvature

Proposition 8.41 *Suppose that (G, H) denotes a Riemannian symmetric pair such that G is connected and the Lie algebra $\mathfrak{h}$ of H has zero center. Then the quotient space G/H has zero curvature if and only if G is isomorphic to $\mathbb{E}^n \ltimes K$, where K is a closed subgroup of the rotation group $O(\mathbb{E}^n)$.*

Proof We have already seen that the semi-direct products $G = \mathbb{E}^n \ltimes K$ with K a closed subgroup of the rotation group $O(\mathbb{E}^n)$ realizes $\mathbb{E}^n$ as a symmetric space $G/(\{I\} \times K)$. In such a situation, $\mathfrak{p} = \mathbb{E}^n \times \{0\}$ and $\mathfrak{h} = \{0\} \times \mathfrak{k}$, where $\mathfrak{k}$ denotes the Lie algebra of K. The Lie brackets in $\mathfrak{g}$ are of the form

$$[(A_1, B_1), (A_2, B_2)] = ([B_2 A_1] - [B_1 A_2], [B_1, B_2])$$

for all (A_1, B_1) and (A_2, B_2) in $\mathfrak{g}$. Evidently the Lie brackets of elements in $\mathfrak{p}$ are zero, and hence G/H has curvature equal to zero.

Conversely, suppose that (G, H) denotes a Riemannian symmetric pair such that the Lie algebra $\mathfrak{h}$ of H has zero center, and suppose that G/H is a space of zero curvature. It follows from Proposition 8.34 that $\mathfrak{p}$ is a commutative subalgebra of $\mathfrak{g}$, hence a Euclidean space $\mathbb{E}^n$. The mapping $B \to ad(B)$ is a Lie algebra representation of $\mathfrak{h}$ into the Lie algebra of skew-symmetric matrices $so_n(\mathbb{R})$. Let $\mathfrak{k} = \{ad(B) : B \in \mathfrak{h}\}$. The mapping $\phi : \mathfrak{g} \to \mathfrak{p} \times \mathfrak{k}$ defined by $\phi(A + B) = (A, -ad(B)), A \in \mathfrak{p}, B \in \mathfrak{h}$ is a Lie algebra isomorphism onto the semi-direct product $\mathbb{E}^n \ltimes \mathfrak{k}$. It is easy to see that $\mathfrak{k}$ is the Lie algebra of Ad_H, and that Ad_H is a subgroup of $O(\mathfrak{p})$ since it leaves the Euclidean norm on $\mathfrak{p}$ invariant.

Let $\hat{G}$ denote the semi-direct product $\mathbb{E}^n \ltimes Ad_H$. It remains to show that G is isomorphic to $\hat{G}$. Let $\mathcal{F}$ denote the family of right-invariant vector fields $X(g) = gA$ such that $A \in \mathfrak{p}$. If P denotes the orbit of $\mathcal{F}$ through the identity of G, then P is an Abelian subgroup of G. It can be proved that $g \in G$ can be written as $g = ph$ for some $p \in P$ and $h \in H$, as a consequence of the fact that $ad(\mathfrak{h})(\mathfrak{p}) = \mathfrak{p}$. Then, $\phi : G \to \hat{G}$ defined by $\phi(ph) = (p, Ad_h)$ is the desired isomorphism. $\qquad\square$

Consider now the spaces of constant non-zero curvature. There is no loss in generality in assuming that $\lambda = \pm 1$, since this can be accomplished by rescaling the metric on $\mathfrak{p}$. We will show that there are only two simply connected cases: the hyperboloid $\mathbb{H}^n = SO(1, n)/\{1\} \times SO_n(\mathbb{R})$ when $\lambda = -1$, and the sphere $S^n = SO_{n+1}(R)/SO_n(R)$ when $\lambda = 1$.

We will consider both cases simultaneously in terms of a parameter ϵ. Let $SO_\epsilon = SO_{n+1}(R)$ when $\epsilon = 1$ and $SO_\epsilon = SO(1, n)$ when $\epsilon = -1$. In addition, let $S_\epsilon = S^n$ when $\epsilon = 1$ and $S_\epsilon = \mathbb{H}^n$ when $\epsilon = -1$. So $S_\epsilon = G_\epsilon/H$, where $H = \{1\} \times SO_n(R)$.

Let us first show that S_ϵ is a space of constant curvature with the metric induced by $\langle A, B \rangle_\epsilon = -\frac{\epsilon}{2}Tr(AB)$ on the Lie algebra so_ϵ of SO_ϵ. As we have seen earlier in this chapter, so_ϵ consists of matrices $M = \begin{pmatrix} 0 & -\epsilon a^T \\ a & A \end{pmatrix}$, where a is a column vector in $\mathbb{R}^n$, a^T is the corresponding row vector, and A a skew-symmetric $n \times n$ matrix. The Cartan space $\mathfrak{p}_\epsilon$ consists of all matrices $A = \begin{pmatrix} 0 & -\epsilon a^T \\ a & 0_n \end{pmatrix}$, where a is a column vector in $\mathbb{R}^n$ and 0_n the $n \times n$ matrix with zero entries. With each such matrix A, $\hat{A}$ will denote the vector a. In this notation, $\langle A, B \rangle_\epsilon = \hat{A} \cdot \hat{B}$ where $\hat{A} \cdot \hat{B}$ denotes the Euclidean product in $\mathbb{R}^n$.

The Lie algebra $\mathfrak{h}$ of $H = \{1\} \times SO_n(\mathbb{R})$ consists of matrices $0 \oplus B = \begin{pmatrix} 0 & 0^T \\ 0 & B \end{pmatrix}$ with B in $so_n(\mathbb{R})$. The reader can easily verify the following formula:

$$[A_1, A_2] = 0 \oplus \epsilon(\hat{A}_1 \wedge \hat{A}_2), \text{ and } [A, 0 \oplus B] = C \text{ if and only if } \hat{C} = B\hat{A}.$$
$$(8.22)$$

for any A_1 and A_2 in $\mathfrak{p}_\epsilon$ and any B in $so_n(\mathbb{R})$. Then, $-ad^2A(X) = \epsilon X$ for any $A \in \mathfrak{p}_\epsilon, ||\hat{A}|| = 1$, and any $X \in \mathfrak{p}_\epsilon$. Therefore, S_ϵ has constant curvature equal to ϵ.

Let us now realize S_ϵ as the orbit of SO_ϵ through e_0, the column vector in $\mathbb{R}^{n+1}$ with the first coordinate equal to 1 and all other coordinates equal to zero. The geodesic curves $x(t)$ through a point $x_0 = g_0 e_0$ in S_ϵ are given by the formula $x(t) = g_0 e^{tA} e_0$ for $A = \begin{pmatrix} 0 & -\epsilon a^T \\ a & 0_n \end{pmatrix}$ with $||a|| = 1$. An easy calculation yields

$$A^2 = \begin{pmatrix} ||a||^2 & 0 \\ 0 & a \otimes a^T \end{pmatrix}, \text{ and } A^3 = -\epsilon A. \qquad (8.23)$$

Therefore,

$$e^{tA} e_0 = \sum_{k=0}^{\infty} \frac{t^k}{k!} A^k e_0 = e_0 \left(1 - \epsilon \frac{t^2}{2!} + \frac{t^4}{4!} - \epsilon \frac{t^6}{6!} + \cdots \right)$$
$$+ a \left(t - \epsilon \frac{t^3}{3!} + \frac{t^5}{5!} - \epsilon \frac{t^7}{7!} + \cdots \right),$$

with the understanding that a is now embedded in $\mathbb{R}^{n+1}$ with its first coordinate equal to zero. It follows that the geodesic curves which originate at e_0 in $\mathbb{H}^n$ are given by $x(t) = e_0 \cosh t + a \sinh t$, while the geodesic curves which originate at e_0 in S^n are of the form $x(t) = e_0 \cos t + a \sin t$.

Evidently, these geodesics are great circles on the sphere, and "great" hyperbolas on the hyperboloid.

Proposition 8.42 *Suppose that G/H is a symmetric space with curvature $\epsilon = \pm 1$. Then the Lie algebra $\mathfrak{g}_\epsilon$ of G is isomorphic to so_ϵ. If ϕ denotes this isomorphism, and if $\langle , \rangle_\epsilon$ denotes the sub-Riemannian metric on $\mathfrak{p}_\epsilon$ then $\langle A_1, A_2 \rangle_\epsilon = \langle \phi(A_1), \phi(A_2) \rangle_\epsilon = \frac{-\epsilon}{2} Tr(\phi(A_1)\phi(A_2))$ for any A_1 and A_2 in $\mathfrak{p}_\epsilon$.*

Proof For simplicity of notation we will omit the dependence on the sign of the curvature, except in the situations where it really matters. The lemma below contains the essential ingredients required for the proof.

Lemma 8.43 *Let $A_1, A_2, \ldots, A_n$ denote an orthonormal basis in $\mathfrak{p}$. Then,*

1. $[A_i, [A_j, A_k]] = 0$ *for distinct indices i, j, k.*
2. $\{[A_i, A_j], 1 \leq j < i \leq n\}$ *is a basis for $\mathfrak{h}$.*

Proof It follows from (8.21) that $[A_i, [A_j, A_k]]$ is orthogonal to A_i. Since A_i, A_j, A_k are mutually orthogonal, $ad^2(A_k)(A_i) = -\epsilon A_i$ and $ad^2(A_k)(A_j) = -\epsilon A_j$. Then,

$$\langle [A_i, [A_j, A_k]], A_k \rangle = \langle [A_i, [[A_j, A_k], A_k] = -\epsilon \langle A_i, A_k \rangle = 0. \tag{8.24}$$

Therefore, $[A_i, [A_j, A_k]]$ is orthogonal to A_k. An identical argument shows that $[A_i, [A_j, A_k]]$ is orthogonal to A_j. Additionally,

$$\begin{aligned}
ad^2 A_i([A_j, [A_k, A_i]]) &= [A_i, [A_i, [A_j, [A_k, A_i]]]] \\
&= -[A_i, [[A_k, A_i], [A_i, A_j]]] - [A_i, [A_j, [[A_k, A_i], A_i]]] \\
&= [[A_i, A_j], [A_i, [A_k, A_i]] + [[A_k, A_i], [[A_i, A_j], A_i]] - [A_i, [A_j, [A_k, A_i], A_i]]] \\
&= [[A_i, A_j], \epsilon A_k] + [[A_k, A_i], \epsilon A_j] + [A_i, [A_j, \epsilon A_k]] \\
&= 2\epsilon [A_i, [A_j, A_k]]. \tag{8.25}
\end{aligned}$$

This yields $ad^2 A_i([A_j, [A_j, A_k]]) = -2ad^2 A_i([A_j, [A_k, A_i]]) = -4\epsilon [A_i, [A_j, A_k]]$. After permuting indices in (8.25), first with respect to i and j, and then in respect to j and k, we get

$$\begin{aligned}
ad^2(A_j([A_i, [A_k, A_j]])) &= 2\epsilon [A_j, [A_i, A_k]], \\
ad^2 A_k([A_i, [A_j, A_k]]) &= 2\epsilon [A_k, [A_i, A_j]]. \tag{8.26}
\end{aligned}$$

These relations imply

$$\begin{aligned}
ad^2 A_j([A_i, [A_j, A_k]]) &= -2\epsilon [A_j, [A_i, A_k]] \text{ and} \\
ad^2 A_k([A_i[A_j, A_k]) &= 2\epsilon [A_k, [A_i, A_j]]. \tag{8.27}
\end{aligned}$$

Since the curvature is constant,

$$\begin{aligned}
ad^2 A_j([A_i, [A_j, A_k]]) &= -4\epsilon [A_i, [A_j, A_k]], \\
ad^2 A_k([A_i, [A_j, A_k]]) &= -4\epsilon [A_i, [A_j, A_k]]. \tag{8.28}
\end{aligned}$$

Equations (8.27) and (8.28) yield

$$2[A_i[A_j, A_k]] = [A_j, [A_i, A_k]] = [A_k, [A_j, A_i]],$$

which further implies that

$$4[A_i, [A_j, A_k]] = [A_j, [A_i, A_k]] + [A_k, [A_j, A_i]] = [A_i, [A_j, A_k]],$$

and therefore, $[A_i, [A_j, A_k]] = 0$.

To show the second part, assume that $\sum_{i>j}^n a_{ij}[A_i, A_j] = 0$. Then,

$$0 = \sum_{\substack{i>j}}^n a_{ij}[A_k, [A_i, A_j]] = \epsilon \left(\sum_{k>j} -a_{kj} A_j + \sum_{i>k} a_{ik} A_i \right),$$

by the previous part. Therefore, $a_{kj} = 0, k > j$ and $a_{ik} = 0, i > k$ for any $k = 1, \ldots, n$. But, this implies that $A_{ij} = 0$, for all i, j. The fact that $[\mathfrak{p}, \mathfrak{p}] = \mathfrak{h}$ implies that $\{[A_i, A_j], i > j\}$ is a basis for $\mathfrak{h}$. $\square$

To show that $\mathfrak{g}_\epsilon$ is isomorphic to so_ϵ we will introduce the coordinates relative to the basis $A_1, \ldots, A_n, [A_i, A_j], i > j$ in $\mathfrak{g}_\epsilon$. Any $A \in \mathfrak{p}_\epsilon$ can be written as $A = \sum_{i=1}^n a_i A_i$, and any $B \in \mathfrak{h}$ can be written as $B = \sum_{i>j}^n a_{ij}[A_i, A_j]$. These coordinates define a matrix $\phi(A + B) = \begin{pmatrix} 0 & -\epsilon a^T \\ a & A \end{pmatrix}$, $A = (\epsilon a_{ij})$ in so_ϵ.

The correspondence $A + B \to \phi(A + B)$ is one to one and onto so_ϵ. We leave it to the reader to verify that ϕ is a Lie algebra automorphism.

The Cartan space consisting of matrices $\begin{pmatrix} 0 & -\epsilon a^T \\ a & 0_n \end{pmatrix}$ is isomorphic to $\mathfrak{p}_\epsilon$.

Evidently, $\langle A_1, A_2 \rangle = \frac{\epsilon}{2} Tr(\phi(A_1), \phi(A_2))$ for any elements A_1 and A_2 in $\mathfrak{p}_\epsilon$. Therefore, (G_ϵ, H_ϵ) and $(SO_\epsilon\{1\}, SO_n(R))$ are isometric.

9

Affine-quadratic problem

Let us now return to the symmetric Riemannian pairs (G, K) and the Cartan decompositions $\mathfrak{g} = \mathfrak{p} \oplus \mathfrak{k}$. In the previous chapter we investigated the relevance of the left-invariant distributions with values in $\mathfrak{p}$ for the structure of G and the associated quotient space G/H. In particular, we showed that the controllability assumption singled out Lie group pairs (G, K) in which the Lie algebraic conditions of Cartan took the strong form, namely,

$$[\mathfrak{p}, \mathfrak{p}] = \mathfrak{k}, \ [\mathfrak{p}, \mathfrak{k}] = \mathfrak{p}. \tag{9.1}$$

In this chapter we will consider complementary variational problems on G defined by a positive-definite quadratic form $Q(u, v)$ in $\mathfrak{k}$ and an element $A \in \mathfrak{p}$. More precisely, we will consider the left-invariant affine distributions $\mathcal{D}(g) = \{g(A + X) : X \in \mathfrak{k}\}$ defined by an element $A \in \mathfrak{p}$. Each affine distribution $\mathcal{D}$ defines a natural control problem in G,

$$\frac{dg}{dt} = g(t)(A + u(t)), \tag{9.2}$$

with control functions $u(t)$ taking values in $\mathfrak{k}$.

We will be interested in the conditions on A that guarantee that any two points of G can be connected by a solution of (9.1), and secondly, we will be interested in the solutions of (9.1) which transfer an initial point g_0 to a given terminal point g_1 for which the energy functional $\frac{1}{2} \int_0^T Q(u(t), u(t)) \, dt$ is minimal. We will refer to this problem as the *affine-quadratic problem*.

In what follows, we will use $\langle u, v \rangle$ to denote the negative of the Killing form $Kl(u, v) = Tr(ad(u) \circ ad(v))$ for any u and v in $\mathfrak{g}$. Since the Killing form is negative-definite on $\mathfrak{k}$, the restriction of $\langle \, , \rangle$ to $\mathfrak{k}$ is positive-definite, and can be used to define a bi-invariant metric on $\mathfrak{k}$. This metric will be used as a bench mark for the affine-quadratic problems. For that reason we will express the quadratic form $Q(u, v)$ as $\langle Q(u), v \rangle$ for some self-adjoint linear mapping

Q on $\mathfrak{k}$ which satisfies $\langle Q(u), u \rangle > 0$ for all $u \neq 0$ in $\mathfrak{k}$. Then, $Q = I$ yields the negative of the Killing form, i.e., the bi-invariant metric on $\mathfrak{k}$.

The transition from the quadratic form to the linear mapping can be justified formally as follows: any non-degenerate and symmetric quadratic form on $\mathfrak{k}$ induces a mapping $\tilde{Q} : \mathfrak{k} \to \mathfrak{k}^*$ defined by $\tilde{Q}(u)(v) = (\tilde{Q}(u), v) = Q(u, v), v \in \mathfrak{k}$. Here, (ℓ, v) denotes the natural pairing $\ell(v)$ between $\ell \in \mathfrak{g}^*$ and $v \in \mathfrak{g}$. Then the mapping Q is defined by $\langle Q(u), v \rangle = (\tilde{Q}(u), v)$ for all $v \in \mathfrak{k}$. Let us also note that any non-degenerate, symmetric form Q on $\mathfrak{k}$ has a dual form Q^* on $\mathfrak{k}^*$ defined by

$$Q^*(k_1, k_2) = (k_2, \tilde{Q}^{-1}(k_1)), k_1 \in \mathfrak{k}^*, k_2 \in \mathfrak{k}^*. \qquad (9.3)$$

It is easy to verify that $Q^*(k_1, k_2) = Q(u, v)$ when $k_1 = \tilde{Q}(u)$ and $k_2 = \tilde{Q}(v)$. In terms of the mapping Q, $Q^*(k_1, k_2) = \langle L_1, Q^{-1}(v) \rangle$, after the identification of $k_1 \in \mathfrak{k}^*$ with $L_1 \in \mathfrak{k}$. In particular, $Q^*(k, k) = \langle L, Q^{-1}(L) \rangle$, for all $k \in \mathfrak{k}^*$ with $(k, X) = \langle L, X \rangle, X \in \mathfrak{k}$.

Before going into further details associated with the above optimal problem, let us first comment on the degenerate case $A = 0$. When $A = 0$, the reachable set of (9.1) from the group identity is equal to K and the reachable set from any other point g_0 is equal to the right coset $g_0 K$. It also follows that any curve in K is a solution of (9.1). Hence the preceding problem essentially reduces to a left-invariant geodesic problem on K of finding the curves of minimal length in K where the length is given by $\int_0^T \sqrt{Q(u(t), u(t))}\, dt$. Then the Maximum Principle identifies

$$H(k) = \frac{1}{2} Q^*(k, k)$$

as the appropriate Hamiltonian. To recapitulate briefly, each extremal control $u(t)$ must satisfy

$$h_{u(t)}(k(t)) = -\frac{1}{2} Q(u(t), u(t)) + (k(t), u(t)) \geq -\frac{1}{2} Q(v, v) + (k(t), v), v \in \mathfrak{k}$$

along each extremal curve $k(t) \in \mathfrak{k}^*$. This implies that $u(t) = \tilde{Q}^{-1}(k)$. Therefore, $h_{u(t)} = \frac{1}{2} Q^*(k(t), k(t)) = H$.

The Hamiltonian equations are then given by

$$\frac{dg}{dt} = g(t)\tilde{Q}^{-1}(k(t)), \frac{dk}{dt} = -ad^* dH(k(t))(k(t)) = -ad^* \tilde{Q}^{-1}(k(t))(k(t)),$$

or in equivalent form on $K \times \mathfrak{g}$ as

$$\frac{dg}{dt} = g(t)Q^{-1}(L(t)), \frac{dL}{dt} = [dH(L), L(t)] = [Q^{-1}(L(t)), L(t)].$$

In the canonical case $Q = I$ the preceding equations reduce to $\frac{dg}{dt} = g(t)L(t)$, $\frac{dl}{dt} = 0$. In this situation, the geodesics are the left-translates of the one-parameter groups $\{e^{tL} : t \in \mathbb{R}\}$ for $L \in \mathfrak{k}$.

Remark 9.1 The above shows that the exponential map is surjective on a compact Lie group K, because the identity can be connected to any other terminal point by a curve of minimal length. But then, the curve of minimal length is the projection of an extremal curve, hence it is geodesic.

Let us now return to the general case, under the assumption that G is semi-simple and that the strong Cartan conditions (9.1) hold. Condition $[\mathfrak{p}, \mathfrak{p}] = \mathfrak{k}$ implies that $\{p \in \mathfrak{p} : adk(p) = 0, k \in \mathfrak{k}\} = 0$. This fact, in turn, implies that Ad_H admits no fixed non-zero points in $\mathfrak{p}$. Additionally, the strong Cartan conditions imply that

$$\mathfrak{g} = \mathfrak{g}_1 \oplus \mathfrak{g}_2 \cdots \oplus \mathfrak{g}_m, \tag{9.4}$$

where each factor $\mathfrak{g}_i$ is a simple ideal of the form $\mathfrak{g}_i = \mathfrak{p}_i + [\mathfrak{p}_i, \mathfrak{p}_i]$ as described by Proposition 8.10 of the previous chapter.

Definition 9.2 An element A in $\mathfrak{p}$ is called regular if $\{X \in \mathfrak{p} : [A, X] = 0\}$ is an abelian subalgebra in $\mathfrak{p}$.

It is easy to prove that the projection of a regular element A on each factor $\mathfrak{g}_i$ in (4) is non-zero.

Proposition 9.3 *The affine system (9.2) is controllable whenever A is an element in $\mathfrak{p}$ such that its projection A_i on $\mathfrak{g}_i$ is not zero for each $i = 1, \ldots, m$.*

For the proof we shall borrow some concepts from differential geometry. Let $\langle \, , \, \rangle$ denote the Ad_K invariant Euclidean inner product on $\mathfrak{p}$ and let S^n denote the unit sphere in $\mathfrak{p}$. A subset S of S^n is said to be *convex* if the angle between any two points of S is less than π and if any two points of S can be connected by an arc of a great circle in S^n that is entirely contained in S [Eb, p. 65].

The dimension of a convex subset S of S^n is the largest integer k such that a k-dimensional disk can be smoothly embedded in S (relative to the canonical metric on S^n). The interior $int(S)$ of S is the union of all smoothly imbedded k-disks smoothly imbedded in S. The boundary S of $\partial(S)$ is equal to $\bar{S} - int(S)$, where $\bar{S}$ denotes the topological closure of S.

If x_1 and x_2 are any points of S^n let $d(x_1, x_2)$ denote the spherical distance between them and let $d(x, X) = inf\{d(x, y) : y \in X\}$ for any closed subset X of S^n.

If S is a convex subset of S^n with a non-empty boundary then there exists a unique point $s(S)$ in $int(S)$ that yields the maximum of the function

$f(x) = d(x, \partial(S))$. This point is called *the soul* of S. Finally, if ϕ is an isometry of S^n that leaves S invariant, then $s(S)$ is a fixed point of ϕ, i.e., $\phi(s(S)) = s(S)$ [Ch; Eb].

With these notions at our disposal we turn to the proof of the proposition.

Proof   We will first show that the Lie algebra generated by $\Gamma = \{A+X; X \in \mathfrak{h}\}$ is equal to $\mathfrak{g}$. There is no loss in generality in assuming that $\mathfrak{g}$ is simple to begin with, since the argument could be reduced to each simple factor by showing that the Lie algebra generated by the projection of Γ on each factor $\mathfrak{g}_i$ is equal to $\mathfrak{g}_i$.

Evidently $\mathfrak{h} \in Lie(\Gamma)$. Therefore, it suffices to show that $\mathfrak{p} \subset Lie(\Gamma)$. Let V denote the linear subspace spanned by $\bigcup_{j=0}^{\infty} ad^j \mathfrak{k}(A) = \bigcup_{j=0}^{\infty} ad^j B(A) : B \in \mathfrak{k}\}$, and let $V^{\perp}$ denote its orthogonal complement in $\mathfrak{p}$ relative to the Killing form. Both V and $V^{\perp}$ are $ad(\mathfrak{k})$ invariant. If $X \in V$ and $Y \in V^{\perp}$ then $Kl(\mathfrak{k}, [X, Y]) = Kl([\mathfrak{k}, X], Y) = 0$ because $[\mathfrak{k}, X] \subseteq V$. Hence $[X, Y] = 0$. It follows that $[V, V^{\perp}] = 0$. Therefore, $V + [V, V]$ is an ideal in $\mathfrak{g}$ and a subset of $Lie(\Gamma)$. But then it must be equal to $\mathfrak{g}$ since $V \neq 0$.

The rest of the proof uses the notion of the Lie saturate $LS(\Gamma)$ introduced in the previous chapters. Let C denote the closure of the positive, convex hull generated by all elements of the form $\{Ad_X^k(A) : X \in \mathfrak{k}, k = 0, 1, \dots \}$. Then C is an Ad_K invariant subset of $LS(\Gamma)$. The fact that $Lie(\Gamma) = \mathfrak{g}$ implies that C has a non-empty interior in $\mathfrak{g}$.

Suppose that C is not equal to $\mathfrak{p}$. Then there exists a vector $v \in \mathfrak{p}$ such that $\langle v, int(C) \rangle > 0$, i.e., C lies on one side of the hyperplane $\langle v, x \rangle = 0$. Let S^n denote the unit sphere in $\mathfrak{p}$ relative to the Ad_K invariant metric $\langle , \rangle$ and let S be equal to $S^n \cap int(C)$.

Then S is a convex subset of S^n with a non-empty interior in S^n that is invariant under Ad_K. If $s(S)$ denotes the soul of S then it follows from above that $s(S)$ is a fixed point of Ad_K, which is not possible since Ad_K acts irreducibly on $\mathfrak{p}$.[1] $\square$

To each affine problem there is a "shadow affine problem" on the semi-direct product $G_s = \mathfrak{p} \ltimes H$ because the Lie algebra $\mathfrak{g}$, as a vector space, admits two Lie bracket structures: the original in $\mathfrak{g}$ and the other induced by the semi-direct product. To be more explicit, note that the Lie algebra $\mathfrak{g}_s$ of G_s consists of pairs (A, B) in $\mathfrak{p} \times \mathfrak{h}$ with the Lie bracket

$$(A_1, B_1), (A_2, B_2)]_s = (adB_1(A_2) - adB_2(A_1), [B_1, B_2]).$$

---

[1] I am grateful to P. Eberlein for suggesting a proof of Proposition 9.4 based on the concept of the soul of a convex subset of the sphere.

If (A, B) in $\mathfrak{p} \ltimes \mathfrak{h}$ is identified with $A + B$ in $\mathfrak{p} + \mathfrak{k}$ then

$$(A_1 + B_1), (A_2 + B_2)]_s = [B_1, A_2] - [B_2, A_1] + [B_1, B_2].$$

Hence, $[\mathfrak{p}, \mathfrak{p}]_s = 0$, $[\mathfrak{p}, \mathfrak{h}]_s = [\mathfrak{p}, \mathfrak{h}]$, $[\mathfrak{h}, \mathfrak{h}]_s = [\mathfrak{h}, \mathfrak{h}]$. Thus vector space $\mathfrak{g}$ is the underlying vector space for both Lie algebras $\mathfrak{g}$ and $\mathfrak{g}_s$.

Proposition 9.4 *The shadow system $\frac{dg}{dt} = g(t)(A + U(t))$, $U(t) \in \mathfrak{h}$ is controllable in G_s whenever Ad_H acts irreducibly of $\mathfrak{p}$.*

Proof Since Ad_H acts irreducibly on $\mathfrak{p}$, $\{Ad_h(A) : h \in H\}$ has a non-empty interior in $\mathfrak{p}$. Therefore, $\mathfrak{p} \subset Lie(\Gamma)$, where $\Gamma = \{A + X : X \in \mathfrak{h}\}$ and consequently, $Lie(\Gamma) = \mathfrak{g}_s$. For the rest of the proof we can either mimic the proof in Proposition 9.3, or use the result in [BJ], which says that the semigroup S generated by $\{e^{tX} : t \geq 0, X \in \Gamma\}$ is equal to G_s whenever $Lie(\Gamma) = \mathfrak{g}_s$. $\square$

Proposition 9.5 *Suppose that system (9.2) is controllable. Then both an affine-quadratic problem on G and its shadow problem on $\mathfrak{p} \ltimes K$ admit optimal solutions for each pair of boundary points $g(0) = g_0$ and $g(T) = g_1$, that is, for each pair of points g_0 and g_1 there exists an interval $[0, T]$ and a control $u(t)$ in $L^2([0, T])$ that generates a trajectory $g(t)$ of minimal energy $\frac{1}{2} \int_0^T \langle Q(u(t), u(t)) \rangle dt$ among all other trajectories that satisfy $g(0) = g_0$ and $g(T) = g_1$.*

Proof Let $\mathcal{T}(g_0, g_1, T)$ denote the set of trajectories $g(t)$ of (9.2) generated by the controls in $L^\infty([0, T])$ that satisfy $g(0) = g_0$ and $g(T) = g_1$. Let T be big enough that $\mathcal{T}(g_0, g_1, T)$ is not empty. Let α denote the infimum of $\int_0^T \langle Q(u(t)), u(t) \rangle dt$ over all controls $u(t)$ in $L^\infty([0, T])$ that generate trajectories in $\mathcal{T}(g_0, g_1, T)$.

Let $L^2([0, T])$ denote the Hilbert space of measurable curves $u(t)$ in $\mathfrak{k}$ that satisfy $\int_0^T \langle Q(u(t)), u(t) \rangle dt < \infty$. Then any control $u_0(t)$ that generate a trajectory in $\mathcal{T}(g_0, g_1, T)$ defines a closed ball $B = \{u \in L^2([0, T]) : \int_0^T \langle Q(u(t)), u(t) \rangle dt \leq \int_0^T \langle Q(u_0(t)), u_0(t) \rangle dt\}$ in $L^2([0, T])$.

Let $\{u_n\}$ denote a sequence in B such that $\alpha = \lim \int_0^T \langle Q(u_n(t)), u_n(t) \rangle dt$. Since closed balls in a Hilbert space are weakly compact, $\{u_n\}$ contains a weakly convergent subsequence. For simplicity of notation, we will assume that $\{u_n\}$ itself is weakly convergent. Let u_∞ denote the weak limit of $\{u_n\}$.

But then the trajectories $g_n(t)$ of (9.2) that are generated by the controls u_n converge uniformly to the trajectory g_∞ that is generated by u_∞ [Jc, p. 118]. Since the convergence is uniform, g_∞ belongs to $\mathcal{T}(g_0, g_1, T)$.

We will complete the proof by showing that g_∞ is the optimal trajectory, i.e., we will show that $\alpha = \int_0^T \langle Q(u_\infty(t)), u_\infty(t) \rangle dt$. To begin with,

weak convergence implies that $\int_0^T \langle Q(u_\infty(t)), u_\infty(t) \rangle \, dt = \lim \int_0^T \langle Q(u_\infty(t)),$
$u_n(t) \rangle \, dt$. Secondly,

$$\int_0^T \langle Q(u_\infty(t)), u_n(t) \rangle \, dt \leq$$

$$\left(\int_0^T \langle Q(u_\infty(t)), u_\infty(t) \rangle \, dt \right)^{\frac{1}{2}} \left(\int_0^T \langle Q(u_n(t)), u_n(t) \rangle \, dt \right)^{\frac{1}{2}}.$$

Together, they imply that $\int_0^T \langle Q(u_\infty(t)), u_\infty(t) \rangle \, dt \leq (\int_0^T \langle Q(u_\infty(t)), u_\infty(t) \rangle \, dt)^{\frac{1}{2}} \alpha^{\frac{1}{2}}$, or that $\int_0^T \langle Q(u_\infty(t)), u_\infty(t) \rangle \, dt \leq \alpha$. $\square$

Remark 9.6 In general the maximum principle is not valid for L^2 controls. Fortunately, that is not the case here. In the class of affine-quadratic systems, optimal trajectories generated by L^2 controls do satisfy the Maximum Principle [Sg; Tr].

9.1 Affine-quadratic Hamiltonians

Let us now turn to the Maximum Principle for the appropriate Hamiltonians associated with the affine problems. To preserve the left-invariant symmetries, the cotangent bundle T^*G will be trivialized by the left-translations and considered as the product $G \times \mathfrak{g}^*$. Then $\mathfrak{g}^* = \mathfrak{p}^* \oplus \mathfrak{k}^*$, where $\mathfrak{p}^*$ is identified with the annihilator $\mathfrak{k}^0 = \{\ell \in \mathfrak{g} : \ell(X) = 0, X \in \mathfrak{k}\}$ and $\mathfrak{k}^*$ with the annihilator $\mathfrak{p}^0 = \{\ell \in \mathfrak{g} : \ell(X) = 0, X \in \mathfrak{p}\}$.

Furthermore, each $\ell \in \mathfrak{g}^*$ will be identified with $L \in \mathfrak{g}$ via $\langle L, X \rangle = \ell(X)$ for all $X \in \mathfrak{g}$. Then $\mathfrak{p}^*$ is identified with $\mathfrak{p}$ and $\mathfrak{k}^*$ is identified with $\mathfrak{k}$. The above implies that $\ell = \ell_\mathfrak{p} + \ell_\mathfrak{k}$, $\ell_\mathfrak{p} \in \mathfrak{p}^*$ and $\ell_\mathfrak{k} \in \mathfrak{k}^*$ is identified with $L = L_\mathfrak{p} + L_\mathfrak{k}$ where $L_\mathfrak{p} \in \mathfrak{p}$ and $L_\mathfrak{k} \in \mathfrak{k}$. Since $\mathfrak{p}$ and $\mathfrak{k}$ are orthogonal relative the Killing form, $L_\mathfrak{p}$ and $L_\mathfrak{k}$ are orthogonal relative to $\langle , \rangle$.

Under these identifications the Hamiltonian lift of the cost-extended system

$$\frac{dx}{dt} = \frac{1}{2} \langle Q(u), u \rangle, \quad \frac{dg}{dt} = g(t)(A + u(t)),$$

is given by

$$h_u(t)(L) = \lambda \frac{1}{2} \langle Q(u), u \rangle + \langle L_\mathfrak{p}, A \rangle + \langle L_\mathfrak{k}, u(t) \rangle, \quad \lambda = 0, -1.$$

The Maximum Principle then yields the Hamiltonian

$$H = \frac{1}{2} \langle Q^{-1}(L_\mathfrak{k}), L_\mathfrak{k} \rangle + \langle A, L_\mathfrak{p} \rangle \tag{9.5}$$

defined by $u = Q^{-1}(L_{\mathfrak{k}})$ for $\lambda = -1$. The abnormal extremals ($\lambda = 0$) are the integral curves of

$$h_{u(t)}(L) = \langle A, L_{\mathfrak{p}} \rangle + \langle L_{\mathfrak{k}}, u(t) \rangle, \tag{9.6}$$

subject to the constraint $L_{\mathfrak{k}} = 0$.

Therefore, the normal extremals are the solutions of

$$\frac{dg}{dt} = g(A + Q^{-1}(L_{\mathfrak{k}})), \quad \frac{dL}{dt} = [Q^{-1}(L_{\mathfrak{k}}) + A, L]. \tag{9.7}$$

The projection on $\mathfrak{g}$ then can be written in expanded form as

$$\frac{dL_{\mathfrak{k}}}{dt} = [Q^{-1}(L_{\mathfrak{k}}), L_{\mathfrak{k}}] + [A, L_{\mathfrak{p}}], \quad \frac{dL_{\mathfrak{p}}}{dt} = [Q^{-1}(L_{\mathfrak{k}}), L_{\mathfrak{p}}] + [A, L_{\mathfrak{k}}]. \tag{9.8}$$

The abnormal extremals are the solutions of

$$\frac{dg}{dt} = g(t)(A + u(t)), \quad \frac{dL}{dt} = [u(t), L(t)]], L_{\mathfrak{k}} = 0. \tag{9.9}$$

On the semi-direct product $\mathfrak{g}_s = \mathfrak{p} \rtimes \mathfrak{k}$, the passage from $\mathfrak{g}_s^*$ to its equivalent representation on $\mathfrak{g}_s$ leads to slightly different equations because the quadratic form $\langle \, , \, \rangle$ is not invariant relative to the semi-direct bracket. On semi-direct products the Hamiltonian equation $\frac{dl}{dt}(X) = -ad^*(dh(l)(l(t)(X)$ corresponds to $\left\langle \frac{dL}{dt}, X \right\rangle = - \langle L, [dh, X]_s \rangle$, for any left-invariant Hamiltonian h. This equation implies that $\left\langle \frac{dL_{\mathfrak{p}}}{dt}, X_{\mathfrak{p}} \right\rangle + \left\langle \frac{dL_{\mathfrak{k}}}{dt}, X_{\mathfrak{k}} \right\rangle = -\langle L_{\mathfrak{p}}, [dh_{\mathfrak{p}}, X_{\mathfrak{k}}] + [dh_{\mathfrak{k}}, X_{\mathfrak{p}}] \rangle + \langle L_{\mathfrak{k}}, [dh_{\mathfrak{k}}, X_{\mathfrak{k}}] \rangle$, or

$$\left\langle \frac{dL_{\mathfrak{p}}}{dt}, X_{\mathfrak{p}} \right\rangle + \left\langle \frac{dL_{\mathfrak{k}}}{dt}, X_{\mathfrak{k}} \right\rangle = \langle [dh_{\mathfrak{k}}, L_{\mathfrak{k}}] + [dh_{\mathfrak{p}}, L_{\mathfrak{p}}], X_{\mathfrak{k}} \rangle + \langle [dh_{\mathfrak{k}}, L_{\mathfrak{p}}], X_{\mathfrak{p}} \rangle.$$

Therefore,

$$\frac{dg}{dt} = g(t)dh(L), \quad \frac{dL_{\mathfrak{h}}}{dt} = [dh_{\mathfrak{h}}, L_{\mathfrak{h}}] + [dh_{\mathfrak{p}}, L_{\mathfrak{p}}], \quad \frac{dL_{\mathfrak{p}}}{dt} = [dh_{\mathfrak{h}}, L_{\mathfrak{p}}].$$

It follows that both the semi-simple and semi-direct Hamiltonian equations can be amalgamated into a single equation

$$\frac{dL_{\mathfrak{h}}}{dt} = [dh_{\mathfrak{h}}, L_{\mathfrak{h}}] + [dh_{\mathfrak{p}}, L_{\mathfrak{p}}], \quad \frac{dL_{\mathfrak{p}}}{dt} = [dh_{\mathfrak{h}}, L_{\mathfrak{p}}] + s[dh_{\mathfrak{p}}, L_{\mathfrak{h}}], \tag{9.10}$$

with the understanding that the parameter s is equal to 0 in the semi-direct case and equal to 1 in the semi-simple case. Of course, it is also understood that $g(t)$ evolves in the appropriate group according to the equation $\frac{dg}{dt} = g(dh)$.

It follows that the integral curves of the affine-quadratic Hamiltonian H are the solutions of the following system of equations:

$$\frac{dg}{dt} = g(t)(A + Q^{-1}(L_{\mathfrak{k}})),$$

$$\frac{dL_{\mathfrak{k}}}{dt} = [Q^{-1}(L_{\mathfrak{k}}), L_{\mathfrak{k}}] + [A, L_{\mathfrak{p}}], \quad \frac{dL_{\mathfrak{p}}}{dt} = [Q^{-1}(L_{\mathfrak{k}}), L_{\mathfrak{p}}] + [A, L_{\mathfrak{k}}]. \tag{9.11}$$

Let us now address the abnormal extremals (equation (9.9). But first, we will need to introduce additional notations: $\mathfrak{k}_A$ will denote the subalgebra of $\mathfrak{k}$ consisting of $X \in \mathfrak{k}$ such that $[A, X] = 0$, and K_A will denote the subgroup in K generated by $\{e^{tX} : X \in \mathfrak{k}_A, t \in R\}$. Then,

Proposition 9.7 *Suppose that A is a regular element in $\mathfrak{p}$. Then abnormal extremal curves $(g(t), L_{\mathfrak{k}}, L_{\mathfrak{p}})$ are the solutions of $\frac{dg}{dt} = g(t)(A + u(t))$ subject to the constraints*

$$L_{\mathfrak{k}} = 0, [A, L_{\mathfrak{p}}] = 0, [L_{\mathfrak{p}}, u(t)] = 0.$$

Consequently, $L_{\mathfrak{p}}$ is constant. If $L_{\mathfrak{p}}$ is regular, then $g(t) = g(0)e^{At}h(t)$, where $h(t)$ denotes the solution of $\frac{dh}{dt}(t) = -h(t)u(t)$, $h(0) = I$. If $g(t)$ is optimal, then $h(t)$ is the curve of minimal length relative to the metric $\langle Q(u), u \rangle$ in K_A that connects I to $h(T)$.

Proof Suppose that $(g(t), L_{\mathfrak{p}}(t), L_{\mathfrak{h}}(t))$ is an abnormal extremal curve generated by a control $u(t)$ in $\mathfrak{k}$. Then equations (9.9) imply that $L_{\mathfrak{k}} = 0$ and $\frac{dL_{\mathfrak{p}}}{dt} = [u, L_{\mathfrak{p}}]$, $[A, L_{\mathfrak{p}}] = 0$. This means that $L_{\mathfrak{p}}(t)$ belongs to the maximal abelian subalgebra $\mathcal{A}$ in $\mathfrak{p}$ that contains A. Therefore, $\frac{dL_{\mathfrak{p}}}{dt}$ also belongs to $\mathcal{A}$.

If B is an arbitrary element of $\mathcal{A}$, then

$$Kl\left(\frac{dL_{\mathfrak{p}}}{dt}, B\right) = Kl([u(t), L_{\mathfrak{p}}], B) = Kl(u(t), [L_{\mathfrak{p}}, B]) = 0.$$

Since the Killing form is non-degenerate on $\mathcal{A}$, $\frac{dL_{\mathfrak{p}}}{dt} = 0$, and therefore $L_{\mathfrak{p}}(t)$ is constant. This proves the first part of the proposition.

To prove the second part assume that $L_{\mathfrak{p}}$ is regular. Then $[A, L_{\mathfrak{p}}] = 0$ implies that $[L_{\mathfrak{p}}, [A, u(t)]] = 0$. Since $L_{\mathfrak{p}}$ is regular and belongs to $\mathcal{A}$, $[A, u(t)]$ also belongs to $\mathcal{A}$. It then follows that $[A, u(t)] = 0$ by the argument identical to the one used in the preceding paragraph. But then

$$g(t) = g(0)e^{At}h(t),$$

where $h(t)$ is the solution of $\frac{dh}{dt}(t) = h(t)u(t)$ with $h(0) = I$. Since $[A, u(t)] = 0$, $h(t) \in K_A$. If $g(t)$ is to be optimal, then $h(t)$ is a curve in K_A of shortest length $\int_0^T \sqrt{\langle Q(u(t), u(t) \rangle} \, dt$. $\square$

Corollary 9.8 *If $g(t)$ is an optimal trajectory that is a projection of an abnormal extremal generated by a control $u(t)$ that satisfies $[A, u(t)] = 0$, then $g(t)$ is also the projection of a normal extremal curve associated with Q.*

Proof The projection $g(t)$ of an abnormal extremal curve that projects onto an optimal trajectory is of the form $g(t) = g_0 e^{At} h(t)$ with $h(t)$ the curve of shortest length in K_A that satisfies the given boundary conditions. Hence $h(t)$ is the projection of an extremal curve in $T^* K_A = K_A \times \mathfrak{k}_A^*$. That is, there exists a curve $L_\mathfrak{k}(t)$ in $\mathfrak{k}_A$ such that $h(t)$ satisfies

$$\frac{dh}{dt} = h(t)(Q^{-1}(L_\mathfrak{k})), \quad \frac{dL_\mathfrak{k}}{dt} = [Q^{-1}(L_\mathfrak{k}), L_\mathfrak{k}].$$

Let $L_\mathfrak{p}(t)$ denote the solution of $\frac{dL_\mathfrak{p}}{dt} = [Q^{-1}(L_\mathfrak{k}), L_\mathfrak{p}]$ such that $[L_\mathfrak{p}(0), A] = 0$. Then,

$$\frac{d}{dt}([A, L_\mathfrak{p}(t)]) = [A, [u(t), L_\mathfrak{p}]] = [[A, L_\mathfrak{p}(t)], u(t)].$$

It follows that $[A, L_\mathfrak{p}(t)]$ is the solution of a linear equation that is equal to zero at $t = 0$. Hence $[A, L_\mathfrak{p}(t)] = 0$ for all t. But then, $g(t), L(t) = L_\mathfrak{k}(t) + L_\mathfrak{p}(t)$ are the solutions of equations (9.8) and (9.11), hence $g(t)$ is the projection of a normal extremal curve. $\qquad\square$

Remark 9.9 The above proposition raises an interesting question: is every optimal trajectory of an arbitrary affine-quadratic problem the projection of a normal extremal curve? It seems that $G = SL_n(R)$ is a good testing ground for this question. In this situation there are plenty of abnormal extremal curves but it is not clear exactly how they relate to optimality. My own guess is that every optimal solution is the projection of a normal extremal curve.

9.2 Isospectral representations

Let us now return to the normal extremal curves equations (9.8) and (9.11). The projection of these equations on either $\mathfrak{g}$ or $\mathfrak{g}_s$ is given by

$$\frac{dL_\mathfrak{k}}{dt} = [Q^{-1}(L_\mathfrak{k}), L_\mathfrak{k}] + [A, L_\mathfrak{p}], \quad \frac{dL_\mathfrak{p}}{dt} = [Q^{-1}(L_\mathfrak{k}), L_\mathfrak{p}] + s[A, L_\mathfrak{k}], s = 0, 1$$
$$(9.12)$$

Equations (9.12) can be regarded also as the Hamiltonian equation on coadjoint orbits in $\mathfrak{g}$ (resp. $\mathfrak{g}_s$) associated with the Hamiltonian $H = \frac{1}{2}\langle Q^{-1}(L_\mathfrak{k}), L_\mathfrak{k}\rangle + \langle A, L_\mathfrak{p}\rangle$.

Definition 9.10 An affine-quadratic Hamiltonian will be called isospectral if its Hamiltonian system (9.12) admits a spectral representation of the form

$$\frac{dL_\lambda}{dt} = [M_\lambda, L_\lambda],$$

$$M_\lambda = Q^{-1}(L_{\mathfrak{k}}) - \lambda A, \ L_\lambda = L_{\mathfrak{p}} - \lambda L_{\mathfrak{k}} + (\lambda^2 - s)B,$$
(9.13)

for some matrix B that commutes with A.

Proposition 9.11 *Suppose that $L(\lambda)$ is a parametrized curve in a Lie algebra $\mathfrak{g}$ that is a solution of $\frac{dL}{dt}(\lambda) = [M(\lambda), L(\lambda)]$ for another curve $M(\lambda)$ in $\mathfrak{g}$. Then each eigenvalue of $L(\lambda)$ is a constant of motion.*

Proof Let $g(t)$ be a solution of $\frac{dg}{dt} = g(t)M_\lambda$ in G. Then,

$$\frac{d}{dt}g^{-1}(t)L_\lambda(t)g(t) = g^{-1}(t)([L_\lambda, M_\lambda] + \frac{dL_\lambda}{dt})g(t) = 0,$$

along the solutions of $\frac{dL}{dt}(\lambda) = [M(\lambda), L(\lambda)]$. Hence, the spectrum of $L_\lambda(t)$ is constant. $\square$

Functions of the eigenvalues of $L(\lambda)$ are called spectral invariants of $L(\lambda)$. It follows that each spectral invariant of $L(\lambda)$ is constant along the solutions of the differential equation for $L(\lambda)$. In particular, the spectral invariants of $L_\lambda = L_{\mathfrak{p}} - \lambda L_{\mathfrak{k}} + (\lambda^2 - s)B$ are constant along the solutions of (9.12).

Proposition 9.12 *The spectral invariants of $L_\lambda = L_{\mathfrak{p}} - \lambda L_{\mathfrak{k}} + (\lambda^2 - 1)B$ Poisson commute with each other relative to the semi-simple Lie algebra structure, while the spectral invariants of $L_\lambda = L_{\mathfrak{p}} - \lambda L_{\mathfrak{k}} + \lambda^2 B$ Poisson commute relative to the semi-direct product structure.*

To address the proof, it will be advantageous to work in $\mathfrak{g}^*$ rather than in $\mathfrak{g}$. Recall that the splitting $\mathfrak{g} = \mathfrak{p} \oplus \mathfrak{k}$ induces the splitting of $\mathfrak{g}^*$ as the sum of the annihilators $\mathfrak{p}^0 \oplus \mathfrak{k}^0$, in which case $\mathfrak{p}^0$ is identified with $\mathfrak{k}^*$ and $\mathfrak{k}^0$ is identified with $\mathfrak{p}^*$. The affine Hamiltonian $H = \frac{1}{2}\langle Q^{-1}(L_{\mathfrak{k}}), L_{\mathfrak{k}}\rangle + \langle A, L_{\mathfrak{p}}\rangle$ is given by $H(p + k) = Q^*(k, k) + (p, A)$ with

$$\frac{d\ell}{dt} = -ad^*dH(\ell)(\ell)$$

equal to its Hamiltonian equations on each coadjoint orbit in $\mathfrak{g}^*$ (respectively, on each coadjoint orbit of $\mathfrak{g}_s^*$).

Definition 9.13 Functions f and h on $\mathfrak{g}^*$ which Poisson commute, i.e., $\{f, h\} = 0$, are said to be in involution. Functions on $\mathfrak{g}^*$ that are in involution with any other function on $\mathfrak{g}^*$ are called invariant or Casimirs.

Invariant functions can be defined alternatively as the functions f that satisfy $ad^* df(\ell) = 0$ for any $\ell \in \mathfrak{g}^*$, which is the same as saying that they are constant on each coadjoint orbit. On semi-simple Lie algebras, coadjoint orbits are identified with the adjoint orbits and the above is the same as saying that $[df, L] = 0$ for each $L \in \mathfrak{g}$. Hence the spectral invariants of L correspond to invariant functions on $\mathfrak{g}^*$

The proof below is a minor adaptation of the arguments presented in [Pr] and [Rm].

Proof Let $T : \mathfrak{g}^* \to \mathfrak{g}^*$ be defined by $T(p+k) = \frac{1}{\lambda} p - k + \mu b$ for $p \in \mathfrak{k}^*, k \in \mathfrak{h}^*$. Here, b is a fixed element of $\mathfrak{p}^*$ and λ and μ are parameters. Thus T is induced by the spectral matrix $L_\lambda = \frac{1}{\lambda} L_\mathfrak{p} - L_\mathfrak{k} + \mu B$ with $\mu = \frac{\lambda^2 - s}{\lambda}$. It follows that $T^{-1} = \lambda p - k - \lambda \mu b$, and hence, T is a diffeomorphism. This diffeomorphism extends to the Poisson form $\{\,,\}$ on $\mathfrak{g}^*$ according to the formula

$$\{f, g\}_{\lambda,\mu}(\xi) = (T\{f, g\})(\xi) = \{T^{-1}f, T^{-1}g\}(T(\xi)).$$

A simple calculation shows that

$$\{f, g\}_{\lambda,\mu} = -\lambda^2 \{f, g\} - \lambda\mu \{f, g\}_b - (1 - \lambda^2)\{f, g\}_s, \tag{9.14}$$

where $\{f, g\}_b = \{f, g\}(b)$ and $\{f, g\}_s$ is the Poisson bracket relative to the semi-direct product Lie bracket structure.

For the semi-direct Poisson bracket $\{\,,\}_s$ the analogous Poisson bracket $\{f, g\}_{s\,\lambda,\mu}(\xi) = (T \circ \{f, g\}_s)(\xi)$ takes on a slightly different form:

$$\{f, g\}_{s\lambda,\mu} = -\{f, g\}_s - \lambda\mu \{f, g\}_{s\,b}. \tag{9.15}$$

If f is any invariant function relative to $\{\,,\}$, then $f_{\lambda,\mu} = T \circ f$ satisfies $\{f_{\lambda,\mu}, g\}_{\lambda,\mu} = 0$ for any function g on $\mathfrak{g}^*$ and any parameters λ and μ. In the case that g is another invariant function, then $f_{\lambda,\mu}$ and $g_{\lambda,\mu} = T \circ g$ satisfy

$$\left\{ f_{\lambda_1,\mu_1}, g_{\lambda_2,\mu_2} \right\}_{\lambda_1,\mu_1} = \left\{ f_{\lambda_1,\mu_1}, g_{\lambda_2,\mu_2} \right\}_{\lambda_2,\mu_2} = 0.$$

for all $\lambda_1, \mu_1, \lambda_2, \mu_2$. The same applies to the semi-direct Poisson bracket $\{\,,\}_s$ and its invariant functions.

Suppose now that $\mu = \frac{\lambda^2 - 1}{\lambda}$. It follows from (9.14) that

$$\{f, g\}_{\lambda,\mu} = -\lambda^2 \{f, g\} - (1 - \lambda^2)(\{f, g\}_s - \{f, g\}_b).$$

Therefore,

$$0 = \frac{1}{\lambda_1^2 - 1} \left\{ f_{\lambda_1, \mu_1}, g_{\lambda_2, \mu_2} \right\}_{\lambda_1, \mu_1} - \frac{1}{\lambda_2^2 - 1} \left\{ f_{\lambda_1, \mu_1}, g_{\lambda_2, \mu_2} \right\}_{\lambda_2, \mu_2}$$

$$= \frac{\lambda_1^2 - \lambda_2^2}{\left(1 - \lambda_1^2\right)\left(1 - \lambda_2^2\right)} \left\{ f_{\lambda_1, \mu_1}, g_{\lambda_2, \mu_2} \right\}.$$

Since λ_1, and λ_2 are arbitrary $\left\{ f_{\lambda_1, \mu_1}, g_{\lambda_2, \mu_2} \right\} = 0$.

This argument proves the first part of the proposition because $\lambda T = L_p - \lambda L_{\mathfrak{k}} + (\lambda^2 - 1)B$, after the identifications $p \to L_p$, $k \to L_{\mathfrak{k}}$ and $b \to B$, and $\mu = \frac{\lambda^2 - 1}{\lambda}$.

To prove the second part let $\lambda = \mu$. Then $\lambda T = L_p - \lambda L_{\mathfrak{k}} + \lambda^2 B$, after the above identifications. Relative to the semi-direct structure,

$$0 = \frac{1}{\lambda_1^2} \left\{ f_{\lambda_1, \mu_1}, g_{\lambda_2, \mu_2} \right\}_{s \, \lambda_1 \mu_1} - \frac{1}{\lambda_2^2} \left\{ f_{\lambda_1, \mu_1}, g_{\lambda_2, \mu_2} \right\}_{s \, \lambda_2, \mu_2}$$

$$= \left(\frac{1}{\lambda_2^2} - \frac{1}{\lambda_1^2} \right) \left\{ f_{\lambda_1, \mu_1}, g_{\lambda_2, \mu_2} \right\}_s.$$

Therefore, $\left\{ f_{\lambda_1, \mu_1}, g_{\lambda_2, \mu_2} \right\}_s = 0.$ □

Evidently the isospectral representation yields a number of constants of motion for the Hamiltonian H all in involution with each other, and this begs the question whether these integrals of motion are sufficient in number to guarantee complete integrability of H. Before addressing this question in some detail, however, let us first address the question of the existence of isospectral representations. In this regard we then have the following:

Proposition 9.14 *An affine Hamiltonian $H = \frac{1}{2}\langle Q^{-1}(L_{\mathfrak{k}}), L_{\mathfrak{k}} \rangle + \langle A, L_p \rangle$ is isospectral if and only if $[Q^{-1}(L_{\mathfrak{k}}), B] = [L_{\mathfrak{k}}, A]$ for some matrix $B \in \mathfrak{p}$ that commutes with A.*

Proof   If $[Q^{-1}(L_{\mathfrak{k}}), B] = [L_{\mathfrak{k}}, A]$, then

$$\frac{dL_\lambda}{dt} = \frac{dL_p}{dt} - \lambda \frac{dL_{\mathfrak{k}}}{dt} = [Q^{-1}(L_{\mathfrak{k}}), L_p] + s[A, L_{\mathfrak{k}}] - \lambda([Q^{-1}(L_{\mathfrak{k}}), L_{\mathfrak{k}}] + [A, L_p])$$

$$= [Q^{-1}(L_{\mathfrak{k}}), L_p - \lambda L_{\mathfrak{k}}] + s[A, L_{\mathfrak{k}}] - \lambda[A, L_p]$$

$$= [Q^{-1}(L_{\mathfrak{k}}) - \lambda A, L_p - \lambda L_{\mathfrak{k}}] - \lambda^2[A, L_{\mathfrak{k}}] + s[A, L_{\mathfrak{k}}]$$

$$= [M_\lambda, L_\lambda] - (\lambda^2 - s)[Q^{-1}(L_{\mathfrak{k}}), B] + (\lambda^2 - s)[L_{\mathfrak{k}}, A] = [M_\lambda, L_\lambda].$$

Conversely, if $\frac{dL_\lambda}{dt} = [M_\lambda, L_\lambda]$, then

$$\frac{dL_\mathfrak{k}}{dt} = [Q^{-1}(L_\mathfrak{k}), L_\mathfrak{k}] + [A, L_\mathfrak{p}] - \lambda(\lambda^2 - s)[A, B],$$

$$\frac{dL_\mathfrak{p}}{dt} = [Q^{-1}(L_\mathfrak{k}), L_\mathfrak{p}] + (\lambda^2 - s)([Q^{-1}(L_\mathfrak{k}), B] + \lambda^2[A, L_\mathfrak{k}]).$$

Therefore, $[A, B] = 0$ and $[Q^{-1}(L_\mathfrak{k}), B] = [L_\mathfrak{k}, A]$. $\square$

Corollary 9.15 *The canonical case $Q = I$ is isospectral. The spectral matrix L_λ is given by $L_\lambda = L_\mathfrak{p} - \lambda L_\mathfrak{k} + (\lambda^2 - s)A$, i.e., $B = A$.*

Proposition 9.16 *Let $H = \frac{1}{2}\langle Q^{-1}(L_\mathfrak{k}), L_\mathfrak{k}\rangle + \langle A, L_\mathfrak{p}\rangle$ be an isospectral Hamiltonian. Then every solution of the corresponding homogeneous Hamiltonian system $\frac{dL_\mathfrak{k}}{dt} = [Q^{-1}(L_\mathfrak{k}), L_\mathfrak{k}]$ is the projection of a solution of the affine-quadratic system (9.12).*

Proof Let B be a matrix that satisfies $[Q^{-1}(L_\mathfrak{k}), B] = [L_\mathfrak{k}, A]$ with $[A, B] = 0$ and let $L_\mathfrak{k}(t)$ denote a solution of the homogeneous system on $\mathfrak{k}$. Then $L(t) = L_\mathfrak{k}(t) + L_\mathfrak{p}$ with $L_\mathfrak{p} = sB$ is a solution of (9.12). $\square$

Corollary 9.17 *The projections $L_\mathfrak{k}(t)$ of the solutions of $L = L_\mathfrak{p} + L_\mathfrak{k}$ of an isospectral system (9.12) with $L_\mathfrak{p} = sB$ are the solutions of the following spectral equation:*

$$\frac{dL_\mathfrak{k}}{dt} = [Q^{-1}(L_\mathfrak{k}) - \lambda A, L_\mathfrak{k} - \lambda B]. \tag{9.16}$$

Conversely, every Hamiltonian system on $\mathfrak{k}$ that admits a spectral representation (9.16) gives rise to an isospectral affine Hamiltonian $H = \frac{1}{2}\langle Q^{-1}(L_\mathfrak{k}), L_\mathfrak{k}\rangle + \langle A, L_\mathfrak{p}\rangle$.

Corollary 9.18 *The spectral invariants of $L_\mathfrak{k} - \lambda B$ are in involution with each other.*

Proof $L_\lambda = L_\mathfrak{p} - \lambda L_\mathfrak{k} + (\lambda^2 - s)B = -\lambda L_\mathfrak{k} + \lambda^2 B = \lambda(L_\mathfrak{k} - \lambda B)$. Hence the spectral invariants of $L_\mathfrak{k} - \lambda B$ are a subset of the ones generated by the spectral matrix L_λ for the affine system. $\square$

9.3 Integrability

Our presentation of isospectral representations makes an original contact with the existing theory of integrable systems on Lie algebras inspired by the seminal work of S. V. Manakov on the integrability of an n-dimensional

Euler's top, a left-invariant quadratic Hamiltonian on the space of $n \times n$ skew-symmetric matrices subject to some restrictions (which we will explain in complete detail later on in the text) [Mn]. This study originated an interest in the spectral matrix $L - \lambda B$. Motivated by this work of Manakov, A. S. Mishchenko and A. T. Fomenko [FM] considered quadratic Hamiltonians H on a complex semi-simple Lie group G that admit a spectral representation $\frac{dL}{dt} = [dH - \lambda A, L - \lambda B]$ for some elements A and B in the Lie algebra $\mathfrak{g}$ of G. More specifically, they considered the set of functions $\mathcal{F}$ that are functionally dependent on the shifts $f(\ell + \lambda b)$, where functions f range over the functions that are constant on the coadjoint orbits in $\mathfrak{g}^*$, and they asked the question, when is a given quadratic Hamiltonian H in $\mathcal{F}$? Here b is a fixed element in $\mathfrak{g}^*$ and λ is an arbitrary real number. When H is an element of $\mathcal{F}$, then each element of $\mathcal{F}$ is an integral of motion for the Hamiltonian flow of H on the coadjoint orbits in $\mathfrak{g}^*$, since the elements of $\mathcal{F}$ are in involution with each other relative to the Poisson bracket $\{\,,\,\}$ on $\mathfrak{g}^*$.

In this remarkable paper [FM], the authors showed first that H is in $\mathcal{F}$ if and only if there exists a vector $A \in \mathfrak{g}$ such that $b \circ adA = 0$ and

$$b \circ ad\, dH(\ell) = \ell \circ adA \text{ for all } \ell \in \mathfrak{g}^*. \qquad (9.17)$$

If the dual $\mathfrak{g}^*$ is identified $\mathfrak{g}$ via the Killing form $\langle\,,\,\rangle b$, then ℓ is identified with L and b is identified with $B \in \mathfrak{g}$, and the preceding condition is equivalent to

$$\langle A, [B, X] \rangle = 0 \text{ and } \langle B, [dH(L), X] \rangle = \langle L, [A, X] \rangle, X \in \mathfrak{g},$$

which is the same as $[dH, B] = [L, A]$ and $[A, B] = 0$. Secondly, they showed that $\mathcal{F}$ is an involutive family and that it contains n functionally independent functions where $2n$ is the dimension of the orbit, i.e., they showed that H is integrable, in the sense of Liouville, on each generic coadjoint orbit. Having in mind applications to an n-dimensional Euler's top, they applied these results to compact real forms to show that Manakov's top is a particular case of integrable tops defined by sectional operators.

The following case is a prototype of the general situation; not only does it illustrate the importance of this study for the theory of tops, but it also provides a constructive procedure for generating affine-quadratic Hamiltonians that admit a spectral matrix $L_\lambda = L_\mathfrak{p} - \lambda L_\mathfrak{k} + (\lambda^2 - s)B$, for some B.

9.3.1 $SL_n(\mathbb{C})$ and its real forms

Let $\mathfrak{g} = sl_n(\mathbb{C})$ and let $\mathfrak{h}$ denote the commutative sub-algebra consisting of all diagonal matrices in $\mathfrak{g}$. We will adopt the following notations: $e_1, \ldots, e_n$ will denote the standard basis in $\mathbb{R}^n$, and $e_i \otimes e_j$ will denote the matrix given by

$(e_i \otimes e_j)e_k = \delta_{jk}e_i, k = 1, \ldots, n$. In addition, $\langle \, , \, \rangle$ will denote the trace form $\langle A, B \rangle = -\frac{1}{2}Tr(AB)$. It follows that $\{e_i \otimes e_j : i \neq j\}$ is a basis for the matrices in $\mathfrak{g}$ whose diagonal entries are zero.

If A is a matrix in $\mathfrak{h}$ with its diagonal entries $a_1, \ldots, a_n$ then

$$adA(e_i \otimes e_j) = (a_j - a_i)e_i \otimes e_j.$$

In the language of Lie groups, $\alpha : \mathfrak{h} \to \mathbb{C}$ given by $\alpha(A) = (a_j - a_i)$ is called a root and the complex line through $X_\alpha = e_i \otimes e_j$ is called the root space corresponding to α, that is, X_α satisfies $adA(X_\alpha) = \alpha(A)X_\alpha$ for each $A \in \mathfrak{h}$. Then $-\alpha$ is also a root with $X_{-\alpha} = e_j \otimes e_i$. If $H_\alpha = [X_\alpha, X_{-\alpha}]$, then $[X_\alpha, X_{-\alpha}] = e_j \otimes e_j - e_i \otimes e_i$, and therefore $H_\alpha \in \mathfrak{h}$.

More generally, a Cartan subalgebra of a semi-simple Lie algebra $\mathfrak{g}$ is a maximal abelian subalgebra $\mathfrak{h}$ of $\mathfrak{g}$ such that $ad(h)$ is semi-simple for each $h \in \mathfrak{h}$. It is known that every semi-simple complex Lie algebra $\mathfrak{g}$ contains a Cartan subalgebra $\mathfrak{h}$ and that

$$\mathfrak{g} = \mathfrak{h} \oplus \sum_\alpha \mathfrak{g}^\alpha, \alpha \in \Delta,$$

where Δ denotes the set of roots generated by $\mathfrak{h}$ [HI]. On $sl_n(\mathbb{C})$ every matrix with distinct eigenvalues $\lambda_1, \ldots, \lambda_n$ is called regular. It is well known that the set of matrices which commute with a regular matrix A is a Cartan subalgebra $\mathfrak{h}$. In the basis of eigenvectors of A, $\mathfrak{h}$ consists of all diagonal matrices having zero trace, and that brings us to the situation above.

It follows that $\{H_\alpha, X_\alpha, X_{-\alpha}, \alpha \in \Delta\}$ is a basis for $sl_n(\mathbb{C})$. It is called a Weyl basis. It is known that every semi-simple Lie algebra admits a Weyl basis [HI].

With this terminology at our disposal let us now return to the work of Mishchenko, Trofimov, and Fomenko, [FM; FT; FT1]. Let $\mathfrak{g}_0$ denote the real vector space spanned by

$$\{iH_\alpha, i(X_\alpha + X_{-\alpha}), (X_\alpha - X_{-\alpha})\}, \alpha \in \Delta$$

Evidently, $\mathfrak{g}_0$ consists of matrices $A + iB$ with A and B matrices having real entries, A skew-symmetric and B symmetric. Therefore, $\mathfrak{g}_0$ is a Lie algebra equal to su_n. It follows that $sl_n(\mathbb{C}) = \mathfrak{g}_0 + i\mathfrak{g}_0$, that is, $\mathfrak{g}_0$ is a real form for $\mathfrak{g}$. It is called a compact normal form for $\mathfrak{g}$ (since SU_n is compact).

Let us note that $\mathfrak{g}_0 = \mathfrak{k} \oplus \mathfrak{p}$, with $\mathfrak{k} = so_n(R)$, and $\mathfrak{p} = \{iX : X \in sl_n(R) : X^T = X\}$, is the Cartan decomposition induced by the automorphism $\sigma(g) = (g^{-1})^T, g \in SU_n$. Suppose now that A is a regular diagonal matrix in $\mathfrak{p}$, that is, suppose that the diagonal entries $ia_1, \ldots, ia_n$ of A are distinct. Then,

$$adA(i(e_i \otimes e_j + e_j \otimes e_i) = -(a_j - a_i)(e_i \otimes e_j - e_j \otimes e_i),$$
$$adA(e_i \otimes e_j - e_j \otimes e_i) = (a_j - a_i)i(e_i \otimes e_j + e_j \otimes e_i).$$

Therefore,

$$adA(i(X_\alpha + X_{-\alpha}) = i\alpha(A)(X_\alpha - X_{-\alpha}), adA(X_\alpha - X_{-\alpha}) = i\alpha(A)i(X_\alpha + X_{-\alpha}),$$

for $\alpha \in \Delta$. It follows that

$$ad^{-1}A \circ adB(i(X_\alpha + X_{-\alpha}) = \frac{\alpha(B)}{\alpha(A)}(i(X_\alpha + X_{-\alpha}),$$

$$ad^{-1}A \circ adB((X_\alpha - X_{-\alpha}) = \frac{\alpha(B)}{\alpha(A)}((X_\alpha - X_{-\alpha}),$$

for any other diagonal element B in $\mathfrak{p}$.

Let $Q = ad^{-1}A \circ adB$ restricted to $\mathfrak{k} = so_n(R)$. If $X \in \mathfrak{k}$, then $X = \sum_{i,j}^n x_{ij}(e_i \wedge e_j) = \sum_\alpha x_\alpha(X_\alpha - X_{-\alpha})$. Therefore,

$$Q(X) = \sum_\alpha \frac{\alpha(B)}{\alpha(A)}x_\alpha = \sum_{i,j} \frac{b_j - b_i}{a_j - a_i}x_{ij},$$

hence Q is a linear mapping on $\mathfrak{k}$. Relative to the trace form, $\langle X, Q(X) \rangle = \sum_{i,j}^n \frac{b_j - b_i}{a_j - a_i}x_{ij}^2$. Moreover,

$$[B, Q^{-1}(X)] = adB \circ ad^{-1}B \circ adA(X) = adA(X),$$

for all X. Therefore, $[Q^{-1}(X), B] = [X, A], X \in \mathfrak{k}$ and hence the affine-quadratic Hamiltonian $H = \frac{1}{2}\langle Q^{-1}(L_\mathfrak{k}), L_\mathfrak{k}\rangle + \langle A, L_\mathfrak{p}\rangle$ is isospectral.

The quadratic form $\langle Q(X, X)$ is positive-definite if and only if all the quotients $\frac{b_j - b_i}{a_j - a_i}$ are positive. For this reason, we need to relax the condition that the matrices A and B have zero trace. This change is inessential for the bracket condition $[Q^{-1}(X), B] = [X, A]$, but it allows for greater flexibility in choosing the ratios $\frac{b_j - b_i}{a_j - a_i}$.

For if A and B are diagonal matrices in u_n instead of su_n that satisfy the above bracket condition, then $A_0 = A - \frac{Tr(A)}{n}I$ and $B_0 = b - \frac{Tr(B)}{n}I$ are diagonal matrices with zero trace that satisfy $[Q^{-1}(X), B_0] = [X, A_0]$.

There are many choices of A and B in u_n that produce positive quadratic forms. One such choice, which makes a contact with the top of Manakov, is given by $b_i = a_i^2, i = 1, \ldots, n$. For then, $\frac{b_j - b_i}{a_j - a_i} = a_i + a_j$. Another choice, which relates to Jacobi's geodesic problem on an ellipsoid (as will be shown later on in the text), occurs when A is a matrix with positive entries and $B = -A^{-1}$.

Then $\frac{b_j - b_i}{a_j - a_i} = \frac{-\frac{1}{a_j} + \frac{1}{a_i}}{a_j - a_i} = \frac{1}{a_i a_j}$. Such a case will be called elliptic. One could

also take $b_i = a_i^3$. Then, $\frac{b_j - b_i}{a_j - a_i} = \frac{\frac{1}{a_j} - \frac{1}{a_i}}{a_j - a_i} = a_j^2 + a_i a_j + a_i^2$.

Let us conclude this section by saying that $sl_n(R)$ is another real form for $sl_n(\mathbb{C})$. When realized as the real vector space spanned by $\{H_\alpha, (X_\alpha + X_{-\alpha}), (X_\alpha - X_{-\alpha}, \alpha \in \Delta\}$, $sl_n(R)$ lends itself to the same procedure as in the case of su_n. The sectional operators in this setting induce isospectral affine-quadratic Hamiltonians under the same conditions as in the previous case, except that the matrices A and B are diagonal with real entries, rather than matrices with imaginary diagonal entries. Both of these two real forms figure prominently in the theory of integrable systems as will be demonstrated in the subsequent chapters.

The above is a special case of the following general situation.

Proposition 9.19 *Suppose that $H = \frac{1}{2}\langle Q-1(L_\mathfrak{k}), L_\mathfrak{k}\rangle + \langle A, L_\mathfrak{p}\rangle$ is an affine-quadratic Hamiltonian that admits a spectral representation*

$$\frac{dL_\lambda}{dt} = [M_\lambda, L_\lambda], L_\lambda = L_\mathfrak{p} - L_\mathfrak{k} + (\lambda^2 - s)B, [A, B] = 0,$$

where B is a regular element. Let $\mathfrak{k}_A = \{X \in \mathfrak{k} : [X, A] = 0\}$. Then $\mathfrak{k}_A$ is a Lie subalgebra of $\mathfrak{k}$ that is invariant under Q, and Q restricted to the orthogonal complement $\mathfrak{k}_A^\perp$ is given by $Q = ad^{-1}A \circ adB$.

It will be convenient first to prove the following lemma.

Lemma 9.20 $\{X \in \mathfrak{k} : [X, A] = 0\} = \{X \in \mathfrak{k} : [X, B] = 0\}$.

Proof Let $\mathfrak{h} = \{X \in \mathfrak{p} : [X, A] = 0\}$ and let $\mathfrak{k}_B = \{X \in \mathfrak{k} : [X, B] = 0\}$. Then,

$$\langle X, [A, \mathfrak{k}_B]\rangle = \langle [X, A], \mathfrak{k}_B\rangle = 0, X \in \mathfrak{h}.$$

Therefore, $[A, \mathfrak{k}_B]$ belongs to the orthogonal complement of $\mathfrak{h}$ in $\mathfrak{p}$. But,

$$[B, [A, \mathfrak{k}_B]] = -[\mathfrak{k}_B, [B], A]] - [[A, [\mathfrak{k}_B, B]] = 0.$$

Regularity of B implies that the set of elements in $\mathfrak{p}$ that commute with B coincides with the set of elements in $\mathfrak{p}$ that commute with A, since $[A, B] = 0$. Therefore, $[B, [A, \mathfrak{k}_B]] = 0$ implies that $[A, \mathfrak{k}_B] \subset \mathfrak{h}$. It follows that $[A, \mathfrak{k}_B] = 0$. Similar argument shows that $[B, \mathfrak{k}_A] = 0$. □

Let us now come to the proof of Proposition 9.19.

Proof The isospectral condition $[Q^{-1}(L_\mathfrak{k}), B] = [L_\mathfrak{k}, A]$ is the same as $[L_\mathfrak{k}, B] = [Q(L_\mathfrak{k}), A]$. Hence $adB(L_\mathfrak{k}) = adA(Q(L_\mathfrak{k}))$ for all $L_\mathfrak{k} \in \mathfrak{k}$. If $L_\mathfrak{k} \in \mathfrak{k}_A$ then $L_\mathfrak{k} \in \mathfrak{k}_B$, therefore $adB(L_\mathfrak{k}) = 0$, and hence $Q(L_\mathfrak{k}) \in \mathfrak{k}_A$. This argument shows that $\mathfrak{k}_A$ is invariant for Q.

When $L_{\mathfrak{k}}$ is an element of $\mathfrak{k}_A^\perp$ then $adB(L_{\mathfrak{k}}] \neq 0$ for $L_{\mathfrak{k}} \neq 0$, Since $adB(L_{\mathfrak{k}}) \in \mathfrak{h}^\perp$ and adA is injective on $\mathfrak{h}^\perp$, $ad^{-1}A \circ adB(L_{\mathfrak{k}}) = Q(L_{\mathfrak{k}}), L_{\mathfrak{k}} \in \mathfrak{k}_A^\perp$.
□

Corollary 9.21 *If A and B are regular elements in a Cartan subalgebra in $\mathfrak{p}$ and if $\mathfrak{k}_A = 0$ then $H = \frac{1}{2}\langle Q^{-1}(L_{\mathfrak{k}}), L_{\mathfrak{k}}\rangle + \langle A, L_{\mathfrak{p}}\rangle$ with $Q = ad^{-1}A \circ adB$ is isospectral.*

Let us now return to the main theme, the existence of integrals of motion associated with the affine-quadratic Hamiltonians on semi-simple Lie that admit isospectral representations. To include the shadow Hamiltonians on the semi-direct product $G_s = \mathfrak{p} \rtimes K$, it will be necessary to think of $\mathfrak{g}$ as a double Lie algebra equipped with two sets of Lie brackets, the semi-simple $[\,,\,]$ and the semi-direct $[\,,\,]_s$. As already remarked earlier, the two types of Lie brackets give rise to two Poisson structures on the dual of $\mathfrak{g}$. The following definition applies to both situations.

Definition 9.22 A family of functions $\mathcal{F}$ on the dual of a Lie algebra $\mathfrak{g}$ of a Lie group G is said to be complete if it contains an involutive subfamily $\mathcal{F}_0$ that is Liouville integrable on each coadjoint orbit of G as well.

In regard to complete integrability of isospectral affine-quadratic Hamiltonians, let $\mathcal{F}$ denote the class of functions functionally dependent on the shifts $f(p - \lambda k + (\lambda^2 - 1)b)$ of invariant functions f on $\mathfrak{g}^*$, where p, k and b denote the corresponding elements of $L_{\mathfrak{p}}, L_{\mathfrak{k}}$ and B in $\mathfrak{g}$. Analogously, $\mathcal{F}_s$ will denote the shifts $f(p - \lambda k + \lambda^2 b)$ of invariant functions f on $\mathfrak{g}_s^*$.

To each of the above families of functions we will adjoin the Hamiltonians generated by the left-invariant vector fields that take values in $\mathfrak{k}_A$. These functions are of the form $f(p, q) = q(X), X \in \mathfrak{k}_A$. We will use $\mathcal{F}_A$ to denote their functional span. Then we have the following proposition.

Proposition 9.23 *If g is a function in $\mathcal{F}_A$, then $\{f, g\} = 0$ for any function f in $\mathcal{F}$. The same applies to functions in $\mathcal{F}_s$, i.e., $\{f, g\}_s = 0$.*

Proof Let us adopt the notations used in the proof of Proposition 9.20. If f is an invariant function on $\mathfrak{g}^*$, let $f_{\lambda,\mu}(p, k) = f\left(\frac{1}{\lambda} - k + \mu b\right)$. If $\{\,,\,\}_{\lambda,\mu}$ is as defined in Proposition 9.12 then

$$0 = \{f_{\lambda,\mu}, g\}_{\lambda,\mu} = -\lambda^2 \{f_{\lambda,\mu}, g\} - \lambda\mu\{f_{\lambda,\mu}, g\}_b - (1 - \lambda^2)\{f_{\lambda,\mu}, g\}_s$$

for any function g on $\mathfrak{g}^*$. If g is a function in $\mathcal{F}_B$ then $\lambda\mu\{f_{\lambda,\mu}, g\}_b = 0$. It follows that

$$\lambda^2\{f_{\lambda,\mu}, g\}_s - \{f_{\lambda,\mu}, g\}) - \{f_{\lambda,\mu}, g\}_s.$$

Since λ is arbitrary, $\{f_{\lambda,\mu}, g\}_s = \{f_{\lambda,\mu}, g\} = 0$. □

The methods of Mischenko and Fomenko could be extended to the affine Hamiltonians to show that $\mathcal{F} \cup \mathcal{F}_A$ is complete on the generic orbits in $\mathfrak{g}^*$. Apparently, $\mathcal{F}_s \cup \mathcal{F}_A$ is complete relative to the semi-direct Poisson structure. [Bv; Pr], although the proofs do not seem to be well documented in the existing literature.

So every isospectral affine-quadratic Hamiltonian H that belongs to $\mathcal{F} \cup \mathcal{F}_A$ is integrable in the Liouville sense on each coadjoint orbit in $\mathfrak{g}^*$ provided that H is in involution with the functions in $\mathcal{F}_A$, and the same applies to the corresponding shadow Hamiltonian on $\mathfrak{g}_s^*$. This means that every isospectral affine-quadratic Hamiltonian is integrable when $\mathfrak{k}_A = 0$.

In the general case when $\mathfrak{k}_A \neq 0$, H will be in involution with the Hamiltonians in $\mathcal{F}_A$ whenever the left-invariant fields with values in $\mathfrak{k}_A$ are symmetries for H. If X is a symmetry for the affine-quadratic problem then its Hamiltonian lift $h_X(g, L) = \langle L, X(g)\rangle$. Poisson commutes with the affine Hamiltonian H as a consequence of Noether's theorem (Proposition 6.25 in Chapter 6).

Proposition 9.24 *Each left-invariant vector field $X(g) = gX$ with $X \in \mathfrak{k}_A$ is a symmetry for the canonical affine-quadratic problem $\frac{dg}{dt} = g(t)(A + u(t))$, $u(t) \in \mathfrak{k}$, $Q = I$*

Proof Let $g(t)$ be a solution of $\frac{dg}{dt} = g(t)(A + u(t))$ and let $h_\lambda(t) = \exp \lambda X g(t) = g(t)e^{\lambda X}$. Then,

$$\frac{dh_\lambda}{dt} = h_\lambda(t)(e^{-\lambda X} A(e^{\lambda X} + e^{\lambda X} u(t)(e^{\lambda X}))$$

$$= h_\lambda(t)(A + v_\lambda(t)),$$

with $v_\lambda(t) = e^{\lambda X} u(t)(e^{\lambda X}$. Since the Killing form is Ad_K invariant, $\langle v_\lambda, v_\lambda\rangle = \langle u, u\rangle$. □

Corollary 9.25 *Every canonical affine Hamiltonian is completely integrable on the coadjoint orbits of G in $\mathfrak{g}$ and the same applies to its shadow problem on the coadjoint orbits of G_s in $\mathfrak{g}_s$.*

In contrast to the existing literature on integrable systems on Lie algebras, in which functions on $\mathfrak{g}^*$ are implicitly regarded as functions on the coadjoint orbits in $\mathfrak{g}_s^*$ and their integrability is interpreted accordingly, the Hamiltonians in the present context are associated with left-invariant variational problems over differential systems in G, and as such, they are regarded as functions on the entire cotangent bundle. Consequently, they will be integrable in the sense of Liouville only when they are a part of an involutive family of functions on T^*G that contains n independent functions, where n is equal to the dimension of G.

Evidently, this notion of integrability is different from integrability on coadjoint orbits. For Hamiltonian systems on coadjoint orbits, the Casimirs and the Hamiltonians of the right-invariant vector fields are irrelevant, apart the fact that they figure in the dimension of the coadjoint orbit, but that is not the case for the Hamiltonians on the entire cotangent bundle of the group.

At this point we remind the reader that the Hamiltonian lift of any right-invariant vector field $X(g) = Xg$, in the left trivialization of the cotangent bundle, is given by $h_X((\ell, g) = \ell(g^{-1}Xg)$. Saying that each such lift h_X is constant along the flow of a left-invariant Hamiltonian H is equivalent to saying that the projection $\ell(t)$ of the integral curve of $\vec{H}$ on $\mathfrak{g}^*$ evolves on a coadjoint orbit. Since $\{h_X, h_Y\} = h_{[X,Y]}$, the maximum number of independent integrals of motion h_X defined by right-invariant vector fields X in involution with each other is independent of H. This observation brings us to the Cartan algebras.

Recall that a Cartan subalgebra of a semi-simple Lie algebra $\mathfrak{g}$ is a maximal abelian subalgebra $\mathfrak{h}$ of $\mathfrak{g}$ such that $ad(h)$ is semi-simple for each $h \in \mathfrak{h}$. Every complex semi-simple Lie algebra has a Cartan subalgebra and any two Cartan subalgebras are isomorphic [Hl, p. 163]. For real Lie algebras there may be several non-conjugate Cartan subalgebras but they all have the same dimension. This dimension is called the rank of $\mathfrak{g}$.

An element A of $\mathfrak{g}$ is called regular if the dimension of $ker(adA)$ is equal to the rank of $\mathfrak{g}$. On a semi-simple Lie algebra each coadjoint orbit that contains a regular element has dimension equal to $\dim(\mathfrak{g}) - rank(\mathfrak{g})$. These orbits are generic. In fact, every element of the Cartan algebra $\mathfrak{h}$ that contains A is orthogonal to the tangent space of the coadjoint orbit at A relative to the Killing form. So elements in the Cartan algebra do not figure in the integrability on coadjoint orbits, but they do in the integrability on the cotangent bundle of G. The same applies to the invariant functions on $\mathfrak{g}$.

It is important to note that integrability on the level of the Lie algebra implies solvability by quadrature on G, because once the extremal control $u(t)$ is known, the remaining equation $\frac{dg}{dt} = g(t)(A + u(t))$ reduces to a time-varying equation on G, which, at least in principle, is solvable in terms of a suitable system of coordinates on G. We will get a chance to see in Chapter 14 that the "good" coordinates on G make use of vector fields in the Cartan algebra in $\mathfrak{g}$ defined by A.

10

Cotangent bundles of homogeneous spaces as coadjoint orbits

This chapter provides an important passageway from Lie algebras to cotangent bundles that connects isospectral Hamiltonians with integrable Hamiltonians systems on these cotangent bundles. We will presently show that the cotangent bundles of certain manifolds M, such as the spheres, the hyperboloids, and their isometry groups $SO_n(R)$ and $SO(1, n)$, can be represented as the coadjoint orbits on $sl_n(R)$, in which case the restriction of the isospectral integrals of H to these coadjoint orbits provides a complete family of integrals of motion for the underlying variational problem on the base manifold.

Recall the classification of semi-simple Lie algebras into compact, non-compact, and Euclidean types. On semi-simple Lie groups of compact type each coadjoint orbit is compact, and cannot be the cotangent bundle of any manifold. In fact, the coadjoint orbits of compact simple Lie groups have been classified by A. Borel and are known to be compact Kähler manifolds [Bo]. Remarkably, the coadjoint orbits of Lie algebras of non-compact type are all cotangent bundles of flag manifolds [GM]. We will show below that the same is true for the coadjoint orbits relative to the semi-direct products.

We will exploit the fact that the dual of any Lie algebra g that admits a Cartan decomposition carries several Poisson structures which account for a greater variety of coadjoint orbits. Our discussion, motivated by some immediate applications, will be confined to the vector space V_{n+1} of $(n + 1) \times (n + 1)$ matrices with real entries of zero trace, but a similar investigation could be carried out on any semi-simple Lie algebra with a Cartan decomposition $\mathfrak{g} = \mathfrak{p} \oplus \mathfrak{k}$.

10.1 Spheres, hyperboloids, Stiefel and Grassmannian manifolds

We will let $\langle\,,\,\rangle$ denote the quadratic form $\langle A, B \rangle = -\frac{1}{2}Tr(AB)$, A, B in V_{n+1}. Since V_{n+1} carries the canonical Lie algebra $sl_{n+1}(R)$ with its Killing form equal to $2(n+1)Tr(AB)$, $\langle\,,\,\rangle$ is a negative scalar multiple of the Killing form on $sl_{n+1}(R)$ and hence, inherits all of its essential properties. In particular, $\langle\,,\,\rangle$ is non-degenerate and invariant, in the sense that $\langle A, [B, C] \rangle = \langle [A, B], C \rangle$ for any matrices A, B, C in $sl_{n+1}(R)$. It is also positive-definite on $\mathfrak{k} = so_{n+1}(R)$.

As remarked before, V_{n+1} has two distinct Poisson structures: the first associated with the dual of $sl_{n+1}(R)$ with its standard Lie–Poisson bracket and the second associated with the dual of the semi-direct product $S_{n+1} \oplus_s so_{n+1}(R)$, where S_{n+1} denotes the spaces of symmetric matrices in V_{n+1}. In addition to these two structures, we will also include the pseudo-Riemannian decompositions of Cartan type in V_{n+1} and the induced Poisson structures. The pseudo-Riemannian decompositions are due to the automorphism

$$\sigma(g) = D(g^T)^{-1}D^{-1}, g \in SL_{p+q}(R),$$

defined by a diagonal matrix D with its first p diagonal entries equal to 1 and the remaining q diagonal entries equal to -1.

It is easy to check that σ is involutive and that the subgroup K of fixed points of σ is equal to $SO(p, q), p+q = n+1$, the group that leaves the quadratic form

$$(x, y)_{p,q} = \sum_{i=1}^{p} x_i y_i - \sum_{n-p+1}^{n+1} x_i y_i \tag{10.1}$$

invariant. The decomposition $V_{n+1} = S_{p,q} \oplus \mathfrak{k}$ induced by the tangent map σ_* consists of the Lie algebra $\mathfrak{k} = so(p, q)$ and the Cartan space

$$S_{p,q} = \left\{ P \in V_{n+1} : P = \begin{pmatrix} A & B \\ -B^T & C \end{pmatrix}, A = A^T, C = C^T \right\}. \tag{10.2}$$

The Cartan space $S_{p,q}$ can be also described as the set of matrices P in V_{p+q} which are symmetric relative to the quadratic form $(x, y)_{p,q}$, i.e., $(Px, y)_{p,q} = (x, Py)_{p,q}$. In what follows it will be convenient to regard the Riemannian metric $(x, y) = \sum_{i=1}^{n+1} x_i y_i$ as the limiting case of the pseudo-Riemannian case with $q = 0$.

The symmetric pair $(SL_{p+q}(R), SO(p, q))$ for $q \neq 0$ is strictly pseudo-Riemannian because the subgroup of the restrictions of $Ad_{SO(p,q)}$ to S_{p+q} is not a compact subgroup of $Gl(S_{p+q})$ (since the trace form is indefinite on S_{p+q}). Nevertheless, this automorphism endows V_{n+1} with another semi-direct product Lie algebra structure, namely $S_{p,q} \oplus_s so(p, q)$.

So V_{n+1} has three distinct Poisson structures, the first induced by the semi-simple Lie brackets, and the other two induced by the semi-direct products corresponding to the Riemannian and pseudo-Riemannian cases. In each case, V_{n+1} is foliated by the coadjoint orbits. We will presently show that the cotangent bundles of many symmetric spaces can be realized as the coadjoint orbits on V_{n+1}, and that will make a link with mechanical and geometric systems of the underlying spaces.

In the semi-simple case, the coadjoint orbits can be identified with the adjoint orbits via the Killing form. As a consequence, each adjoint orbit is symplectic and hence is even dimensional. The adjoint orbit through any matrix P_0 can be identified with the quotient $SL_{n+1}(R)/St(P_0)$, where $St(P_0)$ denotes the stationary group of P_0, consisting of the matrices g such that $gP_0g^{-1} = P_0$. The Lie algebra of $St(P_0)$ consists of the matrices $X \in sl_{n+1}(R)$ which commute with P_0.

The orbits of maximal dimension, called generic, correspond to matrices P_0 whose stationary group is of minimal dimension. It can be proved that each generic orbit is generated by a matrix P_0 that belongs to some Cartan subalgebra of $sl_{n+1}(R)$. Since Cartan algebras in $sl_{n+1}(R)$ are n-dimensional, generic orbits in $sl_{n+1}(R)$ are $n(n-1)$-dimensional. A more detailed study of adjoint orbits in $sl_{n+1}(R)$ is sufficiently complicated, and would take us away from our main objectives, so will not pursue it here except in the simplest cases. Instead, we will turn attention to the coadjoint orbits relative to the semi-direct products.

The following lemma describes the coadjoint orbits on any semi-simple Lie algebra $\mathfrak{g}$ that admits a Cartan decomposition $\mathfrak{p} \oplus \mathfrak{k}$.

Lemma 10.1 *Let G_s denote the semi-direct product $\mathfrak{p} \ltimes K$ and let $\langle\,,\,\rangle$ denote any scalar multiple of the Killing form. Suppose that $l = Ad^*_{g^{-1}}(l_0)$, for some $\ell_0 \in \mathfrak{g}^*$ and some $g = (X, h) \in G_s$. Suppose further that $l_0 \longrightarrow L_0 = P_0 + Q_0$, and $l \longrightarrow L = P + Q$ are the correspondences defined by the quadratic form $\langle\,,\,\rangle$ with P_0 and P in $\mathfrak{p}$ and Q_0 and Q in $\mathfrak{k}$. Then*

$$P = Ad_h(P_0), \text{ and } Q = [Ad_h(P_0), X] + Ad_h(Q_0). \tag{10.3}$$

Proof If $Z = U + V$ is an arbitrary point of $\mathfrak{g}$ with $U \in \mathfrak{p}$ and $V \in \mathfrak{h}$, then

$$Ad_{g^{-1}}(Z) = \frac{d}{ds}(g^{-1}(sU, e^{sV})g)|_{s=0}$$

$$= \frac{d}{ds}(-Ad_{h^{-1}}(X) + Ad_{h^{-1}}(sU + e^{sV}(X)e^{-sV}), h^{-1}e^{sV}h)|_{s=0}$$

$$= Ad_{h^{-1}}(U + [X, V]) + Ad_{h^{-1}}(V).$$

Hence,

$$l(Z) = l_0(Ad_{g^{-1}}(Z) = \langle P_0, Ad_{h^{-1}}(U + [X, V]) \rangle + \langle Q_0, Ad_{h^{-1}}(V) \rangle$$

$$= \langle Ad_h(P_0), U \rangle + \langle Ad_h(P_0), [X, V] \rangle + \langle Ad_h(Q_0), V \rangle$$

$$= \langle Ad_h(P_0), U \rangle + \langle [Ad_h(P_0), X], V \rangle + \langle Ad_h(Q_0), V \rangle = \langle P, U \rangle + \langle Q, V \rangle.$$

Since U and V are arbitrary,

$$P = Ad_h(P_0), \quad Q = [Ad_h(P_0), X] + Ad_h(Q_0),$$

and (10.3) follows. □

Consider now the coadjoint orbit through a symmetric matrix P_0 in the semi-direct product $S_{n+1} \oplus_s so_{n+1}(R)$, where S_{n+1} denotes S_{p+q} with $p = n + 1$, $q = 0$. For simplicity, we will dispense with the constraint that P_0 has trace zero. Any symmetric matrix P can be written as $P = P_0 + \frac{1}{n+1} Tr(P)I$, where P_0 has zero trace. Then the coadjoint orbit through P_0 is essentially the same as the coadjoint orbit through P, they differ by a constant multiple of the identity.

If a and b are any points in $\mathbb{R}^{n+1}$ then $P = a \otimes b$ will denote the rank one matrix defined by $Px = (x, b)a$ for any $x \in \mathbb{R}^{n+1}$. It is easy to verify that $a \otimes b + b \otimes a$ is symmetric while $a \wedge b = a \otimes b - b \otimes a$ is skew-symmetric. In particular, $a \otimes a$ is symmetric for any $a \in \mathbb{R}^{n+1}$.

Proposition 10.2 *Let $a_1, \ldots, a_r, r > 1$ be any orthonormal vectors in $\mathbb{R}^{n+1}$. Then the coadjoint orbit through $P_0 = \sum_{i=1}^{r} a_i \otimes a_i$ relative to the semi-direct bracket in $\mathfrak{g}_s$ is the tangent bundle of the oriented Grassmannian $Gr(n+1, r) = SO_{n+1}(R)/SO_r(R) \times SO_{n+1-r}(R)$.*

The coadjoint orbit through $P_0 = \sum_{i=1}^{r} \lambda_i(a_i \otimes a_i)$ with distinct numbers $\lambda_1, \ldots, \lambda_r$ is the tangent bundle of the positively oriented Stiefel manifold $St(n + 1, r)$.

Proof Let V_0 denote the linear span of $a_1, \ldots, a_r$ oriented so that $a_1, \ldots, a_r$ is a positively oriented frame. With each matrix P_h in (10.3) given by $P_h = \sum_{i=1}^{r} h(a_i) \otimes h(a_i)$, with $h \in SO_{n+1}(R)$, we will associate the vector space V_h spanned by $h(a_1), \ldots, h(a_r)$ and then identify P_h with the reflection R_h around V_h.

This reflection is defined by $R_h x = x$ for x in V_h and $R_h x = -x$ for x in the positively oriented orthogonal complement of V_h. Every r-dimensional vector subspace V of $\mathbb{R}^{n+1}$ admits a positively oriented orthonormal frame $b_1, \ldots, b_r$ which can be identified with V_h for some $h \in SO_{n+1}(R)$, because $SO_{n+1}(R)$ acts transitively on the space of r-dimensional spaces that are positively oriented relative to V_0.

In the identification with the reflections R_h, the action of $SO_{n+1}(R)$ on the vector spaces translates into the adjoint action $R_h = h^{-1}R_0h, h \in SO_{n+1}(R)$. The isotropy group of R_0 in $SO_{n+1}(R)$ is equal to $SO_r(R) \times SO_{n+1-r}(R)$. Hence, the orbit through R_0 is diffeomorphic with $G(n + 1, r)$, which in turn is diffeomorphic with the matrices P in expression (10.3).

The tangent space at V_h can be realized as the vector spaces of skew-symmetric matrices Ω obtained by differentiating curves $R(\epsilon)$ in the space of reflections of V_h. Each such curve $R(\epsilon)$ is a curve in $SO_{n+1}(R)$ that satisfies $R^2(\epsilon) = I$. Therefore, $R\frac{dR}{d\epsilon} + \frac{dR}{d\epsilon}R = 0$. If $R\frac{dR}{d\epsilon} = R(\epsilon)\Omega(\epsilon)$, then the preceding yields

$$R\Omega R = -\Omega. \tag{10.4}$$

This means that

$$R\Omega(x) = -x, \, x \in V_h, \text{ and } R\Omega x = x, \, x \in V_h^\perp. \tag{10.5}$$

Hence, $\Omega(V_h) \subseteq V_h^\perp$ and $\Omega(V_h^\perp) \subseteq V_h$. If $c_1, \ldots, c_{n+1-r}$ denotes any orthonormal basis in $V_h^\perp$, then

$$\Omega = \sum_{j=1}^{n+1-r} \sum_{k=1}^{r} \Omega_{jk} c_j \wedge b_k \text{ where } \Omega_{jk} = (c_j, \Omega b_k). \tag{10.6}$$

It follows that the tangent space at V_h and the linear span of matrices $c_j \wedge b_k$ are isomorphic But this linear span is also generated by the matrices Q in equation (10.3) since

$$[P, X] = \sum_{i=1}^{r} [b_i \otimes b_i, X] = \sum_{i=1}^{r} [b_i \wedge Xb_i]$$

$$= \sum_{i,j=1}^{r} \langle Xb_i, b_j \rangle b_i \wedge b_j + \sum_{j=1}^{n+1-r} \sum_{k}^{r} (c_j, Xb_k) c_j \wedge b_k$$

$$= \sum_{j=1}^{n+1-r} \sum_{k}^{r} (c_j, Xb_k) c_j \wedge b_k,$$

because $\langle Xb_i, b_j \rangle b_i \wedge b_j + \langle Xb_j, b_i \rangle b_j \wedge b_i = 0$ for any symmetric matrix X. This proves the first part of the proposition.

Suppose now that $P_0 = \sum_{i=1}^{r} \lambda_i(a_i \otimes a_i)$ with $\lambda_i \neq \lambda_j, i \neq j$. The correspondence between points $p = (b_1, \ldots, b_r)$ of the Steifel manifold $St(n + 1, r)$ and matrices $P = \sum_{i=1}^{r} \lambda_i(b_i \otimes b_i)$ is one to one. Then $Ad_h(P_0) = \sum_{i=1}^{r} \lambda_i(h(a_i) \otimes h(a_i))$ corresponds to the point $hp_0 = h(a_1, \ldots, a_r)$. That is, the adjoint orbit $\{Ad_h(P_0); h \in SO_{n+1}(R)\}$ corresponds to the orbit $\{hp_0 : h \in SO_{n+1}(R)\}$. The action of $SO_{n+1}(R)$ on $St(n+1, r)$ by the matrix multiplications from the right identifies $St(n+1, r)$ as the homogenous space $SO_{n+1}(R)/SO_{n+1-r}(R)$.

The tangent space at $(b_1, \ldots, b_r) = h(a_1, \ldots, a_r)$ consists of vectors of the form $h\Omega(a_1, \ldots, a_r)$ for some skew-symmetric matrix Ω in the linear span of $b_j \wedge b_k, 1 \leq j, k \leq r$ and $b_j \wedge c_k, j = 1, \ldots, r, k = 1, \ldots, n+1-r$, where $c_1, \ldots, c_{n+1-r}$ stands for any orthonormal basis in the orthogonal complement of the vector space spanned by $b_1, \ldots, b_r$.

As X varies over all symmetric matrices having zero trace, matrices $Q = [P, X]$ span the same space as the Ω's in the above paragraph. The following argument demonstrates the validity of this statement.

Matrices $(b_i \otimes b_j) + (b_j \otimes b_i), (c_i \otimes b_j) + (b_i \otimes c_j), (c_i \otimes c_j) + (c_j \otimes c_i)$ form a basis for the space of symmetric matrices, and hence, any symmetric matrix X can be written as the sum

$$X = \sum_{i,j}^{r} \langle Xb_i, b_j \rangle ((b_i \otimes b_j) + (b_j \otimes b_i))$$

$$+ \sum_{i=1}^{r} \sum_{j=1}^{n+1-r} \langle Xb_i, c_j \rangle ((b_i \otimes c_j) + (c_j \otimes b_i))$$

$$+ \sum_{i,j}^{n+1-r} \langle Xc_i, c_j \rangle ((c_i \otimes c_j) + (c_j \otimes c_i)).$$

Then,

$$\sum_{i=1}^{r} [\lambda_i (b_i \otimes b_i), (b_j \otimes b_k) + b_k \otimes b_j)]$$

$$= \sum_{i=1}^{r} \lambda_i (\delta_{ij}(b_k \otimes b_i) + \delta_{ik}(b_j \otimes b_i) - \delta_{ij}(b_i \otimes b_k) - \delta_{ik}(b_i \otimes b_j))$$

$$= \lambda_j(b_k \wedge b_j) + \lambda_k(b_j \wedge b_k) = (\lambda_j - \lambda_k)(b_k \wedge b_j),$$

$$\sum_{i=1}^{r} [\lambda_i (b_i \otimes b_i), (b_j \otimes c_k) + (c_k \otimes b_j)]$$

$$= \sum_{i=1}^{r} \lambda_i (\delta_{ij}(c_k \otimes b_i) - (b_i \otimes c_k)) = \lambda_j(c_k \otimes b_j), \text{ and}$$

$$\sum_{i=1}^{r} [\lambda_i (b_i \otimes b_i), (c_j \otimes c_k) + (c_k \otimes c_j)] = 0$$

yields $Q = [P, X] = \sum_{j,k}^{r} (b_j, Xb_k)(\lambda_j - \lambda_k)(b_k \wedge b_j) + \sum_{j=1}^{r} \sum_{k=1}^{n+1-r} (c_k, Xb_j)$ $\lambda_j (c_k \wedge b_j)$. $\qquad\qquad\qquad\qquad\qquad\qquad\qquad\qquad\qquad\qquad\qquad$ $\square$

Proposition 10.3 *The canonical symplectic form on each of St$(n + 1, r)$ and Gr$(n + 1, r)$ agrees with the symplectic form of the corresponding coadjoint orbits.*

Proof The symplectic form on coadjoint orbits is given by $\omega_L(V_1, V_2) = \langle L, [V_2, V_1] \rangle$, where V_1 and V_2 are tangent vectors at L. Every tangent vector V at a point $L = P + Q$ is of the form

$$V = [P, A] + [Q, A] + [P, B], \qquad (10.7)$$

for some skew-symmetric matrix A and some symmetric matrix B.

The argument is simple: if $(X(t), h(t))$ denotes any curve in G_s such that $\frac{dh}{dt}(0) = A$ and $\frac{dX}{dt}(0) = B_0$, then $P(t) = Ad_h(P)$, and $Q = [Ad_h(P), X] + Ad_h(Q_0)$ is the corresponding curve on the coadjoint orbit through $P + Q$. Therefore, $\frac{dP}{dt}(0) = [P, A]$, $\frac{dQ}{dt}(0) = [[P, A], X] + [P, B_0] = [Q, A] + [P, [A, X] + B_0]$ with the aid of Jacobi's identity. This yields (10.7) with $B = [A, X] + B_0$.

Suppose now that $P = \sum_{i=1}^{r} x_i \otimes x_i$ and $Q = [P, X] = \sum_{i=1}^{r} x_i \wedge y_i$ is a point of the coadjoint orbit through $P_0 = \sum_{i=1}^{r} a_i \otimes a_i$ corresponding to some orthonormal vectors $a_1, \ldots, a_r$ in $\mathbb{R}^{n+1}$. That is, $x_i = h(a_i)$ for some $h \in SO_{n+1}(R)$ and $y_i = \sum_{j=1}^{n+1-r} \langle Xx_i, c_j \rangle c_j$ for each $i = 1, \ldots, r$, where $c_1, \ldots, c_{n+1-r}$ is any orthonormal extension of $x_1, \ldots, x_r$. Then

$$\dot{P} = \sum_{i=1}^{r} x_i \otimes \dot{x}_i + \dot{x}_i \otimes x_i, \dot{Q} = \sum_{i=1}^{r} \dot{x}_i \wedge y_i + x_i \wedge \dot{y}_i \qquad (10.8)$$

are tangent vectors at P, Q in the tangent bundle of $Gr(n + 1, r)$ for any vectors $\dot{x}_1, \ldots, \dot{x}_r$ and $\dot{y}_1, \ldots, \dot{y}_r$ that satisfy $(x_i, \dot{x}_j) + (\dot{x}_i, x_j) = 0$ and $(x_i, \dot{y}_j) + (\dot{x}_i, y_j) = 0$. The canonical symplectic form on the cotangent (tangent) bundle of $Gr(n + 1, r)$ is given by

$$\sum_{i=1}^{r} (\dot{z}_i, \dot{y}_i) - (\dot{x}_i, \dot{w}_i) \qquad (10.9)$$

for any tangent vectors $(\dot{x}_1, \ldots, \dot{x}_r, \dot{y}_1, \ldots, \dot{y}_r)$ and $(\dot{z}_1, \ldots, \dot{z}_r, \dot{w}_1, \ldots, \dot{w}_r)$ at (P, Q). The coadjoint symplectic form is given by

$$\omega_{P+Q}(V_1, V_2) = \langle L, [B_2, A_1] + [A_2, B_1] + [A_2, A_1] \rangle \qquad (10.10)$$

for any tangent vectors $V_i = [L, A_i + B_i], i = 1, 2$. To equate these two expressions, let $\dot{P} = [P, A]$ and $\dot{Q} = [Q, A] + [P, B]$ for some matrices A and B with A skew-symmetric and B symmetric. Then,

$$\sum_{i=1}^{r} x_i \otimes \dot{x}_i + \dot{x}_i \otimes x_i = \sum_{i=1}^{r}[x_i \otimes x_i, A] = \sum_{i=1}^{r} x_i \wedge Ax_i + Ax_i \wedge x_i$$

implies that

$$\dot{x}_i = Ax_i - \sum_{j=1}^{r}(Ax_j, x_i)x_j. \tag{10.11}$$

Similar calculation yields

$$\dot{y}_i = Ay_i - Bx_i + \sum_{j=1}^{r}(Bx_ix_j)x_j. \tag{10.12}$$

The following lemma will be useful for the calculations below.

Lemma 10.4 *Suppose that B is a symmetric and A a skew-symmetric matrix. Then,*

$$\langle x \otimes x, [A,B] \rangle = (Bx, Ax), \langle x \wedge y, [A_2, A_1] \rangle = (A_2 x, A_1 y) - (A_1 x, A_2 y).$$

Proof First note that $Tr(x \otimes y) = \sum_{i=0}^{n} x_i(y, e_i) = (x, y)$. Then,

$$\langle x \otimes x, [A,B] \rangle = \frac{1}{2} Tr(x \otimes [A,B]x)$$

$$= \frac{1}{2}(Tr(x \otimes BAx - x \otimes ABx)) = \frac{1}{2}((x, BAx) - (x, ABx))$$

$$= \frac{1}{2}((Bx, Ax) + (Ax, Bx)) = (Bx, Ax).$$

The second formula follows by an analogous argument.   □

With these notations behind us, let

$$\dot{x}_i = Ax_i - \sum_{j=1}^{r}(A_1 x_j, x_i)x_j, \dot{y}_i = A_1 y_i - B_1 x_i + \sum_{j=1}^{r}(Bx_ix_j)x_j,$$

$$\dot{z}_i = A_2 x_i - \sum_{j=1}^{r}(A_2 x_j, x_i)x_j, \dot{w}_i = A_2 y_i - B_2 x_i + \sum_{j=1}^{r}(Bx_ix_j)x_j$$

denote the tangent vectors at $P = \sum_{i=1}^{r} x_i \otimes x_i, Q = \sum_{i=1}^{r} x_i \wedge y_i$ defined by the matrices $B_1 + A_1$ and $B_2 + A_2$ in $S_{n+1} \oplus so_{n+1}(R)$. Then,

$$\omega_{P+Q}(V_1, V_2) = \langle L, [B_2, A_1] + [A_2, B_1] + [A_2, A_1] \rangle$$

$$= \sum_{i=1}^{r} \langle x_i \otimes x_i, [B_2, A_1] + [A_2, B_1] \rangle + \langle x_i \wedge y_i, [A_2, A_1] \rangle$$

$$= \sum_{i=1}^{r} (A_1 x_i, B_2 x_i) - (A_2 x_i, B_1 x_i) + (A_2 x_i, A_1 y_i) - (A_1 x_i, A_2 y_i)$$

$$= \sum_{i=1}^{r} (A_1 x_i, -\dot{w}_i + A_2 y_i + \sum_{j=1}^{r} (B_2 x_i, x_j) x_j) - (A_2 x_i, -\dot{y}_i + A_1 y_i$$

$$+ \sum_{j=1}^{r} (B_1 x_i, x_j) x_j) + (A_2 x_i, A_1 y_i) - (A_1 x_i, A_2 y_i)$$

$$= \sum_{i=1}^{r} (A_1 x_i, -\dot{w}_i) + (A_2 x_i, \dot{y}_i) = \sum_{i=1}^{r} (\dot{z}_i, \dot{y}_i) - (\dot{x}_i, \dot{w}_i),$$

because $\sum_{i,j=1}^{r}(B_1 x_i, x_j)(A_2 x_i, x_j) = \sum_{i,j=1}^{r}(B_2 x_i, x_j)(A_1 x_i, x_j) = 0$.

The calculation on the Stiefel manifold is quite analogous and will be omitted. $\square$

Corollary 10.5   *The generic coadjoint orbit of G_s through a non-singular symmetric matrix P_0 is equal to the tangent bundle of $SO_{n+1}(R)$ realized as the Stiefel manifold $St(n + 1, n + 1)$ while the coadjoint orbit through rank one symmetric matrix $P_0 = a \otimes a$ is equal to the tangent bundle of the sphere $S^n = \{x : ||x|| = ||a||\}$.*

Next consider analogous orbits through a point P in S_{p+q} defined by the pseudo Riemannian symmetric pair $(SL_{p+q}(R), SO(p, q))$. In contrast to the Riemannian case, where each symmetric matrix admits an orthonormal basis of eigenvectors, matrices P in $S_{p,q}$ may have complex eigenvalues, which accounts for different types.

The hyperbolic rank one matrices are matrices of the form $x \otimes Dx$ for some vector $x \in \mathbb{R}^{n+1}$, where D is the diagonal matrix with its first p diagonal entries equal to 1, and the remaining q diagonal entries equal to -1. These matrices will be denoted by $x \otimes_{p,q} x$, while $x \wedge_{p,q} y$ will denote the matrix $x \otimes_{p,q} y - y \otimes_{p,q} x$.

It follows that $(x \otimes x)_{p,q} u = (x, u)_{p,q} x$, and therefore,

$$(x \otimes_{p,q} x) u, v)_{p,q} = (x, u)_{p,q}(x, v)_{p,q} = (u, (x \otimes_{p,q} x) v)_{p,q}.$$

Since the trace of $(x \otimes_{p,q} x)$ is equal to $(x, x)_{p,q}$, $P = (x \otimes_{p,q} x) - \frac{1}{n+1}(x, x)_{p,q} I$ is in $S_{p,q}$. A similar calculation shows that $Q = x \wedge_{p,q} y$ belongs to $so(p, q)$.

The coadjoint orbits relative to $G_{p,q} = S_{p,q} \rtimes SO(p,q)$ can be identified as the tangent bundles of the "pseudo-Stiefel" and "pseudo-Grassmannians" in a manner similar to the Riemannian setting. We will not pursue these investigations in any detail here, except to note the extreme cases, the coadjoint orbit through $P_0 = a \otimes_{p,q} a$ and the coadjoint orbit through $P_0 = \sum_{i=1}^{n+1} \lambda_i a_i \otimes_{p,q} a_i, \lambda_i \neq \lambda_j, i \neq j$. As in the Riemannian case, we will omit the zero-trace requirement.

Proposition 10.6   *The coadjoint orbit of the semi-direct product $G_{p,q} = S_{p,q} \rtimes SO(p,q)$ through $P_0 = a \otimes_{p,q} a$ is equal to $\{x \otimes_{p,q} x + x \wedge_{p,q} y : (x,x)_{p,q} = (a,a)_{p,q}, (x,y)_{p,q} = 0\}$. The latter is symplectomorphic to the cotangent bundle of the hyperboloid $\mathbb{H}^{p,q} = \{(x,y) : (x,x)_{p,q} = (a,a)_{p,q}, (x,y)_{p,q} = 0\}$. The symplectic form on the coadjoint orbit coincides with the canonical symplectic form ω on $\mathbb{H}^{p,q}$ given by*

$$\omega_{x,y}((\dot{x}_1, \dot{y}_{p,q}), (\dot{x}_2, \dot{y}_2)) = (\dot{x}_1, \dot{y}_2)_{p,q} - (\dot{x}_2, \dot{y}_1)_{p,q},$$

where $(\dot{x}_1, \dot{y}_1)$ and $(\dot{x}_2, \dot{y}_2)$ denote tangent vectors at (x,y).

Proposition 10.7   *Let $P_0 = \sum_{i=1}^{p} \lambda_i (a_i \otimes_{p,q} a_i) + \sum_{i=1}^{q} \mu_i (b_i \otimes_{p,q} b_i)$ for vectors $a_1, \ldots, a_p, b_1, \ldots, b_q$ that satisfy $(a_i, a_j)_{p,q} = \delta_{ij}, (a_i, b_j)_{p,q} = 0, (b_i, b_j)_{p,q} = -\delta_{ij}$ and some distinct numbers $\lambda_1, \ldots, \lambda_p$ and $\mu_1, \ldots, \mu_q$. The coadjoint orbit of the semi-direct product $G_{p,q} = S_{p,q} \rtimes SO(p,q)$ through P_0 is symplectomorphic to the cotangent bundle of $SO(p,q)$.*

The proofs are similar to the proofs in Propositions 10.2 and 10.3 and will be omitted. The case $p = 1, q = n$ is particularly interesting for the applications, for then the hyperboloid $\mathbb{H}^{1,n}$ defined by a vector a that satisfies $(a,a)_{1,n} = 1$ coincides with the standard hyperboloid $\mathbb{H}^n = \left\{ x \in \mathbb{R}^{n+1} : x_1^2 - \sum_{i=2}^{n+1} x_i^2 = 1 \right\}$. Its realization as a coadjoint orbit allows for conceptually clear explanations of the mysterious connections between the problem of Kepler and the geodesics of space forms [Ms1; Os].

This connection brings us to an important general remark about the modifications of the affine problem in the case that the quadratic form $\langle , \rangle$ is indefinite on $\mathfrak{k}$, as is on $\mathfrak{k} = so(p,q)$ where the Killing form is indefinite. In these situations the problem of minimizing $\frac{1}{2} \int_0^T \langle u, u \rangle \, dt$ over the trajectories of $\frac{dg}{dt} = g(t)((A + u(t))$ in $G_{p,q}$ with the controls $u(t)$ in $so(p,q)$ is not well defined since the cost is not bounded below. Of course, the affine system is still controllable for any regular drift A. So any pair of points in $G_{p,q}$ can be connected by a trajectory in some finite time T, but a priori there is no guarantee that the infimum of the cost over such trajectories will be finite.

Nevertheless, the affine Hamiltonian $H = \frac{1}{2}\langle L_{\mathfrak{k}}, L_{\mathfrak{k}} \rangle + \langle A, L_{\mathfrak{p}} \rangle$ is well defined and corresponds to the critical value of the cost, rather than to its minimum. That is to say, the endpoint map $u \rightarrow \left(\frac{1}{2} \int_0^T \langle u(t), u(t) \rangle \right) dt, g(T))$ is singular at the contol $u(t) = L_{\mathfrak{k}}(t)$ associated with an integral curve $(g(t), L(t))$ of the Hamiltonian vector field $\vec{H}$.

As we have already remarked above, the semi-simple adjoint orbits are also the cotangent bundles. For the sake of comparison with the sphere, the proposition below describes such an orbit through rank one matrices.

Proposition 10.8   *The adjoint orbit through $P_0 = a \otimes a - \frac{1}{n+1}I$, $(a, a) = 1$, is symplectomorphic to the cotangent bundle of the real projective space $\mathbb{P}^{n+1}$.*

Proof   Let S denote the orbit $gP_0g^{-1}, g \in SL_{n+1}(R)$. If $R \in SO_{n+1}(R)$ then $RP_0R^{-1} = Ra \otimes Ra - \frac{1}{n+1}I$. Since $SO_{n+1}(R)$ acts transitively on the unit sphere in $\mathbb{R}^{n+1}$, a can be replaced by e_0.

It follows that $g(e_0 \otimes e_0 - \frac{1}{n+1}I)g^{-1} = g(e_0 \otimes e_0)g^{-1} - \frac{1}{n+1}I$, hence S is diffeomorphic to the orbit $\{g(e_0 \otimes e_0))g^{-1} : g \in SL_{n+1}(R)\}$. The reader can readily verify that

$$g(e_0 \otimes e_0)g^{-1} = ge_0 \otimes (g^T)^{-1}e_0.$$

This relation shows that the the Euclidean inner product $x \cdot y$ is equal to 1 whenever $x = ge_0$ and $y = (g^T)^{-1}e_0$. Hence the orbit $\{g(e_0 \otimes e_0))g^{-1} : g \in SL_{n+1}(R)\}$ is confined the set $\{x \otimes y : x \in \mathbb{R}^{n+1}, y \in \mathbb{R}^{n+1}, (x \cdot y) = 1\}$. But these sets are equal since every pair of points x and y such that $x \cdot y = 1$ can be realized as $x = ge_0$ and $y = g^{T-1}e_0$ for some matrix $g \in SL_{n+1}(R)$. Then the set of matrices $\{x \otimes y : x \cdot y = 1\}$ can be identified with the set of lines $\{\alpha x, \frac{1}{\alpha}y), x \cdot y = 1\}$, and the later set is diffeomorphic to the cotangent bundle of the real projective space $\mathbb{P}^{n+1}$. $\square$

10.2 Canonical affine Hamiltonians on rank one orbits: Kepler and Newmann

We will now return to the canonical affine Hamiltonian $H = \frac{1}{2}\langle L_{\mathfrak{h}}, L_{\mathfrak{h}} \rangle + \langle A, L_{\mathfrak{p}} \rangle$ and consider its restrictions to the coadjoint orbits that correspond to the cotangent bundles of spheres and hyperboloids. This means that $L_{\mathfrak{p}_\epsilon}$ is restricted to matrices $x \otimes_\epsilon x - \frac{\|x\|_\epsilon^2}{n+1}I$, where $\|x\|_\epsilon^2 = (x, x)_\epsilon$, and $L_{\mathfrak{h}_\epsilon}$ is restricted to matrices $x \wedge_\epsilon y$, $(x, y)_\epsilon = 0$. Then,

$$\langle A, L_{\mathfrak{p}_\epsilon} \rangle = -\frac{1}{2} Tr \left(A \left(x \otimes_\epsilon x - \frac{\|x\|_\epsilon^2}{n+1} I \right) \right) = -\frac{1}{2} Tr(A(x \otimes_\epsilon x)) = -\frac{1}{2}(Ax, x)_\epsilon$$

$$\langle L_{\mathfrak{h}_\epsilon}, L_{\mathfrak{h}_\epsilon} \rangle = -\frac{1}{2} Tr(x \wedge_\epsilon y)^2 = Tr(\|x\|_\epsilon^2 y \otimes_\epsilon y + \|y\|_\epsilon^2 x \otimes_\epsilon x) = \|x\|_\epsilon^2 \|y\|_\epsilon^2.$$

Hence H reduces to

$$H = \frac{1}{2} \|x\|_\epsilon^2 \|y\|_\epsilon^2 - \frac{1}{2}(Ax, x)_\epsilon. \tag{10.13}$$

The Hamiltonian equations

$$\frac{dL_{\mathfrak{h}_\epsilon}}{dt} = [A, L_{\mathfrak{p}_\epsilon}], \quad \frac{dL_{\mathfrak{p}_\epsilon}}{dt} = [L_{\mathfrak{h}_\epsilon}, L_{\mathfrak{p}_\epsilon}]$$

reduce to

$$\frac{d}{dt} \left(x \otimes_\epsilon x - \frac{\|x\|_\epsilon^2}{n+1} I \right) = \|x\|_\epsilon^2 (x \otimes_\epsilon y + y \otimes_\epsilon x), \quad \frac{d}{dt}(x \wedge_\epsilon y) = x \wedge_\epsilon Ax. \tag{10.14}$$

It follows that

$$\dot{x} \otimes_\epsilon x + x \otimes_\epsilon \dot{x} - \frac{2\|x\|_\epsilon (x, \dot{x})_\epsilon}{n+1} I = \|x\|_\epsilon^2 (x \otimes_\epsilon y + y \otimes_\epsilon x), \quad \dot{x} \wedge_\epsilon y + x \wedge_\epsilon \dot{y}$$

$$= x \wedge_\epsilon Ax.$$

On the "sphere" $\|x\|_\epsilon^2 = h^2$, $(x, \dot{x})_\epsilon = 0$, and $(x, \dot{y})_\epsilon = -(\dot{x}, y)_\epsilon$, and

$$\left(\dot{x} \otimes_\epsilon x + x \otimes_\epsilon \dot{x} - \frac{2\|x\|_\epsilon (x, \dot{x})_\epsilon}{n+1} I \right) x = (\|x\|_\epsilon^2 (x \otimes_\epsilon y + y \otimes_\epsilon x))x.$$

Therefore $\dot{x} = h^2 y$. The remaining equation $(\dot{x} \wedge_\epsilon y + x \wedge_\epsilon \dot{y})x = (x \wedge_\epsilon Ax)x$ yields $\dot{y} = Ax - \frac{1}{h^2}((Ax, x)_\epsilon + h^2 \|y\|_\epsilon^2)x$. It follows that equations 10.14 reduce to

$$\dot{x} = h^2 y, \dot{y} = Ax - \frac{1}{h^2}((Ax, x)_\epsilon + h^2 \|y\|_\epsilon^2)x. \tag{10.15}$$

If we replace A by $-A$ and take $h = 1$ then equations (10.15) become

$$\dot{x} = y, \dot{y} = -Ax + ((Ax, x)_\epsilon - \|y\|_\epsilon^2)x. \tag{10.16}$$

Equations (10.16) with $\epsilon = 1$ correspond to the Hamiltonian equations for the motions of a particle on the sphere S^n under a quadratic potential $\frac{1}{2}(Ax, x)$ [Ms4]. This system originated with C. Newmann in 1856 [Nm] and has been known ever since as the Newmann system on the sphere. It is a point of departure for J. Moser in his book on integrable Hamiltonian systems [Ms2].

We will presently show that all integrals of motion found in Moser's book are simple consequences of the spectral representation discussed in the previous chapter. For $\epsilon = -1$ equations (10.16) coincide with the Hamiltonian equations on the hyperboloid $\mathbb{H}^n$ for the motion of a particle on $\mathbb{H}^n$ in the presence of a quadratic potential $(Ax, x)_{-1}$ and the same can be said for the mechanical system on the projective space on the semi-simple orbit. Of course, the integration procedure in all cases is essentially the same, as will soon be made clear.

Remarkably, even the degenerate case $A = 0$ sheds light on mechanical systems; we will presently show that it provides natural connections between Kepler's system and the Riemannian geodesics on spaces of constant curvature [Ms2; O2].

10.3 Degenerate case and Kepler's problem

When $A = 0$, the Hamiltonian H reduces to $H = \frac{1}{2}||x||_\epsilon^2 ||y||_\epsilon^2$, and equations (10.15) reduce to

$$\dot{x} = ||x||_\epsilon^2 y, \ \dot{y} = -||y||_\epsilon^2 x. \tag{10.17}$$

It will be convenient to revert to the notations employed earlier, and use SO_ϵ to denote the group $SO_{n+1}(R)$ when $\epsilon = 1$, and $SO(1, n)$ when $\epsilon = -1$. It is clear that our Hamiltonian H is invariant under SO_ϵ. The associated moment map Λ is given by

$$\Lambda = x \wedge_\epsilon y. \tag{10.18}$$

Evidently, Λ is constant along the solutions of (10.17). The solutions of (10.17) satisfy

$$\ddot{x} + ||x||_\epsilon^2 ||y||_\epsilon^2 x = 0.$$

On any sphere $||x||_\epsilon^2 = ||a||_\epsilon^2$ and energy level $H = h^2$, the solutions are given by

$$x(t) = a \cos th + b \sin th,$$

with a and b constant vectors that satisfy $||a||_\epsilon^2 = ||b||_\epsilon^2 = h^2$, $(a, b)_\epsilon = 0$. When $h^2 ||a||_\epsilon^2 = 1$, then $||\frac{dx}{dt}||_\epsilon^2 = 1$ and $\frac{D_{x(t)}}{dt}\left(\frac{dx}{dt}\right) = 0$, therefore $x(t)$ is a geodesic on the sphere $||x||_\epsilon^2 = ||a||_\epsilon^2$. It follows that each solution curve traces a geodesic circle with speed h on either the sphere $||x||^2 = ||a||^2$ when $\epsilon = 1$ or the hyperboloid $||x||_{-1}^2 = ||a||_{-1}^2$ when $\epsilon = -1$. In the latter case the geodesic is confined to the region

$$\left(\frac{dx_0}{dt}\right)^2 - \sum_{i=1}^n \left(\frac{dx_i}{dt}\right)^2 > 0.$$

In the complementary open region, $H = -h^2$, the solutions are given by

$$x(t) = a \cosh ht + b \sinh ht.$$

These solutions are also geodesics on each hyperboloid $||x||_\epsilon^2 = ||a||_\epsilon^2$, but now they trace hyperbolas with speed h. The remaining case, $H = 0$, generates horocycles on each hyperboloid $||x||_\epsilon^2 = ||a||_\epsilon^2$. These solutions are given by

$$x(t) = at + b.$$

10.3.1 Kepler's problem

Let us now digress briefly into Kepler's problem. Kepler's problem concerns the motion of a planet around an immovable planet in the presence of the gravitational force. On the basis of the empirical evidence, Kepler made two fundamental observations about the motion of the moon around the earth, known today as Kepler's laws. The first law of Kepler states that the moon moves on an elliptic orbit around the earth, and the second law of Kepler states that the position vector of the moving planet relative to the stationary planet sweeps out equal areas in equal time intervals as the planet orbits around the stationary planet.

The main motivation for this diversion, however, is not so much driven by this classic knowledge but rather by a remarkable discovery that the solutions of Kepler's problem are intimately related to the geometry of spaces of constant curvature. This discovery goes back to A. V. Fock's paper of 1935 [Fk] in which he reported that the symmetry group for the motions of the hydrogen atom is $O_4(R)$ for negative energy, $E^3 \rtimes O_3(R)$ for zero energy, and $O(1, 3)$ for positive energy. It is then not altogether surprising that similar results apply to the problem of Kepler, since the energy function for Kepler's problem is formally the same as the energy function for the hydrogen atom.

This connection between the problem of Kepler and the geodesics on the sphere was reported by J. Moser in 1970 [Ms1], and even earlier, by G. Györgyi in 1968 [Gy], while similar results for positive energies and the geodesics on spaces of negative constant curvature were reported later by Y. Osipov [O1; O2]. As brilliant as these contributions were, they, nevertheless, did not offer any explanations for these enigmatic connections between planetary motions and geodesics on space forms. This enigma later inspired V. Guillemin and S. Sternberg to take up the problem of Kepler in a larger geometric context with Moser's observation as the focal point for this work [GS], but this attempt did not give any clues for the basic mystery. Our aim is to show that the affine Hamiltonian lifts this mystery behind the Kepler's problem.

Let us first recall the essential facts. It was not until Newton that a mathematical foundation was laid out from which Kepler's observations could be deduced mathematically. According to Newton, the gravitational force of attraction exerted on a planet of mass m by another planet of mass M is given by

$$F = -kMm\frac{q}{\|q\|^3},$$

where k denotes the gravitational constant and q denotes the position vector of the first planet relative to the second planet. This formulation of the force presupposes that the planets are immersed in a three-dimensional Euclidean space $\mathbb{E}^3$ with an origin O so that the position vector q is equal to the difference of the position vectors $\vec{OP_2} - \vec{OP_1}$, and that the distance between the planets is expressed by the usual formula

$$\|q\|^2 = \|\vec{OP_2} - \vec{OP_1}\|^2.$$

For the problem of Kepler one of the planets is assumed stationary, in which case the origin of $\mathbb{E}^3$ is placed at its center. Then $q = \vec{OP}$, where $\vec{OP}$ denotes the position vector of the moving planet, and the movements of the planet are governed by the equation

$$\frac{d^2q}{dt^2} = -kmM\frac{q}{\|q\|^3}, \tag{10.19}$$

according to Newton's second law of motion.

Equation (10.19) can be expressed in Hamiltonian form

$$\frac{dq}{dt} = mp, \quad \frac{dp}{dt} = -kM\frac{q}{\|q\|^3}. \tag{10.20}$$

In the Hamiltonian formalism the gravitational force is replaced by the potential energy $V(q) = -kM\frac{q}{\|q\|}$ and the equations of motion are generated by the energy function $E(q,p) = \frac{1}{2}m^2\|p\|^2 + V(q)$, in the sense that equation (10.2) coincides with

$$\frac{dq}{dt} = \frac{\partial E}{\partial p}, \quad \frac{dp}{dt} = -\frac{\partial E}{\partial q}.$$

The solutions of the above equations are easily described in terms of the energy E and the angular momentum $L = q \times p$, both of which are constant along each solution of equation (10.2). The simplest case occurs when $L = 0$. Then the initial position $q(0)$ is colinear with the initial momentum $p(0)$. In such a case the solutions of (10.2) remain on the line defined by these initial conditions.

The angular momentum is not equal to zero whenever the initial conditions are not colinear. Then each solution $(q(t), p(t))$ of (10.2) is confined to the plane spanned by the initial vectors $q(0)$ and $p(0)$, a consequence of the conservation of the angular momentum, and traces a conic in this plane; a hyperbola when $E > 0$, a parabola when $E = 0$, and an ellipse when $E < 0$.

To make our earlier claim more compelling, we will consider the n-dimensional Kepler problem given by the Hamiltonian

$$E = \frac{1}{2}\|p\|^2 - \frac{1}{\|q\|}$$

in the phase space $\mathbb{R}^n/0 \times \mathbb{R}^n$ corresonding to the normalized constants $m = kM = 1$, and the associated differential equations

$$\frac{dq}{dt} = p, \frac{dp}{dt} = -\frac{1}{\|q\|^3}q. \tag{10.21}$$

Lemma 10.9   *Each of $L = q \wedge p$ and $F = Lp - \frac{q}{\|q\|}$ is constant along any solution of (10.21).*

The proof is simple and will be omitted.

It follows that L is a direct generalization of the angular momentum since

$$x \times (q \times p) = (x \cdot p)q - (x \cdot q)p = L(x).$$

The other conserved quantity F is called *the Runge–Lenz vector*, or *the eccentricity vector*. The length of F is called the *eccentricity* for reasons that will become clear below (see also [An] for more historical details).

Lemma 10.10   *Let $\|L\|^2 = -\frac{1}{2}Tr(L^2)$ denote the standard trace metric on $so_n(R)$. Then,*

(a) $\|L\|^2 = \|q\|^2\|p\|^2 - (q \cdot p)^2$.
(b) $\|F\|^2 = 2\|L\|^2E + 1$.

Proof   Let $e_1, \ldots, e_n$ denote the standard basis in $\mathbb{R}^n$. Then

$$\|L\|^2 = -\frac{1}{2}\sum_{i=1}^n e_i \cdot L^2 e_i$$

$$= -\frac{1}{2}\sum_{i=1}^n e_i \cdot ((p_i((q \cdot p)q - \|q\|^2p) - q_i(\|p\|^2q - (q \cdot p)p))$$

$$= \|q\|^2\|p\|^2 - (q \cdot p)^2.$$

To prove (b), let $e = \frac{1}{\|q\|}q$. Then $F = \|p\|^2q - (p \cdot q)p - e = (\|p\|^2\|q\| - 1)e - (p \cdot q)p$. It follows that

$$F \cdot e = \frac{\|L\|^2}{\|q\|} - 1 \text{ and } F \cdot p = -e \cdot p.$$

Then,

$$\|F\|^2 = F \cdot F = (\|p\|^2\|q\| - 1)F \cdot e - (p \cdot q)F \cdot p$$

$$= (\|p\|^2\|q\| - 1)\left(\frac{\|L\|^2}{\|q\|} - 1\right) + (p \cdot q)e \cdot p$$

$$= (\|p\|^2\|q\| - 1)\frac{\|L\|^2}{\|q\|} + \frac{1}{\|q\|}(-\|L\|^2 + \|p\|^2\|q\|^2) - (\|p\|^2\|q\| - 1)$$

$$= (\|p\|^2\|q\| - 2)\frac{\|L\|^2}{\|q\|} + 1 = 2E\|L\|^2 + 1.$$

$\square$

Proposition 10.11 *Let* $(q(t), p(t))$ *denote the solution of (10.21) that originates at* (q_0, p_0) *at* $t = 0$. *Then* $q(t)$ *evolves on a line through the origin if and only if* q_0 *and* p_0 *are colinear, that is, whenever* $L = 0$.

In the case that q_0 *and* p_0 *are not colinear, then* $q(t)$ *remains in the plane* P *spanned by* q_0 *and* p_0, *where it traces an ellipse when* $E(q_0, p_0) < 0$, *a parabola when* $E(q_0, p_0) = 0$ *and a hyperbola when* $E(q_0, p_0) > 0$.

Proof If q_0 and p_0 are colinear then $L = 0$ and therefore $q(t)$ and $p(t)$ are colinear. But then $F = -e$, and hence, e is constant. To prove the converse reverse the steps.

Assume now that q_0 and p_0 are not colinear and let P denote the plane spanned by q_0 and p_0. Since L is constant, $q(t)$ and $p(t)$ are not colinear for all t. It follows that for any x in the kernel of L both $x \cdot q(t)$ and $x \cdot p(t)$ are equal to zero. But this implies that both $q(t)$ and $p(t)$ are in P for all t.

Evidently the Runge–Lenz vector is in P. Let ϕ denote the angle between F and q. Then,

$$\|L\|^2 - \|q\| = F \cdot q = \|F\|\|q\| \cos\phi,$$

and therefore

$$\|q\| = \frac{\|L\|^2}{1 + \|F\| \cos\phi}. \tag{10.22}$$

It follows that $q(t)$ traces an ellipse when $\|F\| < 1$, a parabola when $\|F\| = 1$, and a hyperbola when $\|F\| > 1$. Since $\|F\|^2 = 2\|L\|^2 E + 1$, $E < 0$ if and only if $\|F\| < 1$, $E = 0$ if and only if $\|F\| = 1$, and $E > 1$ if and only if $\|F\| > 1$. $\square$

With this background at our disposal let us now connect Kepler to the affine Hamiltonians. Let us first establish some additional notations that will facilitate the transition to equations (10.21). We will use $S_\epsilon(h)$ to denote the Euclidean sphere $||x||^2 = h^2$ for $\epsilon > 0$ and the hyperboloid $(x,x)_\epsilon = h^2, x_0 > 0$ when $\epsilon < 0$. The cotangent bundle of $S_\epsilon(h)$ will be identified with the tangent bundle via the quadratic form $(\,,\,)_\epsilon$. Points of the tangent bundle will be represented by the pairs $(x,y) \in \mathbb{R}^{n+1} \times \mathbb{R}^{n+1}$ that satisfy $||x||_\epsilon = h^2$ and $(x,y)_\epsilon = 0$. For each $x \in S_\epsilon(h)$ the stereographic projection p in $\mathbb{R}^n$ is given by

$$\lambda(x - he_0) + he_0 = (0,p) \text{ with } \lambda = \frac{h}{h - x_0}.$$

The inverse map $x = \Phi_\epsilon(p)$ is given by

$$x_0 = \frac{h(||p||^2 - \varepsilon h^2)}{||p||^2 + \varepsilon h^2}, \text{ and } \bar{x} = (x_1, \ldots, x_n) = \frac{2\epsilon h^2}{||p||^2 + \varepsilon h^2}p. \qquad (10.23)$$

It follows that $\mathbb{R}^n \cup \{\infty\}$ is the range under the stereographic projection of $S_\epsilon(h)$ in the Euclidean case and the annulus $\{p \in \mathbb{R}^n : ||p||^2 > h^2\} \cup \{\infty\}$ in the hyperbolic case.

Let $\left(\frac{\partial \Phi_\epsilon}{\partial p}\right)$ denote the Jacobian matrix with the entries $\left(\frac{\partial x_i}{\partial p_j}\right), i = 0, \ldots, n, j = 1, \ldots, n$. Then,

$$dx = \left(\frac{\partial \Phi_\epsilon}{\partial p}\right) dp =$$

$$\left(\frac{4\varepsilon h^3}{(||p||^2 + \epsilon h^2)^2}p \cdot dp, \frac{2\epsilon h^2}{||p||^2 + \varepsilon h^2}dp - \frac{4\epsilon h^2 p \cdot dp}{(||p||^2 + \epsilon h^2)^2}p\right), \qquad (10.24)$$

that is,

$$dx_0 = \frac{4\varepsilon h^3}{(||p||^2 + \varepsilon h^2)^2}p \cdot dp \text{ and } d\bar{x} = \frac{2\epsilon h^2}{||p||^2 + \varepsilon h^2}dp - \frac{4\epsilon h^2 p \cdot dp}{(||p||^2 + \varepsilon h^2)^2}p.$$

Assume that the cotangent bundle of $\mathbb{R}^n$ is identified with its tangent bundle $\mathbb{R}^n \times \mathbb{R}^n$ via the Euclidean inner product $(\cdot)$, and let (p,q) denote the points of $\mathbb{R}^n \times \mathbb{R}^n$. Let $q = \Psi(x,y)$ denote the mapping such that

$$(dx, y)_\epsilon = (dp \cdot \Psi(x,y)), \qquad (10.25)$$

for all (x,y) with $x \in S_\epsilon$ and $(x,y)_\epsilon = 0$. It follows that

$$\sum_{j=1}^{n} y_0 \frac{\partial x_0}{\partial p_j}dp_j + \epsilon \sum_{i=1}^{n}\sum_{j=1}^{n} y_i \frac{\partial x_i}{\partial p_j}dp_j = \sum_{j=1}^{n}\left(y_0 \frac{\partial x_0}{\partial p_j} + \epsilon \sum_{i=1}^{n} y_i \frac{\partial x_i}{\partial p_j}\right)dp_j$$

$$= \sum_{j=1}^{n} q_j dp_j = q \cdot dp.$$

Therefore,

$$q_j = \frac{\partial x_0}{\partial p_j} y_0 + \epsilon \sum_{i=1}^{n} y_i \frac{\partial x_i}{\partial p_j}, j = 1, \ldots, n. \tag{10.26}$$

Then (10.24) yields

$$q = \frac{2h^2}{||p||^2 + \epsilon h^2} \left(\frac{2\epsilon h y_0}{||p||^2 + \epsilon h^2} p + \bar{y} - \frac{2(\bar{y} \cdot p)}{||p||^2 + \epsilon h^2} p \right), \bar{y} = (y_1, \ldots, y_n). \tag{10.27}$$

Hence,

$$\frac{||p||^2 + \epsilon h^2}{2h^2}(q \cdot p) = \frac{2\epsilon h y_0}{||p||^2|| + \epsilon h^2}||p||^2 - \frac{||p||^2 - \epsilon h^2}{||p||^2 + \epsilon h^2}(\bar{y} \cdot p). \tag{10.28}$$

Since y is orthogonal to x, $(\bar{y} \cdot p) = -\frac{y_0}{2h}(||p||^2 - \epsilon h^2)$. Therefore,

$$y_0 = \frac{1}{h} q \cdot p, \bar{y} = \frac{||p||^2 + \varepsilon h^2}{2h^2} q - \frac{q \cdot p}{h^2} p. \tag{10.29}$$

The transformation $(p, q) \in \mathbb{R}^n \times \mathbb{R}^n \to (x, y) \in TS_\epsilon(h)$ is a symplectomorphism since it pulls back the Liouville form $(dx, y)_\epsilon$ on $S_\epsilon(h)$ onto the Liouville form $(dp \cdot q)$ in $\mathbb{R}^n$ (symplectic forms are the exterior derivative of the Liouville forms). It then follows from (10.24) and (10.29) that

$$||dx||_\epsilon^2 \frac{4h^4 \varepsilon}{(||p||^2 + \varepsilon h^2)^2}||dp||^2, ||y||_\epsilon^2 = \varepsilon \frac{(||p||^2 + \varepsilon h^2)^2}{4h^4}||q||^2. \tag{10.30}$$

To pass to the problem of Kepler, write the Hamiltonian $H = \frac{1}{2}||x||_\epsilon^2||y||_\epsilon^2$ in the variables (p, q). It follows that

$$H = \frac{1}{2}h^2\varepsilon \frac{(||p||^2 + \varepsilon h^2)^2}{4h^4}||q||^2 = \frac{1}{2}\varepsilon \frac{(||p||^2 + \varepsilon h^2)^2}{4h^2}||q||^2.$$

The corresponding flow is given by

$$\frac{dp}{ds} = \frac{\partial H}{\partial q} = \varepsilon \frac{(||p||^2 + \varepsilon h^2)^2}{4h^2}q, \frac{dq}{ds} = -\frac{\partial H}{\partial p} = -\varepsilon \frac{||p||^2 + \varepsilon h^2}{2h^2}||q||^2 p.$$

On energy level $H = \frac{\varepsilon}{2h^2}$, $\frac{(||p||^2 + \varepsilon h^2)^2}{4}||q||^2 = 1$ and the preceding equations reduce to

$$\frac{dp}{ds} = \varepsilon \frac{q}{h^2||q||^2}, \frac{dq}{ds} = -\varepsilon \frac{||q||}{h^2}p. \tag{10.31}$$

After the reparametrization $t = -\frac{\varepsilon}{h^2} \int_0^s ||q(\tau)||d\tau$, equations (10.31) become

$$\frac{dp}{dt} = \frac{dp}{ds}\frac{ds}{dt} = -\frac{q}{||q||^3}, \frac{dq}{dt} = \frac{dq}{ds}\frac{ds}{dt} = p.$$

Since $\frac{(\|p\|^2 + \varepsilon h^2)^2}{4}\|q\|^2 = 1$,

$$E = \frac{1}{2}\|p\|^2 - \frac{1}{\|q\|} = \frac{1}{2\|q\|}(\|p\|^2\|q\| - 2)$$

$$= \frac{1}{2\|q\|}(2 - \varepsilon h^2\|q\| - 2) = -\frac{1}{2}\varepsilon h^2. \tag{10.32}$$

So $E < 0$ in the spherical case, and $E > 0$ in the hyperbolic case.

The Euclidean case $E = 0$ can be obtained by a limiting argument in which ε is regarded as a continuous parameter which tends to zero.

To explain in more detail, let $w_0 = \lim_{\epsilon \to 0} x_0$ and $w = \lim_{\epsilon \to 0} \frac{1}{2\epsilon h^2}\bar{x}$. It follows that (10.23) and (10.24) have limiting values

$$w_0 = h, \ w = \frac{1}{\|p\|^2}p, \ dw_0 = 0, dw = \frac{1}{\|p\|^2}dp - 2\frac{p \cdot dp}{\|p\|^4}p. \tag{10.33}$$

The transformation $p \to w$ with $w = \frac{1}{\|p\|^2}p$ is the inversion about the circle $\|p\|^2 = 1$ in the affine hyperplane $w_0 = h$, and $\|dw\|^2 = \frac{1}{\|p\|^4}\|dp\|^2$ is the corresponding transformation of the Euclidean metric $\|dp\|^2$. The Hamiltonian H_0 associated with this metric is equal to $\frac{1}{2}\frac{\|p\|^4}{4}\|q\|^2$. This Hamiltonian can also be obtained as the limit of $(\frac{h^2}{\epsilon})\frac{1}{2}\frac{(\|p\|^2 + \epsilon h^2)^2}{4h^2}\|q\|^2$ when $\epsilon \to 0$. On energy level $H = \frac{1}{2}$, $\|p\|^2\|q\| = 2$, and therefore $E = 0$ by the calculation in (10.32). Of course, the solutions of (10.17) tend to the Euclidean geodesics as ϵ tends to zero. Consequently, $w(t) = \lim_{\epsilon \to 0} \frac{1}{2h^2\epsilon}(\bar{x}(t))$ is a solution of $\frac{d^2w}{dt^2} = 0$, and hence, is a geodesic corresponding to the standard Euclidean metric.

The angular momentum $L = q \wedge p$ and the Runge–Lenz vector $F = Lp - \frac{q}{\|q\|}$ for Kepler's problem correspond to the moment map Λ (10.18) according to the following proposition.

Proposition 10.12 *Let* $x = x_0 e_0 + \bar{x}$ *and* $y = y_0 e_0 + \bar{y}$. *On energy level* $H = \frac{\epsilon}{2h^2}$,

$$L = (\bar{y} \wedge_\epsilon \bar{x}) \text{ and } F = h(y_0(e_0 \wedge \bar{x})_\varepsilon - x_0(e_0 \wedge \bar{y})_\varepsilon)e_0.$$

Proof If we identify $\bar{x}$ and $\bar{y}$ with their projections on $\mathbb{R}^n$, then

$$\bar{x} = \frac{2\epsilon h^2}{\|p\|^2 + \epsilon h^2}p, \ \bar{y} = \frac{\|p\|^2 + \epsilon h^2}{2h^2}q - \frac{(q \cdot p)}{h}p$$

according to (10.23) and (10.29). Therefore,

$$(\bar{y} \wedge_\epsilon \bar{x}) = \epsilon\left(\frac{\|p\|^2 + \epsilon h^2}{2h^2}q - \frac{(q \cdot p)}{h}p \wedge \frac{2\epsilon h^2}{\|p\|^2 + \epsilon h^2}p\right) = q \wedge p = L.$$

Then,

$$h(y_0(e_0 \wedge \bar{x})_\varepsilon - x_0(e_0 \wedge \bar{y})_\varepsilon)e_0 = h(x_0\bar{y} - y_0\bar{x})$$

$$= -h\left(\frac{1}{h}q \cdot p\right)\frac{2\varepsilon h^2}{||p||^2 + \varepsilon h^2}p + \left(\frac{h^2(||p||^2 - \varepsilon h^2)}{||p||^2 + \varepsilon h^2}\right)$$

$$\times \left(\frac{||p||^2 + \varepsilon h^2}{2h^2}q - \frac{q \cdot p}{h^2}p\right)$$

$$= -(q \cdot p)p + \frac{||p||^2 - \varepsilon h^2}{2}q = (q \wedge p)p - ||p||^2 q + \frac{||p||^2 - \varepsilon h^2}{2}q$$

$$= (q \wedge p)p - \frac{||p||^2 + \epsilon h^2}{2}q.$$

The preceding expression reduces to F on $H = \frac{\epsilon}{2h^2}$ because $\frac{(||p||^2 + \varepsilon h^2)}{2}||q|| = 1$. $\square$

To make the correspondence complete, let us mention that the stereographic projections of the great circles on the sphere trace ellipses in the plane spanned by the initial data q_0 and p_0, while the projections of the hyperbolas on the hyperboloid trace hyperbolas in the same plane. The Euclidean geodesics trace parabolas. We will leave these details to the reader.

10.4 Mechanical problem of C. Newmann

Let us now return to the full affine Hamiltonian $H = \frac{1}{2}\langle L_{\mathfrak{h}_\epsilon}, L_{\mathfrak{h}_\epsilon}\rangle + \langle A, L_{\mathfrak{p}_\epsilon}\rangle$ with A a regular element in $\mathfrak{p}_\epsilon$. When $\epsilon = 1$ then $\mathfrak{p}_\epsilon$ is equal to the space symmetric matrices relative to the Euclidean inner product. Every symmetric matrix is conjugate to a diagonal matrix with real entries. Then A is regular if and only if all the diagonal entries are distinct, that is, if and only if A has distinct eigenvalues.

For $\epsilon = -1$, A is symmetric relative to the Lorentzian quadratic form $x_0 y_0 - \sum_{i=1}^n x_i y_i$. Then every matrix A in $\mathfrak{p}_\epsilon$ is of the form $\begin{pmatrix} a_0 & -a^T \\ a & A_0 \end{pmatrix}$, where $a \in \mathbb{R}^n$ and A_0 is a symmetric $n \times n$ matrix. The matrix $\begin{pmatrix} 0 & -a^T \\ a & 0_n \end{pmatrix}$ can be written as $a \wedge_\epsilon e_0$, where now a is embedded in $\mathbb{R}^{n+1}$ with $a_0 = 0$. Hence, every matrix in $\mathfrak{p}_\epsilon$ is of the form $A = a \wedge_\epsilon e_0 + S$ with $S = \begin{pmatrix} a_0 & 0 \\ 0 & A_0 \end{pmatrix}$, a_0 a real number, and A_0 a symmetric $n \times n$ matrix.

There are two distinct possibiities. Either a is zero, in which case A_0 can be diagonalized. In this situation, A is regular if the eigenvalues of A_0 are distinct and different from a_0. Or, $a \neq 0$. In this situation A cannot be conjugated to a diagonal matrix by the elements of $SO(1, n)$, because

$$g(a \wedge_\epsilon e_0 + S)g^{-1} = g(a) \wedge_\epsilon g(e_0) + gSg^{-1} = b(g) \wedge_\epsilon e_0 + S(g)$$

with $b(g) = (g(a) \wedge_\epsilon g(e_0))e_0$. Hence, $b(g) \neq 0$ for any $g \in SO(1, n)$.

The above implies that regular elements may belong to different conjugacy classes. For instance, the matrix $A = \begin{pmatrix} D_0 & 0 \\ 0 & D \end{pmatrix}$, where D_0 is a 2×2 matrix $\begin{pmatrix} \alpha & -\beta \\ \beta & \alpha \end{pmatrix}$ with α and β are real numbers and D a diagonal $(n-1) \times (n-1)$ matrix with non-zero diagonal real entries, is regular whenever $\alpha^2 + \beta^2 \neq 0$ and the remaining diagonal entries are all distinct, but is not conjugate to any diagonal matrix.

Remarkably on the coadjoint orbits through rank one matrices, the restriction of H is completely integrable, and the required integrals of motion are easily obtained from the spectral representation $L_\lambda = L_\mathfrak{p} - \lambda L_\mathfrak{k} + \lambda^2 A$. The text below contains the necessary details.

Consider first the coadjoint orbits in $\mathfrak{g}_\epsilon$. It follows from Section 10.2 that

$$L_{\mathfrak{p}_\epsilon} = \left\{ x \otimes_\epsilon x - \frac{||x||_\epsilon^2}{n+1} I, \ ||x||_\epsilon^2 = ||x_0||^2 \right\}, \ L_{\mathfrak{h}_\epsilon} = \{ x \wedge_\epsilon y : (x, y)_\epsilon = 0 \}.$$

The zero trace requirement is inessential for the calculations below and will be disregarded. Additionally, A will be replaced by $-A$ and L_λ will be rescaled by dividing by $-\lambda^2$ to read $L_\lambda = -\frac{1}{\lambda^2} L_{\mathfrak{p}_\epsilon} + \frac{1}{\lambda} L_{\mathfrak{h}_\epsilon} + A$.

The spectrum of L_λ is then given by

$$0 = Det(zI - L_\lambda) = Det(zI - A)Det\left(I - (zI - A)^{-1} \left(-\frac{1}{\lambda^2} L_{\mathfrak{p}_\epsilon} + \frac{1}{\lambda} L_{\mathfrak{h}_\epsilon} \right) \right).$$

It follows that on $Det(zI - A) \neq 0$, $0 = Det(zI - L_\lambda)$ whenever

$$0 = Det\left(I - (zI - A)^{-1} \left(-\frac{1}{\lambda^2} L_\mathfrak{p} + \frac{1}{\lambda} L_\mathfrak{k} \right) \right).$$

Matrix $M = I - (zI - A)^{-1} \left(-\frac{1}{\lambda^2} L_\mathfrak{p} + \frac{1}{\lambda} L_\mathfrak{k} \right)$ is of the form

$$M = I + \frac{1}{\lambda^2} R_z x \otimes_\epsilon x - \frac{1}{\lambda} (R_z x \otimes_\epsilon y - R_z y \otimes_\epsilon x), \ R_z = (zI - A)^{-1}.$$

Lemma 10.13

$$Det(M) = \frac{1}{\lambda^2} ((R_z x, x)_\epsilon + (R_z x, x)_\epsilon (R_z y, y)_\epsilon - (R_z x, y)_\epsilon^2) + 1.$$

Proof Let V be the linear span of $\{x, y, R_z x, R_z y\}$ and let $V^\perp$ denote the orthogonal complement relative to $(,)_\epsilon$. Since $M(V) \subseteq V$ and M is equal to the identity on $V^\perp$, the determinant of M is equal to the determinant of the restriction of M to V. Then

$$Mx = x + \frac{1}{\lambda^2}||x||^2_\epsilon R_z x + \frac{1}{\lambda}||x||^2_\epsilon R_z y, \ My = y - \frac{1}{\lambda}||y||^2_\epsilon R_z x,$$

$$MR_z x = R_z x + \frac{1}{\lambda^2}(R_z x, x)_\epsilon R_z x - \frac{1}{\lambda}(R_z x, y)_\epsilon R_z x + \frac{1}{\lambda}(R_z x, x)_\epsilon R_z y,$$

$$MR_z y = R_z y + \frac{1}{\lambda^2}(R_z x, y)_\epsilon R_z x - \frac{1}{\lambda}(R_z x, y)_\epsilon R_z x + \frac{1}{\lambda}(R_z x, x)_\epsilon R_z y.$$

The corresponding matrix is given by

$$\begin{pmatrix} 1 & 0 & 0 & 0 \\ 0 & 1 & 0 & 0 \\ \frac{1}{\lambda^2}||x||^2_\epsilon & -\frac{1}{\lambda}||y||^2_\epsilon & 1 + \frac{1}{\lambda^2}(R_z x, x)_\epsilon - \frac{1}{\lambda}(R_z x, y)_\epsilon & \frac{1}{\lambda^2}(R_z x, y)_\epsilon - \frac{1}{\lambda}(R_z y, y)_\epsilon \\ \frac{1}{\lambda}||x||^2_\epsilon & 0 & \frac{1}{\lambda}(R_z x, x)_\epsilon & 1 + \frac{1}{\lambda}(R_z x, y)_\epsilon \end{pmatrix}.$$

The determinant of this matrix is equal to

$$\left(1 + \frac{1}{\lambda^2}(R_z x, x)_\epsilon - \frac{1}{\lambda}(R_z x, y)_\epsilon\right)\left(\frac{1}{\lambda}(R_z x, y)_\epsilon + 1\right)$$

$$- \frac{1}{\lambda}(R_z x, x)_\epsilon \left(\frac{1}{\lambda^2}(R_z x, y)_\epsilon \ \frac{1}{\lambda}(R_z y, y)_\epsilon\right)$$

$$= \frac{1}{\lambda^2}((R_z x, x)_\epsilon + (R_z x, x)_\epsilon (R_z y, y)_\epsilon - (R_z x, y)^2_\epsilon) + 1.$$

$\square$

Lemma 10.14 *Function* $F(z) = (R_z x, x)_\epsilon + (R_z x, x)_\epsilon (R_z y, y)_\epsilon - (R_z x, y)^2_\epsilon, z \in \mathbb{R}$ *is an integral of motion for H.*

Proof It follows from above that $Det(I - W_z) = 0$ if and only if $F(z) = -\lambda^2$. Since $z(zI - A)^{-1} = \left(I - \frac{1}{z}A\right)^{-1}$, $\lim_{z \to \pm\infty}(z(zI - A)^{-1} = I$. Therefore,

$$\lim_{z \to \pm\infty} zF(z) = ||x||^2_\epsilon ||y||^2_\epsilon + ||x||^2_\epsilon.$$

This argument shows that $F(z)$ takes negative values on some open interval. Hence, $F(z)$ is an integral of motion on that interval. But then $F(z)$ is an integral of motion for all z since F is an analytic function of z on $Det(zI - A) \neq 0$. $\square$

Function F is a rational function with poles at the eigenvalues of the matrix A. Hence, $F(z)$ is an integral of motion for H if and only if the residues of F are constants of motion for H.

In the Euclidean case, the eigenvalues of A are real and distinct, since A is symmetric and regular. Hence, there is no loss in generality in assuming that A is diagonal. Let $\alpha_1, \ldots, \alpha_{n+1}$ denote its diagonal entries. Then

$$F(z) = \sum_{k=0}^{n} \frac{F_k}{z - \alpha_k},$$

where $F_0, \ldots, F_n$ denote the residues of F. It follows that

$$F(z) = \sum_{k=0}^{n} \frac{x_k^2}{z - \alpha_k} + \sum_{k=0}^{n} \sum_{j=0}^{n} \frac{x_k^2 y_j^2}{(z - \alpha_k)(z - \alpha_j)} - \left(\sum_{k=0}^{n} \frac{x_k y_k}{z - \alpha_k} \right)^2$$

$$= \sum_{k=0}^{n} \frac{x_k^2}{z - \alpha_k} + \sum_{k=0}^{n} \sum_{j=0, j \neq k}^{n} \frac{x_k^2 y_j^2}{(z - \alpha_k)(z - \alpha_j)}$$

$$- 2 \sum_{k=0}^{n} \sum_{j=0, j \neq k}^{n} \frac{x_k y_k x_j y_j}{(z - \alpha_k)(z - \alpha_j)}. \tag{10.34}$$

Hence,

$$F_k = \lim_{z \to \alpha_k} (z - \alpha_k) F(z)$$

$$= x_k^2 + \sum_{j=0, j \neq k}^{n} \frac{x_j^2 y_k + x_k^2 y_j^2}{(\alpha_k - \alpha_j)} - 2 \sum_{j=0, j \neq k}^{n} \frac{x_k y_k x_j y_j}{(\alpha_k - \alpha_j)}$$

$$= x_k^2 + \sum_{j=0, j \neq k}^{n} \frac{(x_j y_k - x_k y_j)^2}{(\alpha_k - \alpha_j)}, \, k = 0, \ldots, n.$$

The preceding calculation yields the following proposition.

Proposition 10.15 *Each residue* $F_k = x_k^2 + \sum_{j=0, j \neq k}^{n} \frac{(x_j y_k - x_k y_j)^2}{(\alpha_k - \alpha_j)}, k = 0, \ldots, n$ *is an integral of motion for the Newmann's spherical system (10.16) with $\epsilon = 1$. Moreover, functions $F_0, \ldots, F_n$ are in involution.*

Proof The Poisson bracket relative to the orbit structure coincides with the canonical Poisson bracket on $\mathbb{R}^{n+1} \times \mathbb{R}^{n+1}$. □

These results coincide with the ones reported in [Ms2; Rt2], but our derivation and the connection with the affine-quadratic problem is original.

Similar results hold on the hyperboloid. When A is a diagonal matrix with distinct diagonal entries, then

$$(R_z x, y)_\epsilon = \sum_{i=0}^{n} \frac{v_i w_i}{z - \alpha_i},$$

where $v_0 = x_0, w_0 = y_0$ and $v_i = ix_i, w_i = iy_i, i = 1. \ldots, n$ Therefore, the residues F_k are exactly as in (10.34) with x and y replaced by v and w. Hence,

$$F_k = v_k^2 + \sum_{j=0, j \neq k}^{n} \frac{(v_j w_k - v_k w_j)^2}{(\alpha_k - \alpha_j)}, k = 0, \ldots, n$$

are the integrals of motion for the Newmann's problem on the hyperboloid. In the case that $A = \begin{pmatrix} D_0 & 0 \\ 0 & D \end{pmatrix}$ with $D_0 = \begin{pmatrix} \alpha & -\beta \\ \beta & \alpha \end{pmatrix}$ and D a diagonal with diagonal matrix with diagonal entries $\alpha_2, \ldots, \alpha_n$, then define

$$v_0 = \frac{1}{\sqrt{2}} (x_0 + ix_1), v_1 = \frac{1}{\sqrt{2}} (x_0 - ix_1), v_i = ix_i, i = 2 \ldots, n,$$

$$w_0 = \frac{1}{\sqrt{2}} (y_0 + iy_1), w_1 = \frac{1}{\sqrt{2}} (y_0 - iy_1), w_i = iy_i, i = 2, \ldots, n.$$

Then,

$$(R_z x, y)_{-1} = \left((z - A)^{-1} x, y \right)_{-1}$$

$$= \frac{1}{(z - \alpha^2) + \beta^2} (z - \alpha)(x_0 y_0 - x_1 y_1) - \beta (x_0 y_1 + x_1 y_0))$$

$$- \sum_{j=2}^{n} \frac{1}{z - \alpha_j} x_j y_j$$

$$= \frac{1}{z - (\alpha + i\beta)} v_0 w_0 + \frac{1}{z - (\alpha - i\beta)} v_1 w_1 + \sum_{j=2}^{n} \frac{1}{z - \alpha_j} v_j w_j$$

$$= \sum_{j=0}^{n} \frac{1}{z - \alpha_j} v_j w_j,$$

provided that we identify $\alpha_0 = \alpha + i\beta$ and $\alpha_1 = \alpha - i\beta$.

Therefore, the spectral function $F(z)$ has the same form as in the spherical case:

$$F(z) = \sum_{k=0}^{n} \frac{v_k^2}{z - \alpha_k} + \sum_{k=0}^{n} \sum_{j=0}^{n} \frac{v_k^2 w_j^2}{(z - \alpha_k)(z - \alpha_j)} - \left(\sum_{k=0}^{n} \frac{v_k w_k}{z - \alpha_k} \right)^2.$$

It follows that

$$F(z) = \sum_{k=0}^{n} \frac{F_k}{z - \alpha_k}, \quad F_k = v_k^2 + \sum_{j=0, j \neq k}^{n} \frac{(v_j w_k - v_k w_j)^2}{(\alpha_k - \alpha_j)}, k = 0, \ldots, n .$$

An easy calculation shows that $\bar{F}_0 = F_1$ and that each $F_k, k \geq 2$ is real valued. Therefore,

$$Re(F_0), \; Im(F_0), F_2, \ldots, F_n \tag{10.35}$$

are n independent integrals of motion.

The integration technique based on the use of elliptic coordinates is intimately tied to another famous problem in the theory of integrable systems, Jacobi's geodesic problem on the ellipsoid. We will defer this connection to the next chapter.

In the meantime we would like to contrast this class of integrable systems with another famous integrable system, the Toda system.

10.5 The group of upper triangular matrices and Toda lattices

A Toda lattice is a system of n particles on the line in motion under an exponential interaction between its nearest neighbors. This mechanical system is at the core of the literature on integrable systems, partly for historical reasons, as the first mechanical system whose integrals of motion were found on a coadjoint orbit of a Lie algebra, but mostly because of the methodology which seemed to carry over to other Lie algebras with remarkable success [Pr; RS]. Its magical power led to a new paradigm in the theory of integrable systems based on the use of R-matrices. This paradigm provides an abstract procedure to construct large families of functions whose members are in involution with each other, and is looked upon in much of the current literature as an indispensable tool for discovering integrable systems [Pr; RT]. Our approach to the Toda system is completely different and makes no use of double Lie algebras.

Rather than starting with a particular Hamiltonian on the space of symmetric matrices, as is commonly done in the literature on Toda systems, we will start with a geodesic problem on the group G of upper triangular matrices, and along the way discover the integrals of motion for a Toda lattice.

The basic setting is as follows: G will denote the subgroup of matrices in $SL_n(R)$ consisting of matrices $g = (g_{ij})$ such that $g_{ij} = 0$ for $i < j$, and $\mathfrak{g}$ will denote its Lie algebra, i.e., the algebra of all upper triangular matrices of trace zero. Then $\langle A, B \rangle = Tr(AB^T)$ is positive-definite on $\mathfrak{g}$, since $\langle A, A \rangle = \sum_{i \geq j}^n a_{ij}^2$ for any matrix $A = (a_{ij})$ in $\mathfrak{g}$.

We will use $\mathfrak{g}_0$ to denote the space of upper triangular matrices having zero diagonal part. Then $\mathfrak{g}_0$ is an ideal in $\mathfrak{g}$, and its orthogonal complement $\mathfrak{g}_0^\perp$ in $\mathfrak{g}$ consists of all diagonal matrices having zero trace.

In order to make contact with Toda lattices, we will introduce the weighted quadratic form $(A, B) = \langle D_A, D_B \rangle + \frac{1}{2}\langle A_0, B_0 \rangle$, where D_A and D_B are the diagonal parts of A and B, and A_0 and B_0 are the projections of A and B onto $\mathfrak{g}_0$. Then $||A||^2$ will denote the induced norm (A, A). This norm can be used to define the length $\int_0^T ||U(t)||\, dt$ and the energy $E = \frac{1}{2}\int_0^T ||U(t)||^2\, dt$ of any curve $g(t)$ in G where $U(t) = g^{-1}(t)\frac{dg}{dt}(t)$.

The associated Riemannian problem is left-invariant, hence the corresponding Hamiltonian is a function on $\mathfrak{g}^*$ when T^*G is represented as $G \times \mathfrak{g}^*$. A distinctive feature of this set-up, however, is that $\mathfrak{g}^*$ can be identified with the space of symmetric matrices in $sl_n(R)$. The identification depends on the choice of a complementary subalgebra in $sl_n(R)$ and goes as follows.

Any upper triangular matrix U can be expressed as the sum of a symmetric and a skew-symmetric matrix. In fact,

$$U = \frac{1}{2}\left(U + U^T\right) + \frac{1}{2}\left(U - U^T\right).$$

Conversely, any symmetric matrix S defines an uper triangular matrix S_+ with its entries equal to s_{ij} for $i \leq j$, and zero otherwise. It follows that $S_+ + S_+^T = S - S_d$, where S_d denotes the diagonal part of S. Therefore, $S_+ = \frac{1}{2}(S_+ + S_+^T) + \frac{1}{2}(S_+ - S_+^T)$ can be written as

$$2S_+ = S - S_d + (S_+ - S_-), \tag{10.36}$$

where $S_- = S_+^T$. These decompositions show that $sl_n(R) = so_n(R) \oplus \mathfrak{g}$, because any matrix in $sl_n(R)$ is the sum of a symmetric and a skew-symmetric matrix.

The decomposition $sl_n(R) = so_n(R) \oplus \mathfrak{g}$ induces a decomposition of the dual $sl_n(R)^*$ in terms of the annihilators $so_n^0(R) = \{\ell \in sl_n(R)^* : \ell|_{so_n(R)} = 0\}$ and $\mathfrak{g}^0 = \{\ell \in sl_n(R)^* : \ell|_{\mathfrak{g}} = 0\}$. Then $\mathfrak{g}^*$ will be identified with $so_n^0(R)$, and $so_n^*(R)$ with $\mathfrak{g}^0$.

When $sl_n^*(R)$ is identified with $sl_n(R)$ via the trace form $\langle\,,\,\rangle$, then $so_n^0(R)$ is identified with matrices $L \in sl_n(R)$ such that $\langle L, X \rangle = 0$ for all $X \in so_n(R)$. That is, $\mathfrak{g}^*$ is identified with the orthogonal complement of $sl_n(R)$ relative to the trace form, which is exactly the space of symmetric matrices having zero trace.

Remark 10.16 The above reasoning shows that the dual of any Lie algebra which is transversal to $so_n(R)$, such as the Lie algebra of lower triangular matrices, can be represented by the symmetric matrices. Hence the identification of $\mathfrak{g}^*$ with the symmetric matrices is not canonical.

In this identification the Hamiltonian lift of any left-invariant vector field $X_A(g) = gA$ is identified with the symmetric matrix $S_A = \frac{1}{2}(A + A^T)$ via the formula $h_A(\xi) = \langle S, \frac{1}{2}(A + A^T) \rangle$. In particular, $h_{ij} = h_{E_{ij}}$ is identified with the the matrix $S_{ij} = \frac{1}{2}(E_{ij} + E_{ji})$ and h_D is identified with D, where D is a diagonal matrix and $E_{ij} = e_i \otimes e_j, i < j, i = 1, \ldots, n, j = 1, \ldots, n$. Matrices $E_{ij}, i < j$ forms an orthonormal basis for $\mathfrak{g}_0$ relative to $\langle , \rangle$. Let $D_1, \ldots, D_{n-1}$ denote any orthonormal basis in $\mathfrak{g}_0^\perp$ (the two forms agree on $\mathfrak{g}_0^\perp$). Then any curve $g(t) \in G$ is a solution of

$$\frac{dg}{dt} = g(t) \left(\sum_{i=1}^{n-1} u_i(t) D_i + \sum_{i<j}^{n} u_{ij}(t) E_{ij} \right) \tag{10.37}$$

and its energy associated on an interval $[0, T]$ is given by

$$E = \frac{1}{2} \int_0^T \left(\sum_{i=1}^{n-1} u_i^2(t) + \frac{1}{2} \sum_{i<j} u_{ij}^2(t) \right) dt.$$

It follows that the Hamiltonian lift of the energy-extended system (10.37) is given by

$$h_{u(t)}(S) = -\frac{1}{2} \left(\sum_{i=1}^{n-1} u_i^2(t) + \frac{1}{2} \sum_{i<j} u_{ij}^2(t) \right) + \sum_{i=1}^{n-1} u_i(t) h_i(S) + \sum_{i<j} h_{ij}(S) u_{ij}(t)$$

$$= -\frac{1}{2} \left(\sum_{i=1}^{n-1} u_i^2(t) + \frac{1}{2} \sum_{i<j} u_{ij}^2(t) \right) + \sum_{i=1}^{n-1} u_i S_i + \sum_{i<j} u_{ij} S_{ij}.$$

According to the Maximum Principle the extremal control functions are given by

$$u_i = h_i = S_i, i = 1, \ldots, n - 1, u_{ij} = 2h_{ij} = 2S_{ij}, i < j, \tag{10.38}$$

and the maximal Hamiltonian H is given by

$$H(S) = \frac{1}{2} \left(\sum_{i=1}^{n-1} h_i^2(S) + 2 \sum_{i<j} h_{ij}^2(S) \right) = \frac{1}{2} \left(\sum_{i=1}^{n-1} S_i^2 + 2 \sum_{i<j} S_{ij}^2 \right) = \frac{1}{2} Tr(S^2).$$
$$\tag{10.39}$$

It remains to express the solution curves $\xi(t)$ of the Hamiltonian system

$$\frac{dg}{dt} = g(t) dH(\xi(t)), \frac{d\xi}{dt} = -ad^*(dH(\xi(t)))(\xi(t)), \tag{10.40}$$

in terms of symmetric matrices. To do so, note first that

$$dH = U = \sum_{i=1}^{n-1} S_i D_i + 2 \sum_{i<j} S_{ij} E_{ij} = 2S_+ - S_d. \tag{10.41}$$

Consequently, $\frac{1}{2}(U + U^T) = S$. Secondly,

$$[U, X] + [U, X]^T = \frac{1}{2}([U + U^T, X - X^T] + [U - U^T, X + X^T])$$

for any upper triangular matrices U and X.

Return now to the Hamiltonian equation (10.40). We have

$$\left\langle \frac{dS}{dt}, \frac{1}{2}(X + X^T) \right\rangle = \frac{d\xi}{dt}(X) = -ad^*(dH(\xi(t)))(\xi(t))(X) = -\xi([dH, X])$$

$$= -\left\langle S, \frac{1}{2}([dH, X] + [dH, X]^T) \right\rangle$$

$$= -\frac{1}{2}\left\langle S, \frac{1}{2}\left([U + U^T, X - X^T] + [U - U^T, X + X^T]\right) \right\rangle$$

$$= -\frac{1}{2}\left\langle S, [S, X - X^T] + \frac{1}{2}[U - U^T, X + X^T] \right\rangle$$

$$= -\frac{1}{2}\left\langle \left[S, \frac{1}{2}(U - U^T)\right], X + X^T \right\rangle$$

$$= \frac{1}{2}\left\langle [S_+ - S_-, S], X + X^T \right\rangle.$$

The above implies that (10.40) is equivalent to

$$\frac{dS}{dt} = [M, S], M = S_+ - S_-. \tag{10.42}$$

This equation coincides with the famous Lax pair equation associated with the Toda lattice [Pr]. It follows that the spectral invariants of S are the integrals of motion for (10.42). These invariants are usually expressed in terms of symmetric functions $I_j = Tr(S^j), j = 1, 2, \ldots$.

To make an explicit contact with Toda system, consider the restriction of (10.42) to the coadjoint orbit through

$$S_0 = \begin{pmatrix} 0 & 1 & 0 & \cdots & & 0 \\ 1 & 0 & 1 & 0\cdots & & 0 \\ 0 & \ddots & 0 & \ddots & & 0 \\ 0 & \cdots & 1 & 0 & & 1 \\ 0 & \cdots & 0 & 1 & & 0 \end{pmatrix}.$$

This orbit consists of matrices gS_0g^{-1} such that $(gS_0g^{-1})^T = gS_0g^{-1}$, where g ranges over all upper triangular matrices of determinant 1. It is easy to verify that any such symmetric matrix is of the form

$$S_0 = \begin{pmatrix} b_1 & a_1 & 0 & \cdots & & 0 \\ a_1 & b_2 & a_2 & 0\cdots & & 0 \\ 0 & \ddots & \ddots & \ddots & & 0 \\ 0 & \cdots & a_{n-2} & b_{n-1} & a_{n-1} \\ 0 & \cdots & & 0 & a_{n-1} & b_n \end{pmatrix} \qquad (10.43)$$

with $a_1, \ldots, a_{n-1}$ and $b_1, \ldots, b_n$ arbitrary numbers such that $\sum_{i=1}^{n} b_i = 0$. Let M denote the set of matrices given by (10.43).

Proposition 10.17 *The restriction of equations (10.42) to M is given by*

$$\frac{db_1}{dt} = -2a_1^2, \frac{db_k}{dt} = 2(a_{k-1}^2 - a_k^2), \frac{db_n}{dt} = 2a_{n-1}^2, k = 2, \ldots, n-1,$$

$$\frac{da_k}{dt} = a_k(b_{k+1} - b_k), k = 1, \ldots, n-1. \qquad (10.44)$$

Proof If S is as in (10.43), then let B denote its diagonal part and let $A = S_+ - B$. Then equation (10.42) reduces to

$$\frac{dA}{dt} = [A, D] - [A^T, D], \frac{dB}{dt} = 2[A, A^T].$$

Equations (10.43) follow by a simple calculation of these Lie brackets.   □

Recall now that each coadjoint orbit is symplectic. On M, it suffices to calculate the Poisson brackets of the coordinates $a, \ldots, a_{n-1}$ and $b_1, \ldots, b_n$.

Proposition 10.18 *The Poisson brackets on M satisfy the following table:*

$$\{a_i, a_j\} = \{b_i, b_j\} = 0, \{b_j, a_j\} = -a_j, \{b_{j+1}, a_j\} = a_j, \{b_j, a_i\} = 0, |i - j| > 2. \qquad (10.45)$$

Proof We will only do $\{b_i, a_j\}$. The rest follow by similar calculations.

$$\{b_i, a_j\}(S) = \left\langle S, \frac{1}{2} \left([e_i \otimes e_i, e_j \otimes e_{j+1}] + [e_i \otimes e_i, e_j \otimes e_{j+1}]^T\right)\right\rangle.$$

The above expression is zero unless $i = j$ or $i = j + 1$. In the first case $\{b_j, a_j\} = -a_j$ and in the second $\{b_{j+1}, a_j\} = a_j$.   □

To relate the above structure to the Hamiltonian system associated with a Toda system of $n - 1$ particles, let

$$a_i = e^{q_i - q_{i+1}}, p_i = -b_i, i = 1, \ldots, n-1.$$

Then (q_i, p_i) can be identified with the position and the momentum of the ith particle. The case $q_n = q_1$ corresponds to the periodic Toda system, and $q_n = 0$ to the open system on a line. An easy calculation based on (10.45) shows that

$$\{p_i, p_j\} = \{q_i, q_j\} = 0, \ \{q_i, p_j\} = \delta_{ij}.$$

Thus $q_1, \ldots, q_{n-1}, p_1, \ldots, p_{n-1}$ are the canonical coordinates in $\mathbb{R}^{n-1} \times \mathbb{R}^{n-1}$. Therefore, the transformation $(q, p) \to (a, b)$ is symplectic and the Hamiltonian $H = \frac{1}{2} Tr(S^2)$ is transformed into the Hamiltonian

$$H = \frac{1}{2} \sum_{i=1}^{n-1} p_i^2 + \sum_{i=1}^{n-1} e^{q_i - q_{i+1}}$$

associated with the Toda mechanical system.

It seems that this Hamiltonian system has very little in common with the isospectral Hamiltonian systems encountered before. Nevertheless, any attempt to comparisons raises an interesting question: does any left-invariant quadratic Hamiltonian on a solvable Lie algebra admit a spectral curve of the Manakov type?

11

Elliptic geodesic problem on the sphere

Let us now return to the general construction of isospectral systems based on the use of sectional operators in the unitary algebra u_n (Chapter 9). Recall that the linear operator $Q = ad^{-1}A \circ adB$ on $so_n(R)$ defined by two diagonal matrices A and B with diagonal entries $a_1, \ldots, a_n$ of A, and $b_1, \ldots, b_n$ conforms to

$$Q(X) = \sum_\alpha \frac{\alpha(B)}{\alpha(A)} x_\alpha = \sum_{i,j} \frac{b_j - b_i}{a_j - a_i} x_{ij}, \tag{11.1}$$

for any skew-symmetric matrix $X = (x_{ij})$. Relative to the trace form $\langle X, Y \rangle = -\frac{1}{2}Tr(XY)$, $\langle X, Q(X) \rangle = \sum_{i,j}^n \frac{b_j - b_i}{a_j - a_i} x_{ij}^2$. Hence, Q defines a positive-definite quadratic form $\langle Q(X, X) \rangle$ whenever all the quotients $\frac{b_j - b_i}{a_j - a_i}$ are positive. Since

$$[B, Q^{-1}(X)] = adB \circ ad^{-1}B \circ adA(X) = adA(X),$$

the affine-quadratic Hamiltonian $H = \frac{1}{2}\langle Q^{-1}(L_{\mathfrak{k}}), L_{\mathfrak{k}} \rangle + \langle A, L_{\mathfrak{p}} \rangle$ is isospectral.

We will now focus on the elliptic case given by $B = -A^{-1}$, and leave the top of Manakov given by $b_i = a_i^2, i = 1, \ldots, n$ to the next chapter. In the elliptic case, $\frac{b_j - b_i}{a_j - a_i} = \frac{\frac{1}{a_i} - \frac{1}{a_j}}{a_j - a_i} = \frac{1}{a_i a_j}$, and

$$Q(X) = ad^{-2}A(X) = \sum_{ij} \frac{1}{a_i a_j} x_{ij} = DXD,$$

where $D = A^{-1}$. Conversely, any diagonal matrix D with positive entries can be identified with symmetric diagonal matrices $A = D^{-1}$ and $B = -D$, in which case $ad^{-1} \circ adB(X) = DXD, X \in so_n(R)$, and consequently,

$$[D, Q^{-1}(X)] = [D, D^{-1}XD^{-1}] = adB \circ ad^{-1}B \circ adA(X) = [D^{-1}, X].$$

198

It follows that $A = D - Tr(D)I$ defines an affine Hamiltonian

$$H = \frac{1}{2}\langle D^{-1}L_{\mathfrak{h}}D^{-1}, L_{\mathfrak{h}}\rangle + \langle A, L_{\mathfrak{p}}\rangle \tag{11.2}$$

on $SL_n(R)$ whose Hamiltonian equations

$$\frac{dg}{dt} = g(t)(A + D^{-1}L_{\mathfrak{h}}D^{-1}),$$

$$\frac{dL_{\mathfrak{h}}}{dt} = [D^{-1}L_{\mathfrak{h}}D^{-1}, L_{\mathfrak{h}}] + [A, L_{\mathfrak{p}}], \frac{dL_{\mathfrak{p}}}{dt} = [D^{-1}L_{\mathfrak{h}}D^{-1}, L_{\mathfrak{p}}] + s[A, L_{\mathfrak{h}}],$$

$$\tag{11.3}$$

admit an isospectral representation

$$\frac{dL_{s,\lambda}}{dt} = [\Omega_\lambda, L_{s,\lambda}],$$

with $L_{s,\lambda} = L_{\mathfrak{p}} - \lambda L_{\mathfrak{k}} - (\lambda^2 - s)D$, and $\Omega_\lambda = D^{-1}L_{\mathfrak{k}}D^{-1} - \lambda D^{-1}$. As usual, the zero trace constraint could be disregarded, and equations (11.3) can be written as

$$\frac{dg}{dt} = g(t)(A + D^{-1}L_{\mathfrak{h}}D^{-1}),$$

$$\frac{dL_{\mathfrak{h}}}{dt} = [D^{-1}L_{\mathfrak{h}}D^{-1}, L_{\mathfrak{h}}] + [D^{-1}, L_{\mathfrak{p}}], \frac{dL_{\mathfrak{p}}}{dt} = [D^{-1}L_{\mathfrak{h}}D^{-1}, L_{\mathfrak{p}}] + s[D^{-1}, L_{\mathfrak{h}}]$$

$$\tag{11.4}$$

Morcover,

$$H = \frac{1}{2}\langle D^{-1}L_{\mathfrak{h}}D^{-1}, L_{\mathfrak{h}}\rangle + \langle A, L_{\mathfrak{p}}\rangle = \frac{1}{2}\langle D^{-1}L_{\mathfrak{h}}D^{-1}, L_{\mathfrak{h}}\rangle + \langle D^{-1}, L_{\mathfrak{p}}\rangle,$$

because $\langle A, L_{\mathfrak{p}}\rangle = \langle D^{-1}, L_{\mathfrak{p}}\rangle$, when $L_{\mathfrak{p}}$ is a symmetric matrix with zero trace. Consequently,

Proposition 11.1 *The spectral invariants of* $L_{\mathfrak{p}} - \lambda L_{\mathfrak{h}} - (\lambda^2 - s)D$ *are common constants of motion for both the Hamiltonians* $H_1 = \frac{1}{2}\langle L_{\mathfrak{h}}, L_{\mathfrak{h}}\rangle - \langle D, L_{\mathfrak{p}}\rangle$ *and* $H = \frac{1}{2}\langle D^{-1}L_{\mathfrak{h}}D^{-1}, L_{\mathfrak{h}}\rangle + \langle D^{-1}, L_{\mathfrak{p}}\rangle.$

11.1 Elliptic Hamiltonian on semi-direct rank one orbits

Let us now consider the elliptic-affine Hamiltonian on the semi-direct product $G_s = \mathfrak{p} \ltimes SO_{n+1}(R)$ restricted to the coadjoint orbit through the rank one matrix

$L_0 = x_0 \otimes x_0 - -\frac{1}{n+1}I$, $||x_0|| = 1$. As we have seen in the previous chapter, this coadjoint orbit is equal to

$$\left\{ L_p + L_\mathfrak{h} : L_p = x \otimes x - \frac{||x||^2||}{n+1}I, L_\mathfrak{h} = x \wedge y, ||x||^2 = 1, x \cdot y = 0 \right\},$$

and is diffeomorphic the cotangent bundle of the unit sphere under the correspondence $(x, y) \to x \otimes x - \frac{1}{n+1}I + x \wedge y$.

Then, the solutions of (11.4) evolve according to $L_p(t) = x(t) \otimes x(t) - \frac{1}{n+1}I$ and $L_\mathfrak{k} = x(t) \wedge y(t)$. Since equations (11.4) are unaltered if the zero trace constraint is dropped, L_p will be taken as $L_p = x(t) \otimes x(t)$. Then equations (11.4) reduce to

$$\frac{d}{dt}(x \wedge y) = [D^{-1}(x \wedge y)D^{-1}, x \wedge y] + [D^{-1}, x \otimes x],$$
$$\frac{d}{dt}(x \otimes x) = [D^{-1}(x \wedge y)D^{-1}, x \otimes x]. \tag{11.5}$$

The Hamiltonian in (11.2) can be written as

$$H = \frac{1}{2}\left((D^{-1}y \cdot y) - \frac{(D^{-1}x \cdot y)^2}{(D^{-1}x \cdot x)} - 1 \right)(D^{-1}x \cdot x), \tag{11.6}$$

as a consequence of the following calculations:

$$\langle D^{-1}L_\mathfrak{k}D^{-1}, L_\mathfrak{k} \rangle = \langle D^{-1}x \wedge yD^{-1}, x \wedge y \rangle$$
$$= (D^{-1}x \cdot x)(D^{-1}y \cdot y) - (D^{-1}x \cdot y)^2,$$
$$\langle D^{-1}, L_p \rangle = -\frac{1}{2}(x \cdot D^{-1}x).$$

Proposition 11.2 *On energy level $2H = 0$, equations (11.5) correspond to*

$$\frac{dx}{dt} = (D^{-1}x \cdot x)\left(D^{-1}y - \frac{(D^{-1}x \cdot y)}{(D^{-1}x \cdot x)}D^{-1}x \right),$$
$$\frac{dy}{dt} = (D^{-1}x \cdot x)\left(\frac{(D^{-1}x \cdot y)}{(D^{-1}x \cdot x)}D^{-1}y - \frac{(D^{-1}x \cdot y)^2}{(D^{-1}x \cdot x)^2}D^{-1}x - x \right), \tag{11.7}$$

under the correspondence $(x \otimes x, x \wedge y) \to (x, y)$.

We first assemble several auxiliary formulas that are needed for the proof.

Lemma 11.3 *(a) $D^{-1}(x \wedge y)D^{-1} = D^{-1}x \wedge D^{-1}y$, and $[D^{-1}, x \otimes x] = x \wedge D^{-1}x$.*

(b) $[D^{-1}x \wedge D^{-1}y, x \wedge y] = (D^{-1}y \cdot y)(D^{-1}x \wedge x) + (D^{-1}x \cdot x)(D^{-1}y \wedge y)$
$$- (D^{-1}x \cdot y)(D^{-1}y \wedge x + D^{-1}x \wedge y).$$

(c) $[D^{-1}x \wedge D^{-1}y, x \otimes x] = (D^{-1}x \cdot x)(D^{-1}y \otimes x + x \otimes D^{-1}y)$
$$- (D^{-1}y \cdot x)(D^{-1}x \otimes x + x \otimes D^{-1}x).$$

Proof Since $x = (x \otimes x)x$,

$$\dot{x} = \frac{d}{dt}(x \otimes x)x = \left(\frac{d}{dt}x \otimes x\right)x + (x \otimes x)\dot{x} = [D^{-1}x \wedge yD^{-1}, x \otimes x]x$$

$$= ((D^{-1}x \cdot x)(D^{-1}y \otimes x + x \otimes D^{-1}y) - (D^{-1}y \cdot x)(D^{-1}x \otimes x + x \otimes D^{-1}x))x$$

$$= (D^{-1}x \cdot x)D^{-1}y - (D^{-1}x \cdot y)D^{-1}x$$

$$= (D^{-1}x \cdot x)\left(D^{-1}y - \frac{(D^{-1}x \cdot y)}{(D^{-1}x \cdot x)}D^{-1}x\right).$$

To derive the differential equation for y, first note that $x \cdot y = 0 = x \cdot \dot{x}$. Therefore,

$$[D^{-1}(x \wedge y)D^{-1}, x \wedge y]x + [D^{-1}, x \otimes x]x = \left(\frac{d}{dt}(x \wedge y)\right)x$$

$$= (\dot{x} \wedge y)x + (x \wedge \dot{y})x = (x \cdot \dot{y})x - \dot{y} = -(\dot{x} \cdot y)x - \dot{y}.$$

It follows that

$$-\dot{y} = [D^{-1}(x \wedge y)D^{-1}, x \wedge y]x + [D^{-1}, x \otimes x]x + (y \cdot \dot{x})x. \qquad (11.8)$$

Then, $[D^{-1}, x \otimes x]x = (D^{-1}x \cdot x)x - D^{-1}x$, and

$$[D^{-1}x \wedge D^{-1}y, x \wedge y]x$$

$$= (D^{-1}y \cdot y)D^{-1}x - (D^{-1}x \cdot y)D^{-1}y - ((D^{-1}y \cdot y)(D^{-1}x \cdot x) - (D^{-1}x \cdot y)^2)x$$

$$= (D^{-1}y \cdot y)D^{-1}x - (D^{-1}x \cdot y)D^{-1}y - (2H + (D^{-1}x \cdot x))x.$$

On energy level $H = 0$,

$$(D^{-1}y \cdot y) - \frac{(D^{-1}x \cdot y)^2}{(D^{-1}x \cdot x)} = 1, \text{ and } (y \cdot \dot{x})x = (D^{-1}x \cdot x)x,$$

and

$$[D^{-1}x \wedge D^{-1}y, x \wedge y]x = (D^{-1}y \cdot y)D^{-1}x - (D^{-1}x \cdot y)D^{-1}y - (D^{-1}x \cdot x)x.$$

After the substitutions in (11.8) we get

$$\dot{y} = (D^{-1}x \cdot y)D^{-1}y - (D^{-1}y \cdot y) - 1)D^{-1}x - x(D^{-1}x \cdot x)$$

$$= (D^{-1}x \cdot x)\left(\frac{(D^{-1}x \cdot y)}{D^{-1}x \cdot x}D^{-1}y - \frac{(D^{-1}x \cdot y)^2}{(D^{-1}x \cdot x)^2}D^{-1}x - x.\right)$$

$\square$

Equations (11.7) can be reparametrized by $s = \int (D^{-1}x(t) \cdot x(t))\, dt$ to read

$$
\begin{aligned}
\frac{dx}{ds} &= \frac{dx}{dt}\frac{dt}{ds} = D^{-1}y - \frac{(D^{-1}x \cdot y)}{(D^{-1}x \cdot x)}D^{-1}x, \\
\frac{dy}{ds} &= \frac{dy}{dt}\frac{dt}{ds} = \frac{(D^{-1}x \cdot y)}{(D^{-1}x \cdot x)}\left(D^{-1}y - \frac{(D^{-1}x \cdot y)}{(D^{-1}x \cdot x)}D^{-1}x\right) - x.
\end{aligned}
\tag{11.9}
$$

We will presently show that equations (11.9) are the Hamiltonian equations associated with the geodesic problem on the sphere corresponding to the elliptic metric $\langle Ax, x\rangle$. For that reason we will refer to these equations as *the elliptic geodesic equations*.

Proposition 11.4 *Elliptic geodesic equations are completely integrable on the sphere S^n with*

$$
F_k = x_k^2 + \sum_{j=0,\, j\neq k}^{n} \frac{(x_j y_k - x_k y_j)^2}{(d_k - d_j)},\ k = 0,\dots,n,
\tag{11.10}
$$

an involutive system of first integrals.

Proof Proposition 11.1 implies that the canonical affine Hamiltonian $H = \frac{1}{2}\langle L_{\mathfrak{h}}, L_{\mathfrak{h}}\rangle - \langle D, L_{\mathfrak{p}}\rangle$ and the elliptic affine Hamiltonian have the same isospectral integrals of motion. On the cotangent bundle of the sphere the Hamiltonian equations for H coincide with the Newmann mechanical system. Since the functions F_k are integrals of motion for H, they are also integrals of motion for the elliptic problem. $\qquad\square$

Let us now verify that equations (11.9) do correspond to the Hamiltonian system associated with the geodesic problem on the sphere corresponding to the elliptic metric $\langle Ax, x\rangle$ as claimed in the earlier paragraph. More generally, let us derive the Hamiltonian equations on the cotangent bundle of a quadric surface $N = \{x \in \mathbb{R}^{n+1} : (A^{-1}x \cdot x) = 1\}$, defined by a positive-definite matrix A, corresponding to the geodesic problem relative to the metric $\langle \dot{x}, \dot{x}\rangle = (D\dot{x} \cdot \dot{x})$ defined by another positive-definite matrix D.

The classical theory tells us how to obtain Hamiltonians in a canonical system of coordinates but does not say anything how to obtain the Hamiltonians in terms of the redundant coordinates of the ambient space. Since the need for the latter arises in many applications, we will digress briefly into the theory of constrained Hamiltonian systems relevant for the correct formulation of the Maximum Principle in the extraneous coordinates of the ambient space.

11.2 The Maximum Principle in ambient coordinates

Let us begin with some general observations. If N is a submanifold of M then the tangent bundle TN is a sub-bundle of the tangent bundle TM, but the cotangent bundle T^*N is not a sub-bundle of T^*M. Even in vector spaces, there is no canonical way to extend linear functions on N into the space of linear functions on M. When the ambient manifold has an additional structure, such as a Riemannian structure, then the cotangent vectors in T^*M can be identified with the tangent vectors in TM via the Riemannian metric, and then covectors in T^*N can be identified with vectors in T^*M.

In these situations the cotangent bundle T^*N of a k-dimensional submanifold is typically expressed as the zero set

$$G_1 = G_2 = \cdots = G_{2(n-k)} = 0,$$

of some functions $G_1, \ldots, G_{2(n-k)}$ in T^*M. For instance, the cotangent bundle of the n-dimensional unit sphere $||x||^2 = 1$ is identified with $\{(x, y) \in \mathbb{R}^{n+1} \times \mathbb{R}^{n+1} : ||x||^2 = 1, (x, y) = 0\}$. More generally, the cotangent bundle of a quadric surface

$$N = \{x \in \mathbb{R}^{n+1} : (x \cdot A^{-1}x) = 1\}$$

relative to the metric $\langle \dot{x}, \dot{y} \rangle = (D\dot{x}, \dot{y})$ in $\mathbb{R}^{n+1}$ is identified with

$$\left\{(x, y) \in \mathbb{R}^{2(n+1)} : (A^{-1}x, x) = 1, (DA^{-1}x, y) = 0\right\},$$

and the cotangent bundle of the unit tangent bundle $N = \{(x, y) \in \mathbb{R}^{n+1} \times \mathbb{R}^{n+1}; ||y||^2 = 1\}$ is identified with $\{(x, y, p, q) \in \mathbb{R}^{4(n+1)} : ||y||^2 = 1, (y, q) = 0\}$. The metric $\langle \dot{x}, \dot{y} \rangle = (D\dot{x}, \dot{y})$ defined by a positive definite matrix D will be called elliptic.

We will now explain how to adapt the Maximum Principle to situations in which an optimal control problem on N is expressed in terms of the ambient coordinates of M, and also explain how to express the associated Hamiltonian equations in these coordinates. The main difficulty is visible even in the simplest geodesic problems, such as the geodesic problem on the sphere. For when the velocity vector $\frac{dx}{dt}$ on the unit sphere was expressed in terms of the standard frame $E_1, \ldots, E_{n+1}$ in $\mathbb{R}^{n+1}$, then the induced "control" system $\frac{dx}{dt} = \sum_{i=1}^{n+1} u_i(t)E_i$ is bound by the constraint $(u(t), x(t)) = 0$ in order that the trajectory $x(t)$ remains on the sphere. But the Maximum Principle is not readily applicable to the situation in which the control depends on the state variables – the proof of the Maximum Principle requires piecewise constant perturbations of an optimal trajectory, which in this case are not available. Even if we attempted to circumvent this problem by choosing a local frame

$E_1, \ldots, E_n$ of vector fields in $\mathbb{R}^{n+1}$ that are tangent to the sphere, such as, for instance, $E_1 = A_1 x, \ldots, E_n = A_n x$, with $A_1, \ldots, A_n$ skew-symmetric matrices, the cotangent bundle of the sphere would not be invariant for the resulting Hamiltonian system without additional constraints.

The following proposition is fundamental.

Proposition 11.5 *Suppose that N is a submanifold of a manifold M such that its cotangent bundle of N is given by $G_1 = G_2 = \ldots G_{2(n-k)} = 0$, for some functions $G_1, \ldots G_k$ on T^*M that are functionally independent on T^*N. Let V be a vector field in M that is tangent to N and let V_N denote the restriction of V to N. Then the Hamiltonian lift of V_N is the restriction of the Hamiltonian vector field $\vec{h}$ to T^*N corresponding to*

$$h = h_V + \lambda_1 G_1 + \lambda_2 G_2 + \cdots + \lambda_k G_{2(n-k)},$$

where the multipliers $\lambda_1, \ldots, \lambda_k$ are chosen so that $\{h, G_1\} = \cdots = \{h, G_{2(n-k)}\} = 0$.

Proof Let $(x_1, \ldots, x_k, p_1, \ldots, p_k)$ denote a symplectic coordinate system on T^*N and let $(x_1, \ldots, x_n, p_1, \ldots, p_n)$ denote its extension to a symplectic basis in T^*M. Then

$$G_1 = x_{k+1}, \ldots, G_{n-k} = x_n, G_{n-(k+1)} = p_{k+1}, \ldots, G_{2(n-k)} = p_n$$

is the most natural choice of functions to define T^*N. Since the coordinates are symplectic, $\{G_i, G_j\} = \delta_{i,k+i}$. This implies that $\lambda_i = -\{h_V, p_{i+k}\}$ and $\lambda_{i+k} = \{h_V, x_{i+k}\}$, where $h_V = \sum_{i=1}^{n} p_i V_i$. It is now an easy verification to show that the restriction of $\vec{h}$ to T^*N where $h = h_V + \sum_{i=1}^{2k} \lambda_i G_i$ coincides with $\frac{dx_i}{dt} = V_i$, $\frac{dp_i}{dt} = -\sum_{j=1}^{k} p_j \frac{\partial V_j}{\partial x_i}$. We will leave to the reader to show that the restriction of $\vec{h}$ to T^*N is independent of the particular choice of functions that are used to define T^*N. □

The following example illustrates some subtleties behind the proposition.

Example 11.6 Vector field $V(x, y) = (y, Ay)$ is tangent to $N = \{(x, y) \in \mathbb{R}^{2(n+1)}; ||y||^2 = 1\}$ for any skew-symmetric matrix A. The Hamiltonian lift of V to T^*M is given by $h_V = p \cdot y + q \cdot Ay$ and the integral curves of $\vec{h}_V$ are the solutions of

$$\frac{dx}{dt} = y, \quad \frac{dy}{dt} = Ay, \quad \frac{dp}{dt} = 0, \quad \frac{dq}{dt} = -p + Aq. \tag{11.11}$$

The cotangent bundle of N is defined by the constraints $G_1 = ||y||^2 - 1$, $G_2 = y \cdot q$. Since $\{h_V, G_2\} = y \cdot p$ is not equal to zero on T^*N, T^*N is not invariant under the restriction of $\vec{h}_V$ to T^*N.

However, the Hamiltonian field associated with the modified lift $h = h_V + \lambda_1 G_1 + \lambda_2 G_2$ is tangent to T^*N whenever the multipliers λ_1 and λ_2 chosen so that $\{h_V, G_1\} = \{h_V, G_2\} = 0$. These relations yield

$$\lambda_1 = -\frac{\{h_V, G_2\}}{\{G_1, G_2\}} = -\frac{1}{2} y \cdot p, \ \lambda_2 = -\frac{\{h_V, G_1\}}{\{G_2, G_1\}} = 0.$$

It follows that

$$\frac{dx}{dt} = y, \frac{dy}{dt} = Ay, \frac{dp}{dt} = 0, \frac{dq}{dt} = -p + Aq + (y \cdot p)y \qquad (11.12)$$

are the equations for the Hamiltonian vector field $\vec{h}$ on $G_1 = G_2 = 0$.

Suppose now that we have selected a system of coordinates on N. For instance, $x = (x_0, \ldots, x_n) \in \mathbb{R}^{n+1}$ and $\bar{y} = (y_1, \ldots, y_n)$ on the sphere $||y||^2 = 1$ so that $y_0 = \sqrt{1 - ||\bar{y}||^2}$. Then V_N is described by the following equations:

$$\frac{dx_0}{dt} = \sqrt{1 - ||\bar{y}||^2}, \frac{d\bar{x}}{dt} = \bar{y}, \frac{d\bar{y}}{dt} = y_0 a + \bar{A}\bar{y},$$

where $A = \begin{pmatrix} 0 & -a^T \\ a & \bar{A} \end{pmatrix}$. If $(\xi_0, \xi_1, \ldots \xi_n, \eta_1, \ldots, \eta_n)$ denote the dual coordinates in T^*N then the Hamiltonian lift of V_N is given by $h = \xi_0\sqrt{1 - ||\bar{y}||^2} + \bar{\xi} \cdot \bar{y} + \eta \cdot (y_0 a + \bar{A}\bar{y})$ with

$$\frac{dx_0}{dt} = \sqrt{1 - ||\bar{y}||^2}, \frac{d\bar{x}}{dt} = \bar{y}, \frac{d\bar{y}}{dt} = y_0 a + \bar{A}\bar{y},$$
$$\frac{d\xi}{dt} = 0, \frac{d\eta}{dt} = \frac{\xi_0}{\sqrt{1 - ||\bar{y}||^2}}\bar{y} - \bar{\xi} + \frac{\eta \cdot a}{\sqrt{1 - ||\bar{y}||^2}}\bar{y} + \bar{A}\eta, \qquad (11.13)$$

the Hamiltonian equations of the lift of V_N in these coordinates.

To see that equations (11.12) and (11.13) are the same, use

$$\sum_{i=0}^{n} \xi_i dx_i + \sum_{i=1}^{n} \eta_i dy_i = \sum_{i=0}^{n} p_i dx_i + q_i dy_i. \qquad (11.14)$$

It follows that $\bar{\xi} = p$. Since $y_0 = \sqrt{1 - ||\bar{y}||^2}$, $dy_0 = -\frac{1}{y_0}\sum_{i=1}^{n} y_i dy_i$. Moreover, $y \cdot q = 0$ implies that $q_0 y_0 = -\bar{q} \cdot \bar{y}$, where $\bar{q} = (q_1, \ldots, q_n)$. Therefore, $\eta = \bar{q} - \frac{q_0}{y_0}\bar{y}$. But then, $q_0 y_0 = -\eta \cdot \bar{y} - \frac{q_0}{y_0}||\bar{y}||^2$, which further implies that $q_0 = -y_0(\eta \cdot \bar{y})$. These calculations show that, on T^*N,

$$\bar{q} = \eta - (\eta \cdot \bar{y})\bar{y}, q_0 = -(\eta \cdot \bar{y})\sqrt{1 - ||\bar{y}||^2}. \qquad (11.15)$$

Relations (11.15) transform equation $\frac{dq}{dt} = -p + Aq + (p \cdot y)y$ in (11.12) onto the equation $\frac{d\eta}{dt} = \frac{\xi_0}{\sqrt{1-||\bar{y}||^2}}\bar{y} - \bar{\xi} + \frac{\eta \cdot a}{\sqrt{1-||\bar{y}||^2}}\bar{y} + \bar{A}\eta$ in (11.13). We leave these details to the reader.

We will now consider an optimal problem defined by the following general situation:

(a) a control system

$$\frac{dz}{dt} = F(z(t), u(t)), u(t) \in U \tag{11.16}$$

on a manifold M with U a subset of $\mathbb{R}^m$;
(b) a submanifold N such that its cotangent bundle is given by $G_1 = G_2 = \ldots G_{2(n-k)} = 0$ for some functionally independent functions $G_1, \ldots G_{2(n-k)}$ on T^*M;
(c) additional constraints

$$U_1(z, u) = \ldots, U_l(z, u) = 0, \tag{11.17}$$

with $U_1, \ldots U_l$ smooth functions on $M \times U$ such that the restriction of (11.16) to N with controls $u(t)$ in U and subject to these constraints results in a control system on N

$$\frac{dz}{dt} = F(z(t), u(t)), u(t) \in U, U_1(z, u) = \cdots = U_l(z, u) = 0, \tag{11.18}$$

We aim to obtain the appropriate Hamiltonians associated with the problem of minimizing a functional $\int_0^T f(z(t), u(t)) \, dt$ over the trajectories $(z(t), u(t))$ of system (11.18) that satisfy the given boundary conditions $z(0) = z_0$ and $z(T) = z_1$.

The geodesic problem on a quadric $N = \{x \in \mathbb{R}^{n+1} : A^{-1}x \cdot x = 1\}$ induced by an elliptic metric (x, Dx) is considered as a special case of the above situation. If $x(0) \in N$, then $x(t)$ will remain in N provided that $u(t)$ satisfies the constraint $U_1(x) = A^{-1}x \cdot u = 0$. The geodesic problem then can be formulated as the problem of finding the minimum of the functional $\frac{1}{2} \int_0^T u(t) \cdot Du(t) \, dt$ over the solutions $\frac{dx}{dt} = u(t)$ subject to $U_1 = A^{-1}x(t) \cdot u(t) = 0$.

Control system (11.18) together with the cost f lifts to the cost-extended Hamiltonians h_u on T^*M of the form

$$h_{u,\mu}(\xi) = -\mu f(z, u) + \xi(F(z, u) + \lambda_1 G_1(\xi)$$
$$+ \cdots + \lambda_{2(n-k)} G_{2k}(\xi), \xi \in T_z^*M, \mu = 0, 1,$$

with $\lambda_1, \ldots, \lambda_k$ chosen so that $\{h_u, G_1\} = \cdots = \{h_u, G_{2(n-k)}\} = 0$ for any $u \in U$.

It then follows that $z(t) = \pi(\xi(t))$ is a trajectory of (11.16) generated by $u(t)$, where π denote the natural projection $\pi(\xi) = z$. Moreover, $\vec{h}_{u,\mu}$ is tangent to T^*N whenever the control $u(t)$ satisfies the constraints $U_1 = \cdots = U_l = 0$.

The Maximum Principle states that an optimal trajectory $z(t)$ generated by a control $u(t)$ is the projection of an extremal curve $\xi(t)$ of $\vec{h}_{u(t),\mu}$ restricted to $G_1 = \cdots = G_{2(n-k)} = 0$, subject to further conditions that

$$h_{u(t),\mu}(\xi(t)) \geq h_{v,\mu}(\xi(t)), \tag{11.19}$$

for all $v \in U$ that satisfy $U_1(\pi(\xi(t)), v)) = \cdots = U_l(\pi(\xi(t)), v) = 0$.

For the geodesic problem $\frac{1}{2}\int_0^T u(t) \cdot Du(t)\, dt$ over the quadric surface $A^{-1}x \cdot x = 1$,

$$G_1 = (x \cdot A^{-1}x) - 1 = 0, G_2 = (x \cdot A^{-1}p) = 0,$$

and

$$h_{u,\mu}(x,p) = \mu\frac{1}{2}(Du \cdot u) + p \cdot u + \lambda_1 G_1 + \lambda_2 G_2.$$

If we let $h_u^0 = -\mu\frac{1}{2}(Du \cdot u) + p \cdot u$, then

$$\{h_u, G_1\} = \{h_u^0, G_1\} + \lambda_2\{G_2, G_1\}, \quad \{h_u, G_2\} = \{h_u^0, G_2\} + \lambda_1\{G_1, G_2\}.$$

It follows that

$$\{h_u^0, G_1\} = -2u \cdot A^{-1}x, \quad \{h_u^0, G_2\} = -u \cdot A^{-1}p, \quad \{G_1, G_2\} = 2A^{-1}x \cdot A^{-1}x.$$

Hence,

$$\lambda_1 = -\frac{1}{\{G_1, G_2\}}\{h_u^0, G_2\} = \frac{1}{2}\left(\frac{u \cdot A^{-1}p}{A^{-1}x \cdot A^{-1}x}\right),$$

$$\lambda_2 = -\frac{1}{\{G_2, G_1\}}\{h_u^0, G_1\} = -\frac{u \cdot A^{-1}x}{A^{-1}x \cdot A^{-1}x}.$$

Indeed, the integral curves of $\vec{h}_u$ restricted to $G_1 = G_2 = 0$ are given by

$$\frac{dx}{dt} = \frac{\partial h_u}{\partial p} = u, \frac{dp}{dt} = -\frac{\partial h_u}{\partial x} = -\lambda_1 A^{-1}x.$$

Conversely, every solution that originates in T^*N remains in T^*N for all t for any control $u(t)$ that satisfes $u \cdot A^{-1}x = 0$.

Extremal controls attain the maximum for

$$h_u^0 = \lambda\frac{1}{2}(u \cdot Du) + p \cdot u \tag{11.20}$$

on $G_1 = G_2 = 0$, subject to the constraint $u \cdot A^{-1}x = 0$.

For abnormal extremals the above optimization reduces to maximizing a linear function $p(t) \cdot u$ on $G_1 = G_2 = 0$ subject to $u \cdot A^{-1}x = 0$. This means that $p(t) = 0$, which contradicts the non-degeneracy condition of the Maximum Principle. Therefore abnormal extremals are ruled out.

It follows that the extremal controls are to be found among the critical points of $-\frac{1}{2}(u \cdot Du) + p \cdot u - \alpha_0(u \cdot A^{-1}x)$ for some multiplier α_0, that is, they are the solutions of $-Du + p - \alpha_0 A^{-1}x = 0$. This equation is easily solvable:

$$u = D^{-1}(p - \alpha_0 A^{-1}x) \text{ and } \alpha_0 = \frac{A^{-1}x \cdot D^{-1}p}{D^{-1}A^{-1}x \cdot A^{-1}x}. \qquad (11.21)$$

This choice of control determines the Hamiltonian $h = H + \lambda_1 G_1 + \lambda_2 G_2$, where $H = \frac{1}{2}(D^{-1}(p - \alpha_0 A^{-1}x) \cdot p - \alpha_0 A^{-1}x)$. An easy calculation shows that

$$H = \frac{1}{2}((D^{-1}p \cdot p) - \frac{(D^{-1}p \cdot A^{-1}x)^2}{(D^{-1}A^{-1}x \cdot A^{-1}x)}). \qquad (11.22)$$

The extremal curves are the solutions of

$$\frac{dx}{dt} = \frac{\partial H}{\partial p} = D^{-1}(p - \alpha_0 A^{-1}x),$$

$$\frac{dp}{dt} = -\frac{\partial H}{\partial x} - 2\lambda_1 \frac{\partial G_1}{\partial x} = \alpha_0(A^{-1}(D^{-1}(p - \alpha_0 A^{-1}x)) - 2\lambda_1 A^{-1}x. \quad (11.23)$$

The geodesics are the projections of the extremal curves on energy level $H = \frac{1}{2}$, that is, they satisfy an additional condition:

$$(D^{-1}p \cdot p)(D^{-1}A^{-1}x \cdot A^{-1}x) - (D^{-1}p \cdot A^{-1}x)^2 = D^{-1}A^{-1}x \cdot A^{-1}x. \quad (11.24)$$

Let us single out some relevant cases.

1. The geodesic problem on the ellipsoid In this classic case, initiated by C. Jacobi, the ambient metric is Euclidean. Hence, $D = I$, $\alpha_0 = \frac{A^{-1}x \cdot p}{A^{-1}x \cdot A^{-1}x} = 0$, and $\lambda_1 = \frac{1}{2} \frac{p \cdot A^{-1}p}{A^{-1}x \cdot A^{-1}x}$. Then equations (11.23) reduce to

$$\frac{dx}{dt} = p, \frac{dp}{dt} = -\frac{p \cdot A^{-1}p}{A^{-1}x \cdot A^{-1}x} A^{-1}x. \qquad (11.25)$$

The preceding equations called *Jacobi's equations*, also appear in the writings of J. Moser [Ms1; Ms2].

2. The elliptic problem on the sphere Here the ambient metric is defined by a positive-definite matrix D and $A = I$. Then the Hamiltonian (11.24) can be rewritten as

$$(D^{-1}p \cdot p) - \frac{(D^{-1}p \cdot x)^2}{(D^{-1}x \cdot x)} = 1. \qquad (11.26)$$

Furthermore,

$$\lambda_1 = \frac{1}{2}D^{-1}(p - \alpha_0 x) \cdot p = \frac{1}{2}\left(D^{-1}p \cdot p - \frac{(x \cdot D^{-1}p)^2}{D^{-1}x \cdot x}\right) = \frac{1}{2}, \alpha_0 = \frac{D^{-1}x \cdot p}{D^{-1}x \cdot x}.$$

The Hamiltonian system is given by

$$\frac{dx}{dt} = D^{-1}p - \frac{D^{-1}x \cdot p}{D^{-1}x \cdot x}D^{-1}x, \quad \frac{dp}{dt} = \frac{D^{-1}x \cdot p}{D^{-1}x \cdot x}\left(D^{-1}p - \frac{D^{-1}x \cdot p}{D^{-1}x \cdot x}D^{-1}x\right) - x,$$

$$(11.27)$$

which agrees with equations (11.9) in the previous section when $p = y$.

3. The pseudo-Riemannian metric and the hyperboloid Consider now the case $A^{-1} = -D$, where D is a diagonal matrix with its diagonal entries $(-1, 1, \ldots, 1)$. Then $(x \cdot A^{-1}x) = -x_0^2 + \sum_{i=1}^{n} x_i^2$. Hence N is the hyperboloid $\mathbb{H}^n = \{x \in \mathbb{R}^{n+1} : x_0^2 - \sum_{i=1}^{n} x_i^2 = 1\}$ with its hyperbolic metric $\dot{x} \cdot D\dot{x} = -\dot{x}_0^2 + \sum_{i=1}^{n} \dot{x}_i^2$. In this case, $\alpha_0 = x \cdot p$ and the extremal control u is given by $u = D^{-1}p + \alpha_0 \, x$. The Hamiltonian equations (11.23) reduce to

$$\dot{x} = u, \quad \dot{p} = -\alpha_0 Du + 2\lambda_1 Dx.$$

It follows that

$$\dot{u} = D^{-1}(-\alpha_0 Du + 2\lambda_1 Dx) + \alpha_0 u + \dot{\alpha}_0 x = (2\lambda_1 + \dot{\alpha}_0)x.$$

Furthermore,

$$2\lambda_1 + \dot{\alpha}_0 = 2\lambda_1 - \alpha_0 Du \cdot x + 2\lambda_1 Dx \cdot x + u \cdot p = u \cdot p,$$

since $Dx \cdot x = -1$ and $Du \cdot x = 0$. But $u \cdot p = D^{-1}p \cdot p + (x \cdot p)^2 = 2H$. It follows that the geodesics are the solutions of $\dot{x} = u$, $\dot{u} = x$, that is, $x(t) = x_0 \cosh t + \dot{x}_0 \sinh t$

This procedure for getting the appropriate Hamiltonians in the extraneous systems of coordinates is equally applicable for the problems of mechanics. For instance, if $V(x)$ as any potential function on N, then the corresponding Lagrangian L is given by $L = \frac{1}{2}(Du \cdot u - V(x)$. Then the cost-extended Hamiltonian lift h_u^0 is given by $h_u^0 = -L + p \cdot u$. Proceeding in the same manner as above leads to the same expression for the extremal control $u(t)$ as given by (11.21). The multiplier λ_2 is the same as before, but λ_1 is modified: it is given by $\alpha_1 = \frac{1}{2}\left(\frac{u \cdot A^{-1}p}{A^{-1}x \cdot A^{-1}x}\right) - \frac{dV}{dx} \cdot x$. This leads to the Hamiltonian

$$h = \frac{1}{2}(D^{-1}(p - \alpha_0 A^{-1}x) \cdot (p - \alpha_0 A^{-1}x)) + V(x) + \lambda_1 G_1 + \lambda_2 G_2.$$

$$(11.28)$$

The Hamiltonian equations are then given by the following equations:

$$\frac{dx}{dt} = D^{-1}(p - \alpha_0 A^{-1}x), \quad \frac{dp}{dt} = -\frac{dV}{dx} - \lambda_1 x. \qquad (11.29)$$

In particular, for the mechanical problem of Newmann on the sphere, $D = A = I$ and $V(x) = \frac{1}{2}(Cx \cdot x)$, where C is a symmetric matrix. In this situation, $\alpha_0 = 0$ and $\lambda_1 = ||p||^2 - Cx \cdot x)$. Then equations (11.29) reduce to

$$\frac{dx}{dt} = p, \quad \frac{dp}{dt} = -Cx - (||p||^2 - Cx \cdot x)x. \tag{11.30}$$

These equations agree with the ones obtained in the previous chapter.

Constrained Hamiltonian equations go back to P. Dirac [D]. Fedorov and Jovanovich used his techniques to obtain the right Hamiltonians for the geodesic problems on Steifel manifolds realized as the submanifolds of an ambient Euclidean space [FJ]. J. Moser dealt with ambient coordinates in a somewhat ad-hoc manner. For instance, in deriving the Hamiltonian equations for the problem of Jacobi and the mechanical problem of Newmann, Moser begins by constraining the canonical Hamiltonian $H = \frac{1}{2}||p||^2$ in the ambient spaces so that the corresponding flow leaves the cotangent bundle of the relevant submanifold invariant [Ms2]. In contrast to our methodology based on the Maximum Principle, where the lifted Hamiltonians in the ambient space are first constrained and then the optimization is carried out relative to the controls, Moser optimizes the lifted Hamiltonians over the unconstrained control functions first and then constrains the resulting Hamiltonian so that its the flow leaves the cotangent bundle of the submanifold invariant. For the geodesic problem of Jacobi and the mechanical problem of Newmann both methods lead to the same equations, because the multiplier α_0 is equal to 0, but in general that need not be the case.

For instance on a submaniofld N defined by a matrix A with a metric defined by a matrix D, the canonical Hamiltonian H in the ambient space is equal to $\frac{1}{2}(D^{-1}p \cdot p)$. The constrained Hamiltonian is given by $\mathcal{H} = H + \lambda_1 G_1 + \lambda_2 G_2$, where G_1 and G_2 are the same as before. The conditions $\{\mathcal{H}, G_1\} = \{\mathcal{H}, G_2\} = 0$ yield

$$\lambda_1 = \frac{D^{-1}p \cdot A^{-1}p}{A^{-1}x \cdot A^{-1}x}, \quad \lambda_2 = -\frac{D^{-1}p \cdot A^{-1}x}{A^{-1}x \cdot A^{-1}x}.$$

Then the corresponding flow on T^*N is given by

$$\frac{dx}{dt} = D^{-1}p + \lambda_2 A^{-1}x, \quad \frac{dp}{dt} = -2\lambda_1 A^{-1}x - \alpha_2 A^{-1}p. \tag{11.31}$$

Equations (11.31) are different from equations (11.23).

11.3 Elliptic problem on the sphere and Jacobi's problem on the ellipsoid

Proposition 11.7 *The Hamiltonian systems that correspond to the elliptic problem on the sphere and the geodesic problem on the ellipsoid are symplectomorphic.*

Proof We need to show that (11.25) and (11.27) are symplectomorphic when $D = A$. Let us use (x, y) for the coordinates of the cotangent bundle of the sphere and (q, p) for the coordinates of the cotangent bundle of the ellipsoid $E = q \cdot A^{-1}q - 1 = 0$. In these coordinates, the systems in question are given by

$$\frac{dx}{dt} = u, \ \frac{dy}{dt} = \alpha u - x, \text{ and } \frac{dq}{dt} = p, \ \frac{dp}{dt} = -\frac{A^{-1}p \cdot p}{A^{-1}q \cdot A^{-1}q}A^{-1}q, \quad (11.32)$$

where $u = A^{-1}(y - \alpha x)$ and $\alpha = \frac{A^{-1}x \cdot y}{A^{-1}x \cdot x}$.

Let Φ denote the mapping from the cotangent bundle of the sphere to the cotangent bundle of E defined by

$$q = A^{\frac{1}{2}}x, \ p = A^{-\frac{1}{2}}(y - \alpha x) = A^{\frac{1}{2}}u \quad (11.33)$$

Let $\theta = \sum_{i=0}^{n} p_i dq_i = p \cdot dq, dq \cdot A^{-1}q = 0$, denote the Liouville canonical form on T^*E. Then,

$$\Phi^*\theta = A^{-\frac{1}{2}}(y - \alpha x)A^{-\frac{1}{2}}dx = y \cdot dx - \alpha x dx = y \cdot dx,$$

because $0 = dq \cdot A^{-1}q = A^{\frac{1}{2}}dx \cdot A^{-1}A^{\frac{1}{2}}x = dx \cdot x$. Since Φ^* takes the Liouville form on T^*E to the Liouville form on T^*S^n, it also takes the canonical symplectic form on T^*E to the canonical symplectic form on T^*S_n and hence is a symplectomorphism.

Going back to equations (11.32) we get the following relations:

$$\frac{du}{dt} = -A^{-1}x\left(1 + \frac{d\alpha}{dt}\right)$$

$$1 + \frac{d\alpha}{dt} = \frac{u \cdot u}{A^{-1}x \cdot x}.$$

Then,

$$\frac{dq}{dt} = A^{\frac{1}{2}}\frac{dx}{dt} = A^{\frac{1}{2}}u = p,$$

$$\frac{dp}{dt} = A^{\frac{1}{2}}\frac{du}{dt} = -\left(\frac{u \cdot u}{A^{-1}x \cdot x}\right)A^{-1}x = -\frac{A^{-1}p \cdot p}{A^{-1}q \cdot A^{-1}q}A^{-1}q,$$

and thus Φ_* takes the Hamiltonian flow on the sphere onto the Hamiltonian flow on the ellipsoid. $\qquad\qquad\square$

Proposition 11.8 *Jacobi's equations are completely integrable. Functions*

$$G_k = p_k^2 + \sum_{j=1, j\neq k}^{n+1} \frac{(q_j p_k - q_k p_j)^2}{(a_k - a_j)}, k = 1, \ldots, (n+1) \qquad (11.34)$$

are constants of constants of motion in involution with each other.

Proof We have shown that the functions

$$F_k = x_k^2 + \sum_{j=0, j\neq k}^{n} \frac{(x_j y_k - x_k y_j)^2}{(a_k - a_j)}, k = 0, \ldots, n$$

are an involutive family and are integrals of motion for both the elliptic-geodesic problem on the sphere and the Newmann's mechanical problem (Proposition 11.2 and Proposition 10.15, Chapter 10). We have also shown that the above integrals of motion are the residues of the function

$$F(z) = (R_z x \cdot x) + (R_z x \cdot x)(R_z y \cdot y) - (R_z x \cdot y)^2, R_z = (zI - A)^{-1},$$

see Lemma 10.14 in Section 10.4 of Chapter 10. We will now show that functions (11.34) are the residues of the pull-back of F under the symplecto-morphism Φ. First note that F remains unchanged if the variable y is replaced by $y + \alpha x$ with α an arbitrary number. Since $A^{\frac{1}{2}} p = y - \frac{(A^{-1} x \cdot y)}{A^{-1} x \cdot x} x$, we may replace y by $A^{\frac{1}{2}} p$ and x by $A^{-\frac{1}{2}} q$. In addition, note that

$$p \cdot p = A^{-1} y \cdot y - \frac{(A^{-1} x \cdot y)^2}{(A^{-1} x \cdot x)} = 1.$$

Then,

$$1 + R_z y \cdot y = 1 + R_z A p \cdot p = 1 + \sum_{k=1}^{n+1} \frac{a_k p_k^2}{z - a_k} = \sum_{k=1}^{n+1} p_k^2 + \frac{a_k p_k^2}{z - a_k}$$

$$= z \sum_{k=1}^{n+1} \frac{p_k^2}{z - a_k} = z R_z p \cdot p,$$

$$R_z x \cdot x = R_z A^{-1} q \cdot q = \sum_{k=1}^{n+1} \frac{q_k^2}{a_k (z - a_k)}$$

$$= \frac{1}{z} \sum_{k=1}^{n+1} \frac{q_k^2}{a_k} + \frac{q_k^2}{z - a_k} = \frac{1}{z}(1 + R_z q \cdot q),$$

and

$$R_z x \cdot y = R_z q \cdot p.$$

It follows that

$$F(z) = (R_z p \cdot p)(1 + R_z q \cdot q) - (R_z q \cdot p)^2 \tag{11.35}$$

is constant along the solutions of Jacobi's equations. A calculation identical to the one used for the Newmann's system shows that

$$G_k = p_k^2 + \sum_{j=1, j \neq k}^{n+1} \frac{(q_j p_k - q_k p_j)^2}{(a_k - a_j)}, k = 1, \ldots, (n+1)$$

are the residues of F, and hence are integrals of motion for Jacobi's equations. □

In the literature on integrable systems, the integrals of motion for the Newmann's problem are related to the integrals of motion for the problem of Jacobi through the transformation of H. Knörrer that transforms the Newmann's equations on energy level $H = 0$ onto the equations of Jacobi [Kn; Ms2]. Our exposition bypasses this remarkable and somewhat enigmatic transformation. Instead it derives its results from the fact that the Hamiltonian systems associated with the Newmann's problem and the elliptic problem on the sphere are the restrictions of affine Hamiltonian systems that admit the same spectral representation. Alternatively, the Hamiltonian corresponding to the elliptic geodesic problem on the sphere could be associated with the Maupertuis metric on energy level $H = 0$ [AKN]. This observation may explain why the elliptic problem on the sphere and the mechanical problem of Newmann have the same integrals of motion, and since the geodesic problem on the sphere is essentially the same as the Jacobi's problem on the ellipsoid, it then may also explain the origins behind the enigmatic integrals of motion in (11.34).

Analogous investigations on rank two orbits would reveal that both the Newmann problem and the elliptic problem are also integrable on the Stiefel and the Grassmannian manifolds, a fact already observed in [FJ] for the problem of Newmann. A more detailed study along these lines would undoubtedly result in beautiful extensions of the classical theory initiated by Jacobi.

11.4 Elliptic coordinates on the sphere

Let us now address the existence of a suitable system of coordinates in which the elliptic Hamiltonian equations for the elliptic problem can be integrated by quadrature. Jacobi, in his famous treatise on Dynamics, states that the main difficulty in integrating differential equations lies precisely in the absence of any systematic way of finding a system of coordinates that is suitable for resolving the given equations. Instead, he proposed an opposite approach largely inspired by his discovery of elliptic coordinates, of finding a class of problems which are solvable in a given system of coordinates [Jb, Ch. 26]. Jacobi's method of integrating the geodesic equations of the ellipsoid led to the discovery of a method of integration, known today as the method of separation of variables [LL; Wh]. It is one of the principal methods of integrating the equations of mathematical physics [Pr].

Below we will show that it the geodesic elliptic problem on the sphere is integrable in the elliptic system of coordinates. To introduce the elliptic coordinates, let $(\lambda_0 \ldots, \lambda_n)$ denote the zeros of

$$\alpha(z) = \sum_{k=0}^{n} \frac{a_k x_k^2}{z - a_k} + 1, \tag{11.36}$$

which is associated with a non-singular diagonal matrix A with its diagonal terms $0 < a_0 < a_1 < \cdots < a_n$. Evidently, α is a monotone decreasing function in each interval $(-\infty, a_0), (a_i, a_{i+1}), i = 0, \ldots, n, (a_n, \infty)$. In the interval $(-\infty, a_0)$, α starts from 1 at $-\infty$ and ends at $-\infty$ at a_0. Hence the first zero is in the interval $(-\infty, a_0)$. Each subsequent zero λ_i belongs to the interval (a_i, a_{i+1}) since $\alpha(z)$ starts at ∞ at $z = a_i$, and decreases to $-\infty$ at $z = a_{i+1}$. In the interval (a_n, ∞) α is positive, and hence, this interval contains no zeros of α.

Proposition 11.9   *Let $m(z) = \prod_{i=0}^{n}(z - \lambda_i)$. Then*

$$a_k x_k^2 = \frac{m(a_k)}{a'(a_k)} = \frac{\prod_{j=0}^{n}(a_k - \lambda_j)}{\prod_{j \neq k}^{n}(a_k - a_j)}, \quad k = 0, \ldots, n, \tag{11.37}$$

where $a(z) = \prod_{i=0}^{n}(z - a_i)$.

Proof $\alpha(z)$ is a rational function and hence is defined by its zeros and poles. It follows that $F(z$ can be written as $c\frac{m(z)}{a(z)}$, where $c = \lim_{z \to \infty} \alpha(z)$. Since $\lim_{z \to \infty} \alpha(z) = 1$,

$$\sum_{k=0}^{n} \frac{a_k x_k^2}{z - a_k} + 1 = \frac{m(z)}{a(z)}.$$

But then

$$a_k x_k^2 = \lim_{z \to a_k} (z - a_k) \left(\sum_{j=0}^{n} \frac{a_j x_j^2}{z - a_j} + 1 \right) = \lim_{z \to a_k} (z - a_k) \frac{m(z)}{a(z)} = \frac{m(a_k)}{a'(a_k)},$$

because $a'(a_k) = \lim_{z \to a_k} \frac{a(z)}{z - a_k}$. $\square$

If x is a point on the unit sphere then the first zero λ_0 is zero. The remaining zeros $\lambda_1, \ldots, \lambda_n$ are called *elliptic coordinates* of x. It follows that

$$x_k = \pm \sqrt{\frac{m(a_k)}{a_k a'(a_k)}}, k = 0, \ldots, n.$$

The following proposition, which originated with Jacobi [Jc], is essential for some computations below.

Proposition 11.10 *Let $g(z)$ be a polynomial of degree n having n real roots $a_1, \ldots, a_n$ and let $f(z)$ be any polynomial of degree less than n. Then*

$$\frac{f(z)}{g(z)} = \sum_{k=1}^n \frac{f(a_k)}{g'(a_k)(z - a_k)}, \text{ and } \lim_{z \to \infty} z \frac{f(z)}{g(z)} = \sum_{k=1}^n \frac{f(a_k)}{g'(a_k)}$$

Proof $\frac{f(z)}{g(z)} = \sum_{j=1}^n \frac{A_k}{z - a_k}$ by the partial fractions expansion. Then

$$A_k = \lim_{z \to a_k} (z - a_k) \frac{f(z)}{g(z)} = \frac{f(a_k)}{g'(a_k)}.$$

The second equality readily follows from the first expression. $\qquad\square$

Corollary 11.11 $\sum_{k=1}^n \frac{f(a_k)}{g'(a_k)} = 0$ *if* $\deg(f) < n - 1$, *and* $\sum_{k=1}^n \frac{f(a_k)}{g'(a_k)} = 1$ *if* $\deg(f) = n - 1$.

Both of these statements follow directly from the second formula in the preceding proposition.

Since each point x on S^n is parametrized by $\lambda_1, \ldots, \lambda_n$, each directional derivative $\frac{\partial x}{\partial \lambda_j}$ defines a vector field on S^n. Their essential properties are given by:

Proposition 11.12 (a) $\frac{\partial x}{\partial \lambda_j} = \frac{1}{2}(\lambda_j I - A)^{-1} x, j = 1, \ldots, n.$

(b) $\left(A \frac{\partial x}{\partial \lambda_i} \cdot \frac{\partial x}{\partial \lambda_j} \right) = -\frac{1}{4} \frac{m'(\lambda_j)}{a(\lambda_j)} \delta_{ij}.$

Proof Use logarithmic differentiation in (11.37). Since $\ln(a_k x_k^2) = \ln m(a_k) - \ln a'(a_k)$, $\frac{2}{a_k x_k^2} a_k x_k \frac{\partial x_k}{\partial \lambda_j} = -\frac{1}{a_k - \lambda_j}$. Therefore, $\frac{\partial x_k}{\partial \lambda_j} = \frac{1}{2}(\lambda_j - a_k)^{-1} x_k$. This proves part (a).

Then,

$$\left(A \frac{\partial x}{\partial \lambda_i} \cdot \frac{\partial x}{\partial \lambda_j} \right) = \frac{1}{4} \sum_{k=0}^n a_k x_k^2 (\lambda_i - a_k)^{-1} (\lambda_j - a_k)^{-1}.$$

When $i \neq j$,

$$\sum_{k=0}^{n} a_k x_k^2 (\lambda_i - a_k)^{-1} (\lambda_j - a_k)^{-1}$$

$$= \frac{1}{(\lambda_i - \lambda_j)} \sum_{k=0}^{n} \frac{a_k x_k^2}{a_k - \lambda_i} - \frac{a_k x_k^2}{a_k - \lambda_j} = \frac{1}{(\lambda_i - \lambda_j)}(1 - 1) = 0.$$

For $i = j$, we need to differentiate $\sum_{k=0}^{n} \frac{a_k x_k^2}{z - a_k} = \frac{m(z)}{a(z)} + 1$. It follows that

$$-\sum_{k=0}^{n} \frac{a_k x_k^2}{(z - a_k)^2} = \frac{m'(z)}{a(z)} - m(z)\frac{a'(z)}{a^2(z)}. \tag{11.38}$$

The substitution $z = \lambda_j$ in (11.38) gives $\sum_{k=0}^{n} \frac{a_k x_k^2}{(\lambda_j - a_k)^2} = -\frac{m'(\lambda_j)}{a(\lambda_j)}$. Therefore,

$$\left(A\frac{\partial x}{\partial \lambda_j} \cdot \frac{\partial x}{\partial \lambda_j} \right) = \frac{1}{4} \sum_{k=0}^{n} a_k x_k^2 (\lambda_j - a_k)^{-2} = -\frac{1}{4}\frac{m'(\lambda_j)}{a(\lambda_j)}.$$

$\square$

It follows that $X_1(\lambda) = \frac{\partial x}{\partial \lambda_1}, \ldots, X_n(\lambda) = \frac{\partial x}{\partial \lambda_n}$ is an orthogonal frame on S^n, relative to the elliptic metric, hence any tangent vector $\dot{x}$ can be written as $\dot{x} = \sum_{i=1}^{n} X_i(\lambda)\dot{\lambda}_i$. Then, the elliptic metric $\frac{1}{2}(A\dot{x}, \dot{x})$ can be expressed in terms of the elliptic coordinates via the formula

$$\frac{1}{2}(A\dot{x} \cdot \dot{x}) = -\frac{1}{8} \sum_{j=1}^{n} \frac{m'(\lambda_j)}{a(\lambda_j)}\dot{\lambda}_j^2 = \frac{1}{2}(\dot{\lambda} \cdot M\dot{\lambda}),$$

where M is a diagonal matrix with its diagonal entries $-\frac{1}{4}\frac{m'(\lambda_1)}{a(\lambda_1)}, \ldots, -\frac{1}{4}\frac{m'(\lambda_n)}{a(\lambda_n)}$. After an easy calculation, which we will skip, we obtain the corresponding Hamiltonian

$$H = -\frac{1}{2}\left(4\sum_{j=1}^{n} \frac{m'(\lambda_j)}{a(\lambda_j)}\left(\frac{a(\lambda_j)}{m'(\lambda_j)} \right)^2 \mu_j^2 \right) = -2\sum_{j=1}^{n} \frac{a(\lambda_j)}{m'(\lambda_j)}\mu_j^2,$$

where $\mu = (\mu_1, \ldots, \mu_n)$ denote the dual coordinates given by $\mu = M\dot{\lambda}$. The Hamiltonian equations are then given by the usual formulas:

$$\dot\lambda_j = \frac{\partial H}{\partial \mu_j} = -4\frac{a(\lambda_j)}{m'(\lambda_j)}\mu_j, \dot\mu_j = -\frac{\partial H}{\partial \lambda_j}$$

$$= 2\sum_{k=1}^{n}\mu_k^2\frac{\partial}{\partial\lambda_j}\left(\frac{a(\lambda_k)}{m'(\lambda_k)}\right) = -2\sum_{k\neq j}^{n}\mu_k^2\frac{a(\lambda_k)}{m'(\lambda_k)}\frac{\partial}{\partial\lambda_j}m'(\lambda_k) + 2\mu_j^2\frac{\partial}{\partial\lambda_j}\left(\frac{a(\lambda_j)}{m'(\lambda_j)}\right)$$

(11.39)

$$= 2\sum_{k\neq j}^{n}\frac{1}{\lambda_k-\lambda_j}\mu_k^2\frac{a(\lambda_k)}{m'(\lambda_k)} + 2\mu_j^2\frac{a(\lambda_j)}{m'(\lambda_j)}\sum_{k\neq j}^{n}\frac{1}{\lambda_j-a_k} - 2\mu_j^2 a(\lambda_j)\frac{m''(\lambda_j)}{m'(\lambda_j)^2}.$$

Rather than attempting to separate the variables in (11.39) via the classical methods, we will proceed via the equations

$$\frac{dx}{dt} = u, \frac{dy}{dt} = \alpha u - x, u = A^{-1}(y - \alpha x), \alpha = \frac{A^{-1}x\cdot y}{A^{-1}x\cdot x},$$

and their constants of motion

$$F_k = x_k^2 + \sum_{j=0,j\neq k}^{n}\frac{(x_j y_k - x_k y_j)^2}{(a_k - a_j)}, k = 0,\ldots,n.$$

We will now express the preceding equations in terms of the elliptic coordinates on the level surface $F_k = c_k, k = 0,\ldots,n$, defined by some numbers $c_0,\ldots,c_n$. Such constants must satisfy two auxiliary conditions

$$\sum_{k=0}^{n}c_k = 1, \text{ and } \sum_{k=0}^{n}\frac{c_k}{a_k} = 0,$$

for the following reasons:

$$F(z) = (R_z x\cdot x) + (R_z x\cdot x)(R_z y\cdot y) - (R_z x\cdot y)^2 = \sum_{k=0}^{n}\frac{F_k}{z - a_k}, R_z = (z - A)^{-1},$$

implies that

$$F(0) = -(A^{-1}x\cdot x) + (A^{-1}x\cdot x)(A^{-1}y\cdot y) - (A^{-1}x\cdot y)^2 = 2H = -\sum_{i=0}^{n}\frac{F_k}{a_k}.$$

The elliptic geodesics satisfy $H = 0$, hence, $\sum_{k=0}^{n}\frac{c_k}{a_k} = 0$. Additionally,

$$1 = \sum_{i=0}^{n}x_k^2 = \lim_{z\to\infty}zF(z) = \sum_{i=0}^{n}F_k = \sum_{i=0}^{n}c_k.$$

Let now $b_0,\ldots,b_m$ denote the distinct zeros of $F(z) = \sum_{k=0}^{n}\frac{c_k}{z-a_k}$ and let $b(z) = \prod_{i=0}^{m}(z - b_i)$. The poles of $F(z)$ occur at $z = a_k$. Recall that $a(z) = det(A - zI) = \prod_{i=0}^{n}(z - a_k)$. Then,

$$F(z) = \sum_{k=0}^{n} \frac{c_k}{z - a_k} = c \frac{b(z)}{a(z)},$$

for some constant c. Since

$$1 = \sum_{k=0}^{n} c_k = \lim_{z \to \infty} zF(z) = c \lim_{z \to \infty} \frac{zb(z)}{a(z)},$$

$m = n - 1$ and $c = 1$. Moreover, $\sum_{k=0}^{n} \frac{c_k}{a_k} = 0$ implies that $z = 0$ is one of the zeros of $F(z)$.

Since $c_k = \frac{b(c_k)}{a'(c_k)}, k = 0, \ldots, n$, the correspondence from $\{(c_0, \ldots, c_n) : \sum_{k=0}^{n} c_k = 1, \sum_{k=0}^{n} \frac{c_k}{a_k} = 0\}$ onto the zeros $b_0 = 0, b_1, \ldots, b_{n-1}$ is one to one.

Proposition 11.13 *Let $\lambda(t) = (\lambda_1(t), \ldots, \lambda_n(t))$ denote the elliptic coordinates of $x(t)$ corresponding to a solution $(x(t), y(t))$ of the elliptic system*

$$\frac{dx}{dt} = u, \quad \frac{dy}{dt} = \alpha u - x, u = A^{-1}(y - \alpha x), \alpha = \frac{A^{-1}x \cdot y}{A^{-1}x \cdot x},$$

on the level surface $F_k = c_k, k = 0, \ldots, n$. Then $\lambda(t)$ is a solution of the following system of equations:

$$\frac{d\lambda_j}{dt} = \pm \frac{2}{m'(\lambda_j)} \sqrt{-a(\lambda_j)b(\lambda_j)}, j = 1, \ldots, n. \tag{11.40}$$

Proof Let $u(t) = \sum_{k=1}^{n} P_k(t) \frac{\partial x}{\partial \lambda_k}(x(t), y(t))$ for some functions $P_1(t), \ldots, P_n(t)$. Then,

$$\sum_{k=1}^{n} \frac{\partial x}{\partial \lambda_k} \frac{d\lambda_k}{dt} = \frac{dx}{dt} = u(t) = \sum_{k=1}^{n} P_k(t) \frac{\partial x}{\partial \lambda_k},$$

implies that $\frac{d\lambda_k}{dt} = P_k$ for each k. First, note that

$$R_{\lambda_j}x \cdot x = \sum_{l=0}^{n} \frac{a_l x_l^2}{a_l(\lambda_j - a_l)} = -\frac{1}{\lambda_j} \sum_{l=0}^{n} a_l x_l^2 \left(\frac{1}{a_l} - \frac{1}{a_l - \lambda_j} \right)$$

$$= -\frac{1}{\lambda_j} \left(\sum_{l=0}^{n} x_l^2 - \sum_{l=0}^{n} \frac{a_l x_l^2}{a_l - \lambda_j} \right) = 0.$$

Hence, $F(\lambda_j) = -(R_{\lambda_j} x, y)^2$. Now,

$$R_{\lambda_j} x \cdot y = R_{\lambda_j} x \cdot (y - \alpha x) = \langle R_{\lambda_j} x, A^{-1}(y - \alpha x)\rangle = \langle R_{\lambda_j} x, u\rangle$$

$$= \sum_{k=1}^{n} \left\langle R_{\lambda_j} x, P_k \frac{\partial x}{\partial \lambda_k}\right\rangle = \frac{1}{2}\sum_{k=1}^{n} P_k \langle (\lambda_j - A)^{-1}x, (\lambda_k - A)^{-1}x\rangle$$

$$= \frac{1}{2}\sum_{k=1}^{n} P_k \sum_{l=0}^{n} \frac{a_l x_l^2}{(\lambda_j - a_l)(\lambda_k - a_l)}$$

$$= \frac{1}{2}P_j \sum_{l=0}^{n} \frac{a_l x_l^2}{(\lambda_j - a_l)^2} = -\frac{1}{2}P_j \frac{m'(\lambda_j)}{a(\lambda_j)}.$$

The last equality follows from (11.38) and the equation directly above it. Hence, $F(\lambda_j) = -\frac{1}{4}P_j^2 \left(\frac{m'(\lambda_j)}{a(\lambda_j)}\right)^2 = \frac{b(\lambda_j)}{a(\lambda_j)}$, which yields $P_j = \pm\frac{2}{m'(\lambda_j)}$ $\sqrt{-a(\lambda_j)b(\lambda_j)}$. $\square$

Alternatively, we can take advantage of the relations

$$\sum_{k=0}^{n} \frac{\lambda_k^j}{m'(\lambda_k)} = \delta_{j(n-1)}, \tag{11.41}$$

obtained in the corollary to Proposition 11.10, and write (11.40) as

$$\sum_{k=0}^{n} \frac{\lambda_k^j d\lambda_k}{\sqrt{f(\lambda_k)}} = \pm 2\delta_{j(n-1)}dt, j = 0, \dots, n-1, \tag{11.42}$$

which is obtained by the substitution $m'(\lambda_k)dt = \pm 2\sqrt{-a(\lambda_k)b(\lambda_k)}d\lambda_k$ into (11.41).

Equation (11.41) appears first in Jacobi's book on dynamics in the chapter on Abel's theorem and the solutions of differential equations [Jb, Lecture 30]. It also appears in Moser's book of his treatment of Newmann's problem [Ms2].

Moser points out that equation (11.42) is naturally related to the Jacobi map of the Riemann surface $w^2 = -4f(z), f(z) = a(z)b(z)$. The surface is a hyperelliptic curve of genus n, with branch points at $b_0, a_0, b_1, a_1, \dots, b_{n-1}, a_n$. The Jacobi map is given by

$$\sum_{k=0}^{n} \int_{0,0}^{\lambda_k, w_k} \frac{z^j dz}{2\sqrt{-f(z)}} = s_j, j = 0, \dots, n.$$

In the variables $s_0, \dots, s_n$, the solutions of (11.42) are given by

$$s_j(t) = \delta_{j(n-1)}t + s_j(0), j = 0, \dots, n.$$

In particular, the solutions are quasi-periodic with at most n frequencies.

12

Rigid body and its generalizations

A rigid body that is free to rotate around a fixed point in a three-dimensional Euclidean space under the gravitational force is known as the heavy top in the literature on mechanics. Its equations of motion are described by six parameters $c_1, c_2, c_3, I_1, I_2, I_3$, where c_1, c_2, c_3 are the coordinates of the center of mass of the body relative to the fixed point around which the body rotates, in an orthonormal frame affixed to the body at the center of mass; I_1, I_2, I_3 are the principal moments of inertia determined by the shape of the body. Remarkably, the solutions of the equations of motion are explicitly known in only four cases:

1. The center of mass coincides with the fixed point. This case is known as *the top of Euler.*
2. The principal moments of inertia are all equal and c_1, c_2, c_3 are arbitrary.
3. Two principal moments of inertia are equal and c is along the axis that corresponds to the remaining moment of inertia. This case is known as *the top of Lagrange.*
4. Two principal moments of inertia are equal and are twice the remaining principal moment and c lies in the plane defined by the equal moments of inertia. This case is known as *the top of Kowalewski.*

These cases are known as the integrable tops and it is commonly believed that these are the only cases that are integrable. The search for integrable cases and the integrals of motion was principally responsible for the growth of the theory of integrable systems. For this reason, the top plays a somewhat sacrosanct role in the theory of integrable systems.

Our interest in this subject is twofold. Firstly, we would like to derive its equations of motion in a self-contained manner with a particular interest on the passage to Lie groups and the general theory of integrable systems on coadjoint orbits. Secondly, we are interested in its relations with another famous equation of applied mathematics, Kirchhoff's elastic rod equation.

Kirchhoff in his remarkable paper of 1859 not only wrote the differential equations for the equilibrium configurations of an elastic rod in $\mathbb{R}^3$ subject to bending and twisting torques at its ends, but he also likened the elastic equations to the motions of the heavy top, in a statement known ever since as "Kirchhoff's kinetic analogue" (see [Lv] for the exact reference).

According to Kirchhoff, an elastic rod is modeled by a curve $\gamma(t)$, called the central line of the rod, together with an orthonormal frame $v_1(t), v_2(t), v_3(t)$ along γ that measures the amount of twisting and bending along the rod relative to a fixed reference frame defined by the rod in its unstressed mode. The usual assumptions are that the rod is inextensible, which implies that $\left\| \frac{d\gamma}{dt} \right\| = 1$, i.e., that γ is parametrized by arc length and, secondly, that the frame is adapted to the central line by $v_1(t) = \frac{d\gamma}{dt}(t)$. Kirchhoff's model for the equilibrium configurations of the rod postulates that the equilibrium configurations correspond to the minima of the elastic energy $V = \int_0^L (A_1 u_1^2(t) + A_2 u_2^2(t) + A_3 u_3^2(t))\,dt$, where A_1, A_2, A_3 are constants determined by the physical characteristics of the rod, L is the length of the rod, and $u_1(t), u_2(t), u_3(t)$ are the strains along the rod.

If the relation between the frame and the central line is taken in a more general form as

$$\frac{d\gamma}{dt}(t) = a_1 v_1(t) + a_2 v_2(t) + a_3 v_3(t),$$

for some constants a_1, a_2, a_3, then the configurations of the rod are described by six parameters, $a_1, a_2, a_3, A_1, A_2, A_3$. Remarkably, the Hamiltonian equations that correspond to the elastic problem of Kirchhoff are integrable exactly under the same conditions as the equations of the heavy top under the correspondence

$$(a_1, a_2, a_3) \leftrightarrow (c_1, c_2, c_3) \text{ and } (A_1, A_2, A_3) \leftrightarrow (I_1, I_2, I_3).$$

This phenomenon, along with its many variants together with Lie algebraic extensions of the equations of the top, forms the the main body of this chapter. Let us begin with the Euler top.

12.1 The Euler top and geodesic problems on $SO_n(R)$

Equations of the heavy top are derived under the assumption that the ambient space in which the body is situated is equipped with an orthonormal frame

$\vec{e}_1, \vec{e}_2, \vec{e}_3$ centered at some point O_f of the space. Then each point P of the ambient space can be represented by the vector $\overrightarrow{O_fP}$, and the latter can be represented by the coordinate vector $q = (q_1, q_2, q_3)$ defined by $O_fP = q_1\vec{e}_1 + q_2\vec{e}_2 + q_3\vec{e}_3$. The frame $\vec{e}_1, \vec{e}_2, \vec{e}_3$ is called *fixed* or *absolute*, and q_1, q_2, q_3 are called the *space* coordinates of P.

In addition to the fixed frame, there is another orthonormal frame $\vec{f}_1, \vec{f}_2, \vec{f}_3$, which is rigidly affixed to the body at its stationary point O_b. It is called the *moving* frame. Then each point P of the body can be described by the vector $\overrightarrow{O_bP}$ and the coordinates $Q = (Q_1, Q_2, Q_3)$ given by $O_bP = Q_1\vec{f}_1 + Q_2\vec{f}_2 + Q_3\vec{f}_3$. Coordinates Q_1, Q_2, Q_3 are called the *body* coordinates of P. Since each point of the body remains stationary relative to the moving frame, coordinates Q remain stationary during each motion of the body. It is convenient to place the origin of the fixed frame O_f at O_b.

In what follows, points will be identified with their coordinate vectors; vectors relative to the moving frame will be denoted by small letters and vectors relative to the moving frame will be denoted by capital letters. Then the passage from the body coordinates to the space coordinates is described by a rotation matrix R whose i-th column vector is $\vec{f}_i$, i.e., $\vec{f}_i = \sum_{j=1}^{3} R_{ji}\vec{e}_j$, $j = 1, 2, 3$. Then,

$$\sum_{i=1}^{3} Q_i\vec{f}_i = \sum_{i=1}^{3}\sum_{j=1}^{3} Q_i R_{ji}\vec{e}_j = \sum_{j=1}^{3} q_j\vec{e}_j,$$

implies that $q = RQ$. It is commonly assumed that the frames are oriented so that $Det(R) = 1$, which then implies that $R \in SO_3(R)$.

It follows that any motion of a point P induces a curve $R(t)$ in $SO_3(R)$ so that $q(t) = R(t)Q$. Since the body is rigid, the distance $||q_1 - q_2||$ remains constant during the motions of any two points P_1 and P_2 on the body, that is, $||q_1 - q_2|| = ||Q_1 - Q_2||$. This means that if the motion of one point P of the body is described by a rotation $R(t)$, then the motion of any other point of the body is described by the same rotation $R(t)$.

This remarkable fact that the motions of a rigid body can be mirrored by curves in the rotation group $SO_3(R)$ is linked with another, perhaps even more remarkable fact, that $\mathbb{R}^3$ with its vector product is a Lie algebra, isomorphic to $so_3(R)$ that led to the notion that $SO_3(R) \times so_3(R)$ is a natural configuration space for the motions of the top. This discovery had profound effect on the subsequent interest in the Hamiltonian systems on Lie groups and is largely responsible for the common perseption that a Hamiltonian system on a Lie group is a mechanical system in disguise [RT]. We will come to this point in more detail later on in the text.

In the meantime let us introduce some notations that facilitate the passage between vectors in $\mathbb{R}^3$ and matrices in $so_3(R)$. Matrices

$$U = \begin{pmatrix} 0 & -u_3 & u_2 \\ u_3 & 0 & -u_1 \\ -u_2 & u_1 & 0 \end{pmatrix}$$

in $so_3(R)$ are in one to one correspondence with the vectors

$$\hat{U} = \begin{pmatrix} u_1 \\ u_2 \\ u_3 \end{pmatrix}$$

in $\mathbb{R}^3$. Recall that the vector product in $\mathbb{R}^3$ is given by

$$a \times b = \begin{pmatrix} a_2 b_3 - a_3 b_2 \\ a_3 b_1 - a_1 b_3 \\ a_1 b_2 - a_2 b_1 \end{pmatrix}, a = \begin{pmatrix} a_1 \\ a_2 \\ a_3 \end{pmatrix}, b = \begin{pmatrix} b_1 \\ b_2 \\ b_3 \end{pmatrix}.$$

The reader can easily verify the following relations:

$$Aq = \hat{A} \times q, \ \widehat{[A,B]} = \hat{B} \times \hat{A}, \tag{12.1}$$

for any matrices A and B and any vector q (remember that $[A, B] = BA - AB$). This relation implies that the mapping $U \to \hat{U}$ is a Lie algebra isomorphism between $so_3(R)$ and $\mathbb{R}^3$ with $\mathbb{R}^3$ equipped with the Lie bracket $[a, b] = b \times a$. The above mapping is also an isometry from $so_3(R)$ with its trace metric $\langle A, B \rangle = -\frac{1}{2} Tr(AB)$ onto $\mathbb{R}^3$ with its Euclidean metric (,). On occasions we will also make use of the inverse operation $\check{u} = U$ that assigns

$$U = \begin{pmatrix} 0 & -u_3 & u_2 \\ u_3 & 0 & -u_1 \\ -u_2 & u_1 & 0 \end{pmatrix}$$

to each vector vector $u \in \mathbb{R}^3$.

Let us now return to the motions $q(t) = R(t)Q$. The tangent vector $\frac{dR}{dt}$ can be represented either in the right-invariant form as $\frac{dR}{dt} = A(t)R(t)$ for some skew-symmetric matrix $A(t)$, or it can be represented in the left-invariant form as $\frac{dR}{dt} = R(t)B(t)$. Of course, A and B are related through

$$B(t) = R^T(t)A(t)R(t). \tag{12.2}$$

We will presently show that both of these representations have a role to play in the dynamics of the rigid body. If

$$\hat{A}(t) = \begin{pmatrix} \omega_1(t) \\ \omega_2(t) \\ \omega_3(t) \end{pmatrix} \quad \text{and} \quad \hat{B}(t) = \begin{pmatrix} \Omega_1(t) \\ \Omega_2(t) \\ \Omega_3(t) \end{pmatrix}$$

then $\omega(t) = \omega_1(t)e_1 + \omega_2(t)e_2 + \omega_3(t)e_3$ is called *the angular velocity relative to the fixed frame* and $\Omega(t) = \Omega_1(t)a_1 + \Omega_2(t)a_2 + \Omega_3(t)a_3$ is called *the angular velocity relative to the moving frame*. They are related by the following formula:

$$R(t)\Omega(t) = \omega(t). \tag{12.3}$$

The demonstration is easy:

$$\hat{\Omega} \times q = Bq = (R^{-1}AR)q = R^{-1}(\hat{A} \times Rq) = R^{-1}(\omega \times Rq) = R^{-1}\omega \times q.$$

Since q is arbitrary, $\Omega = R^{-1}\omega$. Furthermore, $\frac{dq}{dt} = ARQ = Aq = \hat{A} \times q = \omega \times q$, and $R^{-1}\frac{dq}{dt} = R^{-1}Aq = BR^{-1}q = BQ = \hat{B} \times Q = \Omega \times Q$. Hence,

$$\omega \times q = R(\Omega \times Q). \tag{12.4}$$

The angular momentum $\lambda(t)$ of a particle of mass m along a path $q(t)$ is defined as $\lambda(t) = \lambda_1(t)e_1 + \lambda_2(t)e_2 + \lambda_3(t)e_3$, where $\lambda(t) = m(q(t) \times \frac{dq}{dt}) = m(q(t) \times (\omega(t) \times q(t)))$. Relative to the moving frame, the angular momentum is given by $\Lambda(t) = R^{-1}(t)\lambda(t)$. Then

$$\Lambda(t) = R^{-1}\lambda = mR^{-1}(q \times (\omega \times q)) = m(R^{-1}q \times R^{-1}(\omega \times q))$$
$$= m(Q \times (\Omega \times Q)).$$

Hence,

$$\lambda = m(q \times (\omega \times q)), \text{ and } \Lambda = m(Q \times (\Omega \times Q)). \tag{12.5}$$

The operator $\Lambda_Q(\Omega) = m(Q \times (\Omega \times Q))$ is called the *inertia tensor* at Q [Ar].

Proposition 12.1 *The inertia tensor Λ_Q is semi-positive-definite. Its kernel consists of the angular velocities colinear with Q.*

Proof We shall use the fact that $\mathbb{R}^3$ with its Euclidean metric is isomorphic and isometric with $so_3(R)$ with the trace metric. Let Ω_1 and Ω_2 be any vectors in $\mathbb{R}^3$. Then

$$(\Lambda_Q(\Omega_1), \Omega_2) = (Q \times (\Omega_1 \times Q), \Omega_2) = ([[Q, \Omega_1], Q], \Omega_2)$$
$$= ([Q, \Omega_1], [Q, \Omega_2]) = ([Q, \Omega_2], [Q, \Omega_1]) = (\Lambda_Q(\Omega_2), \Omega_1).$$

The above shows that $(\Lambda_Q(\Omega), \Omega) = ||[Q, \Omega]||^2 = ||Q \times \Omega||^2$. Evidently, $\Lambda_Q(\Omega) = 0$ if and only if Q and Ω are colinear. $\square$

The kinetic energy T of a particle of mass m along its path $q(t)$ given by $\frac{1}{2}m||\frac{dq}{dt}||^2$. It follows that

$$T = \frac{1}{2}m||\omega \times q||^2 = \frac{1}{2}||R(\omega \times q)||^2 = \frac{1}{2}m||R\omega \times Rq||^2 = \frac{1}{2}||\Omega \times Q||^2.$$
(12.6)

Consider now a finite collection of points on the body having coordinate vectors $Q_1, \ldots, Q_k$. Then the total kinetic energy associated with this ensemble of points is equal to $\frac{1}{2}\sum_{i=1}^k m_i||\frac{dq_i}{dt}||^2 = \frac{1}{2}\sum_{i=1}^k m_i(\Lambda_{Q_i}(\Omega), \Omega)$. This quadratic form in Ω is positive-definite if and only the points $Q_1, \ldots, Q_k$ are not all colinear.

The above sums can be regarded as the the Riemann sums $\frac{1}{2}\sum_{i=1}^k dm_i(\Lambda_{Q_i}(\Omega,), \Omega) = \frac{1}{2}\sum_{i=1}^k (\Lambda_{Q_i}(\Omega), \Omega)\rho(Q_i)\Delta Q_1 \Delta Q_2 \Delta Q_3$ associated with the integral

$$\int \int \int \rho(Q)\langle \Lambda_Q(\Omega), \Omega \rangle dQ_1 dQ_2 dQ_3,$$

where the integration is carried over the body, and where ρ denotes the density of the body so that infinitesimally, $dm - \rho d(vol)$. The integral $\Lambda(\Omega) = \int \int \int \rho \Lambda_Q(\Omega)dQ_1 dQ_2 dQ_3$ is called the *inertia tensor of the body* [AM]. The inertia tensor defines the kinetic energy T of the body with

$$T = \frac{1}{2}(\Lambda(\Omega), \Omega).$$
(12.7)

It follows that T is a positive-definite quadratic form on $\mathbb{R}^3$ whenever the body is not a line. Its eigenvalues I_1, I_2, I_3 are called the *principal moments of inertia*.

Definition 12.2 A positive-definite quadratic form $\langle \mathcal{P}(U), U \rangle, U \in so_3(R)$ will be called an Euler form if there exists a rigid body with its inertia tensor Λ such that $\langle \mathcal{P}(U), U \rangle = (\Lambda(\hat{U}), \hat{U})$ for all $U \in so_3(R)$.

Proposition 12.3 *Every solid rigid body (a body which does not lie in a plane) generates a unique Euler form* $\langle \mathcal{P}(U), U \rangle$ *with* $\mathcal{P}(U) = PU + UP$ *for some positive-definite matrix P.*

We will revert to the earlier notations with $a \otimes b$ rank one matrix defined by $(a \otimes b)x = (b, x)x, x \in \mathbb{R}^3$, and $a \wedge b$ defined by $a \otimes b - b \otimes a$.

Lemma 12.4 $\widehat{a \wedge b} = b \times a$.

The proof is straightforward and will be omitted.

Proof of Proposition 12.3 The inertia tensor $\Lambda_Q(\Omega)$ has an isomorphic representation on $so_3(R)$ given by

$$\mathcal{P}_Q(U) = m[V, [U, V]], \tag{12.8}$$

under the identification $\hat{U} = \Omega$, $\hat{V} = Q$, since

$$\widehat{\mathcal{P}_Q(U)} = m(\widehat{[U, V]} \times \hat{V} = m(\hat{V} \times \hat{U}) \times \hat{V}) = m((Q \times \hat{U}) \times Q) = \Lambda_Q(\hat{U}).$$

Now,

$$(Q \otimes Q)U + U(Q \otimes Q) = -Q \otimes UQ + UQ \otimes Q = UQ \wedge Q.$$

Since $\widehat{UQ \wedge Q} = Q \times UQ = (Q \times \hat{U}) \times Q$, then

$$\mathcal{P}_Q(U) = m((Q \otimes Q)U + U(Q \otimes Q)). \tag{12.9}$$

If $Q_1, \ldots, Q_k$ are any points on the body then $\sum_{i=1}^{k} \mathcal{P}_{Q_i}(U) = \widehat{\sum_{i=1}^{k} \Lambda_{Q_i}(\hat{U})}$.

Matrices $P_k = \sum_{i=1}^{k} Q_i \otimes Q_i$ are positive semi-definite for any choice of the points $Q_1, \ldots, Q_k$. If the points Q_i are linearly independent, then $P_k v = 0$ if and only if v is orthogonal to each point Q_i. When the body is solid, the points $Q_1, \ldots, Q_k$ can always be chosen so that they span $\mathbb{R}^3$. For such a choice of points P_k is positive-definite. Moreover, if Q_{k+1} is added to $Q_1, \ldots, Q_k$ then $(P_{k+1}v, v) = (P_k v, v) + (Q_{k+1}v, Q_{k+1}v)^2 \geq (P_k v, v)$.

As before, the above sums can be regarded as the Riemann sums

$$\sum_{i=1}^{k} ((Q_i \otimes Q_i)U + U(Q_i \otimes Q_i))\rho(Q_i)\Delta Q_i$$

associated with the integral $\mathcal{P}(U) = \int \int \int ((Q \otimes Q)U + U(Q \otimes Q))\rho(Q)dQ$.

It follows that $\widehat{\mathcal{P}(U)} = \Lambda(\hat{U})$. Therefore, $\langle \mathcal{P}(U), U \rangle = (\Lambda(\hat{U}), \hat{U})$ for all $U \in so_3(R)$. Moreover, the matrix P defined by $\int \int \int \rho(Q)(Q \otimes Q)dQ$ is positive-definite. $\qquad\square$

The eigenvalues of $\mathcal{P}$ are the same as the eigenvalues of Λ and, therefore, the principal moments of inertia can be obtained from $\mathcal{P}$. Let D be a diagonal matrix with its diagonal entries d_1, d_2, d_3 such that $P = RDR^T$ for some $R \in SO_3(R)$. Then

$$\mathcal{P}(U) = PU + UP = (RDR^T)U + U(RDR^T) = R(DV + VD)R^T,$$

where $V = R^T UR$. It follows that $R^T \mathcal{P}(U)R = DV + VD$.

Proposition 12.5   *The principal moments of inertia* I_1, I_2, I_3 *are given by*

$$I_1 = d_2 + d_3, I_2 = d_1 + d_3, I_3 = d_1 + d_2.$$

Proof   Let e_1, e_2, e_3 denote the standard basis in $\mathbb{R}^3$ and let

$$A_1 = e_2 \wedge e_3, A_2 = e_3 \wedge e_1, A_3 = e_2 \wedge e_3.$$

It follows that $D(e_i \wedge e_j) + (e_i \wedge e_j)D = (d_i + d_j)e_i \wedge e_j, i \leq 3, j \leq 3$, and hence

$$DA_1 + A_1 D = (d_2 + d_3)A_1, DA_2 + A_2 D = (d_1 + d_3)A_2, DA_3 + A_3 D = (d_1 + d_2)A_3.$$

Hence, $d_i + d_j, i \neq j$ are the eigenvalues of $\mathcal{P}$ with eigenvectors

$$U_1 = Re_2 \wedge Re_3, U_2 = Re_1 \wedge Re_3, U_3 = Re_1 \wedge Re_2.$$

$\square$

Corollary 12.6   *Matrices* U_1, U_2, U_3 *are orthonormal and satisfy the following Lie bracket relations:*

$$[U_1, U_2] = -U_3, [U_1, U_3] = U_2, [U_2, U_3] = -U_1.$$

If $U = u_1 U_1 + u_2 U_2 + u_3 U_3$, *then* $\langle \mathcal{P}(U), U \rangle = I_1 u_1^2 + I_2 u_2^2 + I_3 u_3^2$. *Hence* $\hat{U}_1, \hat{U}_2, \hat{U}_3$ *form the principal axes for the inertia tensor* $(\Lambda(\hat{U}), \hat{U})$.

Corollary 12.7 $\mathcal{P}(U) = DU + UD$ *with* $D = \frac{1}{2}(I_1 + I_2 + I_3)I - diag(I_1, I_2, I_3)$.

Corollary 12.8 $I_1 + I_2 < I_3, I_1 + I_3 < I_2, I_2 + I_3 < I_1$.

These corollaries are evident and do not require proofs. We leave it to the reader to show that the sum of any two principal moments of inertia can be equal to the remaining moment of inertia only if P is semi-positive-definite, that is, only when the body is a planar body

To obtain the equations of motion we will use the Principle of Least Action. This association enables us to treat Euler's top like any other left-invariant geodesic problem on $SO_3(R)$. Then the Maximum Principle readily yields the Hamiltonian

$$H = \frac{1}{2} \left\langle \mathcal{P}^{-1}(L), L \right\rangle = \frac{1}{2} \left(\frac{1}{I_1} l_1^2 + \frac{1}{I_2} l_2^2 + \frac{1}{I_3} l_3^2 \right),$$

where $L = l_1 U_1 + l_2 U_2 + l_3 U_3$, with

$$\frac{dR}{dt} = R(t)\mathcal{P}^{-1}(L), \frac{dL}{dt} = [\mathcal{P}^{-1}(L), L] \qquad (12.10)$$

the corresponding Hamiltonian equations.

The vector $M = \hat{L}$ is called the angular momentum in the literature on mechanics, and the vector $\Omega = \hat{U} = \widehat{\mathcal{P}^{-1}(L)}$ is called the angular velocity. Then equations (12.10) can be written in terms of M and Ω as

$$\frac{dM}{dt} = M \times \Omega. \tag{12.11}$$

Euler's top, much like any other left-invariant optimal control problem on $SO_3(R)$, is completely integrable. The integrals of motion are the Hamiltonian H, the Casimir $\langle L, L \rangle$ and the Hamiltonian of any right-invariant vector field. Stated somewhat differently, each coadjoint orbit on $so_3(R)$ is a the momentum sphere $\langle L, L \rangle = l_1^2 + l_2^2 + l_3^2 = c^2$. The Hamiltonian H defines the energy ellipsoid $H = \frac{l_1^2}{I_1} + \frac{l_2^2}{I_2} + \frac{l_3^2}{I_3}$. Hence the solution curves are the projections of the extremals which lie in the intersection of the momentum sphere with the energy ellipsoid.

Equations of an Euler top are solvable by quadrature. We will demonstrate below that the procedure used to integrate its equations is equally valid for any left-invariant geodesic problem and not just the top.

To carry out the details it will be convenient to relabel the variables:

$$A = I_1, B = I_2, C = I_3, p = \frac{l_1}{I_1}, q = \frac{l_2}{I_2}, r = \frac{l_3}{I_3}.$$

Then p, q, r are the solutions of

$$A\frac{dp}{dt} = (B - C)qr, B\frac{dq}{dt} = (A - C)pr, C\frac{dr}{dt} = (A - B)pq,$$

and are subject to the constraints

$$Ap^2 + Bq^2 + Cr^2 = 2H, \; A^2p^2 + B^2q^2 + C^2r^2 = c^2.$$

For simplicity assume that $A > B > C$. It follows that

$$A(A - C)p^2 + B(B - C)q^2 = c^2 - 2HC.$$

Evidently $c^2 - 2HC$ is non-negative. Therefore,

$$p^2 = \frac{B(B - C)}{A(A - C)}(f^2 - q^2), f^2 = \frac{c^2 - 2HC}{B(B - C)}. \tag{12.12}$$

Similarly,

$$B(A - B)q^2 + C(A - C)r^2 = 2HA - c^2,$$

implies that

$$r^2 = \frac{B(A - B)}{C(A - C)}(g^2 - q^2), g^2 = \frac{2HA - c^2}{B(A - B)}. \tag{12.13}$$

But then

$$\frac{dq}{dt} = \frac{1}{B}(A - C)pr = \sqrt{\frac{(A-B)(B-C)}{AC}(f^2 - q^2)(g^2 - q^2)}. \qquad (12.14)$$

After the substitutions $q = fs$ and $k^2 = \frac{f^2}{g^2}$, the preceding equation becomes

$$\frac{ds}{\sqrt{(1 - s^2)(1 - k^2 s^2)}} = \sqrt{\frac{(B - C)(2HA - c^2)}{ABC}} dt = \alpha dt, \qquad (12.15)$$

where $\alpha = \sqrt{\frac{(B-C)(2HA-c^2)}{ABC}}$. If we now let $\tau = \alpha t$ then $s = sn(\tau)$, where sn is the elliptic function of Legendre. But then

$$p(\tau) = f\sqrt{\frac{B(B - C)}{A(A - C)}}\sqrt{1 - sn^2(\tau)} = f\sqrt{\frac{B(B - C)}{A(A - C)}}cn(\tau)$$

and

$$r(\tau) = g\sqrt{\frac{B(A - B)}{C(A - C)}}\sqrt{1 - k^2 sn^2(\tau)} = g\sqrt{\frac{B(A - B)}{C(A - C)}}dn(\tau),$$

obtained from (12.12) and (12.13).

It remains to integrate the equation $\frac{dR}{dt} = R(t)U(t)$ with $U(t) = p(t)A_1 + q(t)A_2 + r(t)A_3$. To do so, we will take advantage of the fact that the Hamiltonian of every right-invariant vector field is constant along the extremals, which means that $\langle R^T(t)AR(t), L(t)\rangle$ is constant for each skew-symmetric matrix A. But this means that $R(t)L(t)R^T(t)$ is a constant matrix Λ for each extremal curve $(R(t), L(t))$. The rotation matrix $R(t)$ can be always conjugated so that Λ is colinear with one the principal axes. For simplicity we will assume that $\Lambda = cA_3$. We will see that such a choice greatly simplifies the calculations.

We will also choose quaternions over the Euler angles since it is easier to work with the matrices in su_2 rather than with the matrices in $so_3(R)$. For that reason we will identify the principal axes A_1, A_2, A_3 with the skew-symmetric Pauli matrices

$$A_1 = \frac{1}{2}\begin{pmatrix} i & 0 \\ 0 & -i \end{pmatrix}, A_2 = \frac{1}{2}\begin{pmatrix} 0 & 1 \\ -1 & 0 \end{pmatrix}, A_3 = \frac{1}{2}\begin{pmatrix} 0 & i \\ i & 0 \end{pmatrix}.$$

Then $U(t) = \frac{1}{2}\begin{pmatrix} ip(t) & q(t) + ir(t) \\ -q(t) + ir(t) & -ip(t) \end{pmatrix}$ and $L = \frac{1}{2}\begin{pmatrix} iAp & Bq + iCr \\ -Bq + iCr & -iAp \end{pmatrix}$.

We will also identify rotations $R(t)$ with their counterparts $\begin{pmatrix} z & w \\ -\bar{w} & \bar{z} \end{pmatrix}$,

$|z|^2 + |w|^2 = 1$, in SU_2. Then R in SU_2 will be parametrized by angles $\theta_1, \theta_2, \theta_3$ defined by

$$R = e^{\theta_1 A_3} e^{\theta_2 A_1} e^{\theta_3 A_3}.$$

This choice is compatible with the assumption that $\Lambda = cA_3$, as will be made clear below. Then

$$e^{\theta A_1} = \begin{pmatrix} e^{\frac{1}{2}\theta} & 0 \\ 0 & e^{-\frac{1}{2}\theta} \end{pmatrix}, e^{\theta A_3} = \begin{pmatrix} \cos\frac{1}{2}\theta & i\sin\frac{1}{2}\theta \\ i\sin\frac{1}{2}\theta & \cos\frac{1}{2}\theta \end{pmatrix}.$$

The assumption $RLR^T = cA_3$ implies that

$$L = ce^{-\theta_3 A_3} e^{-\theta_2 A_1} A_3 e^{\theta_2 A_1} e^{\theta_3 A_3}$$

$$= \frac{1}{2}c \begin{pmatrix} \sin\theta_2\sin\theta_3 & \sin\theta_2\cos\theta_3 + i\cos\theta_2 \\ -\sin\theta_2\cos\theta_3 + i\cos\theta_2 & -\sin\theta_2\sin\theta_3 \end{pmatrix}.$$

Hence,

$$Ap = \sin\theta_2\sin\theta_3, Bq = \sin\theta_2\cos\theta_3, Cr = \cos\theta_2. \tag{12.16}$$

the remaining angle θ_1 is obtained from the equation of motion $\frac{dR}{dt} = R(t)U(t)$. It follows that

$$R^T \frac{dR}{dt} = \dot{\theta}_1 R^T A_3 R + \dot{\theta}_2 e^{-\theta_3 A_3} A_1 e^{\theta_3 A_3} + \dot{\theta}_3 A_3$$

$$= \frac{1}{c}\dot{\theta}_1 L + \dot{\theta}_2(\cos\theta_3 A_1 - \sin\theta_3 A_2) + \dot{\theta}_3 A_3$$

$$= U(t) = p(t)A_1 + q(t)A_2 + r(t)A_3.$$

The above yields

$$\frac{1}{c}\dot{\theta}_1 Ap + \dot{\theta}_2\cos\theta_3 = p, \frac{1}{c}\dot{\theta}_1 Bq - \dot{\theta}_2\sin\theta_3 = q, \frac{1}{c}\dot{\theta}_1 Cr + \dot{\theta}_3 = r,$$

which together with (12.16) gives

$$\dot{\theta}_1 = c\frac{p\sin\theta_3 + q\cos\theta_3}{Ap\sin\theta_3 + Bq\cos\theta_3} = c\frac{Ap^2 + Bq^2}{A^2p^2 + B^2q^2}. \tag{12.17}$$

Equation (12.17) can be integrated in terms of Jacobi's Θ function. The interested reader can find the details in [Ap]. Let us just remark that in the degenerate case, when two principal moments of inertia are equal to each other, the solutions are expressed in terms of circular (trigonometric) functions.

The fact that the energy function $\frac{1}{2}\langle \mathcal{P}(U), U \rangle$ is of special form $\mathcal{P}(U) = PU + UP$ does not have any particular significance in regard to the above calculations. The special form, however, singles out a particular class of geodesic problems on $SO_n(R), n > 3$ of central importance for the theory of integrable systems, even though, paradoxically, there are no "real" rigid bodies in dimensions larger than 3.

Definition 12.9 The Hamiltonian system

$$\frac{dR}{dt} = R(t)\mathcal{P}^{-1}(L), \frac{dL}{dt} = [\mathcal{P}^{-1}(L), L] \tag{12.18}$$

generated by $H = \frac{1}{2}\langle \mathcal{P}^{-1}(L), L \rangle$ with $\mathcal{P}(U) = PU + UP$ for some positive-definite matrix P is called an n-dimensional Euler's top.

There is no loss in generality in assuming that P is a diagonal matrix; for, if T is a rotation matrix such that TPT^{-1} is a diagonal matrix D, then

$$T\mathcal{P}(U)T^{-1} = \tilde{U}D + D\tilde{U}, \tilde{U} = TUT^{-1}.$$

Therefore, $\langle \mathcal{P}(U), U \rangle = \langle \tilde{U}D + D\tilde{U}, \tilde{U} \rangle$ and $\frac{d\tilde{R}}{dt} = \tilde{R}(\tilde{U})$, where $\tilde{R} = TRT^{-1}$.

It was C. B. Manakov [Mn] who noticed first that an n-dimensional Euler's top admits a spectral representation of the form

$$\frac{dL}{dt} = [U + \lambda P, L + \lambda P^2], U = \mathcal{P}^{-1}(L). \tag{12.19}$$

The proof is remarkably simple: since $L = PU + UP$, $[P, L] = [P, PU + UP] = [P^2, U]$ and therefore,

$$\frac{dL}{dt} = [U, L] = [U, L] + \lambda[P, L] + \lambda[U, P^2] = [U + \lambda P, L + \lambda P^2].$$

Manakov's observation was a catalyst for the general construction of isospectral systems based on the sectional operators in the unitary algebra u_n (Chapter 9). We have seen that any linear operator $\mathcal{P}(U) = PU + UD$, $U \in so_n(R)$ can be realized as $\mathcal{P} = ad^{-1}P \circ adP^2$, in which case $[P, \mathcal{P}(U)] = [P^2, U]$, $U \in so_n(R)$, or $[P, L_{\mathfrak{k}}] = [P^2, \mathcal{P}^{-1}(L_{\mathfrak{k}})]$, $L_{\mathfrak{k}} = \mathcal{P}(U)$. Therefore, the affine Hamiltonian $H(L_{\mathfrak{k}}, L_{\mathfrak{p}}) = \frac{1}{2}\langle \mathcal{P}^{-1}(L_{\mathfrak{k}}), L_{\mathfrak{k}} \rangle + \langle P, L_{\mathfrak{p}} \rangle$ admits an isospectral representation in $sl_n(R)$ with the spectral matrix

$$L_\lambda = L_{\mathfrak{p}} - \lambda L_{\mathfrak{k}} + (\lambda^2 - s)P^2.$$

We will now identify this affine Hamiltonian with a top in the presence of a Newtonian potential.

12.2 Tops in the presence of Newtonian potentials

Real-valued functions of the ambient space can be expressed in terms of the coordinates relative to the fixed frame and then can be interpreted as a source of potential energy for the body. In this context functions in the ambient space are referred to as the Newtonian potential fields [Bg1]. If ϕ is any such function, then

$$V(R) = \int \rho(Q)\phi((Q, R^T e_1), (Q, R^T e_2), (Q, R^T e_3))dQ$$

is the induced potential energy of the body, where ρ is the density of the body, and R the matrix that relates the absolute coordinates to the body coordinates via $q = RQ$ [Bg1]. Then each coordinate q_i is given by $q_i = (Q, R^T e_i) = (Q, \alpha_i)$, where $\alpha_i = R^T e_i, i = 1, 2, 3$. Each α_i is the coordinate vector of e_i relative to the moving frame.

The total energy of the body is given by the Hamiltonian

$$H(R, L) = \frac{1}{2}\langle \mathcal{P}^{-1}(L), L \rangle + V(R),$$

which then is regarded as a function on the cotangent bundle $T^*SO_3(R)$, after the identifications of $T^*SO_3(R)$ with $SO_3(R) \times so_3^*(R)$ via the left-translations and the identification of $so_3^*(R)$ with $so_3(R)$ via the trace form. In contrast to the situations encountered so far, this Hamiltonian is not left-invariant, and hence its Hamiltonian equations contain an extra term, called the external torque, due to the force imparted by V.

The correct equations are obtained by identifying a function H on $G \times \mathfrak{g}^*$ with its Hamiltonian vector field $\vec{H}_{(g,\ell)} = (A(g, \ell), a(g, \ell))$ through the formula

$$\omega_{(g,\ell)}((A(g, \ell), a(g, \ell))), (B, b)) = dH_{(g,\ell)}(B, b),$$

where $\omega_{(g,\ell)}((A_1, l_1), (A_2, l_2)) = l_2(A_1) - l_1(A_2) - \ell([A_1, A_2])$ is the left-invariant form introduced earlier in Chapter 5.

When $H(g, \ell)$ is the sum $H_0(\ell) + V(g)$, then its differential dH at a point (g, ℓ) is equal to the sum $dH_0 + dV$, where dH_0 is the differential of H_0 on $\mathfrak{g}^*$ at ℓ and dV is the differential of V at g. It follows that

$$dH_{g,\ell}(B, 0) = \frac{d}{dt}V(g \exp(Bt))|_{t=0} = dV(gB)$$

and

$$dH_{g,\ell}(0, b) = \frac{d}{dt}H_0(\ell + tb)|_{t=0} = dH_0(b).$$

Since dH_0 is a linear function on $\mathfrak{g}^*$ it is naturally identified with an element of $\mathfrak{g}$. Therefore,

$$A(g, \ell) = dH_0(\ell) \text{ and } a(g, \ell) = -dV \circ L_g - ad^* dH(\ell)(\ell).$$

It follows that the integral curves $(g(t), \ell(t))$ of $\vec{H}$ are the solution curves of

$$\frac{dg}{dt}(t) = g(t)dH_0(\ell) \quad \frac{d\ell}{dt}(t) = -dV \circ L_g - ad^* dH_0(\ell(t))(\ell(t)) \quad (12.20)$$

where dL_g denotes the tangent map of the left translation $L_g(x) = gx$, $x \in G$.

Let us now calculate the directional derivative of functions V in $SO_n(R)$ along the left-invariant vector fields. Let $\alpha_1, \ldots, \alpha_n$ be the orthonormal frame of vectors associated to the rotation matrix $R \in SO_n(R)$ via the relations $\alpha_i = R^T e_i, i = 1, \ldots, n$.

Consider now the curve $\sigma(\epsilon) = Re^{\epsilon X}$ in $SO_n(R)$ defined by a skew-symmetric matrix X. Along this curve, $\alpha_i(\epsilon) = e^{-\epsilon X} R^T e_i = e^{-\epsilon X} \alpha_i$. The directional derivative of V along σ at $\epsilon = 0$ is given by

$$dV(RX) = \sum_{i=1}^{n} \left(\frac{\partial V}{\partial \alpha_i}, \frac{d\alpha_i}{d\epsilon} \right) = \sum_{i=1}^{n} \left(\frac{\partial V}{\partial \alpha_i}, -X\alpha_i \right) = \left\langle \frac{\partial V}{\partial \alpha_i} \wedge \alpha_i, X \right\rangle.$$

Then equations (12.20), after the identification of $\ell \in so_n^*$ with $L \in so_n$ via the trace form $\langle , \rangle$, become

$$\frac{dR}{dt} = R(t)dH_0(L), \quad \frac{dL}{dt} = [dH_0(L), L] + \sum_{i=1}^{n} \alpha_i \wedge \frac{\partial V}{\partial \alpha_i}, \quad (12.21)$$

The term $\sum_{i=1}^{n} \alpha_i \wedge \frac{\partial V}{\partial \alpha_i}$ is the external torque due to the external force $-grad\ V$.

In the particular case that $H(R, L) = \frac{1}{2}\langle \mathcal{P}^{-1}(L), L \rangle + V(R)$, the Hamiltonian equations are given by

$$\frac{dR}{dt} = R(\mathcal{P}^{-1}(L)), \quad \frac{dL}{dt} = [\mathcal{P}^{-1}(L), L] + \sum_{i=1}^{n} \alpha_i \wedge \frac{\partial V}{\partial \alpha_i}. \quad (12.22)$$

On $SO_3(R)$ the preceding equations can be expressed in terms of the vector product as

$$\frac{dM}{dt} = M \times \Omega + \sum_{i=1}^{3} \frac{\partial V}{\partial \alpha_i} \times \alpha_i,$$

$$\frac{d\alpha_i}{dt} = \alpha_i \times \Omega, \quad i = 1, 2, 3. \quad (12.23)$$

In the case of a heavy top, the Newtonian potential $i\phi$ is given by $\phi(q_1, q_2, q_3) = Cq_1$, where C denotes the gravitational constant. In this representation, the fixed frame is oriented so that its first leg is in the direction of the gravitational force. Therefore, the potential energy of a point mass m is given by $Cm(e_1, q) = C\rho(Q)(e_1, RQ)dQ$, which then is integrated over the body to yield

$$V = C \int \rho(Q)(R^T e_1, Q)\, dQ = C(R^T e_1, \int \rho(Q)Q\, dQ) = C(Q_c, \alpha_1),$$

where Q_c is the center of mass of the body defined by $Q_c \int \rho(Q)dQ = \int \rho(Q)QdQ$. Evidently, any heavy top reduces to the top of Euler when the center of gravity coincides with the fixed point of the body.

The heavy top is a particular case of the top in the presence of a linear Newtonian field $\phi(q_1, q_2, q_3) = a_1 q_1 + a_2 q_2 + a_3 q_3$. Then, $V(R) = \sum_{i=1}^{3} a_i(\alpha_i, Q_c)$ and the associated Hamiltonian equations are given by the following:

$$\frac{dR}{dt} = R(t)(\mathcal{P}^{-1}(L)), \quad \frac{dL}{dt} = [\mathcal{P}^{-1}(L), L] + \sum_{i=1}^{3} a_i \alpha_i \wedge Q_c. \tag{12.24}$$

Remarkably, the Hamiltonian system associated with the heavy top can be regarded as an invariant subsystem of the affine Hamiltonian $H = \frac{1}{2}\langle \mathcal{P}^{-1}(L_\ell), L_\ell \rangle + \langle A, L_p \rangle$ on the cotangent bundle of the semi-direct product $G = \mathbb{R}^3 \rtimes SO_3(R)$ through the following identifications.

Each point (x, R) in G can be represented by the matrix $g = \begin{pmatrix} 1 & 0 \\ x & R \end{pmatrix}$.
Then the Lie algebra of G is realized as the space of 4×4 matrices of the form $\begin{pmatrix} 0 & 0 \\ p & L \end{pmatrix}$ with p a column vector in $\mathbb{R}^3$ and L a skew-symmetric 3×3 matrix.

Matrices $L_p = \begin{pmatrix} 0 & 0 \\ p & 0 \end{pmatrix}$ and $L_\ell = \begin{pmatrix} 0 & 0 \\ 0 & L \end{pmatrix}$ correspond to the matrices in the factor spaces $\mathfrak{p}$ and $\mathfrak{k}$ in the Cartan decomposition of $\mathfrak{g}$. Moreover, $[L_p, L_\ell] = \begin{pmatrix} 0 & 0 \\ Lp & 0 \end{pmatrix}$.

Let now $A = \begin{pmatrix} 0 & 0 \\ Q_c & 0 \end{pmatrix}$. Every solution curve $(R(t), L(t))$ of equation (12.24) defines a curve $(g(t), (L_\ell(t), L_p(t)))$ in $G \times \mathfrak{g}$ via the following identifications:

$$g(t) = \begin{pmatrix} 1 & 0 \\ x(t) & R(t) \end{pmatrix}, \quad x(t) = -\int_0^t R(\tau)Q_c\, d\tau,$$

$$L_\ell(t) = \begin{pmatrix} 0 & 0 \\ 0 & L(t) \end{pmatrix}, \quad L_p(t) = \begin{pmatrix} 0 & 0 \\ \sum_{i=1}^{3} a_i \alpha_i(t) & 0 \end{pmatrix}.$$

If we now write $\mathcal{P}^{-1}(L_\ell(t)) = \begin{pmatrix} 0 & 0 \\ 0 & \mathcal{P}^{-1}(L(t)) \end{pmatrix}$, then $g(t), L_\ell(t), L_p(t)$ are the solutions of the following system of equations:

$$\frac{dg}{dt} = g(t)(A + \mathcal{P}(L_\ell(t))), \quad \frac{dL_\ell}{dt} = [\mathcal{P}^{-1}(L_\ell), L_\ell] + \hat{A} \wedge \hat{L}_p, \quad \frac{dL_p}{dt} = [\mathcal{P}^{-1}(L_\ell), L_p],$$

$$\tag{12.25}$$

where $\hat{A} = \begin{pmatrix} 0 \\ -Q_c \end{pmatrix}$ and $\hat{L}_{\mathfrak{p}} = \begin{pmatrix} 0 \\ p \end{pmatrix}$. Equations (12.25) are the

Hamiltonian equations on the semi-direct product $\mathbb{R}^3 \rtimes SO_3(R)$ generated by the Hamiltonian $H = \frac{1}{2}\langle \mathcal{P}^{-1}(L_{\mathfrak{t}}), L_{\mathfrak{t}} \rangle + \langle A, L_{\mathfrak{p}} \rangle$.

This passage to the semi-direct product links the equations of the top with a linear potential with the equations Kirchhoff corresponding to the equilibrium configurations of an elastic rod. We will defer these details to the next chapter.

In the meantime let us shift attention to the top in the presence of a quadratic Newtonian field $\phi(q_1, q_2, q_3) = \frac{1}{2}(a_1 q_1^2 + a_1 q_2^2 + a_3 q_3^2)$. Then, the potential energy V is given by

$$2V = \sum_{i=1}^{3} a_i \int \rho(Q, \alpha_i)^2 \, dQ.$$

There is no loss in generality in assuming that the moving frame $\vec{f}_1, \vec{f}_2, \vec{f}_3$ coincides with the principal axes of the inertia tensor, in which case the coordinates of a point relative to the moving frame coincide with the coordinates relative to the principal axes. If Λ denotes the inertia tensor of the body and if I_1, I_2, I_3 denote the principal moment of inertia, then

$$I_1(v^{(1)})^2 + I_2(v^{(2)})^2 + I_3(v^{(3)})^2 = (\Lambda(v), v) = \int \rho(||Q||^2 - (v, Q)^2) \, dQ,$$

for any vector v, where $v^{(1)}, v^{(2)}, v^{(3)}$ denote the coordinates of v relative to the principal axes. Therefore, $I_i = \int \rho(||Q||^2 - Q_i^2) \, dQ, i = 1, 2, 3$, and $I_1 + I_2 + I_3 = 2 \int \rho ||Q||^2 \, dQ$.

In addition,

$$I_1(\alpha_i^{(1)})^2 + I_2(\alpha_i^{(2)})^2 + I_3(\alpha_i^{(3)})^2 = (\Lambda(\alpha_i), \alpha_i) = \int \rho(||Q||^2 - (\alpha_i, Q)^2) \, dQ.$$

The above implies that

$$2V = \frac{1}{2}(a_1 + a_2 + a_3)(I_1 + I_2 + I_3) - \sum_{i=1}^{3} a_i(I_1(\alpha_i^1))^2 + I_2(\alpha_i^2) + I_3(\alpha_i^3).$$

$$(12.26)$$

It follows that $\frac{\partial V}{\partial \alpha_i} = -a_i(I_1 \alpha_i^{(1)} e_1 + I_2 \alpha_i^{(2)} e_2 + I_3 \alpha_i^{(3)} e_3)$.

Let us now bring back the Euler form $\mathcal{P}$ (Definition 12.2) and its diagonal representation $\mathcal{P}(U) = DU + UD$, with $D = \frac{1}{2}(I_1 + I_2 + I_3)I - \text{diag}(I_1, I_2, I_3)$ (Corollary 12.7). The reader can readily verify that

$$\frac{\partial V}{\partial \alpha_i} = -a_i \hat{\mathcal{P}}(\check{\alpha}_i) = -a_i(D\check{\alpha}_i + \check{\alpha}_i D),$$

In these notations the external torque is given by $\sum_{i=1}^{3} a_i[D(\check{\alpha}_i) + \check{\alpha}_i D, \check{\alpha}_i]$. Since $[DU + UD, U] = [D, U^2]$, the Hamiltonian equations of the top with the potential V can be written as

$$\frac{dL}{dt} = [\mathcal{P}^{-1}(L), L] + \sum_{i=1}^{3} a_i[\check{\alpha}_i^2, D], \quad \frac{d\check{\alpha}_i}{dt} = [\mathcal{P}^{-1}(L), \check{\alpha}_i], i = 1, 2, 3. \quad (12.27)$$

We now come to an ingenious discovery of O. Bogoyavlensky [Bg1] that system (12.27) is integrable, because it can be represented as the Hamiltonian system on $sl_n(R)$ associated with the isospectral Hamiltonian $H = \frac{1}{2}\langle \mathcal{P}(L_{\mathfrak{k}}, L_{\mathfrak{k}}) + \langle D, L_{\mathfrak{p}} \rangle\rangle$, in which case the spectral matrix yields the extra integrals required for Liouville integrability.

The lemma below singles out two facts that are relevant for the representation on $sl_n(R)$.

Lemma 12.10 *Let λ be a 3×3 skew-symmetric matrix such that $\|\lambda\| = 1$. Then*

$$\lambda^2 = -I + \hat{\lambda} \otimes \hat{\lambda}.$$

If $\frac{d\lambda}{dt} = [\Omega, \lambda]$, then $\frac{d}{dt}\lambda^2 = [\Omega, \lambda^2]$.

The proof is easy and will be omitted. Let now

$$L_{\mathfrak{p}} = -\left(\frac{2}{3}(a_1 + a_2 + a_3)I + \sum_{i=1}^{3} a_i\check{\alpha}_i^2\right) = \frac{1}{3}(a_1 + a_2 + a_3)I - \sum_{i=1}^{3} a_i\lambda_i \otimes \lambda_i.$$

It follows that $L_{\mathfrak{p}}$ is a symmetric matrix of trace zero. If we rename L as $L_{\mathfrak{k}}$ and take $A = D$, then equations (12.27) can be written in the following form:

$$\frac{dL_{\mathfrak{k}}}{dt} = [\mathcal{P}^{-1}(L_{\mathfrak{k}}), L_{\mathfrak{k}}] + [D, L_{\mathfrak{p}}], \quad \frac{dL_{\mathfrak{p}}}{dt} = [\mathcal{P}^{-1}(L_{\mathfrak{k}}), L_{\mathfrak{p}}]. \quad (12.28)$$

Equations (12.28) admit an isospectral representation

$$\frac{dL(\lambda)}{dt} = [\mathcal{P}^{-1}(L_{\mathfrak{k}}) - \lambda D, L(\lambda)], \quad L(\lambda) = -L_{\mathfrak{p}} + \lambda L_{\mathfrak{k}} + \lambda^2 D^2, \quad (12.29)$$

and hence are integrable on each coadjoint orbit in $sl_3(R)$ relative to the action of the semi-direct product $G_s = S \ltimes SO_n(R)$.

The coadjoint orbit through $L_{\mathfrak{p}}(0) = \frac{1}{3}(a_1 + a_2 + a_3)I - \sum_{i=1}^{3} a_i(\lambda_i(0) \otimes \lambda_i(0)), L_{\mathfrak{k}}(0) = 0$ is diffeomorphic to the cotangent bundle of $SO_3(R)$ whenever a_1, a_2, a_3 are all distinct.

The case $a_1 = a_2 = a_3 = a$ is degenerate, in the sense that $\sum_{i=1}^{3} a_i\lambda_i \otimes \lambda_i = aI$, and hence $L_{\mathfrak{p}} = 0$. The orbit is the sphere $\|L_{\mathfrak{k}}(t)\| = $

$L_{\mathfrak{k}}(0)||$. In the remaining case, where only two of a_1, a_2, a_3 are equal, the coadjoint orbit through $L_{\mathfrak{p}}(0)$ is equal to the tangent bundle of the sphere S^2.

This conclusion is essentially a paraphrase of Bogoyavlensky's remarkable paper [Bg1] with some minor diferences. Bogoyavlensky uses I to denote our matrix D and matrix $B = \begin{pmatrix} I_2 I_3 & 0 & 0 \\ 0 & I_1 I_3 & 0 \\ 0 & 0 & I_1 I_2 \end{pmatrix}$ instead of D^2. However, $B = D^2 - cI$, where here I denotes the identity matrix, and $c = \frac{1}{4}(I_1^2 + I_2^2 + I_3^2) - \frac{1}{2}(I_1 I_2 + I_1 I_3 + I_2 I_3)$. Hence B and D^2 differ by a scalar multiple of the identity matrix and hence the Lie brackets with B and D^2 are the same.

In all cases, the required integrals of motion are obtained from the spectral invariants of $L_\lambda = L_{\mathfrak{p}} - \lambda L_{\mathfrak{k}} + \lambda^2 B$. It is easy to verify that $-\frac{1}{2}Tr(L_\lambda^2)$ yields $Tr(\frac{1}{2}L_{\mathfrak{k}}^2 + BL_{\mathfrak{k}})$ as a first integral of motion. This integral of motion is equal to the Hamiltonian $H_1 = \frac{1}{2}\langle L_{\mathfrak{k}}, L_{\mathfrak{k}}\rangle + \langle B, L_{\mathfrak{p}}\rangle$ corresponding to the canonical affine Hamiltonian defined by the drift $B = D^2$. The second independent integral of motion $H_2 = \langle L_{\mathfrak{k}}^2, L_{\mathfrak{p}}\rangle + \langle B, L_{\mathfrak{p}}^2\rangle$ is obtained from $-\frac{1}{2}Tr(L_\lambda^3)$. These integrals of motion are in involution relative to the semi-direct Poisson bracket in $sl_3(R)$. Hence H is Liouville integrable on every coadjoint orbit in $sl_3(R)$ of dimension less or equal to six. Since each level set $H_1 = c_1, H_2 = c_2, H = c_3$ is compact, each trajectory is quasi-periodic on the three-dimensional torus T^3 defined by these level sets.

It is important to note that the passage from the top with a quadratic potential to the affine Hamiltonian on $sl_n(R)$ is very different from the passage to $sl_n(R)$ in the case of the tops with a linear potential. Tops with linear potential are subordinate to the affine Hamiltonians on space forms as we have shown earlier and seem to conform to a completely different paradigm.

13

Isometry groups of space forms and affine systems: Kirchhoff's elastic problem

Our study of affine Hamiltonians brings us to a particularly interesting juncture, the triple $G, K, G/K$, where G is both the orthonormal frame bundle and the isometry group for the underlying symmetric space G/K. In these situations variational problems involving the geometrical invariants of curves, such as the geodesic curvature and the torsion, can be phrased as optimal control problems on G with the corresponding Hamiltonians affine, and hence amenable to the general methodology described earlier. On these groups, Kirchhoff's elastic problem leads to a distinguished class of affine Hamiltonians whose integrable cases exhibit striking similarities with the theory of the heavy tops. However, their study seems to require different tools from those developed in the previous chapters. We will show that the canonical affine systems are the only affine Hamiltonian that admit the isospectral property in this setting. So the search for extra integrals of motion via the spectral curves is hopeless.

We will be able to show that the solutions of the canonical affine Hamiltonian reveal a curious connection between the elastic curves and the motions of an n-dimensional pendulum analogous to the case of non-Euclidean elastica discussed in Chapter 7. This material sets the stage for Kirchhoff's elastic problem and the investigations of its integrable cases in the subsequent chapters.

Our discussion will be confined to the isometry groups of the space forms. We will revert to the notations introduced earlier and use $G = SO_\epsilon$ to denote $SO_{n+1}(R)$ for $\epsilon = 1$, $SO(1, n)$ for $\epsilon = -1$ and $\mathbb{R}^n \rtimes SO_n(R)$ for $\epsilon = 0$. In all these cases, the isotropy group K, the group that fixes the point e_0 is independent of ϵ and is isomorphic to $\{1\} \times SO_n(R)$. Then, S_ϵ will denote the homogeneous space G/K. As we have seen before, S_ϵ is the n-dimensional sphere S^n for $\epsilon = 1$, the hyperboloid $\mathbb{H}^n = \{x \in R^{n+1}; x_0^2 - \sum_{i=1}^n x_i^2 = 1, x_0 > 0\}$ for $\epsilon = -1$, and the affine Euclidean space $\mathbb{E}^n = \{x \in \mathbb{R}^{n+1} : x_0 = 1\}$ for $\epsilon = 0$.

The Lie algebra of G_ϵ will be denoted by so_ϵ with $\mathfrak{p}_\epsilon$ and $\mathfrak{k}$ denoting the factor spaces in the Cartan decomposition of $\mathfrak{g}_\epsilon$. Any X in $\mathfrak{p}_\epsilon$ is of the form $\begin{pmatrix} 0 & -\epsilon x^T \\ x & 0_n \end{pmatrix}$, $x \in \mathbb{R}^n$, and any $Y \in \mathfrak{k}$ is of the form $Y = \begin{pmatrix} 0 & 0 \\ 0 & h \end{pmatrix}$, $h \in so_n(R)$. We recall that

$$[X_1, X_2] = \begin{pmatrix} 0 & 0 \\ 0 & \epsilon(x_1 \wedge x_2) \end{pmatrix}, [X, Y] = \begin{pmatrix} 0 & -\epsilon(hx)^T \\ hx & 0 \end{pmatrix} \qquad (13.1)$$

for any matrices $X_i = \begin{pmatrix} 0 & -\epsilon x_i^T \\ x_i & 0_n \end{pmatrix}$ in $\mathfrak{p}_\epsilon$, and any matrix $Y = \begin{pmatrix} 0 & 0 \\ 0 & h \end{pmatrix}$ in $\mathfrak{k}$. Hence,

$$Ad_g(X) = \begin{pmatrix} 0 & -\epsilon(Rx)^T \\ Rx & 0_n \end{pmatrix} \qquad (13.2)$$

for any $g = \begin{pmatrix} 1 & 0 \\ 0 & R \end{pmatrix}$ in K and any X in $\mathfrak{p}_\epsilon$.

We recall our earlier convention that $\langle , \rangle$ is a negative multiple of the Killing form on a semi-simple Lie algebra $\mathfrak{g}$. In this context, $\langle A, B \rangle = -\frac{1}{2} Tr(AB)$ for any matrices A and B in $\mathfrak{g}_\epsilon$, $\epsilon = \pm 1$. The restriction of this form to $\mathfrak{p}_\epsilon$ negative relative to the canonical metric on $\mathfrak{p}_\epsilon$ when $\epsilon = -1$. To compensate for this inconvenience, it is natural to introduce another form $\langle , \rangle_\epsilon$ defined by

$$\langle X_1 + Y + 1, X_2 + Y_2, \rangle_\epsilon = -\epsilon \frac{1}{2} Tr(X_1 X_2) - \frac{1}{2} Tr(Y_1 Y_2)$$

for any X_1, X_2 in $\mathfrak{p}_\epsilon$. and any Y_1, Y_2 in $\mathfrak{k}$. It follows that $B_i = \begin{pmatrix} 0 & -\epsilon e_i^T \\ e_i & 0_n \end{pmatrix}$, $i = 1, \ldots, e_n$ is an orthonormal basis on $\mathfrak{p}_\epsilon$ relative to $\langle , \rangle_\epsilon$.

The restriction of $\langle , \rangle_\epsilon$ to $\mathfrak{p}_\epsilon$ agrees with the sub-Riemannian length on the space of horizontal curves with values in $\mathfrak{p}_\epsilon$, and the sub-Riemannian length of a horizontal curve agrees with the Riemannian length of the projected curve in G/K.

Let us now consider affine Hamiltonians $H = \frac{1}{2}\langle \mathcal{P}(L_\mathfrak{k}), L_\mathfrak{k} \rangle + \langle A, L_\mathfrak{p} \rangle$ on SO_ϵ with $\mathcal{P}(U) = DU + UD$, where D is a diagonal matrix.

Proposition 13.1 *An affine Hamiltonian H admits an isospectral representation if and only if it is canonical, that is, only when $H = \frac{1}{2}||L_\mathfrak{k}||^2 + \langle A, L_\mathfrak{p} \rangle$.*

Proof If $D = diag(d_0, \ldots, d_n)$ then $\mathcal{P}(U) = DU + UD \in \mathfrak{k}$ for any matrix $U \in \mathfrak{k}$. Let $A = \begin{pmatrix} 0 & -\epsilon a^T \\ a & 0_n \end{pmatrix}$ and $B = \begin{pmatrix} 0 & -\epsilon b^T \\ b & 0_n \end{pmatrix}$ be matrices in $\mathfrak{p}_\epsilon$ such that $[U, A] = [\mathcal{P}(U), B]$ for all $U = \begin{pmatrix} 0 & 0 \\ 0 & u \end{pmatrix}$, $u \in so_n(R)$.

According to (13.2), $[U, A] = [\mathcal{P}(U), B]$ is equivalent to $ua = \mathcal{P}(u)b$, where $\mathcal{P}(u)$ is the matrix in $so_n(R)$ with entries $(d_i + d_j)u_{ij}$. Then $u = u_{ij}e_i \wedge e_j$ implies that $u_{ij}(a_j e_i - a_i e_j) = u_{ij}(d_i + d_j)(b_j e_i - b_i e_j)$. Hence

$$a_j = (d_i + d_j)b_j$$

for any $i \neq j$. Therefore, $\frac{a_j}{b_j} = (d_i + d_j), i \neq j$. But this can hold only if $d_1 = d_2 = \cdots = d_n$. □

Let us now recall that the spectral invariants of $L_\lambda = -L_\mathfrak{p} + \lambda L_\mathfrak{k} + (\lambda^2 - s)A$ together with the Hamiltonian lifts of elements in $\mathfrak{k}_A = \{X \in \mathfrak{k} : [X, A] = 0\}$ form a complete family of functions on each coadjoint orbit in $\mathfrak{g}_\epsilon$ (Proposition 9.23, Chapter 9). The subalgebra $\mathfrak{k}_A$ is not commutative for $n > 3$. Hence the Hamiltonians generated by the elements in $\mathfrak{k}_A$ need not Poisson commute with each other. The maximum number of integrals in involution with each other that these integrals can contribute to the isospectral integrals is equal to the rank of $\mathfrak{k}_A$.

Proposition 13.2 *Let* $\mathfrak{p}_A^\perp = \{X \in \mathfrak{p}_\epsilon : \langle X, A \rangle_\epsilon = 0\}$. *Then adA restricted to* $\mathfrak{p}_A^\perp$ *is an isomorphism onto* $\mathfrak{k}_A^\perp$. *Moreover,* $\mathfrak{k}_A$ *and* $\mathfrak{k}_A^\perp$ *satisfy the Cartan conditions*

$$[\mathfrak{k}_A^\perp, \mathfrak{k}_A] \subseteq \mathfrak{k}_A^\perp, \ [\mathfrak{k}_A^\perp, \mathfrak{k}_A^\perp] \subseteq \mathfrak{k}_A. \tag{13.3}$$

Proof Since $\langle [X, A], \mathfrak{k}_A \rangle = \langle X, [A, \mathfrak{k}_A] \rangle = 0$, $[X, A] \in \mathfrak{k}_A^\perp$. Therefore, adA maps $\mathfrak{p}_\epsilon^\perp$ into $\mathfrak{k}_A^\perp$. On spaces of constant curvature, $[[X, A], A] = -\epsilon ||A||_\epsilon^2 X$ for any $X \in \mathfrak{p}_\epsilon^\perp$. Therefore, $ker(adA) = 0$ on $\mathfrak{p}_\epsilon^\perp$. Since $\mathfrak{p}_\epsilon^\perp$ and $\mathfrak{k}_A^\perp$ are of the same dimension, adA is an isomorphism on $\mathfrak{p}_\epsilon^\perp$.

That $[\mathfrak{k}_A, \mathfrak{k}_A^\perp] \subseteq \mathfrak{k}_A^\perp$ is an immediate consequence of the fact that $\mathfrak{k}_A$ is a subalgebra of $\mathfrak{k}$. To prove the second inclusion, let X and Y be any elements of $\mathfrak{k}_A^\perp$. Let P and Q be the elements in $\mathfrak{p}_A^\perp$ such that $[A, P] = X$ and $[A, Q] = Y$. Then,

$$[A, [[A, P], [A, Q]] = [[Q, A], [A, [A, P]]] - [[P, A], [[Q, A], A]]$$
$$= -\epsilon ||A||^2([[Q, A], P] - [[P, A], Q])$$
$$= -\epsilon ||A||^2[A, [Q, P]] = 0.$$

Therefore $[X, Y] \in \mathfrak{k}_A$. □

Corollary 13.3 *The subalgebra* $\mathfrak{k}_A$ *is isomorphic to* $so_{(n-1)}(R)$.

13.1 Elastic curves and the pendulum

Let us now turn to the affine Hamiltonian $H = \frac{1}{2}||L_{\mathfrak{k}}||^2 + \langle A, L_{\mathfrak{p}} \rangle$ and its Hamiltonian equations

$$\frac{dg}{dt} = g(t)(A + U(t)), \frac{dL}{dt} = [U(t), L(t)]_{\epsilon}, U(t) = L_{\mathfrak{k}}(t). \tag{13.4}$$

Each solution $L_{\mathfrak{k}}(t)$ can be written as $L_{\mathfrak{k}}(t) = L_A(t) + L_A^{\perp}(t)$, with $L_A(t) \in \mathfrak{k}_A$ and $L_A^{\perp}(t) \in \mathfrak{k}^{\perp}$. It follows from above that $L_A(t)$ is constant. The following definition is fundamental for the solutions of the above system.

Definition 13.4 A curve $x(t)$ in G/K with the geodesic curvature $\kappa(t)$ is called elastic if there exists $T > 0$ such that the integral $\frac{1}{2} \int_0^T \kappa^2(s)\, ds$ is minimal among all curves in G/K that satisfy the same tangential conditions as $x(t)$ at $t = 0$ and $t = T$.

Proposition 13.5 *Suppose that the drift vector A satisfies $||A||_{\epsilon} = 1$. Then, every solution $L_{\mathfrak{p}}(t)$ and $L_{\mathfrak{k}}(t)$ of (13.4) that satisfies $L_A(t) = 0$ projects onto an elastic curve in G/K.*

Proof Let $L_{\mathfrak{k}}(t) = L_A(t) + L_A^{\perp}(t)$ be as in above. Since the extremal control $U(t)$ is equal to $L_{\mathfrak{k}}(t)$, the associated energy is equal to

$$\frac{1}{2} \int_0^T ||U(t)||^2 dt = \frac{1}{2} \int_0^T (||L_A||^2 + ||L_A^{\perp}(t)||^2) dt.$$

When $L_A(t) = 0$, then $U(t) = L_A^{\perp}(t)$. We need to show that $||L_A^{\perp}(t)||^2 = \kappa^2(t)$, where $\kappa(t)$ is the geodesic curvature of the projected curve $x(t)$.

Recall that a curve $x(t)$ in S_{ϵ} is parametrized by arc length if and only it is a projection of a horizontal curve $g(t)$, a solution of $\frac{dg}{dt} = g(t)\Lambda(t)$, $\Lambda(t) \in \mathfrak{p}_{\epsilon}$ with $||\Lambda(t)||_{\epsilon} = 1$. Then the covariant derivative of $\frac{dx}{dt}$ is equal to the projection of $g(t)\frac{d\Lambda}{dt}(t)$ on S_{ϵ}, and hence, the geodesic curvature $\kappa(t)$ is given by $||\frac{d\Lambda}{dt}(t)||_{\epsilon}$.

If $g(t)$ denotes the solution curve of (13.4) that projects onto $x(t)$, then $\tilde{g}(t) = g(t)h(t)$, where $h(t)$ is the solution of $\frac{dh}{dt} = -U(t)h(t)$, $h(0) = I$ is a solution of

$$\frac{d\tilde{g}}{dt}(t) = \tilde{g}(t)(h^{-1}(t)Ah(t)).$$

It follows that $\tilde{g}(t)$ is a horizontal line that projects onto $x(t)$. Hence, the geodesic curvature of $x(t)$ is given by

$$\kappa(t) = ||\frac{d}{dt}(h^{-1}(t)Ah(t))|| = ||[U(t), A]||. \tag{13.5}$$

It follows that there is a unique curve $Q(t) \in \mathfrak{p}_\epsilon^\perp$ such that $U(t) = [Q(t), A]$ (Proposition 13.2). Then, $||U||^2 = \langle [Q, A], [Q, A] \rangle = \langle Q, [A, [Q, A]] \rangle = \langle Q, \epsilon Q \rangle = ||Q||_\epsilon^2$, and

$$||[U(t), A]||_\epsilon^2 = |[[[Q(t), A], A]||_\epsilon^2 = || - \epsilon Q||_\epsilon^2 = ||Q||_\epsilon^2.$$

Hence $\kappa^2(t) = ||U(t)||^2$. $\square$

Proposition 13.6 *Every elastic curve $x(t)$ is the projection of a solution curve $g(t), L_\mathrm{p}(t), L_\mathfrak{k}(t)$ of (13.4) with $L_A(t) = 0$.*

Proof Every curve $x(t)$ in S_ϵ is the the projection of a horizontal curve $g(t)$. If $\frac{dg}{dt}(t) = g(t)\Lambda(t)$ then $||\frac{d\Lambda}{dt}(t)||$ is the curvature of $x(t)$ provided that $||\Lambda(t)|| = 1$. Since K acts transitively on the unit sphere in $\mathfrak{p}_\epsilon$ by adjoint action, a fact that readily follows from (13.2), there exists a curve $h(t) \in K$ such that $\Lambda(t) = h^{-1}(t)Ah(t)$. The curve $h(t)$ is a solution of $\frac{dh}{dt} = -U(t)h(t)$ with $U(t) \in \mathfrak{k}_A^\perp$. Then, $\tilde{g}(t) = g(t)h^{-1}(t)$ projects onto $x(t)$ and is a solution of

$$\frac{d\tilde{g}}{dt}(t) = \tilde{g}(t)(A + U(t)),$$

and $\int_0^T \kappa^2(s)\, dt = \int_0^T ||U(s)||^2\, dt$, by the same argument as in Proposition 13.5. If $x(t)$ is an elastic curve, then $\tilde{g}(t)$ is an optimal solution of the above affine system, hence must be a projection of a solution $L_\mathrm{p}(t), L_\mathfrak{k}(t)$ in (7), subject to the transversality condition $L_A(0) = 0$ (dictated by the Maximum Principle). Since $L_A(t)$ is constant, $L_A(t) = 0$. $\square$

Proposition 13.7 *Let $L_\mathrm{p} = \begin{pmatrix} 0 & -\epsilon p^T \\ p & 0_n \end{pmatrix}$, and let $Q = \begin{pmatrix} 0 & -\epsilon q^T \\ q & 0_n \end{pmatrix}$ be such that $L_A^\perp = [Q, A]$ and $\langle Q, A \rangle = 0$. Then the characteristic polynomial of $L_\lambda = L_\mathrm{p} - \lambda L_A^\perp + (\lambda^2 - s)A$ is given by*

$$\mu^4 + c_1 \mu^2 + c_2 = 0, \tag{13.6}$$

where

$$
\begin{aligned}
c_1 &= \lambda^2 ||q||^2 + 2\epsilon(\lambda^2 - s)(a \cdot p) + \epsilon(\lambda^2 - s)^2 + \epsilon ||p||^2, \\
c_2 &= \epsilon\lambda^2 (||q||^2 ||p||^2 - ||q||^2 (a \cdot p)^2 - (q \cdot p)^2).
\end{aligned}
\tag{13.7}
$$

Proof It will be convenient to recall the quadratic form $(x, y)_\epsilon = x_0 y_0 + \epsilon \sum_{i-1}^n x_i y_i$, and the wedge product $(a \wedge b)_\epsilon$ in $\mathbb{R}^{n+1}$, $(a \wedge b)_\epsilon(x) = (x, a)_\epsilon b - (x, a)_\epsilon b$, $x \in \mathbb{R}^{n+1}$. Then matrices $\begin{pmatrix} 0 & -\epsilon p^T \\ p & 0_n \end{pmatrix}$ in $\mathfrak{p}_\epsilon$ can be written as $(p \wedge e_0)_\epsilon$, with the understanding that a in the wedge product is embedded in

$\mathbb{R}^{n+1}$ so that its first coordinate $a_0 = 0$. In these notations the spectral matrix L_λ is given by

$$L_\lambda = (p \wedge e_0)_\epsilon + (\lambda^2 - s)(a \wedge e_0)_\epsilon - \lambda(q \wedge a)_\epsilon, \qquad (13.8)$$

where vectors a, p, q in $\mathbb{R}^{n+1}$ are all orthogonal to e_0, $(a, q)_\epsilon = 0$, and $||a|| = 1$.

It follows that the non-zero spectrum of L_λ is determined by the restriction of L_λ to the four-dimensional space spanned by $\{e_0, a, p, q\}$. Let M_λ denote the matrix of this restriction of L_λ relative to the basis $b_1 = e_0, b_2 = a, b_3 = p$, $b_4 = q$. Then,

$$L_\lambda(e_0) = p + (\lambda^2 - s)a, \; L_\lambda(a) = -\epsilon((a \cdot p) + \epsilon(\lambda^2 - s))e_0 + \epsilon\lambda q,$$
$$L_\lambda(p) = -\epsilon(||p||^2 + (\lambda^2 - s)(a \cdot p))e_0 - \epsilon\lambda((a \cdot p)q - (p \cdot q)a),$$
$$L_\lambda(q) = -\epsilon(p \cdot q)e_0 + \epsilon\lambda||q||^2 a.$$

It follows that

$$M_\lambda = \begin{pmatrix} 0 & -\epsilon((\lambda^2 - s) + (a \cdot p)) & -\epsilon(\lambda^2 - s)(a \cdot p) + ||p||^2) & -\epsilon(p \cdot q) \\ \lambda^2 - s & 0 & \epsilon\lambda(q \cdot p) & \epsilon\lambda||q||^2 \\ 1 & 0 & 0 & 0 \\ 0 & \lambda\epsilon & -\epsilon\lambda(a \cdot p) & 0 \end{pmatrix}.$$

The characteristic polynomial of the above matrix is given by $\mu^4 + c_1\mu^2 + c_2 = 0$, with

$$c_1 = \lambda^2||q||^2 + 2\epsilon(\lambda^2 - s)(a \cdot p) + \epsilon(\lambda^2 - s)^2 + \epsilon||p||^2,$$
$$c_2 = \epsilon\lambda^2(||q||^2||p||^2 - ||q||^2(a \cdot p)^2 - (q \cdot p)^2).$$

$\square$

Since $H = \frac{1}{2}||L_{\mathfrak{t}}^{\perp}||^2 + \langle A, L_{\mathfrak{p}}\rangle = \frac{1}{2}||q||^2 + \epsilon(a \cdot p)$, then

$$c_1 = (\lambda^2 - s)H + \epsilon I_1 + \epsilon(\lambda^2 - s)^2,$$

where $I_1 = \epsilon||L_{\mathfrak{p}}||^2 + s||L_{\mathfrak{h}}||^2 = ||p||^2 + s\epsilon||q||^2$ is the Casimir function on so_ϵ. Note that $s\epsilon = \epsilon$ for $\epsilon = \pm 1, 0$ and $s = 0, 1$.

Then c_2 implies that

$$I_2 = ||q||^2||p||^2 - ||q||^2(a \cdot p)^2 - (q \cdot p)^2$$

is a constant of motion for H. One can easily show that I_2 is equal to the square of the signed volume $a \cdot (p \wedge q)$ of the parallelopiped spanned by a, p, q.

It turns out that H, I_1, I_2 give a complete description of the solutions according to the following proposition.

Proposition 13.8   *Let $\kappa(t)$ and $\tau(t)$ denote the curvature and the torsion of the projection curve $x(t)$ associated with an extremal curve. Then, $\xi(t) = \kappa^2(t)$ is the solution of the following equation:*

$$\left(\frac{d\xi}{dt}\right)^2 = -\xi^3 + 4(H - \epsilon)\xi^2 + 4(I_1 - H^2)\xi - 4I_2, \qquad (13.9)$$

and $(\kappa(t)^2\tau(t))^2 = I_2$.

   Moreover, if $T(t), N(t), B(t)$ denote the Serret–Frenet triad along $x(t)$ then $\frac{dB}{dt}(t)$ is contained in the linear span of $T(t), N(t), B(t)$. Consequently, all the remaining curvatures along $x(t)$ are equal to zero.

Proof   Since $L_A^\perp = [Q(t), A]$, $[L_A^\perp, A] = [[Q, A], A] = -\epsilon Q$. Then $-\epsilon \frac{dQ}{dt} = [[A, L_p], A] = \epsilon(L_p - \langle A, L_p \rangle A)$, and therefore

$$\frac{dQ}{dt} = -(L_p - \langle L_p, A \rangle_\epsilon A). \qquad (13.10)$$

It follows that

$$\frac{d\kappa^2}{dt} = 2\left\langle Q, \frac{dQ}{dt}\right\rangle_\epsilon = -2\langle Q, L_p \rangle_\epsilon = -2(q, p).$$

Hence,

$$\frac{d\xi}{dt}^2 = \left(\frac{d\kappa^2}{dt}\right)^2 = 4(p, q)_\epsilon^2 = 4(\|q\|^2\|p\|^2 - \|q\|^2(a \cdot p)^2 - I_2))$$

$$= 4(\|q\|^2(I_1 - \epsilon\|q\|^2) - \|q\|^2\left(H - \frac{1}{2}\|q\|^2 - I_2\right)$$

$$= 4(\xi(I_1 - \epsilon\xi) - \xi\left(H - \frac{1}{2}\xi\right)^2 - I_2)$$

$$= -\xi^3 + (4H - \epsilon)\xi^2 + 4(I_1 - H^2)\xi - 4I_2.$$

To prove the subsequent statements recall the Serret–Frenet relations

$$\frac{DT}{dt} = \kappa(t)N(t), \quad \frac{DN}{dt} = -\kappa(t)T(t) + \tau(t)B(t). \qquad (13.11)$$

In the present situation $x(t)$ is the projection of a horizontal curve $g(t)$ defined by $\frac{dg}{dt} = g(t)(h^{-1}(t)Ah(t))$, where $\frac{dh}{dt} = -U(t)h(t)$ and $U(t) = [Q(t), A]$. Let $\Lambda^T, \Lambda^N, \Lambda^B$ denote the unit vectors in p_ϵ that project onto the tangent vectors T, N, B along $x(t)$. Then Λ^T, Λ^N and Λ^B satisfy equations (13.11) with ordinary derivatives in place of the covariant derivatives. It follows that $\Lambda^T = h^{-1}(t)Ah(t)$ and

$$\frac{d\Lambda^T}{dt} = h^{-1}(t)[U, A]h(t) = h^{-1}(t)[[Q, A], A]h(t) = -\epsilon h^{-1}(t)Qh(t).$$

Hence, $\Lambda^N = -\epsilon \frac{1}{||Q||_\epsilon} h^{-1}(t) Q h(t)$. Then,

$$-\kappa \Lambda^T + \tau \Lambda^B = \frac{d\Lambda^N}{dt}$$

$$= h^{-1}(t) \left(\epsilon \frac{\langle Q, L_\mathrm{p} \rangle_\epsilon}{||Q||_\epsilon^3} Q - \epsilon \frac{1}{||Q||_\epsilon} (L_\mathrm{p} - \langle A, L_\mathrm{p} \rangle_\epsilon A) - \epsilon \frac{1}{||Q||_\epsilon} [[Q, A], A] \right) h(t)$$

$$= h^{-1}(t) \left(\epsilon \frac{\langle Q, L_\mathrm{p} \rangle_\epsilon}{||Q||_\epsilon^3} Q - \epsilon \frac{1}{||Q||_\epsilon} (L_\mathrm{p} - \langle A, L_\mathrm{p} \rangle_\epsilon A) - ||Q||_\epsilon A \right) h(t),$$

from which we get

$$\tau \Lambda^B = (h^{-1}(t) \left(\epsilon \frac{\langle Q, L_\mathrm{p} \rangle_\epsilon}{||Q||_\epsilon^3} Q - \epsilon \frac{1}{||Q||_\epsilon} (L_\mathrm{p} - \langle A, L_\mathrm{p} \rangle_\epsilon A) \right) h(t).$$

This expression shows that Λ^B is a linear combination of vectors $V_1 = h^{-1}(t) Q h(t)$, $V_2 = h^{-1}(t) L_\mathrm{p} h(t)$ and $V_3 = h^{-1}(t) A h(t)$. Since V_1 is a scalar multiple of Λ^N and $V_3 = \Lambda^T$, their time derivatives are in the linear span of $\Lambda^T, \Lambda^N, \Lambda^B$. The same applies to $\frac{dV_2}{dt}$ as can be readily verified from equations (13.4). Hence, $\frac{d\Lambda^B}{dt}$ is in the linear span of $\Lambda^T, \Lambda^N, \Lambda^B$, and therefore, the Serret–Frenet frame terminates with the binormal vector B. Finally,

$$(\kappa^2 \tau)^2 = \left|\left| \frac{\langle Q, L_\mathrm{p} \rangle_\epsilon}{||Q||_\epsilon} Q - ||Q||_\epsilon (L_\mathrm{p} - \langle A, L_\mathrm{p} \rangle_\epsilon A) \right|\right|_\epsilon^2$$

$$= ||Q||_\epsilon^2 ||L_\mathrm{p}||_\epsilon^2 - \langle Q, L_\mathrm{p} \rangle_\epsilon^2 - \langle A, L_\mathrm{p} \rangle_\epsilon ||Q||_\epsilon^2 = I_2.$$

$\square$

Equation (13.9) admits an alternative version in terms of an angle θ that provides a nice segue to the section on the pendulum. To define θ, first notice that

$$\left(\frac{d\xi}{dt} \right)^2 = -\xi^3 + 4(H - \epsilon)\xi^2 + 4(I_1 - H^2)\xi - 4I_2$$

$$= -\xi((\xi - 2(H - \epsilon))^2 - 4J^2) - 4I_2,$$

where $J^2 = 1 - 2H\epsilon + I_1$. Let $\xi(t) - 2(H - \epsilon) = 2J \cos \theta$. Then,

$$4J^2 (\sin^2 \theta) \dot{\theta}^2 = \frac{d\xi}{dt}^2 = 2((H - \epsilon) + J \cos \theta) 4J^2 \sin^2 \theta - 4I_2.$$

Therefore,

$$\frac{d\theta}{dt} = \pm \sqrt{2H - 2\epsilon + 2J \cos \theta - \frac{I_2}{J^2 \sin^2 \theta}}. \tag{13.12}$$

The case $I_2 = 0$ implies that the torsion τ of the projected curve $x(t)$ is equal to zero. In such a case the elastic curves are confined to the two-dimensional submanifold of S_ϵ determined by $x(0)$ and $\frac{dx}{dt}(0)$. The elastic curves then coincide with the non-Euclidean elasticae defined in Chapter 7.

13.1.1 The pendulum

We have already alluded to the mysterious connection between the affine problems on space forms and the equations of heavy tops. We will presently show that the motions of a mathematical pendulum are intricately related with the elastic curves. Consider now an n-dimensional analogue of the spherical pendulum of unit length suspended at the origin of $\mathbb{R}^n$ and acted upon by the force $\vec{F} = -e_1$, where $e_1, \ldots, e_n$ denotes the standard basis in $\mathbb{R}^n$. The motions of such a pendulum are confined to the unit sphere S^{n-1}.

The sphere S^{n-1} can be realized as the quotient K/K_0, where $K = SO_n(R)$ and K_0 is the isotropy group of e_1, that is, $K_0 e_1 = e_1$. Then, $K_0 = \{1\} \times SO_{n-1}(R)$. Let $\mathfrak{k}_0$ denote the Lie algebra of K_0, and let $\mathfrak{k}_1$ denote the orthogonal complement in $so_n(R)$ relative to the trace form $\langle A, B \rangle = -\frac{1}{2}Tr(AB)$. It follows that $\mathfrak{k} = \mathfrak{k}_0 \oplus \mathfrak{k}_1$ and

$$\mathfrak{k}_1 = \left\{ \begin{pmatrix} 0 & -u^T \\ u & 0 \end{pmatrix}, u \in \mathbb{R}^{n-1} \right\}, \mathfrak{k}_0 = \begin{pmatrix} 0 & 0 \\ 0 & so_{n-1}(R) \end{pmatrix}.$$

Let $\mathcal{D}$ denote the left-invariant distribution with values in $\mathfrak{k}_1$. Following our earlier terminology, we will refer to the integral curves of $\mathcal{D}$ as the horizontal curves. Every curve $q(t)$ on S^{n-1} can be lifted to a horizontal curve $R(t)$, in the sense that $q(t) = R(t)e_1$, and $\frac{dR}{dt} = R(t) \begin{pmatrix} 0 & -u^T(t) \\ u(t) & 0 \end{pmatrix}$ for some curve $u(t)$ in $\mathbb{R}^{n-1}$. Any two such liftings differ by a left multiple by an element in K_0.

The kinetic energy T associated with a path $q(t)$ in S^{n-1} is given by

$$T = \frac{1}{2} \left\| \frac{dq}{dt} \right\|^2 = \frac{1}{2} \left\| \frac{dR}{dt} e_1 \right\|^2 = \frac{1}{2} \left\| R(t) \begin{pmatrix} 0 \\ u(t) \end{pmatrix} \right\|^2 = \frac{1}{2} \|u(t)\|^2.$$

The potential energy, given by the formula $V(q) = -\int_{q_0}^q \vec{F} \cdot \frac{d\sigma}{dt} dt$, where $\sigma(t)$ is a path from q_0 to q, is equal to $V(q) = e_1 \cdot (q - q_0)$. For convenience, we will take $q_0 = -e_1$ in which case, $V = e_1 \cdot q + 1$.

The Principle of Least Action states that each motion $q(t)$ of the pendulum minimizes the action $\int_{t_0}^{t_1} \mathcal{L}\left(q(t), \frac{dq}{dt}\right) dt$ over the paths from $q(t_0)$ to $q(t_1)$ for times sufficiently near each other, where $\mathcal{L}$ denotes the Lagrangian $\mathcal{L} = T - V$.

We will rephrase this principle as an optimal control problem on $SO_n(R)$ in order to get its Hamiltonian by the Maximum Principle.

The Principle of Least Action stated as an optimal control problem consists of finding the minimum of the integral $\int_{t_0}^{t_1} \left(\frac{1}{2}\|u(t)\|^2 - e_1 \cdot Re_1 - 1 \right) dt$ over the horizontal curves in $\mathcal{D}$ that satisfy the boundary conditions $R(t_0) \in \{R : Re_0 = q_0\}, R(t_1) \in \{R : Re_0 = q_1\}$. That is, the boundary manifolds consist of the vertical fibers in $SO_n(R)$ above the points q_0 and q_1 and the horizontal curves are the solutions of $\frac{dR}{dt} = R(t)U(t)$ with the control functions $U(t) =$ $\begin{pmatrix} 0 & -u^T(t) \\ u(t) & 0 \end{pmatrix}, u(t) \in \mathbb{R}^{n_1}$.

The Hamiltonian and the resulting Hamiltonian equations are obtained by an argument essentially identical to the one used for the heavy top. We will not repeat the calculation, which shows that the Hamiltonian is given by

$$\mathcal{H}(Q_0, R) = \frac{1}{2}\langle Q_1, Q_1 \rangle + e_1 \cdot Re_1 + 1,$$

with the corresponding Hamiltonian equations given by

$$\frac{dR}{dt} = R(t)(Q_1(t)), \frac{dQ}{dt}(t) = [Q_1(t), Q(t)] + R^T(t)e_1 \wedge e_1. \tag{13.13}$$

These equations agree with equations (12.24) in Chapter 12 for the top when $a_1 = 1, a_2 = 0, a_3 = 0$.

Since $[Q_1, Q] = [Q, Q_0]$ and since the latter bracket is in $\mathfrak{k}_1$, the projection Q_0 of $Q(t)$ on $\mathfrak{k}_0$ is constant. But this constant is must be zero in order to satisfy the transversality conditions imposed by the Maximum Principle. Therefore, equations (13.13) can be rewritten as

$$\frac{dR}{dt} = R(t)(Q_1(t)), \frac{dQ_1}{dt}(t) = R^T(t)e_1 \wedge e_1, Q_0 = 0. \tag{13.14}$$

Equations (13.14) can be identified with equations (13.4) on the semi-direct product $G = \mathfrak{p}_\epsilon \rtimes SO_n(R), \epsilon = 0$ in the case that $U = L_A^\perp$ and $A = \begin{pmatrix} 0 & 0 \\ e_1 & 0_n \end{pmatrix}$. The identification is the same as the one described in Chapter 12 on the tops: each motion $q(t) = R(t)e_1$ is identified with the matrix $g(t) = \begin{pmatrix} 1 & 0 \\ x(t) & R(t) \end{pmatrix}$ in G where $x(t) = \int_0^t q(\tau) d\tau$. Then,

$$\frac{dg}{dt} = \begin{pmatrix} 0 & 0 \\ \dot{x} & \dot{R} \end{pmatrix} = \begin{pmatrix} 0 & 0 \\ Re_1 & RQ_1 \end{pmatrix} = g(t)(A + L_{\mathfrak{k}}^\perp(t)), L_{\mathfrak{k}}^\perp(t) = \begin{pmatrix} 0 & 0 \\ 0 & Q_1(t) \end{pmatrix}.$$

Secondly, the term $R^T e_1$ will be identified with the matrix $L_p = \begin{pmatrix} 0 & -p(t)^T \\ p(t) & 0_n \end{pmatrix}$

via $p(t) = R^T(t)e_1$. Then,

$$\frac{dL_p}{dt} = \begin{pmatrix} 0 & 0 \\ -Q_1 p(t) & 0 \end{pmatrix} = [L_{\mathfrak{k}}^\perp, L_p]$$

and

$$\frac{dL_{\mathfrak{k}}^\perp}{dt} = \begin{pmatrix} 0 & 0 \\ 0 & -(R^T e_1) \wedge e_1 \end{pmatrix} = [A, L_p(t)].$$

The above can be summarized in the following proposition.

Proposition 13.9 *Every motion $q(t)$ of the pendulum traces the curve of tangent vectors of an elastic curve in the affine Euclidean space $\mathbb{E}^n = \{1\} \times \mathbb{R}^n$.*

13.2 Parallel and Serret–Frenet frames and elastic curves

The results of Proposition 13.8 can be obtained several ways. In this section we will examine two such ways, both of which rely on the lifting to Lie groups. The first, which may be considered a reduced version of the canonical affine-quadratic problem, makes use of the parallel frames, and the second way makes use of the Serret–Frenet frames.

Let us begin with the parallel frames.

Definition 13.10 A curve of ordered orthonormal frames $V_1(t), \ldots, V_n(t)$ over a curve $x(t)$ in S_ϵ is called Darboux if $\frac{dx}{dt} = V_1(t)$.

Darboux frames are adapted to the underlying curve by the requirement that the first lef of the frame is equal to the tangent vector of the curve. Every Darboux curve $V_1(t), \ldots, V_n(t)$ over $x(t)$ is a solution of

$$\frac{dg}{dt} = g(t)(A_1 + U(t)), A_1 = \begin{pmatrix} 0 & -\epsilon e_1^T \\ e_1 & U(t) \end{pmatrix}, U(t) \in so_n(R). \qquad (13.15)$$

In fact, $g(t)e_0 = x(t), g(t)e_i = V_i(t), i = 1, \ldots, n$ and $U_{ij}(t) = \left(\frac{D_x}{dt}(V_i(t)), V_j(t)\right)_\epsilon$, for all $1 \leq i, j \leq n$.

Definition 13.11 A vector $V(t)$ along a curve $x(t) \in S_\epsilon$ that is normal to the tangent vector $\frac{dx}{dt}$ is said to parallel to $x(t)$ if $\frac{D_{x(t)}}{dt}(V(t))$ is parallel to the tangent vector $\frac{dx}{dt}$.

Definition 13.12 A Darboux frame $V_1(t), \ldots, V_n(t)$ is said to be parallel along $x(t)$ if each leg $V_i, i > 1$ is parallel to x.

Proposition 13.13 *Let $x(t)$ be any smooth curve in S_ϵ parametrized by arc length, and let $v_1, v_2, \ldots, v_n$ be any orthonormal set of vectors such that $v_1 = \frac{dx}{dt}(0)$. Then there exists a parallel frame $V_1(t), \ldots, V_n(t)$ along $x(t)$ such that $V_i(0) = v_i, i = 2, \ldots, n$.*

Proof Let K_1 denote the subgroup of K that satisfies $K_1 e_1 = e_1$ in K. It follows that $K_1 = \{1\} \times \{1\} \times SO_{n-1}(R)$. Hence its Lie algebra $\mathfrak{k}_1$ is isomorphic to $so_{n-1}(R)$ with $e_i \wedge e_j, i \geq 2 > j$ its orthonormal base. Then $\mathfrak{k}_1^\perp$, its orthogonal complement in $\mathfrak{k}$, is a Cartan space, in the sense that

$$[\mathfrak{k}_1, \mathfrak{k}_1^\perp] \subseteq \mathfrak{k}_1^\perp, \ [\mathfrak{k}_1^\perp, \mathfrak{k}_1^\perp] \subseteq \mathfrak{k}_1. \tag{13.16}$$

Let $g(t)$ be any Darboux curve in SO_ϵ such that $g(t)e_0 = x(t)$ and $\frac{dx}{dt} = g(t)e_1$. Such a curve can always be constructed by the Gram–Schmidt procedure. Then $g(t)$ is a solution of equation (13.15) for some curve $U(t)$ in $so_n(R)$. We may assume that $g(0)e_i = v_i, i = 2, \ldots, n$, since (13.15) is left-invariant. Let $U_1(t)$ denote the projection of $U(t)$ on $\mathfrak{k}_1$ and let $U_0(t)$ denote the projection of $U(t)$ on the orthogonal complement $\mathfrak{k}_1^\perp$ in $\mathfrak{k}$. It follows that $U_0(t) = \sum_{j=2}^{n} U_{1j} e_1 \wedge e_j$ and $U_1 = \sum_{i \geq 2 > j}^{n} U_{ij} e_i \wedge e_j$.

Let $g_1(t)$ denote the solution of $\frac{dg_1}{dt} = -U_1(t)g_1(t)$ in K_1 that originates at the identity at $t = 0$ and let $h(t) = g(t)g_1(t)$. Then $h(t)e_0 = g(t)e_0 = x(t)$ and $h(t)e_1 = g(t)e_1 = \frac{dx}{dt}$. In addition, $h(0)e_i = g(0)e_i = v_i, i = 2, \ldots, n$ and

$$\frac{dh}{dt} = \frac{dg}{dt}g_1(t) + g(t)\frac{dg_1}{dt} = g(t)(A_1 + U_1(t) + U_0(t))g_1(t) - g(t)U_1(t)g_1(t)$$

$$= h(t)g_1^{-1}(A_1 + U_0(t))g_1(t)$$

$$= h(t)(g_1^{-1}(t)A_1 g_1(t) + g_1^{-1}U_0(t)g_1(t)).$$

Since $K_1 e_1 = e_1$, $g_1^{-1}(t)A_1 g_1(t) = A_1$, and $g_1^{-1}(t)U_0(t)g_1(t)$ is in $\mathfrak{k}_1^\perp$ in view of (13.16). Hence, $g_0^\perp(t)U_0(t)g_0(t) = \sum_{i=2}^{n} u_i(t)e_1 \wedge e_i$ for some functions $u_2(t), \ldots, u_n(t)$.

Then $V_i(t) = h(t)e_i, i = 1, \ldots, n$ is the desired parallel frame since

$$\frac{D_x}{dt}(V_1) = -\sum_{i=2}^{n} u_i(t)V_i \text{ and } \frac{D_x}{dt}(V_i) = u_i(t)V_1(t) = u_i(t)\frac{dx}{dt}, i = 2, \ldots, n.$$

$\square$

Corollary 13.14 *Every solution of the equation*

$$\frac{dg}{dt} = g(t)(A_1 + U(t)), U(t) \in \mathfrak{k}_1^\perp, \tag{13.17}$$

defines a parallel frame $V_i(t) = g(t)e_i, i = 1, \ldots, n$ along the curve $x(t) = g(t)e_0$ and $\|U(t)\|$) is equal to the geodesic curvature $\kappa(t)$ of $x(t)$.

Proof The proof in above shows that $U(t) = \sum_{i=2}^{n} u_i(t)e_1 \wedge e_i$, and that $\frac{D_x}{dt}\left(\frac{dx}{dt}\right) = -\sum_{i=2}^{n} u_i(t)V_i(t)$. Hence,

$$\kappa^2(t) = \left\|\frac{D_x}{dt}\left(\frac{dx}{dt}\right)\right\|_\epsilon^2 = \sum_{i=2}^{n} u_i^2(t) = \langle U(t), U(t)\rangle.$$

$\square$

Corollary 13.15 *Every control system $\frac{dg}{dt} = g(t)(A + U(t)), \|A\|_\epsilon = 1, U(t) \in \mathfrak{k}_A^\perp$ is conjugate to system (13.17) above.*

Proof There exists an $h \in \mathfrak{k}$ such that for any $A \in \mathfrak{p}_\epsilon, \|A\|_\epsilon = 1, Ad_h(A) = A_1$, because K acts transitively on each sphere in $\mathfrak{p}_\epsilon$.

Let $g(t)$ be any solution above and let $\tilde{g}(t) = Ad_h(g(t))$. Then, $\tilde{g}(t)$ satisfies

$$\frac{d\tilde{g}}{dt} = \tilde{g}(t)(Ad_h(A) + Ad_h(U(t))) = \tilde{g}(t)(A_1 + \tilde{U}(t)).$$

It is easy to verify that $Ad_h(\mathfrak{k}_A) = \mathfrak{k}_1$, hence $\langle \tilde{U}, \mathfrak{k}_1\rangle = \langle Ad_h(U), Ad_h(\mathfrak{k}_A)\rangle = \langle U, \mathfrak{k}_A\rangle = 0$. This implies that $\tilde{U} \in \mathfrak{k}_1^\perp$ and $\tilde{g}$ is a solution of (13.17). $\square$

Parallel frames offer the most direct route to elastic curves via the following formulation, referred to as the Euler–Griffiths problem in my earlier publications [Jm].

Definition 13.16 The Euler–Griffiths problem Find a solution $g(t)$ of

$$\frac{dg}{dt} = g(t)(A_1 + U(t)), U(t) \in \mathfrak{k}_1^\perp$$

on an interval $[0, T]$ that satisfies $g(0) \in g_0K_1, g(T) \in g_1K_1$, such that the corresponding control $U(t)$ minimizes $\frac{1}{2}\int_0^T \|U(t)\|_\epsilon^2 \, dt$ among all other solutions that satisfy the same boundary conditions.

The Euler–Griffiths problem then generates a reduced affine Hamiltonian $H = \frac{1}{2}\|L_1^\perp\|^2 + \langle L_\mathfrak{p}, A_1\rangle$, subject to the transversality condition $L_1(0) = 0$, where $L_1(t)$ and $L_1^\perp(t)$ denote the projections onto $\mathfrak{k}_1^\perp$ and $\mathfrak{k}_1$. The Hamiltonian equations are conjugate to system (13.4) with $L_A(t) = 0$. Therefore,

Proposition 13.17 *A curve in S_ϵ is elastic if and only if it is a projection of an integral curve of the reduced Hamiltonian $H = \frac{1}{2}\|L_1^\perp\|^2 + \langle L_\mathfrak{p}, A_1\rangle$ on $L_1 = 0$.*

13.3 Serret–Frenet frames and the elastic problem

Serret-Frenet frames provide an alternative setting for formulating the elastic problem as well as other variational problems involving the geometric invariants of the underlying curve relative to its Riemannian metric.

Let us recall the basic facts. A curve $x(t) \in S_\epsilon$ is called non-singular if the vectors $\frac{dx}{dt}, \frac{D_x^k}{dt^k}\left(\frac{dx}{dt}\right), k = 1, \ldots, n-1$ are linearly independent on an interval $[0, T]$. Every non-singular curve can be lifted to the Serret–Frenet frame via the following procedure.

If $x(t)$ is non-singular, then $\frac{dx}{dt}(t) \neq 0$, hence $x(t)$ can be reparametrized by its arc length. Then $\frac{dx}{ds} = v_1$ defines the first leg $v_1(t)$ of the frame. The projection of the covariant derivative $\frac{D_x}{ds}(v_1(s))$ on the orthogonal complement of $v_1(s)$ defines the second leg of the frame $v_2(s)$ and a function $u_1(s)$ via the formula $\frac{D_x}{ds}(v_1) = u_1(s)v_2(s)$. It follows that $||v_2|| = \frac{1}{\left\|\frac{D_x}{ds}(v_1)\right\|}$, which implies that $u_1(s) = \left\|\frac{D_x}{ds}(v_1)\right\|$. The projection of the covariant derivative of v_2 on the orthogonal complement of the linear span of v_1, v_2 determines the third leg v_3 of the frame. It follows that $\frac{D_x}{dt}(v_2(s)) = -u_1(s)v_1(s) + u_2(s)v_3(s)$, where $u_2(s) = \left\|\frac{D_x}{dt}(v_2(s)) + u_1(s)v_1(s)\right\|$.

This procedure generates a frame $v_1(s), \ldots, v_n(s)$ over each non-singular curve $x(s)$ that is a subject to the following differential conditions:

$$\frac{D_x}{ds}(v_1(s)) = u_1(s)v_2,$$

$$\frac{D_x}{ds}(v_i(s)) - -u_{i-1}v_{i-1} + u_i(s)v_{i+1}(s), i = 2, \ldots, n \quad 1,$$

$$\frac{D_x}{ds}(v_n(s)) = -u_{n-1}(s)v_{n_1}(s),$$

where the functions $u_1(s), \ldots, u_{n-2}(s)$ are positive while the last function $u_{n-1}(s)$ can be of arbitrary sign.

The above system can be identified with a curve $g(t)$ in SO_ϵ with $x(t) = g(t)e_0$ and the remaining columns of g equal to $v_1, \ldots, v_n$. Then $g(t)$ is a solution of

$$\frac{dg}{dt}(A_1 + U(t)), U = \begin{pmatrix} 0 & -u_1(t) & 0 & \cdots & & \cdots \\ u_1(t) & 0 & -u_2(t) & 0 & & \cdots \\ 0 & u_2(t) & 0 & 0 & & \cdots \\ \vdots & \vdots & \vdots & \vdots & & \vdots \\ 0 & \cdots & 0 & -u_{n-2}(t) & & 0 \\ 0 & \cdots & u_{n-2}(t) & 0 & & -u_{n-1}(t) \\ 0 & \cdots & 0 & u_{n-1}(t) & & 0 \end{pmatrix}.$$

$$(13.18)$$

It follows that $u_1(t)$ is the geodesic curvature and $u_2(t)$ is the torsion $\tau(t)$ for three-dimensional curves. In arbitrary dimensions, u_1 is called the first curvature, u_2 the second curvature, and so on, to the last curvature u_{n-1}.

At first glance it might seem natural to formulate the variational problems involving geometric invariants of non-singular curves as an optimal control problem on SO_ϵ with functions $u_1(t), \ldots, u_{n-1}(t)$ playing the role of controls. However, optimal solutions tend to violate the requirement that $u_1 > 0, \ldots, u_{n-2} > 0$, and thus loose the connection with the underlying non-singular curve.

Nevertheless, system (13.18) under the relaxed conditions that the control functions are in $L^2([0,T]$ and satisfy $u_1(t) \geq 0, \ldots, u_{n-2}(t) \geq 0$, and $u_{n-1}(t) \in \mathbb{R}$ is a well-defined control system and any semi-positive-definite quadratic form $Q(u_1, \ldots, u_{n-1})$ induces a natural optimal control problem of finding the minimum of $\int_0^T Q(u_1(t), \ldots, u_{n-1}(t)) \, dt$ over the trajectories that satisfy the given boundary conditions. The most natural choice occurs for

$$Q(u_1(t), \ldots, u_{n-1}(t)) = \frac{1}{2}(u_1^2(t) + u_2^2(t) + \cdots + u_{n-1}^2(t)) \qquad (13.19)$$

where the functional could be interpreted as the energy of the underlying curve.

The problem of minimizing $\frac{1}{2} \int_0^T u_1^2(t) \, dt$ over the solutions $g(t)$ of (13.18) that conform to $g(0) \in g_0 K_1$ and $g(T) \in g_1 K_1$, where K_1 is the isotropy subgroup of e_1, is most commonly associated with the Euler–Griffiths problem, even though, as such, this formulation of the elastic problem is subject to certain technical flaws: for instance, singular curves do not lift to the Serret–Frenet system and hence are not in the competition for the minimum, while some Serret–Frenet curves are generated by the controls that bear no relation to the geometric invariants of the projected curve, such as, for instance, the curves with $u_1(t) = 0$, and $u_2(t) \neq 0$.

We will presently investigate this situation on $SE_3(R), SO(1,3)$ and $SO_4(R)$), the orthonormal frame bundles of three-dimensional space forms. We will resume the notations introduced earlier and use SO_ϵ to denote such groups. In these situations, the factor spaces $\mathfrak{k}$ and $\mathfrak{p}_\epsilon$ corresponding to the Cartan decomposition of the Lie algebra so_ϵ are both three dimensional. Let

$$A_1 = \begin{pmatrix} 0 & 0 \\ 0 & e_3 \wedge e_2 \end{pmatrix}, A_2 = \begin{pmatrix} 0 & 0 \\ 0 & e_1 \wedge e_3 \end{pmatrix}, A_3 = \begin{pmatrix} 0 & 0 \\ 0 & e_2 \wedge e_1 \end{pmatrix},$$

$$B_1 = \begin{pmatrix} 0 & -\epsilon e_1^T \\ e_1 & 0 \end{pmatrix}, B_2 = \begin{pmatrix} 0 & -\epsilon e_2^T \\ e_2 & 0 \end{pmatrix}, B_3 = \begin{pmatrix} 0 & -\epsilon e_3^T \\ e_3 & 0 \end{pmatrix}.$$

Matrices $X \in \mathfrak{p}_\epsilon$ and $Y \in \mathfrak{k}$ can be expressed by their coordinate vectors $\begin{pmatrix} \hat{Y} \\ \hat{X} \end{pmatrix}$ in $\mathbb{R}^6$ relative to the above basis, i.e., $X = x_1 B_1 + x_2 B_2 + x_3 B_3$,

$Y = y_1 A_1 + y_2 A_2 + y_3 B_3$ and

$$\hat{X} = \begin{pmatrix} x_1 \\ x_2 \\ x_3 \end{pmatrix}, \hat{Y} = \begin{pmatrix} y_1 \\ y_2 \\ y_3 \end{pmatrix}.$$

It then follows from (13.1) that the coordinates of the Lie brackets in $\mathfrak{g}_\epsilon$ are expressed in terms of the vector product in $\mathbb{R}^3$ according to the following formulas:

$$\widehat{[X_1, X_2]} = \epsilon \begin{pmatrix} \hat{X}_2 \times \hat{X}_1 \\ 0 \end{pmatrix}, \widehat{[X, Y]} = \begin{pmatrix} 0 \\ \hat{Y} \times \hat{X} \end{pmatrix}, \widehat{[Y_1, Y_2]} = \begin{pmatrix} \hat{Y}_2 \times \hat{Y}_1 \\ 0 \end{pmatrix}.$$

(13.20)

The above basis induces its dual basis $A_1^*, A_2^*, A_3^*, B_1^*, B_2^*, B_3^*$ on $\mathfrak{g}_\epsilon^*$. If $M_1, M_2, M_3, p_1, p_2, p_3$ denote the coordinates of any point $\ell \in \mathfrak{g}^*$ relative to the dual basis, then these coordinate functions conform to the following Poisson bracket table:

Table 13.1

$\{\,,\,\}_\epsilon$	M_1	M_2	M_3	p_1	p_2	p_3
M_1	0	$-M_3$	M_2	0	$-p_3$	p_2
M_2	M_3	0	$-M_1$	p_3	0	$-p_1$
M_3	$-M_2$	M_1	0	$-p_2$	p_1	0
p_1	0	$-p_3$	p_2	0	$-\epsilon M_3$	ϵM_2
p_2	p_3	0	$-p_1$	ϵM_3	0	$-\epsilon M_1$
p_3	$-p_2$	p_1	0	$-\epsilon M_2$	ϵM_1	0

In the identification of $\ell \in \mathfrak{g}_\epsilon$ with $L \in \mathfrak{g}_\epsilon$ via the formula $\langle L, X \rangle = \ell(X), X \in \mathfrak{g}_\epsilon, L$ is given by the following matrix:

$$L = \begin{pmatrix} 0 & -p_1 & -p_2 & -p_3 \\ \frac{1}{\epsilon}p_1 & 0 & -M_3 & M_2 \\ \frac{1}{\epsilon}p_2 & M_3 & 0 & -M_1 \\ \frac{1}{\epsilon}p_3 & -M_2 & M_1 & 0 \end{pmatrix}.$$

(13.21)

In terms of these notations the Serret–Frenet system is given by

$$\frac{dg}{dt} = g(t)(B_1 + u_1(t)A_3 + u_2(t)A_1(t)), u_1(t) \geq 0, u_2(t) \in \mathbb{R}.$$ (13.22)

Over the non-singular curves, $u_1(t) = \kappa(t)$ and $u_2(t) = \tau(t)$ of the underlying curve $x(t) = g(t)e_0$. As we mentioned earlier, (13.22) admits solution curves which are not the lifts of curves. For instance, when $u_1 = 0$ the solutions are given by

$$g(t) = e^{tB_1} e^{\int_0^t u_2(s)\, dsA_1},$$ (13.23)

because $[A_1, B_1] = 0$. Then, $x(t) = g(t)e_0 = e^{B_1t}e_0$. Thus all solutions with $u_1 = 0$ project onto the same geodesic, and $u_2(t)$ bears no relation to the torsion of $x(t)$.

Consider now the problem of minimizing $\frac{1}{2}\int_0^T u_1^2(t)\,dt$ over the solutions of (13.22). According to the Maximum Principle each optimal solution $g(t)$ corresponding to a control $u(t)$ is the projection of an extremal curve $(g(t), L(t))$ associated with the Hamiltonian lift $\mathcal{H}_u(t)(L) = -\lambda\frac{1}{2}u_1^2 + p_1 + u_1(t)M_3 + u_2(t)M_1$, $\lambda = 0, 1$ subject to the condition that

$$\mathcal{H}_{u(t)}(L(t)) \geq -\lambda\frac{1}{2}v_1^2 + p_1 + v_1M_3 + v_2M_1, v_1 \geq 0, v_2 \in \mathbb{R}. \qquad (13.24)$$

The abnormal extremals are the integral curves of $\mathcal{H}_{u(t)}(L) = p_1 + u_1(t)M_3(L) + u_2(t)M_1(L)$ subject to $M_1(L(t)) = M_3(L(t)) = 0$. Then $p_1(L(t)) = 0$, since $\mathcal{H}_{u(t)}(L(t)) = 0$ along abnormal extremals. Further,

$$0 = \{M_1, \mathcal{H}_{u(t)}\}_\epsilon(L(t)) = u_1(t)M_2(L(t)),$$
$$0 = \{M_3, \mathcal{H}_{u(t)}\}_\epsilon(L(t)) = -p_2(L(t)) - u_2(t)M_2(L(t)),$$

and therefore $u_1(t) = 0$ and $u_2(t) = -\frac{p_2(L(t))}{M_2(L(t))}$, since $M_2(L(t)) \neq 0$. ($M_2(L(t)) = 0$ would imply that imply that $L(t) = 0$ contrary to the Maximum Principle.) It then follows that the projection of an abnormal extremal is given by (13.23) and therefore projects onto a geodesic in S_ϵ.

In the case of the normal extremals, condition (13.24) yields $u_1(t) = M_3(L(t))$ and a constraint $M_1(L(t)) = 0$. We now need additional differentiations to resolve this constraint. To begin with, $\{M_1, \mathcal{H}_{u(t)}\}(L(t)) = M_2(L(t))M_3(L(t))$ yields another constraint

$$M_2(L(t))M_3(L(t)) = 0.$$

If $M_3(L(t)) = 0$ then the extremal curve $g(t)$ projects onto a geodesic independently of $u_2(t)$. In the remaining case, $M_3(L(t)) \neq 0$ and $M_2(L(t)) = 0$. Then $0 = \{M_2, \mathcal{H}_{u(t)}\} = p_3(L(t)) + u_2(t)M_3(L(t))$ and $u_2(t) = -\frac{p_3(L(t))}{M_3(L(t))}$. Therefore, the extremal curves are the integral curves of the constrained Hamiltonian

$$H = \frac{1}{2}M_3^2 + p_1 - \frac{p_3}{M_3}M_1, M_3 > 0, \text{ on } M_1 = M_2 = 0. \qquad (13.25)$$

On $M_1 = M_2 = 0$ its equations are given by

$$\frac{dM_3}{dt} = -p_2, \frac{dp_1}{dt} = h_2M_3, \frac{dp_2}{dt} = p_3u_2 - p_1M_3 + \epsilon sM_3, \frac{dp_3}{dt} = p_1 - p_2u_2$$

$$(13.26)$$

Remarkably, $\xi(t) = M_3^2(t)$ satisfies the same equation as the square of the curvature of an elastic curve (equation (13.9)) because

$$\frac{1}{4}\frac{d}{dt}(\xi)^2 = M_3^2 p_2^2 = \xi(I_1 - p_1^2 - p_3^3 - \epsilon M_3^2)$$

$$= \xi\left(I_1 - \left(H - \frac{1}{2}\xi\right)^2\right) - I_2^2 - \epsilon\xi^2$$

$$= -\frac{1}{4}\xi^3 + (H - \epsilon)\xi^2 + (I_1 - H^2)\xi - I_2^2,$$

where

$$(M_1^2 + M_2^2 + M_3^2) + p_1^2 + p_2^2 + p_3^2 = I_1, p_1 M_1 + p_2 M_2 = p_3 M_3 = I_2$$

are the Casimirs restricted to $M_1 = M_2 = 0$. The equation

$$\frac{d\xi}{dt}^2 = -\xi^3 + 4(H - \epsilon)\xi^2 + 4(I_1 - H^2)\xi - 4I_2^2 \qquad (13.27)$$

agrees with the corresponding equation for the curvature of an elastic curve.

Let us now consider the extremals associated with optimizing the energy functional $E = \frac{1}{2}\int_0^T u_1^2(t) + u_2^2(t))\,dt$ over the solutions of (13.22). The normal extremals are the integral curves of $H = \frac{1}{2}(M_1^2 + M_3^2) + p_1$, while the abnormal extremals are the same as in the previous case. It follows that the normal extremals are the solutions of

$$\frac{dg}{dt} = g(B_1 + M_1 A_1 + M_3 A_3), \frac{dM_1}{dt} = M_2 M_3, \frac{dM_2}{dt} = p_3,$$

$$\frac{dM_3}{dt} = -M_1 M_2 - p_2, \frac{dp_1}{dt} = p_2 M_3, \frac{dp_2}{dt} = p_3 M_1 - M_3 p_1 + \epsilon M_3,$$

$$\frac{dp_3}{dt} = -p_2 M_1 - \epsilon M_2. \qquad (13.28)$$

Proposition 13.18 *Every optimal solution $g(t)$ of the above energy problem that is the projection of an abnormal extremal is also the projection of a normal extremal.*

Proof Every abnormal extremal is generated by $u_1 = 0$ and $u_2 = -\frac{p_2}{M_2}$. Their projections on SO_ϵ are given by (13.23). These projections will be optimal only for constant controls u_2.

In the normal case the extremal controls are given by $u_1 = M_3$ and $u_2 = M_1$. It is easy to check that (13.28) admits solutions $M_3 = 0$ and M_1 an arbitrary constant. The remaining variables are given by $\frac{dp_1}{dt} = 0, \frac{dp_2}{dt} = 0, h_3 = 0,$ $\epsilon M_2 = M_1 p_2.$ □

It seems that system (13.28) does not admit any extra integrals of motion beyond the obvious ones – the Casimirs, the Hamiltonian, and the

Hamiltonians of right-invariant vector fields. We will be able to show, using the ides of S. Kowalewski, that system (13.26) does not admit meromorphic solutions, which in turn implies that the equations cannot be integrated in terms of elliptic or hyperelliptic integrals.

13.4 Kichhoff's elastic problem

Let us now return to to the general affine Hamiltonian $H = \frac{1}{2}\langle \mathcal{P}(L_{\mathfrak{k}}), L_{\mathfrak{k}} \rangle + \langle A, L_{\mathfrak{p}} \rangle$ on SO_ϵ. For two-dimensional space forms, the Serret–Frenet systems and the general affine-quadratic problem coalesce into a single system whose solutions project onto the elasticae. As we mentioned, earlier elastic curves have a particularly distinguished history and go back to Daniel Bernoulli, who in 1742 suggested to L. Euler that the differential equation for the equilibrium shape of a thin elastic inextensible beam subject to bending torques at its ends could be found by making the integral of the square of of the curvature along the beam a minimum. Euler, acting on this suggestion, obtained the differential equation for this problem in 1744 and was able to describe its solutions, known since then as *elasticae*, well before the discovery of elliptic functions [Eu; Lv].

The passage from Euler's work on elastica to more general theory of elastic plates and rods required new theoretical concepts, and this new subject matter attracted the attention of some of the best mathematical minds of the ninettenth century (see the Historical introduction to the *Mathematical Theory of Elasticity* by A. E. Love [Lv]). In this formative period of the theory, the work of A. Cauchy in 1822 on stresses and strains was of central importance for the subsequent generalizations of Euler's elastica to spatial rods in which the most notable contribution came from G. Kirchhoff. Kirchhoff, in his remarkable paper of 1859, not only wrote the differential equations for the equilibrium configurations of an elastic rod in $\mathbb{R}^3$ subject to bending and twisting torques at its ends, but also likened these equations to the motions of the heavy top metaphorically as " the kinetic analogue" (see [Lv] for the exact reference).

According to A. E. Love, Kirchhoff considered an elastic rod as a framed curve in $\mathbb{R}^3$ [Lv]. As such, the rod is represented by a parametrized curve $\gamma(t), 0 \le t \le T$ that corresponds to the central line of the rod, and an orthonormal frame $V(t) = (v_1(t), v_2(t), v_3(t))$ defined along $\gamma(t)$ that measures the amount of bending and twisting of the rod. To model inextensible rods, it is assumed that $\left\| \dfrac{d\gamma}{dt}(t) \right\| = 1$, which implies that the parameter t corresponds to the length of γ in the interval $[0, t]$ measured from the initial point $\gamma(0)$. The frames $F(s)$ are assumed adapted to the central line by a constraint of the form

$$\frac{d\gamma}{dt}(t) = v_1(t).$$

It is implicitly assumed that the frames $V(s)$ are measured relative to the fixed frame that corresponds to the rod in its unstressed state. Then the frames $V(s)$ can be identified with elements $R(s)$ in $SO_3(\mathbb{R})$, in which case the frame deformation $\frac{dF}{dt}(t)$ is described by an antisymmetric matrix

$$A(t) = \begin{pmatrix} 0 & -u_3(t) & u_2(t) \\ u_3(t) & 0 & -u_1(t) \\ u_2(t) & u_1(t) & 0 \end{pmatrix}.$$

The functions $u_1(t), u_2(t), u_3(t)$ are called strains, which together with constants c_1, c_2, c_3 reflect the tensile and geometric properties of the rod and define the elastic energy E of the deformed rod given by

$$E = \int_0^L (c_1 u_1^2(t) + c_2 u_2^2(t) + c_3 u_3^2(t)) \, dt.$$

Kirchhoff postulated in 1859 that the equilibrium configurations of the rod subjected to fixed boundary conditions at its ends correspond to the stationary configurations for the elastic energy [Lv, Ch. 7]. Over short lengths Kirchhoff's principle can be paraphrased as a Minimum Principle: *The equilibrium config-urations of the rod correspond to the minima of the elastic energy relative to the configurations that conform to the given boundary conditions.* In such a case Kichhoff's Minimum Principle is equivalent to our affine-quadratic problem on the group of motions of $\mathbb{R}^3$ with the drift vector $B = \begin{pmatrix} 0 & 0 \\ e_1 & 0_n \end{pmatrix}$. Then Kirchhoff's problem can be seen as the shadow problem of the affine-quadratic problem on S_ϵ corresponding to the drift vector $A = \begin{pmatrix} 0 & -\epsilon e_1^T \\ e_1 & 0_n \end{pmatrix}$ and the cost functional $E = \int_0^L (c_1 u_1^2(t) + c_2 u_2^2(t) + c_3 u_3^2(t)) \, dt$.

It follows that a general affine-quadratic problem on three-dimensional space forms is a natural geometric extension of Kirchhoff's elastic model in which $\frac{d\gamma}{dt} = v_1$ is replaced by a more general condition

$$\frac{d\gamma}{dt} = b_1 v_1 + b_2 v_2 + b_3 v_3. \tag{13.29}$$

Then all possible configurations of this generalized elastic rod are described by the equations

$$\frac{dg}{dt} = g(t)(B + U(t)), \quad U(t) = u_1(t)A_1 + u_2(t)A_2 + u_3(t)A_3,$$

with the elastic energy given by $E = \int_0^L (c_1 u_1^2(t) + c_2 u_2^2(t) + c_3 u_3^2(t)\, dt$. Hence Kirchhoff's Principle becomes synonymous with the affine-quadratic problem.

Every optimal solution of the above affine-quadratic problem on SO_ϵ is the projection of a normal extremal curve. This follows from Proposition 9.7 and Corollary 9.8 in Chapter 9. For, if $[B, L_p] = 0$, then L_p and B are colinear (formula (13.1)). If $[L_p, U(t)] = 0$, then $[B, U(t)] = 0$ and Corollary is applicable.

Therefore the equilibrium configurations of this rod are the projections on S_ϵ of the integral curve of the Hamiltonian vector field associated with the affine Hamiltonian

$$H = \frac{1}{2}\langle \mathcal{P}^{-1}(L_t), L_t \rangle + \langle B, L_p \rangle$$

defined by $\mathcal{P}(U) = c_1 u_1 A_1 + c_2 u_2 A_2 + c_3 u_3 A_3$ and $B = b_1 B_1 + b_2 B_2 + b_3 B_3$.

Let us now focus on the associated Hamiltonian equations restricted to the Lie algebra $\mathfrak{g}_\epsilon$:

$$\frac{dL_t}{dt} = [\mathcal{P}^{-1}(L_t), L_t] + [B, L_p], \quad \frac{dL_p}{dt} = [\mathcal{P}^{-1}(L_t), L_p] + s[B, L_t]. \quad (13.30)$$

To make the connection with the equations of the heavy top more transparent, introduce the following notations:

$$\hat{L_t} = M = \begin{pmatrix} M_1 \\ M_2 \\ M_3 \end{pmatrix}, \widehat{\mathcal{P}^{-1}(L_t)} = \Omega = \begin{pmatrix} \frac{1}{c_1}M_1 \\ \frac{1}{c_2}M_2 \\ \frac{1}{c_3}M_3 \end{pmatrix},$$

$$\hat{B} = b = \begin{pmatrix} b_1 \\ b_2 \\ b_3 \end{pmatrix}, \hat{L_p} = p = \begin{pmatrix} p_1 \\ p_2 \\ p_3 \end{pmatrix}.$$

Then (13.30) can be written as

$$\begin{pmatrix} \frac{dM}{dt} \\ 0 \end{pmatrix} = \frac{d\hat{L_t}}{dt} = \begin{pmatrix} M \times \Omega + p \times b \\ 0 \end{pmatrix},$$

$$\begin{pmatrix} 0 \\ \frac{1}{\epsilon}\frac{dp}{dt} \end{pmatrix} = \frac{d\hat{L_p}}{dt} = \begin{pmatrix} 0 \\ \frac{1}{\epsilon}(p \times \Omega) + s(M \times b) \end{pmatrix},$$

or in simpler form as

$$\frac{dM}{dt} = M \times \Omega + p \times b, \quad \frac{dp}{dt} = p \times \Omega + s\epsilon(M \times b). \quad (13.31)$$

The term $s\epsilon$ could be now replaced by a single parameter $\epsilon = \pm 1, 0$.

The reader may easily check that the above equations admit two integrals of motion:

$$\epsilon(M_1^2 + M_2^2 + M_3^2) + p_1^2 + p_2^2 + p_3^2 = I_1, p_1M_1 + p_2M_2 + p_3M_3 = I_2.$$
(13.32)

These integrals of motion are invariant functions on so_ϵ, in the sense that they constant along the flow of any left-invariant Hamiltonian vector field on so_ϵ. In addition to these two integrals, the Hamiltonians of right-invariant vector fields are also constant along (13.30). Since the rank of so_ϵ is two, every left-invariant Hamiltonian on so_ϵ has five independent integrals of motion all in involution with each other. Therefore an affine-quadratic Hamiltonian H is Liouville integrable whenever there is an extra integral of motion.

As we have indicated earlier, equations (13.31) can be identified with the equations of the top in the Euclidean case. In this identification, the drift vector b is identified with the center of gravity of this top, the constants c_1, c_2, c_3 are identified with the principal moments of inertia. Then the term $b \times p(t)$ is identified with the external torque $Q_c \times \sum_{i=1}^{3} a_i\alpha_i$ associated with potential energy $V(R) = -\sum_{i=1}^{3} a_i(\alpha_i, Q_c)$. Under these identifications, $p(t) - \sum_{i=1}^{3} u_i\alpha_i$ with $\alpha_i = R^T(t)e_i$, where $R(t)$ is the rotation matrix associated with the solution matrix $g(t) = \begin{pmatrix} 1 & 0 \\ x(t) & R(t) \end{pmatrix}$. Trerefore, $||p||^2 = a_1^2 + a_2^2 + a_3^2$. Evidently solutions of (13.31) which do not satisfy $||p||^2 = a_1^2 + a_2^2 + a_3^2$ do not project onto the solutions of the associated top.

Let us now turn our attention to the integrable cases in equations (13.30) and (13.31). There are two obvious cases; the degenerate case $b = 0$, and the symmetric case, $c_1 = c_2 = c_3$ and b arbitrary. In the first case, both $||M||^2$ and $||p||^2$ are constant along the solutions of (13.31). In analogy with the top, this case occurs when the center of gravity is at the fixed point, i.e., when there is no external torque. For that reason will refer to this case as the Euler case.

The corresponding equations are

$$\frac{dg}{dt} = g(t)\check{\Omega}(t), \frac{dM}{dt} = M(t) \times \Omega(t), \frac{dp}{dt} = p(t) \times \Omega(t),$$
(13.33)

The equation $\frac{dM}{dt} = M \times \Omega$ corresponds to the Euler top with $\mathcal{P}(U) = c_1u_1A_1 + c_2u_2A_2 + c_3u_3A_3$. It can integrated by the same procedure used in the section on the rigid body and the same applies to the equation for p.

The symmetric case corresponds to the canonical affine Hamiltonian. It is the only affine Hamiltonian on so_ϵ that admits isospectral representation (Proposition 13.1). Since $\frac{dM}{dt} = p \times b$, $I_3 = b_1M_1 + b_2M_2 + b_3M_3$ is the missing integral of motion.

It may be instructive to extract I_3 from the spectral matrix $L_\lambda = L_p + \lambda L_\ell +$
$(\lambda^2 - s)B = \begin{pmatrix} 0 & -p_\lambda \\ \frac{1}{\epsilon}p_\lambda & \lambda\check{M}_\lambda \end{pmatrix}$, $p_\lambda = p + (\lambda^2 - 1)b$, $\check{M}_\lambda = \lambda\check{M}$.

An easy calculation shows that

$$L_\lambda^2 = \begin{pmatrix} -\frac{1}{\epsilon}||p_\lambda||^2 & -p_\lambda^T M_\lambda \\ \frac{1}{\epsilon}\check{M}_\lambda p_\lambda & \check{M}_\lambda^2 - \frac{1}{\epsilon}p_\lambda \otimes p_\lambda \end{pmatrix},$$

and therefore, $\epsilon Tr(L_\lambda^2) = \epsilon||M_\lambda||^2 + ||p_\lambda||^2 = 2(\lambda^2 - 1)H + I_1 + (\lambda^2 - 1)^2||b||^2$.
Then

$$Tr(L_\lambda^4) = ||p_\lambda||^4 - \frac{2}{\epsilon^2}\check{M}_\lambda p_\lambda \cdot p_\lambda^T \check{M}_\lambda + Tr(\check{M}_\lambda^2 - \frac{1}{\epsilon}(p \otimes p_\lambda))^2,$$

where $x \cdot y$ denotes the Euclidean inner product in $\mathbb{R}^n$.

Since the matrix $\check{M}_\lambda$ is skew-symmetric, $\check{M}_\lambda^2 = -||M_\lambda||^2 I + M_\lambda \otimes M_\lambda$. Then,
$\check{M}_\lambda p_\lambda \cdot p_\lambda^T \check{M}_\lambda = -(\check{M}_\lambda^2 p_\lambda \cdot p_\lambda) = ||M_\lambda||^2||p_\lambda||^2 - (M_\lambda \cdot p_\lambda)^2$, and $Tr(\check{M}_\lambda^4) =$
$2||M_\lambda||^4$. The above can be rewritten as

$$Tr(L_\lambda^4) = 2||p_\lambda||^4 + 4\epsilon(||M_\lambda||^2||p_\lambda||^2 - (M_\lambda \cdot p_\lambda)^2) + 2||M_\lambda||^4.$$

But then

$$\frac{1}{2}Tr(L_\lambda^4) = (||p_\lambda||^2 + \epsilon||M_\lambda||^2)^2 - 2\epsilon(M_\lambda \cdot p_\lambda)^2.$$

The above implies that $(M_\lambda \cdot p_\lambda)$ is constant along the solutions of (13.33), or
that

$$\lambda(\lambda^2(M \cdot b) + (M \cdot p) - (M \cdot b))$$

is constant. This fact implies that both $(M \cdot b)$ and $(M \cdot p)$ are constant. Of
course, $(M \cdot p)$ is the second Casimir I_2.

Remarkably, the remaining integrable cases occur precisely the same conditions as the ones known in the theory of tops. The integrable cases occur
when two constants $\{c_1, c_2, c_3\}$ are equal to each other and the drift vector has
a particular relation to the equal constants. The easy case corresponds to the top
of Lagrange; it occurs when $c_i = c_j$ and $b_i = b_j = 0$. For then, there is an extra
symmetry of rotation which accounts for the missing integral of motion. For
instance, when $c_1 = c_2$ and $b = b_3 e_3$, then $R(t) = \begin{pmatrix} \cos s & -\sin s & 0 \\ \sin s & \cos s & 0 \\ 0 & 0 & 1 \end{pmatrix}$
fixes b and leaves the quadratic form $\langle \mathcal{P}(U), U \rangle$ invariant. Hence M_1 is
constant along the solutions of (13.31).

The remaining, and most enigmatic, case occurs when $c_1 = c_2 = 2c_3$ and $b_3 = 0$, as in the case of the top of Sonya Kowalewski. Then

$$|z^2 - b(w - \epsilon b)|^2, z = \frac{1}{2}(M_1 + iM_2), w = p_1 + ip_2, b = b_1 + ib_2 \quad (13.34)$$

is the missing integral of motion. This integral coincides with the one discovered by Kowalewski in her famous paper of 1889 [Kw] when $\epsilon = 0$.

We will show that Kowalewski's integral of motion admits holomorphic extensions when affine Hamiltonians are extended to complex Lie groups. This passage to complex Lie algebras enables a complete classifications of integrable cases based on Kowalewski–Lyapunov criteria, and could be seen as a natural conclusion of the ideas initiated by S. Kowalewski in her seminal paper of 1889 in which she treated all the variables as complex quantities [Kw]. These details will be taken up in the next chapter.

14

Kowalewski–Lyapunov criteria

Affine-quadratic complex Hamiltonians can be introduced in the same manner as the real ones, provided that the passage to the Hamiltonians is done via the critical points rather than optimization. Since the critical points are singular points of the end-point map, the Maximum Principle remains valid in the complex setting and can be used to generate complex Hamiltonians, much in the same manner as in the real case. For our purposes it will be sufficient to limit the discussion to $G = SO_n(\mathbb{C})$, the matrix group that leaves the quadratic form

$$(z, w) = z_1 w_1 + \cdots + z_n w_n$$

in $\mathbb{C}^n$ invariant, and it will be more convenient to work with $SO_{n+1}(\mathbb{C})$ rather than $SO_n(\mathbb{C})$. Here, it is understood that G acts by the matrix multiplication on the column vectors $z = \begin{pmatrix} z_1 \\ \vdots \\ z_n \end{pmatrix}$ in $\mathbb{C}^n$. As in the real case, for each R in G, R^{-1} is equal to the matrix transpose R^T. Therefore, the Lie algebra $\mathfrak{g} = o_n(\mathbb{C})$ of G consists of skew-symmetric $n \times n$ matrices with complex entries.

To make connections with complex affine Hamiltonians, it will be necessary to induce Cartan decompositions of $\mathfrak{g}$ through involutive automorphisms σ. As in the real case, we will use $\sigma : SO_{n+1}(\mathbb{C}) \to SO_{n+1}(\mathbb{C})$ defined by $\sigma(g) = DgD^{-1}$, where D is a diagonal matrix of the form $D = \begin{pmatrix} -I_p & 0 \\ 0 & I_q \end{pmatrix}$, with $p + q = n + 1$, and I_p and I_q the identity matrices of dimensions p and q to provide such a decomposition. Then, $\mathfrak{g} = \mathfrak{p} + \mathfrak{k}$, with $\mathfrak{p} = \{A : \sigma_*(A) = -A\}$, and $\mathfrak{k} = \{A : \sigma_*(A) = A\}$, where σ_* is the tangent map of σ at the group identity, subject to the usual Lie bracket conditions

$$[\mathfrak{p}, \mathfrak{p}] = \mathfrak{k}, [\mathfrak{p}, \mathfrak{k}] \subseteq \mathfrak{p}, [\mathfrak{k}, \mathfrak{k}] \subseteq \mathfrak{k}. \tag{14.1}$$

Matrices X in $\mathfrak{p}$ and Y in $\mathfrak{k}$ are of the form

$$X = \begin{pmatrix} 0 & -B^T \\ B & 0 \end{pmatrix} \text{ and } Y = \begin{pmatrix} A_1 & 0^T \\ 0 & A_2 \end{pmatrix}, \tag{14.2}$$

where B is an arbitrary $p \times q$ matrix with complex entries, 0 is the $p \times q$ matrix with zero entries, and A_1 and A_2 are skew-symmetric matrices with complex entries of dimensions $p \times p$ and $q \times q$. It is easy to verify that the Cartan relations (14.1) take on the strong form

$$[\mathfrak{p}, \mathfrak{p}] = \mathfrak{k}, [\mathfrak{p}, \mathfrak{k}] = \mathfrak{p}, [\mathfrak{k}, \mathfrak{k}] = \mathfrak{k}. \tag{14.3}$$

The above implies that $\mathfrak{k}$ is the Lie algebra of the group K consisting of fixed points of σ. Evidently, $K = \{(g, h); g \in O_p(\mathbb{C}), h \in O_q(\mathbb{C}), Det(gh) = 1\}$.

It follows that $\mathfrak{p}$ and $\mathfrak{k}$ are orthogonal relative to the complex valued trace form $\langle A, B \rangle = -\frac{1}{2} Tr(AB) = \sum_{i>j}^{n+1} A_{ij} B_{ij}$.

The fact that $[\mathfrak{p}, \mathfrak{k}] = \mathfrak{p}$ implies that K acts on $\mathfrak{p}$ by adjoint action. As in the real case, the underlying vector space consisting of $(n + 1) \times (n + 1)$ skew-symmetric complex matrices with zero trace can be regarded both as the Lie algebra of $SO_{n+1}(\mathbb{C})$ and the Lie algebra of the semi-direct product $G_s - \mathfrak{p} \rtimes K$. The identification is done completely analogously to the real case. We briefly recall that the group multiplication in G_s is given by $(A, g)(B, h) = (A + gBg^{-1}, gh)$. Hence, $e = (0, I)$ is the group identity, and $(A, g)^{-1} = (-g^{-1}Ag, g^{-1})$ is the group inverse.

The Lie algebra $\mathfrak{g}_s$ of G_s is the product $\mathfrak{p} \rtimes \mathfrak{k}$ with the Lie bracket structure

$$[(B_1, A_1), (B_2, A_2)]_s = (adB_1(A_2) - adB_2(A_1), [A_1, A_2])$$

for any (B_1, A_1) and (B_2, A_2) in $\mathfrak{p} \times \mathfrak{k}$.

After $\mathfrak{p}$ is identified with $\mathfrak{p} \times \{0\}$ and $\mathfrak{k}$ with $\{0\} \times \mathfrak{k}$, the elements (B, A) in $\mathfrak{p} \times \mathfrak{k}$ are identified with the sums $B + A$ in $\mathfrak{g}$, in which case the Lie bracket in $\mathfrak{g}_s$ can be written as

$$[(B_1, A_1), (B_2, A_2)]_s = ([B_1, A_2] - [B_2, A_1]) + [A_1, A_2]. \tag{14.4}$$

Thus $\mathfrak{g}$ as a vector space carries two Lie algebras: one induced by the semi-simple Lie bracket and the other induced by the semi-direct product Lie bracket. We also recall that s can be regarded as a continuous parameter, a fact that will be relevant for the material below.

We shall also recall the following:

Definition 14.1 A real Lie algebra $\mathfrak{g}_0$ is said a real form for a complex Lie algebra $\mathfrak{g}$ if

$$\mathfrak{g} = \mathfrak{g}_0 + i\mathfrak{g}_0.$$

A complex Lie algebra may have several real forms, as we have already noticed earlier. For our purposes, it will be important to note that both $so_n(\mathbb{R})$ and $so(p,q)$ with $n = p + q$ are real forms for $so_n(\mathbb{C})$. That $so_n(R)$ is a real form for $so_n(\mathbb{C})$ is evident. To show that $so(p,q)$ is a real form, let $V_{p,q}$ denote the real vector space spanned by $ie_0, ie_1, \ldots, ie_{p-1}, e_p, \ldots, e_n$, where $e_0, \ldots, e_n$ denotes the standard basis in $\mathbb{C}^{n+1}$. The restriction of the complex form (z, w) to $V_{p,q}$ is given by

$$(x, y) = -\sum_{j=0}^{p-1} x_j y_j + \sum_{j=p}^{n} x_j y_j.$$

Let $\mathfrak{g}_0$ denote the set of matrices $M = \begin{pmatrix} A_1 & iB^T \\ -iB & A_2 \end{pmatrix}$, where A_1 and A_2 are $p \times p$ and $q \times q$ skew-symmetric matrices with real entries and B an arbitrary $p \times q$ matrix with real entries.

An easy calculation shows that $\mathfrak{g}_0$ is a real Lie algebra. Evidently, $\mathfrak{g}_0$ is a real form for $so_{n+1}(\mathbb{C})$. Every element of $\mathfrak{g}_0$ acts linearly on $V_{p,q}$. Relative to the basis $ie_0, \ldots, ie_{p-1}, e_p, \ldots, e_n$ matrices in $\mathfrak{g}_0$ are of the form $M = \begin{pmatrix} A_1 & B \\ B^T & A_2 \end{pmatrix}$. The mapping $\begin{pmatrix} A_1 & B \\ B^T & A_2 \end{pmatrix} \rightarrow \begin{pmatrix} A_1 & iB^T \\ -iB & A_2 \end{pmatrix}$ is an isomorphism from $so(p,q)$ onto $\mathfrak{g}_0$. It follows that if $B = \begin{pmatrix} 0 & b^T \\ b & 0 \end{pmatrix}$ belongs to to the Cartan space in $so(1,n)$, then $\begin{pmatrix} 0 & ib^T \\ -ib & 0 \end{pmatrix}$ belongs to $\mathfrak{p}$ in $so_{n+1}(\mathbb{C})$.

14.1 Complex quaternions and $SO_4(\mathbb{C})$

The Lie algebra $so_4(\mathbb{C})$ together with its real forms has a particularly rich structure that is of considerable importance for the remainder of this chapter. Partly because of the notational necessity, and partly as a convenience to the reader, the relevant facts will be assembled in the text below (these facts, although well known, are somewhat scattered through the existing literature and difficult to reference accurately). Matrices in $so_4(\mathbb{C})$ formally look the same as in $so_4(R)$, except that the entries are complex numbers. In particular, matrices A and B, $A \in \mathfrak{k}$ and $B \in \mathfrak{p}$ given by

$$A = \begin{pmatrix} 0 & 0 & 0 & 0 \\ 0 & 0 & -a_3 & a_2 \\ 0 & a_3 & 0 & -a_1 \\ 0 & -a_2 & a_1 & 0 \end{pmatrix}, B = \begin{pmatrix} 0 & -b_1 & -b_2 & -b_3 \\ b_1 & 0 & 0 & 0 \\ b_2 & 0 & 0 & 0 \\ b_3 & 0 & 0 & 0 \end{pmatrix} \tag{14.5}$$

are now complex-valued.

Let us now introduce some basic isomorphisms which will simplify the notations and elucidate the structure of Hamiltonian systems. The set of matrices

$$Z = \begin{pmatrix} z_0 + iz_1 & z_2 + iz_3 \\ -z_2 + iz_3 & z_0 - iz_1 \end{pmatrix}, (z_1, z_2, z_3, z_4) \in \mathbb{C}^4$$

will be denoted by $\mathbb{H}(\mathbb{C})$. It is easy to verify that $\mathbb{H}(\mathbb{C})$ is a four-dimensional complex vector space as well as an algebra under matrix multiplication. In fact, $\mathbb{H}(\mathbb{C})$ is an algebra isomorphic to the space of complex quaternions. To make this connection more transparent, write each matrix Z in $\mathbb{H}(\mathbb{C})$ as $Z = z_0 E_0 + z_1 E_1 + z_2 E_2 + z_3 E_3$ where

$$E_0 = \begin{pmatrix} 1 & 0 \\ 0 & 1 \end{pmatrix}, E_1 = \begin{pmatrix} i & 0 \\ 0 & -i \end{pmatrix} E_2 = \begin{pmatrix} 0 & 1 \\ -1 & 0 \end{pmatrix}, E_3 = \begin{pmatrix} 0 & i \\ i & 0 \end{pmatrix}.$$

It is easy to verify that

$$E_1^2 = E_2^2 = E_3^2 = -E_0, \; E_1 E_2 = E_3, \; E_3 E_1 = E_2, \; E_2 E_3 = E_1. \quad (14.6)$$

Therefore, E_0, E_1, E_2, E_3 correspond to the standard quaternionic basis e, i, j, k.

An easy calculation shows that

$$\begin{pmatrix} z_0 + iz_1 & z_2 + iz_3 \\ -z_2 + iz_3 & z_0 - iz_1 \end{pmatrix} \begin{pmatrix} w_0 + iw_1 & w_2 + iw_3 \\ -w_2 + iw_3 & w_0 - iw_1 \end{pmatrix} = \begin{pmatrix} u_0 + iu_1 & u_2 + iu_3 \\ -u_2 + iu_3 & u_0 - iu_1 \end{pmatrix},$$

where

$$u_0 = z_0 w_0 - z_1 w_1 - z_2 w_2 - z_3 w_3, u_1 = z_0 w_1 + w_0 z_1 + z_2 w_3 - z_3 w_2,$$

$$u_2 = z_0 w_2 + w_0 z_2 + z_3 w_1 - z_1 w_3, u_3 = w_0 z_3 + z_0 w_3 + z_1 w_2 - z_2 w_1. \quad (14.7)$$

Therefore, quaternionic multiplication is isomorphic with the matrix multiplication in $\mathbb{H}(\mathbb{C})$.

Quaternions which belong to the linear span of E_1, E_2, E_3 satisfy $z_0 = 0$, or equivalently, are of zero trace. They are called pure quaternions. If $Z = z_1 E_1 + z_2 E_2 + z_3 E_3$ and $W = w_1 E_1 + w_2 E_2 + w_3 E_3$ are any such quaternions then

$$ZW = -(z, w) E_0 + (z_2 w_3 - z_3 w_2) E_1 + (z_3 w_1 - z_1 w_3) E_2 + (z_1 w_2 - z_2 w_1) E_3. \quad (14.8)$$

So the pure part of the product of two pure quaternions corresponds to the quaternion associated with the vector product $z \times w$.

The conjugate Z^* of a quaternion Z is given by

$$Z^* = \begin{pmatrix} z_0 - iz_1 & -z_2 - iz_3 \\ z_2 - iz_3 & z_0 + iz_1 \end{pmatrix} = z_0 E_0 - (z_1 E_1 + z_2 E_2 + z_3 E_3). \quad (14.9)$$

The reader can readily verify that

$$(ZW)^* = W^*Z^*, \text{ and } \frac{1}{2}(ZW^* + WZ^*) = (z, w)E_0. \qquad (14.10)$$

Therefore,

$$ZZ^* = (z, z)E_0.$$

An element z in $\mathbb{C}^4$ is called isotropic if $(z, z) = 0$. The quaternion associated with an isotropic vector satisfies $ZZ^* = 0$, or equivalently, satisfies $Det(Z) = 0$. It follows that a non-isotropic quaternion Z has an inverse $Z^{-1} = \frac{1}{ZZ^*}Z^*$. A complex quaternion Z is said to be a unit quaternion if $ZZ^* = E_0$. It follows that Z is a unit quaternion if and only if $Z \in SL_2(\mathbb{C})$. The above also implies that $Z \in SL_2(\mathbb{C})$ if and only if $Z^* = Z^{-1}$.

A quaternion is real if z_0, z_1, z_2, z_3 are all real numbers. Real quaternions are represented by matrices $Z = \begin{pmatrix} x_0 + ix_1 & x_2 + ix_3 \\ -x_2 + ix_3 & x_0 - ix_1 \end{pmatrix} = \begin{pmatrix} z & w \\ -\bar{w} & \bar{x} \end{pmatrix}$. It follows that real unit quaternions are represented by SU_2.

If g is a complex (real) unit quaternion then gZg^* is a pure complex (real) quaternion for each pure complex (real) quaternion Z since $(gZg^*)^* = -(gZg^*)$. It follows that the group of unit quaternions acts on the space of pure quaternions by the action Φ defined by

$$\Phi_g(Z) = gZg^{-1}, g \in SL_2(\mathbb{C}) \text{ respectively}, g \in SU_2. \qquad (14.11)$$

Since

$$\frac{1}{2}((gZg^*)(gWg^*)^* + (gWg^*)(gZg^*)^*) = \frac{1}{2}(g(ZW^* + WZ^*)g^*) = (z, w)E_0,$$

Φ_g is a rotation in $\mathbb{C}^3$ when $g \in SL_2(\mathbb{C})$, and Φ_g is a rotation in $\mathbb{R}^3$ when $g \in SU_2$. The correspondence $g \to \Phi_g$ is a group homomorphism from $SL_2(\mathbb{C})$ (resp. SU_2) into $SO_3(\mathbb{C})$ (resp. $SO_3(R)$) with the kernel $\pm I$, which shows that $SL_2(\mathbb{C})$ is a double cover of $SO_3(\mathbb{C})$ and that SU_2 is a double cover of $SO_3(R)$.

The above action can be extended to $SL_2(\mathbb{C}) \times SL_2(\mathbb{C})$ (resp. $SU_2 \times SU_2$) via the action

$$\Psi_{(g,h)}(Z) = gZh^{-1}, (g, h) \in SL_2(\mathbb{C}) \times SL_2(\mathbb{C}). \qquad (14.12)$$

One can easily verify that $(g, h) \to \Psi_{(g,h)}$ is a group homomorphism from $SL_2(\mathbb{C}) \times SL_2(\mathbb{C})$ (resp. $SU_2 \times SU_2$) onto $SO_4(\mathbb{C})$ (resp. $SO_4(R)$) with the kernel $\pm(I, I)$.

It will be convenient to assemble the above observations into the following:

Proposition 14.2 (a) *$SL_2(\mathbb{C}) \times SL_2(\mathbb{C})$ is a double cover of $SO_4(\mathbb{C})$ and $SL_2(\mathbb{C})$ is a double cover of $SO_3(\mathbb{C})$. Consequently, $sl_2(\mathbb{C}) \times sl_2(\mathbb{C})$ and $so_4(\mathbb{C})$ are isomorphic, as well as $sl_2(\mathbb{C})$ and $so_3(\mathbb{C})$.*

(b) *$SU_2 \times SU_2$ is a double cover of $SO_4(R)$, and SU_2 is double cover of $SO_3(R)$. Consequently, $su_2 \times su_2$ and $so_4(R)$ are isomorphic, as well as su_2 and $so_3(R)$.*

The explicit isomorphisms are provided by the tangent maps Ψ_* at the the group identity. It follows that $(L_1, L_2) \in sl_2(\mathbb{C}) \times sl_2(\mathbb{C})$ corresponds to the matrix $C \in so_4(\mathbb{C})$ defined by

$$Cz = w, \text{ if and only if } L_1 Z - Z L_2 = W \text{ for all } Z \in \mathbb{H}(\mathbb{C}). \qquad (14.13)$$

If $L_1 = l_1 E_1 + l_2 E_2 + l_3 E_3, L_2 = m_1 E_1 + m_2 E_2 + m_3 E_3$ and if $C = A + B$ with A and B defined by (14.5), that is,

$$A = \begin{pmatrix} 0 & 0 & 0 & 0 \\ 0 & 0 & -a_3 & a_2 \\ 0 & a_3 & 0 & -a_1 \\ 0 & -a_2 & a_1 & 0 \end{pmatrix}, B = \begin{pmatrix} 0 & -b_1 & -b_2 & -b_3 \\ b_1 & 0 & 0 & 0 \\ b_2 & 0 & 0 & 0 \\ b_3 & 0 & 0 & 0 \end{pmatrix},$$

then relations (14.13) imply that

$$l_1 = \frac{1}{2}(a_1 + b_1), l_2 = \frac{1}{2}(a_2 + b_2), l_3 = \frac{1}{2}(a_3 + b_3),$$

$$m_1 = \frac{1}{2}(a_1 - b_1), m_2 = \frac{1}{2}(a_2 - b_2), m_3 = \frac{1}{2}(a_3 - b_3),$$

and conversely,

$$L_1 + L_2 = \begin{pmatrix} ia_1 & a_2 + ia_3 \\ -a_2 + ia_3 & -ia_1 \end{pmatrix}, L_1 - L_2 = \begin{pmatrix} ib_1 & b_2 - ib_3 \\ -b_2 - ib_3 & -ib_1 \end{pmatrix}.$$

$$(14.14)$$

The above shows that curves $g(t) \in SO_4(\mathbb{C})$ that satisfy $g(0) = I$ are in one to one correspondence with curves $(g_1(t), g_2(t)) \in SL_2(\mathbb{C}) \times SL_2(\mathbb{C})$, with $g_1(0) = I, g_2(0) = I$, via the correspondence

$$g(z) = w \Leftrightarrow g_1(t) Z g_2^{-1}(t) = W,$$

where $Z = \begin{pmatrix} iz_1 & z_2 + iz_3 \\ -z_2 + iz_3 & -iz_1 \end{pmatrix}, W = \begin{pmatrix} iw_1 & w_2 + iw_3 \\ -w_2 + iw_3 & -iw_1 \end{pmatrix}$ are

the quaternions defined by z and w in $\mathbb{C}^3$. Moreover, if $\frac{dg}{dt} = g(t)(A(t) + B(t))$ with A and B as above, then

$$\frac{dg_1}{dt} = g_1(t) L_1(t), \frac{dg_2}{dt} = g_2(t) L_2(t), \qquad (14.15)$$

with $L_1(t)$ and $L_2(t)$ defined by (14.14) above.

In a similar manner one can show that a matrix $A = \begin{pmatrix} 0 & -a_3 & a_2 \\ a_3 & 0 & a_1 \\ -a_2 & a_1 & 0 \end{pmatrix}$ in $so_3(\mathbb{C})$ corresponds to the matrix $\frac{1}{2}\begin{pmatrix} ia_1 & (a_2 + ia_3) \\ -(a_2 - ia_3) & -ia_1 \end{pmatrix}$ in $sl_2(\mathbb{C})$.

For our purposes, however, it will be more convenient to use another representation in which the standard basis $A_1, A_2, A_3, B_1, B_2, B_3$ in $so_4(\mathbb{C})$ is represented by the matrices

$$\phi(A_i) = \frac{1}{2}\begin{pmatrix} E_i & 0 \\ 0 & E_i \end{pmatrix}, \phi(B_i) = \frac{1}{2}\begin{pmatrix} 0 & E_i \\ E_i & 0 \end{pmatrix}, i = 1, 2, 3.$$

One can easily check that $[\phi(X), \phi(Y)] = \phi([X, Y])$ for any X and Y in $so_4(\mathbb{C})$. Then matrices $C = A + B$ in $so_4(\mathbb{C})$ with $A \in \mathfrak{p}$ and $B \in \mathfrak{p}$ are represented by the matrices $\begin{pmatrix} \phi(A) & \phi(B) \\ \phi(B) & \phi(A) \end{pmatrix}$, where

$$\phi(A) = \frac{1}{2}\begin{pmatrix} ia_1 & a_2 + ia_3 \\ -a_2 + ia_3 & -ia_1 \end{pmatrix}, \phi(B) = \frac{1}{2}\begin{pmatrix} ib_1 & b_2 + iib_3 \\ -b_2 + ib_3 & -ib_1 \end{pmatrix}.$$

We will not make a notational distinction between matrices C and $\phi(C)$ – it will be clear from the context which representation is used. In particular, $L = \sum_{i=1}^{3} m_i A_i + p_i B_i$ will be written as $L = \begin{pmatrix} M & P \\ P & M \end{pmatrix}$, where

$$M = \frac{1}{2}\begin{pmatrix} im_3 & m_1 + im_2 \\ -m_1 + im_2 & -im_3 \end{pmatrix}, P = \frac{1}{2}\begin{pmatrix} ip_3 & p_1 + ip_2 \\ -p_1 + ip_2 & -ip_3 \end{pmatrix}.$$
$$(14.16)$$

As we commented earlier, matrices M and P in $sl_2(\mathbb{C})$ are isomorphic with matrices $\begin{pmatrix} 0 & -m_3 & m_2 \\ m_3 & 0 & m_1 \\ -m_2 & m_1 & 0 \end{pmatrix}$ and $\begin{pmatrix} 0 & -p_3 & p_2 \\ p_3 & 0 & p_1 \\ -p_2 & p_1 & 0 \end{pmatrix}$ in $so_3(\mathbb{C})$. Hence, the Lie brackets $[M, P]$ are isomorphic with the vector products $p \times m$ with $m = \begin{pmatrix} m_1 \\ m_2 \\ m_3 \end{pmatrix}$ and $p = \begin{pmatrix} p_1 \\ p_2 \\ p_3 \end{pmatrix}$. It will be convenient for further reference to restate the above remarks in the form of a proposition.

Proposition 14.3   *The Lie algebra $so_4(\mathbb{C})$ is isomorphic with the subalgebra of $sl_4(\mathbb{C})$ consisting of matrices $L = \begin{pmatrix} M & P \\ P & M \end{pmatrix}$ with M and P in $sl_2(\mathbb{C})$.*

This seems an opportune place to note another group isomorphism which will be relevant in the subsequent text.

Proposition 14.4 $SL_2(\mathbb{C})$ *is a double cover of* $SO(1, 3)$.

Proof   Let $Z^\dagger = \bar{Z}^*$ for each complex quaternion $Z = \begin{pmatrix} z_0 + iz_1 & z_2 + iz_3 \\ -z_2 + iz_3 & z_0 - iz_1 \end{pmatrix}$,

where $\bar{Z} = \begin{pmatrix} \bar{z}_0 - i\bar{z}_1 & \bar{z}_2 - i\bar{z}_3 \\ -\bar{z}_2 - i\bar{z}_3 & \bar{z}_0 + i\bar{z}_1 \end{pmatrix}$. It follows that $(ZW)^\dagger = W^\dagger Z^\dagger$ for any Z and W.

A matrix Z is called Hermitian if $Z^\dagger = Z$. It follows that Z is Hermitian if and only if $z_0 = \bar{z}_0$ and $z_i = -\bar{z}_i, i = 1, 2, 3$, that is, a Hermitian matrix is any matrix of the form $Z = \begin{pmatrix} x_0 - x_1 & ix_2 - x_3 \\ -ix_2 - x_3 & x_0 + x_1 \end{pmatrix}$ for some real numbers x_0, x_1, x_2, x_3. If Z and W are Hermitian then

$$\frac{1}{2}(Z\bar{W} + W\bar{Z}) = \frac{1}{2}(ZW^* + WZ^*) = (z, w)E_0 = \left(x_0 y_0 - \sum_{i=1}^{3} x_i y_i\right)E_0.$$

Moreover,

$$\frac{1}{2}((gWg^\dagger)\overline{(gZg^\dagger)} + (gZg^\dagger)\overline{(gWg^\dagger)}) = g\frac{1}{2}(Z\bar{W} + W\bar{Z})\bar{g}^\dagger$$

$$= (z, w)gE_0 g^* = (z, w)E_0.$$

Hence the mapping $\Phi_g : Z \to gZg^\dagger$ preserves the Lorentzian form $x_0 y_0 - \sum_{i=1}^{3} x_i y_i$ and hence is an element of $SO(1, 3)$. But this implies that $g \to \Phi_g$ is a group homomorphism from $SL_2(\mathbb{C})$ onto $SO(1, 3)$ with the kernel equal to $\pm I$.   □

Corollary 14.5 $sl_2(\mathbb{C})$ *as a six-dimensional real Lie algebra is isomorphic with* $so(1, 3)$.

Proof   The isomorphism is given by the tangent map Φ_{g*}. The tangent map takes every matrix $U = \begin{pmatrix} iu_1 & u_2 + iu_3 \\ -u_2 + iu_3 & -iu_1 \end{pmatrix}$ in $sl_2(\mathbb{C})$ onto the matrix

$$M = \begin{pmatrix} 0 & -i(u_1 - \bar{u}_1) & -i(u_2 - \bar{u}_2) & -i(u_3 - \bar{u}_3) \\ -i(u_1 - \bar{u}_1) & 0 & -(u_3 + \bar{u}_3) & (u_2 + \bar{u}_2) \\ -i(u_2 - \bar{u}_2) & (u_3 + \bar{u}_3) & 0 & -(u_1 + \bar{u}_1) \\ -i(u_3 + \bar{u}_3) & -(u_2 + \bar{u}_2) & (u_1 + \bar{u}_1) & 0 \end{pmatrix}$$

$$(14.17)$$

in $so(1, 3)$.   □

Matrix M admits a simple interpretation when U is written in the basis of the Hermitian Pauli matrices

$$B_1 = \frac{1}{2}\begin{pmatrix} 1 & 0 \\ 0 & -1 \end{pmatrix}, B_2 = \frac{1}{2}\begin{pmatrix} 0 & -i \\ i & 0 \end{pmatrix}, B_3 = \frac{1}{2}\begin{pmatrix} 0 & 1 \\ 1 & 0 \end{pmatrix},$$

and their skew-Hermitian companions

$$A_1 = iB_1, A_2 = iB_2, A_3 = iB_3.$$

For if $U = \sum_{i=1}^{3} x_i B_i + y_i A_i$ then

$$U = \frac{1}{2}\begin{pmatrix} x_1 + iy_1 & (x_3 + iy_3) + (y_2 - ix_2) \\ (x_3 + iy_3) - (y_2 + ix_2) & -(x_1 + iy_1) \end{pmatrix}$$

$$= \begin{pmatrix} iu_1 & u_2 + iu_3 \\ -u_2 + iu_3 & -iu_1. \end{pmatrix}.$$

Therefore,

$$u_1 = y_1 - ix_1, u_2 = y_2 - ix_2, u_3 = y_3 - ix_3,$$

and

$$-i(u_i - \bar{u}_i) = x_i, \ u_i + \bar{u}_i = y_i, \ i = 1, 2, 3.$$

Hence,

$$M = \begin{pmatrix} 0 & x_1 & x_2 & x_3 \\ x_1 & 0 & -y_3 & y_2 \\ x_2 & -y_3 & 0 & -y_1 \\ x_3 & -y_2 & y_1 & 0 \end{pmatrix}. \tag{14.18}$$

With this background material behind, us we now turn attention to complex Hamiltonians.

14.2 Complex Poisson structure and left-invariant Hamiltonians

The symplectic formalism on cotangent bundles of real manifolds carries over to complex manifolds with essentially no alterations, and in particular applies to the cotangent bundles of Lie groups. As in the real case, the cotangent bundle T^*G (respectively the tangent bundle TG) of a complex Lie group G will be identified with the product $G \times \mathfrak{g}^*$ (respectively $G \times \mathfrak{g}$) via the left translations, and the tangent bundle T^*G will be identified with the products $(G \times \mathfrak{g}^*) \times (\mathfrak{g} \times \mathfrak{g}^*)$, where it is understood that the second factor denotes the tangent vectors at the base point described by the first factor. Relative to this decomposition vector fields on T^*G will be written as pairs $V = ((X(g, \ell), Y(g, \ell))$ with $X(g, \ell) \in \mathfrak{g}$, $Y(g, \ell) \in \mathfrak{g}^*$, where (g, ℓ) denotes the base point in $G \times \mathfrak{g}^*$. In particular, vector fields defined by a left-invariant vector field $X(g) = gA$ and a constant vector field on $Y(\ell) = l$ will be represented by the pairs (A, l). Such vector fields span the tangent space at each point (g, ℓ) and hence form a global frame on $G \times \mathfrak{g}^*$. In this frame the symplectic form ω is given by the same expression as in the real case with

$$\omega_{(g,\ell)} ((A_1, l_1), (A_2, l_2)) = l_2(A_1) - l_1(A_2) - \ell([A_1, A_2]). \tag{14.19}$$

Then each function H on $G \times \mathfrak{g}^*$ generates a Hamiltonian vector field $\vec{H}(g, \ell) = (A(g, \ell), l(g, \ell))$ on $G \times \mathfrak{g}^*$ according to the formula

$$\omega_{(g,\ell)} (A(g, \ell), l(g, \ell)), (\cdot, \cdot)) = dH_{(g,\ell)}.$$

As in the real situation,

$$A(g, \ell) = \partial H_l(g, \ell) \text{ and } l(g, \ell) = -\partial H_g(g, \ell) - ad^* \partial H_l(g, \ell)(\ell),$$

and therefore integral curves $(g(t), \ell(t))$ of $\vec{H}$ are the solution curves of

$$\frac{dg}{dt}(t) = g(t)\partial H_l(g(t), \ell(t)), \frac{d\ell}{dt}(t) = -\partial H_g(g(t), \ell(t)) - ad^* \partial H_l(g(t), \ell(t))(\ell(t)), \tag{14.20}$$

where $ad^*A : \mathfrak{g}^* \to \mathfrak{g}^*$ is defined by $(ad^*A(\ell))(B) = \ell[A, B]$ for all $B \in \mathfrak{g}$.

Definition 14.6 The Poisson bracket $\{H_1, H_2\}$ on $G \times \mathfrak{g}^*$ is defined by $\{H_1, H_2\}(g, \ell) = \omega_{g,\ell}(\vec{H}_1(g, \ell), \vec{H}_2(g, \ell))$ for all (g, ℓ) in $G \times \mathfrak{g}^*$.

Definition 14.7 Functions F on $G \times \mathfrak{g}^*$ are said to be left-invariant if $F(gh, \ell) = F(h, \ell)$ for all g and h in G and all ℓ in $\mathfrak{g}^*$. Hamiltonian vector fields generated by left-invariant functions are also called left-invariant.

It follows that the left-invariant functions on $G \times \mathfrak{g}^*$ are in exact correspondence with the functions on $\mathfrak{g}^*$. The integral curves $(g(t), \ell(t))$ of any left-invariant Hamiltonian vector field $\vec{F}$ are the solution curves of the following differential system:

$$\frac{dg}{dt} = g(t)dF_{\ell(t)}, \quad \frac{d\ell}{dt}(t) = -ad^*(dF_{\ell(t)})(\ell(t)). \tag{14.21}$$

The Poisson bracket $\{F, H\}$ of left-invariant Hamiltonians F and H is left-invariant and is given by

$$\{F, H\}(\ell) = \ell([dF_\ell, dH_\ell].$$

In particular, linear functions $h_A(\ell) = \ell(A)$ on $\mathfrak{g}^*$ defined by an element A in $\mathfrak{g}$ are left-invariant. Then the Poisson bracket of such functions is given by

$$\{h_A, h_B\}(\ell) = \ell([A, B]).$$

Linear functions $h_1, \ldots, h_m$ defined by a basis $A_1, \ldots, A_m$ in $\mathfrak{g}$ coincide with the coordinates of $\ell \in \mathfrak{g}^*$ relative to the dual basis $A_1^*, \ldots, A_m^*$. Consequently, any left-invariant Hamiltonian can be expressed as a function of the variables $h_1, \ldots, h_m$.

On Lie algebras that admit an invariant and non-degenerate quadratic form $\langle\,,\rangle$, elements ℓ in $\mathfrak{g}^*$ can be identified with elements L in $\mathfrak{g}$ via the formula $\langle L, X \rangle = \ell(X), X \in \mathfrak{g}$, and equation (14.21) can be expressed on $\mathfrak{g}$ simply as

$$\frac{dg}{dt} = g(t)dF(L(t)), \quad \frac{dL}{dt} = [dF(L(t)), L(t)]. \tag{14.22}$$

Since the trace form is invariant and non-degenerate on $so_{n+1}(\mathbb{C})$, equation (14.22) is valid for the left-invariant Hamiltonians on $so_{n+1}(\mathbb{C})$. Therefore, the eigenvalues of L are invariant functions on $so_{n+1}(\mathbb{C})$.

Since

$$Det(L - \xi I) = Det((L - \xi I)^T) = Det(-L - \xi I) = Det(-I)Det(L + \xi I),$$

the characteristic polynomial of L is an even polynomial on $so_n(\mathbb{C})$. On $so_4(\mathbb{C})$, the characteristic polynomial is given by

$$Det(L - \xi I) = \xi^4 + ((M, M) + (P, P))\xi^2 + (P, M)^2,$$

which yields two invariant functions $I_1 = (M, M) + (P, P)$ and $I_2 = (P, M)$. Here, we may either take M and P as the projections of L on the factors $\mathfrak{p}$ and $\mathfrak{k}$ in $so_4(\mathbb{C})$, or we may take them as matrices in $sl_2(\mathbb{C})$, corresponding to the representation of $so_4(\mathbb{C})$ as the product $sl_2(\mathbb{C}) \times (sl_2(\mathbb{C})$, with $L = \begin{pmatrix} M & P \\ P & M \end{pmatrix}$ as in Proposition 14.3.

On $so_4(\mathbb{C})$ the projection $\frac{dL}{dt} = [dH, L]$ on $so_4(\mathbb{C})$ associated with the Hamiltonian equation $\frac{dg}{dt} = g(t)(dH), \frac{dL}{dt} = [dH, L]$ can be represented either as

$$\frac{d}{dt}\begin{pmatrix} M & P \\ P & M \end{pmatrix} = \left[\begin{pmatrix} dH_{\mathfrak{k}} & dH_{\mathfrak{p}} \\ dH_{\mathfrak{p}} & dH_{\mathfrak{k}} \end{pmatrix}, \begin{pmatrix} M & P \\ P & M \end{pmatrix}\right],$$

where $dH_{\mathfrak{p}}$ and $dH_{\mathfrak{k}}$ denote the projections of dH on the factors $\mathfrak{p}$ and $\mathfrak{k}$, or as

$$\frac{dM}{dt} = [dH_{\mathfrak{k}}, M] + [dH_{\mathfrak{p}}, P], \quad \frac{dP}{dt} = [dH_{\mathfrak{k}}, P] + [dH_{\mathfrak{p}}, M]. \tag{14.23}$$

Equations (14.23) can be also expressed in terms of the vector product as

$$\frac{dm}{dt} = m \times \frac{\partial H}{\partial m} + p \times \frac{\partial H}{\partial p}, \frac{dp}{dt} = p \times \frac{\partial H}{\partial m} + m \times \frac{\partial H}{\partial p}. \qquad (14.23a)$$

Equations (14.23), when considered as the equations associated with a left-invariant Hamiltonian system on the tangent bundle of the double cover $SL_2(\mathbb{C}) \times Sl_2(\mathbb{C})$, should be adjoined

$$\frac{dg_1}{dt} = g_1(t)(dH_{\mathfrak{k}} + dH_{\mathfrak{p}}), \frac{dg_2}{dt} = g_2(t)(dH_{\mathfrak{k}} - dH_{\mathfrak{p}}) \qquad (14.24)$$

to complete the Hamiltonian system.

14.3 Affine Hamiltonians on $SO_4(\mathbb{C})$ and meromorphic solutions

We will now return to the affine Hamiltonians

$$\mathcal{H} = \frac{1}{2} \left(\frac{M_1^2}{\lambda_1} + \frac{M_2^2}{\lambda_2} + \frac{M_3^2}{\lambda_3} \right) + b_1 p_1 + b_2 p_2 + b_3 p_3, \qquad (14.25)$$

where the variables $M_1, M_2, M_3, p_1, p_2, p_3$ are the complex coordinates of a point ℓ in $so_4^*(\mathbb{C})$ relative to the dual basis of $A_1, A_2, A_3, B_1, B_2, B_3$. The parameters $\lambda_1, \lambda_2, \lambda_3$ will be assumed real and positive and b_1, b_2, b_3 will be assumed either all real or all imaginary. The imaginary choice covers the hyperbolic case. Our immediate objective is to seek the conditions on these parameters so that the solutions of the corresponding Hamiltonian equations

$$\frac{dM}{dt} = M \times \Omega + p \times b \quad \frac{dp}{dt} = p \times \Omega + s(M \times b), s = 0, 1 \qquad (14.26)$$

are meromorphic functions of complex time. The parameter s designates the appropriate Lie algebra, $s = 0$ in the semi-direct case, and $s = 1$ in the semi-simple case.

This was the criterion initiated by S. Kowalewski in her famous paper of 1889 [Kw]. Kowalewski began her investigations in this paper with a remark that the equations of the top of Lagrange admit meromorphic solutions of complex time, and then set out to find other cases that exhibit the same property. In doing so she assumed that all the variables in the equations of the top were complex, except for the principal moments of inertia and the coordinates of the center of mass. Formally, her equations are of the same form as our equations (14.26) with $s = 0$. In the process she discovered the famous relations $\lambda_1 = \lambda_2 = 2\lambda_3$, $b_1 = b_2 = 0$, as the only other case that admits

meromorphic solutions. In the rest of the paper, she then solved the equations
under these conditions in terms of hyperelliptic integrals.

The original paper of Kowalewski began with an implicit assumption that

$$M(t) = \frac{1}{t}M_0 + M_1 + M_2 t + \cdots,$$

$$P(t) = \frac{1}{t^2}P_0 + \frac{1}{t}P_1 + P_2 + P_3 t + \cdots,$$

which in time raised questions about her claim concerning the classification
of meromorphic cases. This gap in Kowalewski's paper was first noticed by
A. A. Markov a few years after publication, and attracted the attention of
several Russian mathematicians of that period. The issue was finally settled
by A. M. Lyapunov in 1894 [Lya], who confirmed the findings of Kowalewski
without making any assumption on the order of poles. In this paper, Lyapunov
considered the linearized equations around a particular meromorphic solution,
and then argued that this linear system must be single valued whenever the
flow is meromorphic.

It may be important to note that neither Kowalweski nor Lyapunov made any
claims about integrability based on the meromorphic properties. Of course,
if the equations are ultimately to be integrated on Abelian varieties in terms
of the elliptic or hyperelliptic integrals, then the solutions will be necessarily
meromorphic functions of complex time. However, it is still unclear if there
are integrable cases of (14.26) that do not share the meromorphic property.
The paper of R. Liouville of 1894 suggests that such situations exist, but I
have found Liouville's arguments difficult to follow [Li].

For the moment we will leave these subtleties aside and proceed with the
classification of meromophic solutions based on the Kowalewski–Lyapunov
criteria (rather than the monodromy arguments based on the branching of
solutions [Zg1; Zg2]). The following lemma is basic.

Lemma 14.8   *Let $p(t)$ and $M(t)$ be any solutions of (14.26), and let μ be any
non-zero number. Then $M(t)$ and $q(t) = \dfrac{1}{\mu}p(t)$ are the solutions of*

$$\frac{dM}{dt} = M \times \Omega + q \times b_\mu, \quad \frac{dq}{dt} = q \times \Omega + \frac{s}{\mu^2}M \times b_\mu,$$

where $b_\mu = \mu b$.

Proof   Evidently $q \times b_\mu = p \times b$, and

$$\frac{dq}{dt} = \frac{1}{\mu}\frac{dp}{dt} = \frac{1}{\mu}\left(p \times \Omega + sM \times b\right) = q \times \Omega + \frac{s}{\mu^2}M \times b_\mu.$$

$\square$

It is a corollary of the preceeding lemma that the meromorphic cases are invariant under the dilations $p \to \frac{1}{\mu} p(t)$ and $b \to \mu b$ when $s = 0$. Therefore, the meromorphic cases depend only on the direction of b. Secondly, the lemma implies that it is sufficient to classify meromorphic cases for small s.

Lemma 14.9 *If (14.26) do not admit meromorphic solutions for $s = 0$, then they do not admit meromorphic solutions for $s = 1$.*

Proof The solutions of an analytic differential system with a parameter depend analytically on the parameter. Therefore each solution $(\hat{M}(t), \hat{P}(t))$ of (14.26) can be developed as

$$M(t) = M_0(t) + s M_1(t) + s^2 M_2(t) + \cdots$$

and

$$p(t) = p_0(t) + s p_1(t) + s^2 p_2(t) + \cdots$$

valid for sufficiently small s. It follows that for these values of s each $M_i(t)$ and $p_i(t)$ must be meromorphic if $(M(t), p(t))$ is to be meromorphic. Since $(M_0(t), P_0(t))$ is a solution of (14.26) with $s = 0$, then it follows from above that $(\hat{M}(t), \hat{P}(t))$ cannot be meromorphic for small s whenever $M_0(t), p_0(t)$ is not meromorphic. But then the same must be true for all s as a consequence of Lemma 14.8. $\square$

Proposition 14.10 *Let $s = 0$ and $b \neq 0$. Then meromorphic solutions require that at least two eigenvalues λ_1, λ_2, λ_3 be equal to each other.*

The proof of this proposition will proceed via several lemmas. Assume first that $\lambda_1, \lambda_2, \lambda_3$ are all distinct.

Lemma 14.11 *Meromorphic solutions require that one of the coordinates of b is equal to zero.*

Proof Introduce the following constants:

$$A = \sqrt{\frac{\lambda_1}{\lambda_3 - \lambda_2}} \, , \quad B = \sqrt{\frac{\lambda_2}{\lambda_1 - \lambda_3}} \, , \quad C = \sqrt{\frac{\lambda_3}{\lambda_2 - \lambda_1}}.$$

Because $\lambda_1, \lambda_2, \lambda_3$ are all distinct, either two of the above constants are real and the remaining constant is imaginary, or two constants are imaginary and the remaining constant is real. That is, vector (A, B, C) is not colinear with any vector having either all real or all imaginary coordinates.

Now consider $M(t) = \dfrac{1}{t}M_0$, where

$$M_0 = \begin{pmatrix} BC\lambda_1 \\ AC\lambda_2 \\ AB\lambda_3 \end{pmatrix} \quad \text{and} \quad \Omega = \frac{1}{t}\begin{pmatrix} \dfrac{1}{\lambda_1}BC\lambda_1 \\ \dfrac{1}{\lambda_2}AC\lambda_2 \\ \dfrac{1}{\lambda_3}AB\lambda_3 \end{pmatrix} = \frac{1}{t}\begin{pmatrix} BC \\ AC \\ AB \end{pmatrix}.$$

Then $\frac{dM}{dt}(t) = -\dfrac{1}{t^2}M_0 = M(t) \times \Omega(t)$. It follows that $M(t)$ together with $p(t) = 0$ is a solution of (14.26) for $s = 0$.

We will now consider the variational equations along this solution. Let $u(\tau)$ and $v(\tau)$ denote arbitrary differentiable curves that satisfy $u(0) = M(1) = M_0$ and $v(0) = p(1) = 0$, and let $M_\tau(t)$ and $p_\tau(t)$ denote the solutions of (14.26) that emanate from $u(\tau)$ and $v(\tau)$ at $t = 1$. If we let

$$U(t) = \left.\frac{d}{d\tau}M_\tau(t)\right|_{\tau=0}, \quad V(t) = \left.\frac{d}{d\tau}P_\tau(t)\right|_{\tau=0}, \quad W(t) = \left.\frac{d}{d\tau}\Omega_\tau(t)\right|_{\tau=0},$$

then $U(t)$ and $V(t)$ are the solutions of the variational equation

$$\begin{aligned} \frac{dU}{dt} &= U \times \Omega + M \times W + V \times b, \\ \frac{dV}{dt} &= V \times \Omega, \end{aligned} \tag{14.27}$$

with the initial data $U(1) = \frac{du}{d\tau}(0)$ and $V(1) = \frac{dv}{d\tau}(0)$.

It follows that $W(t) = \begin{pmatrix} \dfrac{1}{\lambda_1}u_1(t) \\ \dfrac{1}{\lambda_2}u_2(t) \\ \dfrac{1}{\lambda_3}u_3(t) \end{pmatrix}$, where $u_1(t), u_2(t), u_3(t)$ are the

coordinates of $U(t)$. Then $U \times \Omega + M \times W = \frac{1}{t}\Delta U$, where

$$\Delta = \begin{pmatrix} 0 & AB\lambda_3\left(\dfrac{1}{\lambda_3}-\dfrac{1}{\lambda_2}\right) & AC\lambda_2\left(\dfrac{1}{\lambda_3}-\dfrac{1}{\lambda_2}\right) \\ AB\lambda_3\left(\dfrac{1}{\lambda_1}-\dfrac{1}{\lambda_3}\right) & 0 & BC\lambda_1\left(\dfrac{1}{\lambda_1}-\dfrac{1}{\lambda_3}\right) \\ AC\lambda_2\left(\dfrac{1}{\lambda_2}-\dfrac{1}{\lambda_1}\right) & BC\lambda_1\left(\dfrac{1}{\lambda_2}-\dfrac{1}{\lambda_1}\right) & 0 \end{pmatrix}.$$

This implies that equations (14.27) can be written as

$$\frac{dU}{dt} = \frac{1}{t}\Delta U(t) + V(t) \times b, \quad \frac{dV}{dt} = -\frac{1}{t}\Lambda(t)V, \tag{14.28}$$

where Λ denotes the skew-symmetric matrix such that $\hat{\Lambda} = \begin{pmatrix} BC \\ AC \\ AB \end{pmatrix}$.

We need to determine the solutions of this linear system. Consider first the solutions of $\frac{dV}{dt} = -\frac{1}{t}\Lambda V$. Note that

$$(AB)^2 + (AC)^2 + (BC)^2$$

$$= \frac{\lambda_1\lambda_2}{(\lambda_3 - \lambda_2)(\lambda_1 - \lambda_3)} + \frac{\lambda_1\lambda_3}{(\lambda_3 - \lambda_2)(\lambda_2 - \lambda_1)} + \frac{\lambda_2\lambda_3}{(\lambda_1 - \lambda_3)(\lambda_2 - \lambda_1)}$$

$$= \frac{\lambda_1\lambda_2(\lambda_2 - \lambda_1) + \lambda_1\lambda_3(\lambda_1 - \lambda_3) + \lambda_2\lambda_3(\lambda_3 - \lambda_2)}{(\lambda_3 - \lambda_2)(\lambda_1 - \lambda_3)(\lambda_2 - \lambda_1)} = -1.$$

This implies that the characteristic polynomial of Λ is equal to $\xi(\xi^2 - 1)$. Therefore $\xi_1 = 1$, $\xi_2 = -1$, $\xi_3 = 0$ are the eigenvalues of Λ. Let W_1, W_2, W_3 denote the corresponding eigenspaces. It is easy to verify that

$$W_1 = \{w \in \mathbb{C}^3 : \hat{\Lambda} \times w = w\}, \ W_2 = \{w \in \mathbb{C}^3 : \hat{\Lambda} \times w = -w\}, \text{ and } W_3 = \mathbb{C}\hat{\Lambda}.$$

It follows that the eigenspaces W_1 and W_2 are isotropic, in the sense that their vectors satisfy $(w, w) = 0$. These eigenvectors also satisfy $(\hat{\Lambda}, w) = 0$.

If we now express $V(t)$ as $V(t) = w_1(t) + w_2(t) + w_3(t)$ with $w_i(t) \in W_i$, $i = 1, 2, 3$, then equation $\frac{dV}{dt} = -\frac{1}{t}\Lambda(t)V$ becomes

$$\dot{w}_1(t) = -\frac{1}{t}w_1(t) \, , \ \dot{w}_2(t) = \frac{1}{t}w_2(t) \, , \ \dot{w}_3(t) = 0$$

Therefore,

$$w_1(t) = \frac{1}{t}w_1(1) \, , \ w_2(t) = tw_2(1) \, , \ w_3(t) = \text{constant} \, . \tag{14.29}$$

Let us now calculate the eigenvalues of Δ. For that it will be convenient to introduce the following notations:

$$\delta_1 = BC\lambda_1 \left(\frac{1}{\lambda_1} - \frac{1}{\lambda_3} \right) \, , \ \delta_2 = AC\lambda_2 \left(\frac{1}{\lambda_3} - \frac{1}{\lambda_2} \right) \, , \ \delta_3 = AB\lambda_3 \left(\frac{1}{\lambda_3} - \frac{1}{\lambda_2} \right) \, .$$

It follows that

$$\delta_1 = BC\lambda_1 \left(\frac{1}{\lambda_1} - \frac{1}{\lambda_3} \right) = \sqrt{\frac{\lambda_2}{\lambda_1 - \lambda_3}} \sqrt{\frac{\lambda_3}{\lambda_2 - \lambda_1}} \frac{\lambda_3 - \lambda_1}{\lambda_3}$$

$$= -\sqrt{\left(\frac{\lambda_1 - \lambda_3}{\lambda_2 - \lambda_1} \right) \frac{\lambda_2}{\lambda_3}} = -\frac{C\lambda_2}{B\lambda_3},$$

and $BC\lambda_1 \left(\frac{1}{\lambda_2} - \frac{1}{\lambda_1} \right) = \frac{1}{\delta_1}.$

Similar calculations show that

$$\delta_2 = -\frac{C\lambda_1}{A\lambda_3}, \quad AC\lambda_2\left(\frac{1}{\lambda_2} - \frac{1}{\lambda_1}\right) = \frac{1}{\delta_2}, \quad \text{and}$$

$$\delta_3 = -\frac{B\lambda_1}{A\lambda_2}, \quad AB\lambda_3\left(\frac{1}{\lambda_1} - \frac{1}{\lambda_3}\right) = \frac{1}{\delta_3}.$$

Therefore,

$$\Delta = \begin{pmatrix} 0 & \delta_3 & \delta_2 \\ \dfrac{1}{\delta_3} & 0 & \delta_1 \\ \dfrac{1}{\delta_2} & \dfrac{1}{\delta_1} & 0 \end{pmatrix}.$$

Furthermore,

$$\frac{\delta_2}{\delta_1\delta_3} = \left(-\frac{A}{B}\frac{\lambda_2}{\lambda}\right)\left(-\frac{B}{C}\frac{\lambda_3}{\lambda_2}\right)\left(-\frac{C}{A}\frac{\lambda_1}{\lambda_3}\right) = -1,$$

which in turn implies that $\det \Delta = -2$. It follows that the characteristic polynomial of Δ is given by

$$-\xi^3 + 3\xi - 2 = -(1-\xi)^2(\xi+2),$$

and therefore Δ has a double eigenvalue $\xi_{1,2} = 1$, and a single eigenvalue $\xi_3 = -2$.

An easy calculation shows that the eigenspace V_1 corresponding to $\xi = 1$ is given by $V_1 = \{x \in \mathbb{C}^3 : x_1 = \delta_3 x_2 + \delta_2 x_3\}$, and the eigenspace V_{-2}, corresponding to $\xi = -2$, is a line through $(\delta_1\delta_3, -\delta_1, 1)$. Then, $\frac{\lambda_3}{C}(\delta_1\delta_3, -\delta_1, 1) = \left(\frac{\lambda_1}{A}, \frac{\lambda_2}{B}, \frac{\lambda_3}{C}\right)$.

If we now write $U(t) = x(t) + y(t), x(t) \in V_1, y(t) \in V_{-2}$ then equation $\frac{dU}{dt} = \frac{1}{t}\Delta U(t) + V(t) \times b$ in (14.28) can be written as

$$\frac{dx}{dt} = \frac{1}{t}\Delta x(t) + f_1, \frac{dy}{dt} = -\frac{2}{t}y(t) + f_{-2}, \qquad (14.30)$$

where f_1 and f_{-2} denote the projections of $V \times b$ onto V_1 and V_{-2}. This is a linear inhomogeneous equation. Every solution is a sum of a homogeneous solution and a particular solution. The homogeneous solutions are given by

$$x(t) = tx(1), y(t) = \frac{1}{t^2}y(1).$$

Consider now the inhomogeneous equation with $V(t) = w_3$ for an arbitrary point in W_3 (14.29). Since $\hat{\Lambda}$ is not colinear with $bf = w_3 \times \hat{B}$ is not zero. Let f_1 and f_{-2} denote the projections of f on V_1 and V_{-2} as (14.30).

If $f_1 \neq 0$ then the solutions of (14.30) cannot be all single valued for the following reason: the restriction of (14.30) to the eigenspace V_1 is of the form

$$\frac{dx}{dt} = \frac{1}{t}x(t) + f_1 .$$

The preceding equation admits a particular solution $x_p(t)$ is of the form $x_p(t) = t\varphi(t)$ for some function $\varphi(t)$. But then $\frac{dx_p}{dt} = t\varphi' + \varphi(t) = \frac{1}{t}x_p + f_1$, and therefore, $\varphi'(t) = \frac{f_1}{t}$. But then φ is multivalued, and hence x_p is multivalued as well.

Therefore meromorphic solutions require that f belongs to V_{-2}. This condition implies that

$$b_2AB - b_3AC = t\frac{\lambda_1}{A},$$

$$b_3BC - b_1AB = t\frac{\lambda_2}{B},$$

$$b_1AC - b_2BC = t\frac{\lambda_3}{C},$$

for some complex number t. Therefore, $b_2B - b_3C = t(\lambda_3 - \lambda_2)$, and $b_3C - b_1A = t(\lambda_1 - \lambda_3)$, or $\frac{b_2B - b_3C}{b_3C - b_1A} = \frac{\lambda_3 - \lambda_2}{\lambda_1 - \lambda_3}$. This, in turn, implies that

$$Ab_1(\lambda_3 - \lambda_2) + Bb_2(\lambda_1 - \lambda_3) + Cb_3(\lambda_2 - \lambda_1) = 0.$$

Therefore,

$$b_1\sqrt{\lambda_1(\lambda_3 - \lambda_2)} + b_2\sqrt{\lambda_2(\lambda_1 - \lambda_3)} + b_3\sqrt{\lambda_3(\lambda_2 - \lambda_1)} - 0.$$

Either one of the above roots is imaginary and the other two roots are real, or the opposite is true. In either case one of b_1, b_2, b_3 must be zero. $\qquad\square$

Let us now turn to the proof of Proposition 14.10.

Proof of Proposition 14.10 We will assume that $\lambda_1 > \lambda_2 > \lambda_3$ in which case $b_2 = 0$, and

$$b_1\sqrt{\lambda_1(\lambda_3 - \lambda_2)} + b_3\sqrt{\lambda_3(\lambda_2 - \lambda_1)} = 0.$$

Let

$$M(t) = \frac{1}{t}\begin{pmatrix} 0 \\ 2i\lambda_2 \\ 0 \end{pmatrix} \text{ and } p(t) = \frac{1}{t^2}\begin{pmatrix} ih \\ 0 \\ h \end{pmatrix}.$$

The reader may readily verify that $M(t), p(t)$ together with $\Omega(t) = \frac{1}{\lambda}M(t)$ is a solution of equation (14.26) for $s = 0$, where $h = \frac{2\lambda_2}{b}$ and $b = b_3 + ib_1$.

Consider now the variational equation associated with the above flow. It is obtained in a manner similar to the previous case, and we have

$$\frac{dU}{dt} = U \times \frac{1}{t}\begin{pmatrix} 0 \\ 2i \\ 0 \end{pmatrix} + \frac{1}{t}\begin{pmatrix} 0 \\ 2i\lambda_2 \\ 0 \end{pmatrix} \times \begin{pmatrix} \frac{1}{\lambda_1}u_1 \\ \frac{1}{\lambda_2}u_2 \\ \frac{1}{\lambda_3}u_3 \end{pmatrix} + V \times b,$$

$$\frac{dV}{dt} = V \times \frac{1}{t}\begin{pmatrix} 0 \\ 2i \\ 0 \end{pmatrix} + \hat{P} \times \begin{pmatrix} \frac{1}{\lambda_1}u_1 \\ \frac{1}{\lambda_2}u_2 \\ \frac{1}{\lambda_3}u_3 \end{pmatrix}.$$

These equations can be rewritten as

$$\frac{dU}{dt} = \frac{1}{t}\left(U \times \left(\begin{pmatrix} 0 \\ 2i \\ 0 \end{pmatrix} + \begin{pmatrix} 0 \\ 2i\lambda_2 \\ 0 \end{pmatrix} \times \begin{pmatrix} \frac{1}{\lambda_1}u_1 \\ \frac{1}{\lambda_2}u_2 \\ \frac{1}{\lambda_3}u_3 \end{pmatrix} \right) + W \times \hat{B} \right),$$

$$\frac{dW}{dt} = \frac{1}{t}\left(W + W \times \left(\begin{pmatrix} 0 \\ 2i \\ 0 \end{pmatrix} + \begin{pmatrix} ih \\ 0 \\ h \end{pmatrix} \times \begin{pmatrix} \frac{1}{\lambda_1}u_1 \\ \frac{1}{\lambda_2}u_2 \\ \frac{1}{\lambda_3}u_3 \end{pmatrix} \right) \right), \quad W(t) = tV(t).$$

The above differential system breaks up into two independent subsystems:

$$\frac{du_1}{dt} = \frac{1}{t}\left(-2iu_3 + 2i\frac{\lambda_2}{\lambda_3}u_3 + w_2 b_3 \right),$$

$$\frac{du_3}{dt} = \frac{1}{t}\left(2iu_1 - 2i\frac{\lambda_2}{\lambda_1}u_1 - w_2 b_1 \right),$$

$$\frac{dw_2}{dt} = \frac{1}{t}\left(w_2 + h\frac{u_1}{\lambda_1} - ih\frac{u_3}{\lambda_3} \right),$$

and

$$\frac{du_2}{dt} = \frac{1}{t}\left(w_3 b_1 - b_3 w_1 \right),$$

$$\frac{dw_1}{dt} = \frac{1}{t}\left(w_1 + 2iw_3 - \frac{h}{\lambda_2}u_2 \right),$$

$$\frac{dw_3}{dt} = \frac{1}{t}\left(w_3 - 2iw_1 + i\frac{h}{\lambda_2}u_2 \right),$$

which are then described by the matrices

$$E = \begin{pmatrix} 0 & -2i\left(\frac{-\lambda_3+\lambda_2}{\lambda_3}\right) & b_3 \\ 2i\left(\frac{\lambda_1-\lambda_2}{\lambda_1}\right) & 0 & -b_1 \\ h\frac{1}{\lambda_1} & -ih\frac{1}{\lambda_3} & 1 \end{pmatrix} \quad \text{and} \quad F = \begin{pmatrix} 0 & -b_3 & b_1 \\ -\frac{h}{\lambda_2} & 1 & 2i \\ i\frac{h}{\lambda_2} & -2i & 1 \end{pmatrix}.$$

The solutions of these two subsystems must be single valued, if the solutions of (14.26) are to be meromorphic functions of complex time. But the solutions of the preceeding linear systems are single valued only when the eigenvalues of E and F are integers. Therefore, if either of the above matrices has an eigenvalue λ which is not an integer, then differential system (14.26) cannot have only meromorphic solutions.

The reader can readily verify that matrix F has eigenvalues $\lambda = 1, \lambda = 3$, and $\lambda = -2$. However, the imaginary part of the determinant of E is not zero when $\lambda_1 > \lambda_2 > \lambda_3$ unless $b = 0$. Therefore, the eigenvalues of E cannot be all real when $b \neq 0$. $\square$

Let us now come to the classification of meromorphic cases. According to the previous proposition, one coordinate of the drift vector must be zero, and at least two constants $\lambda_i, i = 1, 2, 3$ must be equal to each other. Let us assume that $\lambda_1 = \lambda_2$. If, in addition, we assume that $b_1 = b_2 = 0$, then $M_1(t)$ is constant, as we remarked earlier, and the system is integrable. We will subsequently show that, in that case, the solutions have the meromorphic property.

Now investigate the case that either $b_1 \neq 0$, or $b_2 \neq 0$. Then b can be reduced to $b = (b_1, 0, b_3)$ with $b_1 \neq 0$ by a rotation around the e_3 axis. This rotation does not alter the Hamiltonian, apart from moving the drift to a new position. Under these circumstances we then have the following:

Proposition 14.12 *Assume that $b = (b_1, 0, b_3), b_1 \neq 0$ and that $\lambda_1 = \lambda_2$. Then the solutions of equation (14.22) are not meromorphic functions of complex time unless:*

(i) $\lambda_1 = \lambda_2 = \lambda_3$ (the isospectral case);
(ii) $\lambda_1 = \lambda_2, b_1 = b_2 = 0$ (Kirchhoff–Lagrange case);
(iii) $\lambda_1 = \lambda_2 = 2\lambda_3, b_3 = 0$ (Kirchhoff–Kowalewski case).

As in the proof of the preceding proposition, we will follow the procedure initiated by Lyapunov based on the variational equations around the solutions of (14.26) of the form $M(t) = \frac{1}{t}M_0, p(t) = \frac{1}{t^2}p_0$ for some constant vectors M_0 and p_0 in $\mathbb{C}^3$. We will be able to show that outside of the above three cases these variational equations always admit multi-valued solutions. As a consequence of Lemma 14.9, we will confine the investigations to $s = 0$. Then,

Lemma 14.13 *There are only two types of solutions for (14.26) of the form*

$$M(t) = \frac{1}{t}M_0 = \frac{1}{t}\begin{pmatrix} p \\ q \\ r \end{pmatrix}, \quad P(t) = \frac{1}{t^2}P_0 = \frac{1}{t^2}\begin{pmatrix} f \\ g \\ h \end{pmatrix} \tag{14.31}$$

for some complex constants f, g, h and p, q, r. They are as follows:

Type 1: $p = r = 0$, $q = 2i\lambda_1$, $f = ih$, $g = 0$, $h = \dfrac{2\lambda_1}{b_3 + ib_1}$

Type 2: $p = iq$, $q(2\lambda_3 - \lambda_1) = \dfrac{2b_3}{b_1}\lambda_1\lambda_3$, $r = 2i\lambda_3$, $f = \dfrac{2\lambda_3}{b_1}$,
$g = if$, $h = 0$.

Proof   It follows from (14.23) that

$$-M_0 = [\Omega_0, M_0] + [B, P_0] , \text{ and } - 2P_0 = [\Omega_0, P_0],$$

where $\hat{\Omega}_0 = \begin{pmatrix} \dfrac{1}{\lambda_1}p \\ \dfrac{1}{\lambda_2}q \\ \dfrac{1}{\lambda_3}r \end{pmatrix}$, $\hat{M}_0 = \begin{pmatrix} p \\ q \\ r \end{pmatrix}$, and $\hat{P}_0 = \begin{pmatrix} f \\ g \\ h \end{pmatrix}$.

Therefore,

$$-p = qr\left(\frac{1}{\lambda_3} - \frac{1}{\lambda_2}\right) + gb_3, \qquad -2f = \frac{gr}{\lambda_3} - \frac{hq}{\lambda_2},$$

$$-q = pr\left(\frac{1}{\lambda_1} - \frac{1}{\lambda_3}\right) + hb_1 - fb_3, \qquad -2g = \frac{hp}{\lambda_1} - \frac{fr}{\lambda_3},$$

$$-r = pq\left(\frac{1}{\lambda_2} - \frac{1}{\lambda_1}\right) - gb_1, \qquad -2h = \frac{fq}{\lambda_2} - \frac{gp}{\lambda_1}.$$

The fact that both $(M(t), P(t))$ and $(P(t), P(t))$ are constant along the solutions of (14.26) implies that $pf + qg + rh = 0$ and $f^2 + g^2 + h^2 = 0$. In addition, $(P_0, \Omega_0) = 0$, which further implies that $\dfrac{1}{\lambda_1}pf + \dfrac{1}{\lambda_2}qg + \dfrac{1}{\lambda_3}rh = 0$. The above equations imply that $rh = 0$ whenever $\lambda_1 = \lambda_2$.

Type 1 corresponds to $r = 0$. Then $g = 0$, since b_1 is assumed non-zero, and therefore, $p = 0$. Finally, $f = ih$, $q = fb_3 - hb_1 = ih(b_3 + ib_1)$, and $-2ih = -\dfrac{hq}{\lambda_2}$. It follows that $q = 2i\lambda_2 = 2i\lambda_1$ and $h = \dfrac{2\lambda_1}{b_3 + ib_1}$.

Type 2 corresponds to $h = 0$. Further calculations are straightforward and will be omitted. $\square$

With this lemma at our disposal let us go to the proof of Proposition 14.12.

Proof of Proposition 14.12   Consider the variational equations associated around the trajectories of Lemma 14.13. Let $z(\varepsilon)$ and $w(\epsilon)$ denote arbitrary differentiable curves that satisfy $z(0) = M(1) = M_0$ and $w(0) = P(1) = P_0$, where $M(t)$ and $p(t)$ are the variables in (14.31), and let u and v denote their tangents of z and w when $\epsilon = 0$. Let $M_\varepsilon(t)$ and $p_\varepsilon(t)$ denote the solutions of (14.26) passing through $z(\epsilon)$ and $w(\epsilon)$ at $t = 1$, and let $U(t) = \dfrac{d}{d\varepsilon} M_\varepsilon(t) \Big|_{\varepsilon=0}$,

and let $V(t) = \dfrac{d}{d\varepsilon} P_\varepsilon(t) \Big|_{\varepsilon=0}$.

Then $U(t)$ and $V(t)$ are the solutions of the following equations:

$$\frac{dU}{dt} = U \times \Omega + M \times \Lambda + V \times b,$$

$$\frac{dV}{dt} = V \times \Omega + p \times \Lambda.$$

In this notation, $\Lambda(t) = \begin{pmatrix} \frac{1}{\lambda_1} u_1(t) \\ \frac{1}{\lambda_2} u_2(t) \\ \frac{1}{\lambda_3} u_3(t) \end{pmatrix}$, with $u_1(t), u_2(t), u_3(t)$ the coordinates

of $U(t)$.

Since $M(t) = \dfrac{1}{t} M_0$, $\Omega(t) = \dfrac{1}{t}\Omega_0$ and $P(t) = \dfrac{1}{t^2} P_0$,

$$\frac{dU}{dt} = \frac{1}{t}(U \times \Omega_0 + M_0 \times \Lambda) + V \times b,$$

$$\frac{dV}{dt} = \frac{1}{t}(V \times \Omega_0) + \frac{1}{t^2}(P_0 \times \Lambda).$$

Define a new variable $W = tV(t)$. Then,

$$\frac{dU}{dt} = \frac{1}{t}(U \times \Omega_0 + M_0 \times \Lambda + W \times b),$$

$$\frac{dW}{dt} = \frac{1}{t}(W + W \times \Omega_0 + P_0 \times \Lambda).$$

Following Lyapunov's arguments, we conclude that $U(t)$ and $W(t)$ will be single valued if and only if the eigenvalues of the linear operator Δ,

$$(U, W) \longrightarrow (U \times \Omega_0 + M_0 \times \Lambda + W \times b, (W + W \times \Omega_0 + P_0 \times \Lambda)),$$

are integers.

Indeed, if ξ denotes an eigenvalue of Δ, then the solutions of $\dfrac{dx}{dt} = \dfrac{1}{t}\Delta x(t)$ when restricted to the eigenspace corresponding to ξ are of the form $\dfrac{dx}{dt} = \dfrac{1}{t}\xi x(t)$, and hence, $x(t) = t^\xi x(1)$. But t^ξ is single valued only when ξ is a whole number.

To set the stage for the subsequent calculations observe that $\xi = -2$ is an eigenvalue of Δ. The demonstration is as follows: let $U = M_0$ and let $W = 2P_0$.

Since $M_0 \times \Omega_0 + P_0 \times b = -M_0$, and $P_0 \times \Omega_0 = -2P_0$,

$$M_0 \times \Omega_0 + M_0 \times \Lambda(M_0) + W \times b = 2(M_0 \times \Omega_0) + 2(P_0 \times b) = -2M_0 = -2U$$

and

$$W + W \times \Omega_0 + P_0 \times \Lambda(M_0) = 2P_0 + 2(P_0 \times \Omega_0) + p_0 \times \Omega_0$$
$$= 2P_0 - 4P_0 - 2P_0 = -2(2P_0) = -2(W).$$

Hence $\xi = -2$ is an eigenvalue.

The operator Δ is described by the following matrix:

$$\begin{pmatrix} 0 & \Delta_{32}r & \Delta_{32}q & 0 & b_3 & -b_2 \\ \Delta_{13}r & 0 & \Delta_{13}p & -b_3 & 0 & b_1 \\ \Delta_{21}q & \Delta_{21}p & 0 & b_2 & -b_1 & 0 \\ 0 & -\dfrac{h}{\lambda_2} & \dfrac{g}{\lambda_3} & 1 & \dfrac{r}{\lambda_3} & -\dfrac{q}{\lambda_2} \\ \dfrac{h}{\lambda_1} & 0 & -\dfrac{f}{\lambda_3} & -\dfrac{r}{\lambda_3} & 1 & \dfrac{p}{\lambda_1} \\ -\dfrac{g}{\lambda_1} & \dfrac{f}{\lambda_2} & 0 & \dfrac{q}{\lambda_2} & -\dfrac{p}{\lambda_1} & 1 \end{pmatrix},$$

where

$$\Delta_{32} = \left(\frac{1}{\lambda_3} - \frac{1}{\lambda_2} \right), \Delta_{13} = \left(\frac{1}{\lambda_1} - \frac{1}{\lambda_3} \right) \text{ and } \Delta_{21} = \left(\frac{1}{\lambda_2} - \frac{1}{\lambda_1} \right).$$

Consider now the spectrum of Δ associated with each solution in lemma 14.13.

When $p = r = 0, q = 2i\lambda_1, f = ih, g = 0, h = \dfrac{2\lambda_1}{b_3 + ib_1}$,

$$\Delta = \begin{pmatrix} 0 & 0 & \Delta_{32}q & 0 & b_3 & 0 \\ 0 & 0 & 0 & -b_3 & 0 & b_1 \\ 0 & 0 & 0 & 0 & -b_1 & 0 \\ 0 & -\dfrac{h}{\lambda_2} & 0 & 1 & 0 & -\dfrac{q}{\lambda_2} \\ \dfrac{h}{\lambda_1} & 0 & -\dfrac{f}{\lambda_3} & 0 & 1 & 0 \\ 0 & \dfrac{f}{\lambda_2} & 0 & \dfrac{q}{\lambda_2} & 0 & 1 \end{pmatrix}.$$

Rename the variables $x_1 = u_2, x_2 = w_1, x_3 = w_3, x_4 = u_1, x_5 = u_3, x_6 = w_2$. Then,

$$\dot{x}_1 = \dot{u}_2 = \frac{1}{t}(-b_3 x_2 + b_1 x_3),$$

$$\dot{x}_2 = \dot{w}_1 = \frac{1}{t}\left(-\frac{h}{\lambda_2}x_1 + x_2 - \frac{q}{\lambda_2}x_3\right),$$

$$\dot{x}_3 = \dot{w}_3 = \frac{1}{t}\left(\frac{f}{\lambda_2}x_1 + \frac{q}{\lambda_2}x_2 + x_3\right),$$

and

$$\dot{x}_4 = \dot{u}_1 = \frac{1}{t}(\Delta_{32}q x_5 + b_3 x_6),$$

$$\dot{x}_5 = \dot{u}_3 = \frac{1}{t}(-b_1 x_6),$$

$$\dot{x}_6 = \dot{w}_2 = \frac{1}{t}\left(\frac{h}{\lambda_1}x_4 - \frac{f}{\lambda_3}x_5 + x_6\right).$$

The preceding change of variables decomposes the variational equations into two independent subsystems and, hence, the determinant of Δ is equal to the product of the determinants associated with

$$D_1 = \begin{pmatrix} 0 & -b_3 & b_1 \\ -\dfrac{h}{\lambda_2} & 1 & -\dfrac{q}{\lambda_2} \\ \dfrac{f}{\lambda_2} & \dfrac{q}{\lambda_2} & 1 \end{pmatrix}, \quad \text{and} \quad D_2 = \begin{pmatrix} 0 & \Delta_{32}q & b_3 \\ 0 & 0 & -b_1 \\ \dfrac{h}{\lambda_1} & -\dfrac{f}{\lambda_3} & 1 \end{pmatrix}.$$

It is easy to verify that the characteristic polynomial of D_1 is equal to $p_1(\xi) = -\xi(1-\xi)^2 + 6\xi - 6 = (\xi - 1)(\xi - 3)(\xi + 2)$. The characteristic polynomial $p_2(\xi)$ associated with D_2 is given by

$$p_2(\xi) = \xi^2(1-\xi) + \xi\left(\frac{b_1 ih}{\lambda_3} + \frac{hb_3}{\lambda}\right) - \frac{2ihb_1}{\lambda_3} + \frac{2ihb_1}{\lambda},$$

where λ denotes the common eigenvalue $\lambda_1 = \lambda_2$.

Further, it is easy to check that $\xi = 2$ is a root of p_2, and therefore,

$$p_2(\xi) = (\xi - 2)\left(-\xi(\xi + 1) + \frac{1}{2}\left(\frac{2b_1 h}{\lambda_3} - \frac{2ihb_1}{\lambda}\right)\right).$$

$p_2(\xi)$ has integer roots only when

$$p(\xi) = -\xi(\xi + 1) + \left(\frac{b_1 hi}{\lambda_3} - \frac{2hib_1}{\lambda}\right) = -\xi(\xi + 1) + \frac{(\lambda - \lambda_3)}{\lambda\lambda_3}\frac{b_1(b_1 + ib_3)\lambda}{b_1^2 + b_3^2}$$

has integer roots. Since the product of the roots is equal to $\dfrac{(\lambda - \lambda_3)}{\lambda\lambda_3}$ $\dfrac{b_1(b_1 + ib_3)\lambda}{b_1^2 + b_3^2 i}$, integer roots occur only when $b_3 = 0$ and $\dfrac{\lambda - \lambda_3}{\lambda_3}$ is an integer. The latter occurs only when the ratio $\dfrac{\lambda}{\lambda_3}$ is an integer.

We now pass to the second case, under the assumption that $b_3 = 0$, and that the ratio $\frac{\lambda}{\lambda_3}$ is an integer. Recall that this case is given by

$$p = -iq, \quad q(2\lambda_3 - \lambda_1) = 2\frac{b_3}{b_1}\lambda_1\lambda_3, \quad r = 2i\lambda_3, \quad f = \frac{2\lambda_3}{b_1}, \quad g = if, \quad h = 0.$$

The case $2\lambda_3 - \lambda_1 = 0$, yields Kowalewski's case, and there is nothing further to prove. So assume that $2\lambda_3 - \lambda \neq 0$. Then $b_3 = 0$ implies that $q = 0$, and hence $p = 0$.

The corresponding matrix Δ breaks up into two subsystems as in Type 1. Simply rename the variables $x_1 = u_1$ $x_2 = u_2$, $x_3 = w_3$, and $x_4 = u_3$, $x_5 = w_1$ $x_6 = w_2$. Then $\Delta = \begin{pmatrix} D_1 & 0 \\ 0 & D_2 \end{pmatrix}$ with

$$D_1 = \begin{vmatrix} 0 & \Delta_{32}r & 0 \\ \Delta_{13}r & 0 & b_1 \\ -\dfrac{if}{\lambda} & \dfrac{f}{\lambda} & 1 \end{vmatrix}, \quad D_2 = \begin{vmatrix} 0 & 0 & -b_1 \\ \dfrac{if}{\lambda_3} & 1 & \dfrac{r}{\lambda_3} \\ -\dfrac{f}{\lambda_3} & -\dfrac{r}{\lambda_3} & 1 \end{vmatrix}.$$

The characteristic polynomial $p_1(\xi)$ of D_1 is given by

$$p_1(\xi) = \xi^2(1 - \xi) + \xi\left(\Delta_{32}\Delta_{13}r^2 + \frac{b_1 f}{\lambda}\right) - \Delta_{32}\Delta_{13}r^2 - \frac{\Delta_{32}rifb_1}{\lambda}.$$

It follows that $\Delta_{32}\Delta_{13}r^2 = \dfrac{4(\lambda - \lambda_3)^2}{\lambda^2}$, and $\dfrac{b_1 f}{\lambda} = \dfrac{b_1\lambda_3}{\lambda^2}$. Furthermore,

$$\frac{\Delta_{32}rifb_1}{\lambda} = -\frac{4(\lambda - \lambda_3)\lambda_3}{\lambda^2}.$$

Therefore,

$$p_1(\xi) = \xi^2(1 - \xi) + \frac{\xi}{\lambda^2}(4(\lambda - \lambda_3)^2 + \lambda_1\lambda_3) - \frac{4(\lambda - \lambda_3)(\lambda - 2\lambda_3)}{\lambda^2}.$$

For $p_1(\xi)$ to have integer roots, the quantity $\dfrac{4(\lambda - \lambda_3)(\lambda - 2\lambda_3)}{\lambda^2}$ must be an integer. This quantity is equal to $4\left(1 - \dfrac{1}{n}\right)\left(1 - \dfrac{2}{n}\right)$, where n denotes the ratio $\frac{\lambda_3}{\lambda}$. It is an integer only when $n = 1$, or $n = 2$: $n = 1$ implies that $\lambda_1 = \lambda_2 = \lambda_3$, and $n = 2$ implies that $\lambda_1 = \lambda_2 = 2\lambda_3$. □

14.4 Kirchhoff–Lagrange equation and its solution

The preceding section shows that the only candidates for meromorphic solutions are the isospectral case, $\lambda_1 = \lambda_2 = \lambda_3$, Kirchhoff–Lagrange case, $\lambda_2 = \lambda_3$, $b_2 = b_3 = 0$, and Kirchhoff–Kowalewski case, $\lambda_1 = \lambda_2 = 2\lambda_3$, $b_3 = 0$. Since the isospectral case can always be brought to the Kirchhoff–Lagrange form by a suitable rotation of the drift vector b, the number of candidates reduces to two. Remarkably, each of these two cases admits an extra integral of motion on $\mathfrak{g}$ independent of the Casimirs and the Hamiltonian. Thus our search for meromorphic solutions uncovered the integrable Hamiltonians.

Let us now show that the solutions of $H = \frac{1}{2\lambda}(M_2^2 + M_3^2) + \frac{1}{2\lambda_1}M_1^2 + b_1 p_1$ indeed possesses the meromorphic property (we have switched to $\lambda_2 = \lambda_3$ in order to conform to an earlier presentation in [Jm]). The Kowalewski case is more subtle and will be deferred to the next chapter.

We will confine our attention to the semi-simple case, the case that is not treated in the literature of the top. But even in the semi-direct case the solutions of the affine Hamiltonian system are more general than the solutions of the corresponding top of Lagrange–the equations of the top form a six-dimensional subsystem of the corresponding twelve-dimensional affine-quadratic system $(2dim(T^*SO_3(\mathbb{C})) = dim(T^*G) = 12)$.

We need to solve the following system of equations:

$$\frac{dM_1}{dt} = 0, \ \frac{dM_2}{dt} = M_1 M_3 \left(\frac{1}{\lambda_1} - \frac{1}{\lambda}\right) + p_3 b_1, \ \frac{dM_3}{dt} = M_1 M_2 \left(\frac{1}{\lambda} - \frac{1}{\lambda_1}\right) - p_2 b_1,$$

$$\frac{dp_1}{dt} = \frac{p_2 M_3}{\lambda} - \frac{p_3 M_2}{\lambda}, \ \frac{dp_2}{dt} = \frac{p_2 M_3}{\lambda} - \frac{p_3 M_2}{\lambda}, \ \frac{dp_3}{dt} = \frac{p_1 M_2}{\lambda} - \frac{p_2 M_1}{\lambda_1} - M_2 b_1,$$

$$\tag{14.32}$$

$$\frac{dg}{dt} = g(t)(dH), dH = \sum_{i=1}^{3} b_i B_i + \frac{1}{\lambda_i} M_i A_i.$$

For simplicity of exposition we will assume that these equation are scaled so that $b_1 = 1$.

Rather than carrying out the integration down to $SO_4(\mathbb{C})$, we will carry it out down to its double cover $SL_2(\mathbb{C}) \times SL_2(\mathbb{C})$. For that reason variables $M_1, M_2, M_3, p_1, p_2, p_3$ will be identified with matrices

$$M = \frac{1}{2}\begin{pmatrix} iM_1 & M_2 + iM_3 \\ -M_2 + iM_3 & -iM_1 \end{pmatrix}, P = \frac{1}{2}\begin{pmatrix} ip_1 & p_2 + ip_3 \\ -p_2 + ip_3 & -ip_1 \end{pmatrix}.$$

Accordingly, Ω and B will be given by

$$\Omega = \frac{1}{2}\begin{pmatrix} \frac{i}{\lambda_1}M_1 & \frac{1}{\lambda}(M_2 + iM_3) \\ \frac{1}{\lambda}(-M_2 + iM_3) & -\frac{i}{\lambda}M_1 \end{pmatrix}, \text{ and } B = \frac{1}{2}\begin{pmatrix} i & 0 \\ 0 & -i \end{pmatrix},$$

so that (14.32) becomes

$$\frac{dM}{dt} = [\Omega, M] + [B, P], \frac{dP}{dt} = [\Omega, P] + [B, M],$$

$$\frac{dg_1}{dt} = g_1(t)\Omega_1(t), \frac{dg_2}{dt} = g_2(t)\Omega_2(t), \tag{14.33}$$

where

$$\Omega_1 = \frac{1}{2}\begin{pmatrix} i\left(\frac{1}{\lambda_1}M_1 + 1\right) & \frac{1}{\lambda}(M_2 + iM_3) \\ \frac{1}{\lambda}(-M_2 + iM_3) & -i\left(\frac{1}{\lambda_1}M_1 + 1\right) \end{pmatrix},$$

$$\Omega_2 = \frac{1}{2}\begin{pmatrix} i\left(\frac{1}{\lambda_1}M_1 - 1\right) & \frac{1}{\lambda}(M_2 + iM_3) \\ \frac{1}{\lambda}(-M_2 + iM_3) & -i\left(\frac{1}{\lambda_1}M_1 - 1\right) \end{pmatrix}.$$

It will be advantageous to introduce new variables z_1, z_2, w_1, w_2 defined by

$$z_1 = M_2 + iM_3, \, z_2 = M_2 - iM_3, \, w_1 = p_2 + ip_3, \, w_2 = p_2 - ip_3. \tag{14.34}$$

We leave it to the reader to verify that the new variables conform to the following Poisson bracket table:

Table 14.1

$\{,\}$	p_1	M_1	z_1	z_2	w_1	w_2
p_1	0	0	iw_1	$-iw_2$	iz_1	$-iz_2$
M_1	0	0	iz_1	$-iz_2$	iw_1	$-iw_2$
z_1	$-iw_1$	$-iz_1$	0	$2iM_1$	0	$2ip_1$
z_2	iw_2	iz_2	$-2iM_1$	0	$-2ip_1$	0
w_1	$-iz_1$	$-iw_1$	0	$2ip_1$	0	$2iM_1$
w_2	iz_2	iw_2	$-2ip_1$	0	$-2iM_1$	0

In these variables, $H = \frac{1}{2\lambda_1}M_1^2 + \frac{1}{2\lambda}z_1z_2 + p_1$, and

$$\frac{dz_1}{dt} = -i\left(\frac{1}{\lambda_1} - \frac{1}{\lambda}\right)M_1z_1 - iw_1, \frac{dz_2}{dt} = i\left(\frac{1}{\lambda_1} - \frac{1}{\lambda}\right)M_1z_2 + iw_2,$$

$$\frac{dw_1}{dt} = -\frac{iM_1w_1}{\lambda_1} + iz_1\left(\frac{1}{\lambda}p_1 - 1\right), \frac{dw_2}{dt} = \frac{iM_1w_2}{\lambda_1} - iz_2\left(\frac{1}{\lambda}p_1 - 1\right),$$

$$\tag{14.35}$$

$$\frac{dM_1}{dt} = 0, \frac{dp_1}{dt} = -\frac{i}{2\lambda}(w_2 z_1 - w_1 z_2),$$

are the associated Hamiltonian equations, as can be easily verified through the Poisson brackets in Table 14.1. Alternatively, these equations could have been obtained from equations (14.32) by direct substitutions.

Consider now the integral manifold $V(I_1, I_2, I_3, I_4)$ defined by four constants I_1, I_2, I_3, I_4 where

$$I_1 = \mathcal{H} - \frac{M_1^2}{2\lambda_1} = \frac{1}{2\lambda}z_1 z_2 + p_1, \ I_2 = p_1 M_1 + p_2 M_2 + p_3 M_3$$

$$= p_1 M_1 + \frac{1}{2}z_2 w_1 + \frac{1}{2}z_1 w_2,$$

$$I_3 = p_1^2 + p_2^2 + p_3^2 + (M_1^2 + M_2^2 + M_3^2) - M_1^2 = p_1^2 + z_1 z_2 + w_1 w_2, \ I_4 = M_1.$$

We will presently show that $u = w_2 z_1 - w_1 z_2$, and $\xi = z_1 z_2$ can be identified with points on an elliptic curve determined by the above integral manifold.

Let us first note that

$$\begin{aligned} \frac{d\xi}{dt} &= \frac{d}{dt}z_1 z_2 = i(w_2 z_1 - w_1 z_2) = iu, \\ \frac{d}{dt}(w_1 w_2) &= i\left(\frac{p_1}{\lambda} - 1\right)(w_2 z_1 - w_1 z_2). \end{aligned} \tag{14.36}$$

Along each solution of (14.35), the variables $u(t)$ and $\xi(t)$ evolve as follows:

$$\begin{aligned} u^2(t) &= (w_1(t)z_2(t) - w_2(t)z_1(t))^2 \\ &= (w_1(t)z_2(t) + w_2(t)z_1(t))^2 - 4z_1(t)z_2(t)w_1(t)w_2(t) \\ &= 4(I_2 - p_1(t)M_1)^2 - 4\xi(t)\left(I_3 - p_1^2(t) - \xi(t)\right) \\ &= 4(I_2 - (I_1 - \frac{1}{2\lambda}\xi(t))M_1)^2 - 4\xi(t)\left(I_3 - (I_1 - \frac{1}{2\lambda}\xi(t))^2 - \xi(t)\right). \end{aligned}$$

Hence,

$$u^2(t) = \frac{1}{\lambda^2}\xi^3(t) + A\xi^2(t) + B\xi(t) + C,$$

where

$$A = \frac{M_1^2}{\lambda^2} - 4\left(\frac{I_1}{\lambda} - 1\right), B = 4\left(\frac{M_1}{\lambda}(I_2 - I_1 M_1) - (I_3 - I_1^2)\right), C = 4(I_2 - M_1 I_1)^2.$$

Let Γ denote the elliptic curve $u^2 = \frac{1}{\lambda^2}x^3 + Ax^2 + Bx + C$. This curve is equivalent to the cubic curve of Weierstrass $v^2 = 4s^3 - g_2 s - g_3$ under the transformation $v = 2\lambda u$ and $s = \xi + \frac{1}{3}A\lambda$. Since $\frac{d\xi}{dt}(t) = iu(t)$,

$$\left(\frac{d\xi}{dt}\right)^2 = -u^2(t) = -\left(\frac{1}{\lambda^2}\xi^3(t) + A\xi^2(t) + B\xi(t) + C\right).$$

The above implies that the roots of the polynomial $P(x) = \frac{1}{\lambda^2}x^3 + Ax^2 + Bx + C$ correspond to the stationary points of $\xi(t)$. This observation implies that the quantity $z_1(t)z_2(t)$ is constant whenever it originates on the surface $z_1w_2 - z_2w_1 = 0$. The reader may note that these stationary solutions are confined to the singular points of the algebraic variety $V(I_1, I_2, I_3, I_4)$.

We are now ready to outline the integration procedure. Apart from the equilibrium points described earlier, each solution $\xi(t)$ is an elliptic function and hence meromorphic. Every solution of $\left(\frac{dx}{dt}\right)^2 = -u^2 = -\left(\frac{1}{\lambda^2}x^3 + Ax^2 + Bx + C\right)$ can be expressed in terms of the Weierstrass $\wp$-function. Simply define $s(t) = \xi(2it) + \frac{A\lambda^2}{3}$. Then

$$\left(\frac{ds}{dt}\right)^2 = -4\lambda^2\left(\frac{d\xi}{dt}\right)^2$$
$$= 4(\xi^3 + 4\lambda^2 A\xi^2 + 4\lambda^2 B\xi + 4\lambda^2 C)$$
$$= 4s^3(t) - g_2 s(t) - g_3.$$

Hence, $s(t)$ is a time-shift of $\wp$. For simplicity of exposition, assume that $s(t)$ is time shifted so that the pole is at $t = 0$. Then $\xi(t)$ is an even elliptic function having a double pole at $t = 0$. Then $p_1(t) = I_1 - \frac{1}{2\lambda}\xi(t)$, $p_1(t)$ is also an even elliptic function having a double pole at $t = 0$.

The remaining variables are integrated as follows. Since $I_2 = p_1 M_1 + \frac{1}{2}(z_1w_2 + z_2w_1)$ and $\frac{1}{2}u = \frac{1}{2}(z_1w_2 - z_2w_1)$, $z_1w_2 = (I_2 - p_1 M_1) + \frac{1}{2}u$, and $z_2w_1 = (I_2 - p_1 M_1) - \frac{1}{2}u$.

Assuming that $w_1(t)w_2(t) \neq 0$, then

$$\frac{z_1}{w_1} = \frac{z_1w_2}{w_1w_2} = \frac{(I_2 - p_1 M_1) + \frac{1}{2}u}{I_3 - p_1^2 - \xi}, \frac{z_2}{w_2} = \frac{z_2w_1}{w_2w_1} = \frac{(I_2 - p_1 M_1) - \frac{1}{2}u}{I_3 - p_1^2 - \xi}.$$

Therefore,

$$z_1(t) = w_1(t)\frac{(I_2 - p_1(t)M_1) + \frac{1}{2}u(t)}{I_3 - p_1^2(t) - \xi(t)}, z_2(t) = w_2(t)\frac{(I_2 - p_1(t)M_1) - \frac{1}{2}u(t)}{I_3 - p_1^2(t) - \xi(t)}.$$

Equation $\dfrac{dw_1}{dt} = -i\left(\dfrac{M_1 w_1}{\lambda_1} - z_1\left(\dfrac{1}{\lambda}p_1 - 1\right)\right)$ can be written as

$$\frac{dw_1}{dt} = -iw_1\left(\frac{M_1}{\lambda_1} - \frac{(I_2 - p_1 M_1) + \frac{1}{2}u}{(I_3 - p_1^2 - \xi)}\left(\frac{p_1}{\lambda} - 1\right)\right) = -iw_1(t)f(t),$$

(14.37)

where $f(t) = \dfrac{M_1}{\lambda_1} - \dfrac{(I_2 - p_1 M_1) + \frac{1}{2}u}{(I_3 - p_1^2 - \xi)}\left(\dfrac{p_1}{\lambda} - 1\right)$.

It follows that $w_1(t) = (\exp -i\int_0^t f(z)dz)w_1(0)$, provided that $\int_0^t f(z)dz$ is independent of the path that connects 0 to t. That will be the case whenever f has zero residue. Since u is an odd function it is not obvious that the path integral defined by f is single valued, so we will proceed differently.

Let $M^2 = I_3 - 2\lambda I_1 + \lambda^2$. It is easy to verify that $M^2 = (p_1 - \lambda)^2 + w_1 w_2$. Now introduce two new variables $\theta(t)$ and $\varphi(t)$. The first variable θ is defined through the formula

$$p_1 - \lambda = M\cos\theta, \quad \sqrt{w_1 w_2} = M\sin\theta.$$

(14.38)

Then $\dfrac{dp_1}{dt} = -\dfrac{1}{2\lambda}iu$, implies $M\sin\theta\,\dfrac{d\theta}{dt} = \dfrac{i}{2\lambda}u(t)$. Therefore,

$$
\begin{aligned}
\frac{M_1}{\lambda_1} - f(t) &= \frac{\left(I_2 - p_1 M_1 + \frac{1}{2}u\right)}{w_1 w_2}\left(\frac{p_1}{\lambda} - 1\right) = \frac{\left(I_2 - p_1 M_1 + \frac{1}{2}u\right)M\cos\theta}{\lambda M^2 \sin^2\theta} \\
&= \frac{(I_2 - p_1 M_1)M\cos\theta}{\lambda M^2 \sin^2\theta} + \left(\frac{\lambda}{i}M\sin\theta\,\frac{d\theta}{dt}\right)\frac{M\cos\theta}{\lambda M^2 \sin^2\theta} \\
&= \frac{(I_2 - p_1 M_1)\cos\theta}{\lambda M \sin^2\theta} - i\frac{\cos\theta}{\sin\theta}\left(\frac{d\theta}{dt}\right).
\end{aligned}
$$

The second variable ϕ is defined implicitly by $w_1(t) = M\sin\theta(t)e^{-i\left(\varphi(t) + t\frac{M_1}{\lambda_1}\right)}$.
Then,

$$
\begin{aligned}
-iw_1(t)f(t) = \frac{dw_1}{dt} &= \left(M\cos\theta\,\frac{d\theta}{dt} - i\left(\frac{d\varphi}{dt} + \frac{M_1}{\lambda_1}\right)M\sin\theta\right)e^{i\left(\varphi + \frac{M_1(t)}{\lambda_1}\right)} \\
&= \left(M\cos\theta\,\frac{d\theta}{dt} - i\left(\frac{d\varphi}{dt} + \frac{M_1}{\lambda_1}\right)M\sin\theta\right)\frac{w_1(t)}{M\sin\theta}.
\end{aligned}
$$

This equality implies that

$$-if(t)M\sin\theta = M\cos\theta\,\frac{d\theta}{dt} - i\left(\frac{d\varphi}{dt} + \frac{M_1}{\lambda_1}\right)M\sin\theta,$$

or

$$-i\left(f(t) - \frac{M_1}{\lambda_1}\right) M \sin\theta = M \cos\theta \frac{d\theta}{dt} - i\frac{d\varphi}{dt} M \sin\theta.$$

The substitution of $\dfrac{M_1}{\lambda_1} - f(t) = \dfrac{(I_2 - p_1 M_1)\cos\theta}{\lambda M \sin^2\theta} - i\dfrac{\cos\theta}{\sin\theta}\left(\dfrac{d\theta}{dt}\right)$ into the above equation yields

$$i\left(\frac{(I_3 - p_1 M_1)\cos\theta}{\lambda M \sin^2\theta} - i\frac{\cos\theta}{\sin\theta}\frac{d\theta}{dt}\right) M \sin\theta = M \cos\theta \frac{d\theta}{dt} - i\frac{d\varphi}{dt} M \sin\theta.$$

Therefore,

$$\frac{d\varphi}{dt} = -\frac{(I_3 - p_1 M_1)\cos\theta}{\lambda M \sin^2\theta} = \frac{(I_3 - p_1 M_1)(p_1 - \lambda)}{\lambda\left(M^2 - (p_1 - \lambda)^2\right)}. \tag{14.39}$$

Since p_1 is an even function, the right-hand side of equation (14.39) is an even function, and consequently has zero residue. It follows that

$$\varphi(t) = \int_0^t \frac{(I_3 - p_1(z)M_1)p_1(z) - \lambda}{\lambda(M^2 - (p_1(z) - \lambda)^2}dz$$

is a well-defined meromorphic function. Therefore,

$$w_1(t) = M \sin\theta(t)e^{\left(i\varphi(t) + t\frac{M_1}{\lambda}\right)}$$

is also a meromorphic function of complex time t. The remaining variable $w_2(t)$ is then determined through the relation $w_1(t)w_2(t) = M^2 \sin^2\theta(t)$, which then yields

$$w_2(t) = \frac{M^2 \sin^2\theta}{w_1(t)} = M \sin\theta e^{\left(i\varphi(t) + t\frac{M_1}{\lambda_1}t\right)}.$$

Finally, $z_1(t)$ and $z_2(t)$ are determined through the formulas

$$z_1(t)w_2(t) = (I_3 - p_1(t)M_1) + \frac{1}{2}u(t),$$

$$z_2(t)w_1(t) = (I_3 - p_1(t)M_1) - \frac{1}{2}u(t).$$

In the literature of the top, the real counterparts of θ and φ are known as the nutation and the precession angles [Ar].

It remains to carry out the integration procedure all the way down to the underlying group $G = SL_2(\mathbb{C}) \times SL_2(\mathbb{C})$, that is, it remains to solve

$$\frac{dg_1}{dt}(t) = g_1(t)\Omega_1(t) \text{ and } \frac{dg_2}{dt}(t) = g_2(t)\Omega_2(t),$$

where

$$\Omega_1 = \frac{1}{2}\begin{pmatrix} i\left(\frac{1}{\lambda_1}M_1 + 1\right) & \frac{1}{\lambda}(M_2 + iM_3) \\ \frac{1}{\lambda}(-M_2 + iM_3) & -i\left(\frac{1}{\lambda_1}M_1 + 1\right) \end{pmatrix},$$

$$\Omega_2 = \frac{1}{2}\begin{pmatrix} i\left(\frac{1}{\lambda_1}M_1 - 1\right) & \frac{1}{\lambda}(M_2 + iM_3) \\ \frac{1}{\lambda}(-M_2 + iM_3) & -i\left(\frac{1}{\lambda_1}M_1 - 1\right) \end{pmatrix}.$$

To find the solutions of the above system we proceed via the coordinates on G adapted to the symmetries of the extremal equations. Recall that along each extremal curve $(g(t), L(t))$, $g^{-1}(t)L(t)g(t)$ is constant. This means that both $g_1^{-1}(t)L_1(t)g_1(t)$ $g_2^{-1}(t)L_2(t)g_2(t)$ are constant, where

$$L_1 = \frac{1}{2}\begin{pmatrix} i(M_1 + p_1) & z_1 + w_1 \\ -z_2 - w_2 & -i(M_1 + p_1) \end{pmatrix}, L_2 = \frac{1}{2}\begin{pmatrix} i(M_1 - p_1) & z_1 - w_1 \\ -z_2 + w_2 & -i(M_1 - p_1) \end{pmatrix}.$$

Let

$$g_1^{-1}(t)L_1(t)g_1(t) = \Lambda_1, \text{ and let } g_2^{-1}(t)L_2(t)g_2(t) = \Lambda_2.$$

There is no loss of generality if $\Lambda_1 = K_1E_1$ and $\Lambda_2 = K_2E_1$, for some numbers K_1 and K_2, where

$$E_1 = \begin{pmatrix} i & 0 \\ 0 & -i \end{pmatrix}, E_2 = \begin{pmatrix} 0 & 1 \\ -1 & 0 \end{pmatrix}, E_3 = \begin{pmatrix} 0 & i \\ i & 0 \end{pmatrix},$$

since every solution can be conjugated to such a configuration.

The above choice of coadjoint orbits calls for an adapted system of coordinates $\varphi_1, \varphi_2, \varphi_3$ and ψ_1, ψ_2, ψ_3 according to the formulas

$$g_1 = \exp\left(\phi_1\frac{1}{2}E_1\right)\exp\left(\phi_2\frac{1}{2}E_2\right)\exp\left(\phi_3\frac{1}{2}E_1\right),$$

$$g_2 = \exp\left(\psi_1\frac{1}{2}E_1\right)\exp\left(\psi_2\frac{1}{2}E_2\right)\exp\left(\psi_3\frac{1}{2}E_1\right),$$

Then,

$$L_1(t) = K_1 g_1^{-1}(t)E_1 g_1(t), \quad L_2(t) = K_2 g_2^{-1}(t)E_1 g_2(t)$$

implies that

$$L_1 = K_1 e^{-\frac{1}{2}E_1\varphi_3}e^{-\frac{1}{2}E_2\varphi_2}E_1 e^{\frac{1}{2}E_2\varphi_2}e^{\frac{1}{2}E_1\varphi_3},$$

$$L_2 = K_2 e^{-\frac{1}{2}E_1\psi_3}e^{-\frac{1}{2}E_2\psi_2}E_1 e^{\frac{1}{2}E_2\psi_2}e^{\frac{1}{2}E_1\psi_3}.$$

It follows that

$$L_1(t) = iK_1 \begin{pmatrix} \cos\varphi_2 & e^{-i\varphi_3}\sin\varphi_2 \\ e^{i\varphi_3}\sin\varphi_2 & -\cos\varphi_2 \end{pmatrix},$$

$$L_2(t) = iK_2 \begin{pmatrix} \cos\psi_2 & e^{-i\psi_3}\sin\psi_2 \\ e^{i\psi_3}\sin\psi_2 & -\cos\psi_2 \end{pmatrix},$$

and therefore,

$$M_1 + p_1 = K_1\cos\varphi_2, M_1 - p_1 = K_2\cos\psi_2.$$

Furthermore,

$$z_1 + w_1 = iK_1 e^{-i\varphi_3}\sin\varphi_2 \,, z_2 + w_2 = -iK_1 e^{i\varphi_3}\sin\varphi_2 \,,$$
$$z_1 - w_1 = iK_2 e^{-i\psi_3}\sin\psi_2 \,, z_2 - w_2 = -iK_2 e^{i\psi_3}\sin\psi_2 \,. \tag{14.40}$$

Hence,

$$\sin\phi_2 = K_1^2(z_1 + w_1)(z_2 + w_2), \cos\phi_2 = \frac{1}{K_1}(M_1 + p_1),$$

$$\sin\psi_2 = K_1^2(z_1 - w_1)(z_2 - w_2), \cos\psi_2 = \frac{1}{K_2}(M_1 - p_1),$$

$$e^{2i\phi_3} = -\frac{z_2 + w_2}{z_1 + w + 1}, e^{2i\psi_3} = -\frac{z_2 - w_2}{z_1 - w_1}.$$

The remaining variables ϕ_1 and ψ_1 are obtained as follows:

$$g_1^{-1}\frac{dg_1}{dt} = \frac{1}{2}\left(\dot\varphi_1 g_1^{-1}E_1 g_1 + \dot\varphi_2\left(e^{-\frac{1}{2}E_1\varphi_3}E_2 e^{\frac{1}{2}E_1\varphi_3}\right) + \dot\varphi_3 E_1\right).$$

But $\frac{1}{2}g^{-1}(t)E_1 g_1(t) = \frac{1}{K_1}L_1(t)$, and $e^{-\frac{1}{2}E_1\varphi_3}\frac{1}{2}E_2 e^{\frac{1}{2}E_1\varphi_3} = \frac{1}{2}\begin{pmatrix} 0 & e^{i\varphi_3} \\ -e^{-i\varphi_3} & 0 \end{pmatrix}.$

Hence, $g_1^{-1}\frac{dg_1}{dt} = d\mathcal{H}_1$, with $d\mathcal{H}_1 = \frac{1}{2}\begin{pmatrix} i\left(\dfrac{M_1}{\lambda_1}+1\right) & \dfrac{z_1}{\lambda} \\ -\dfrac{z_2}{\lambda} & -i\left(\dfrac{M_1}{\lambda_1}+1\right) \end{pmatrix}$

leads to

$$\frac{\dot\varphi_1}{2}\begin{pmatrix} \cos\varphi_2 & e^{-i\varphi_3}\sin\varphi_2 \\ e^{i\varphi_3}\sin\varphi_2 & -\cos\varphi_2 \end{pmatrix} + \frac{1}{2}\dot\varphi_2\begin{pmatrix} 0 & e^{-i\varphi_3} \\ -e^{i\varphi_3} & 0 \end{pmatrix} + \frac{\dot\varphi_3}{2}\begin{pmatrix} 1 & 0 \\ 0 & -1 \end{pmatrix} = d\mathcal{H}_1,$$

which in turn yields

$$\dot\varphi_1 e^{-i\varphi_3}\sin\varphi_2 + \dot\varphi_2 e^{-i\varphi_3} = \frac{1}{\lambda}z_1, \dot\varphi_1 e^{i\varphi_3}\sin\varphi_2 - \dot\varphi_2 e^{i\varphi_3} = -\frac{1}{\lambda}z_2.$$

Therefore,

$$\dot{\varphi}_1 \sin \varphi_2 = \frac{1}{\lambda} \left(e^{i\varphi_3} z_1 - e^{-i\varphi_3} z_2 \right),$$

or

$$\dot{\varphi}_1 = \frac{1}{\lambda \sin \varphi_2} \left(\frac{z_1}{e^{-i\varphi_3}} - \frac{z_2}{e^{i\varphi_3}} \right).$$

It follows that

$$\dot{\varphi}_1 = i\frac{K_1}{2\lambda} \left(\frac{z_1}{z_1 + w_1} + \frac{z_2}{z_2 + w_2} \right) = i\frac{K_1}{2\lambda} \left(\frac{2z_1 z_2 + z_1 w_2 - z_2 w_1}{(z_1 + w_1)(z_2 + w_2)} \right).$$

The constants of motion $I_1 = \frac{z_1 z_2}{2\lambda} + p_1$, $I_2 = p_1 M_1 + \frac{1}{2}(z_1 w_1 + z_1 w_2)$, and $I_3 = p_1^2 + z_1 z_2 + z_1 w_2 + z_2 w_1$ imply that

$$\dot{\varphi}_1 = i\frac{K_1}{2\lambda} \left(\frac{4\lambda(I_1 - p_1) + 2(I_2 - p_1 M_1)}{2\lambda(I_2 - p_1 M_1) + I_3 - p_1^2} \right).$$

The right-hand side of the above equation is an even function (since p_1 is even), hence residue free. Therefore, $\varphi_1(t)$ is single valued. An analogous argument applied to $g_2^{-1}(t)\frac{dg_2}{dt} = d\mathcal{H}_2$ yields that $\psi_1(t)$ is also single valued. This completes our integration procedure.

15

Kirchhoff–Kowalewski equation

We will now investigate the Hamiltonian equations corresponding to the remaining case $\lambda_1 = \lambda_2 = 2\lambda_3, b_3 = 0$, under the assumption that the remaining coordinates b_1, b_2 of B are either both real or both imaginary. In this situation the Hamiltonian will be written as

$$\mathcal{H} = \frac{1}{2\lambda}(M_1^2 + M_2^2) + \frac{1}{\lambda}M_3^2 + b_1 p_1 + b_2 p_2,$$

with λ equal to the common value $\lambda_1 = \lambda_2$. We shall refer to it as the Kirchhoff–Kowalewski Hamiltonian. Then

$$\frac{dg}{dt} = g(t)(d\mathcal{H}), \quad \frac{dM}{dt} = [\Omega, M] + [B, P], \quad \frac{dP}{dt} = [\Omega, P] + s[B, M], s = 0, 1$$

(15.1a)

are the associated equations where

$$M = \frac{1}{2}\begin{pmatrix} iM_3 & M_1 + iM_2 \\ -M_1 + iM_2 & -iM_3 \end{pmatrix}, P = \frac{1}{2}\begin{pmatrix} ip_3 & p_1 + ip_2 \\ -p_1 + ip_2 & -ip_3 \end{pmatrix},$$

$$\Omega = \frac{1}{2}\begin{pmatrix} \frac{2i}{\lambda_1}M_3 & \frac{1}{\lambda}(M_2 + iM_3) \\ \frac{1}{\lambda}(-M_2 + iM_3) & -\frac{2}{i}\lambda M_1 \end{pmatrix}, B = \frac{1}{2}\begin{pmatrix} 0 & b_1 + ib_2 \\ -b_1 + ib_2 & 0 \end{pmatrix}.$$

Alternatively, these equations can be written in terms of the vector product as

$$\frac{dg}{dt} = g(t)(d\mathcal{H}), \quad \frac{d\hat{M}}{dt} = \hat{M} \times \hat{\Omega} + \hat{P} \times \hat{B}, \quad \frac{d\hat{P}}{dt} = \hat{P} \times \hat{\Omega} + s\hat{M} \times \hat{B}. \quad (15.1b)$$

Let us now pass to the variables used by Kowalewski in her original paper of 1889

$$z_1 = \frac{1}{2}(M_1 + iM_2), z_2 = \frac{1}{2}(M_1 - iM_2), w_1 = p_1 + ip_2, w_2 = p_1 - ip_2,$$

296

The reader should keep in mind that all the quantities are complex, and therefore z_2 and w_2 are not the complex conjugates of z_1 and w_1. In terms of these variables equations (15.1a) and (15.1b) reduce to

$$\dot{z}_1 = -\frac{1}{\lambda}z_1z_3 + \frac{1}{2}bw_3, \ \dot{z}_2 = \frac{1}{\lambda}z_2z_3 - \frac{1}{2}\bar{b}w_3, \ \dot{z}_3 = -\frac{1}{2}\bar{b}w_1 + \frac{1}{2}bw_2,$$

$$\dot{w}_1 = \frac{2}{\lambda}z_1w_3 - \frac{2}{\lambda}z_3w_1 + sbz_3, \ \dot{w}_2 = -\frac{2}{\lambda}z_2w_3 + \frac{2}{\lambda}z_3w_2 - s\bar{b}z_3, \quad (15.2)$$

$$\dot{w}_3 = \frac{1}{2\lambda}(z_1w_2 - z_2w_1) - s(\bar{b}z_1 - bz_2),$$

where $z_3 = iM_3, w_3 = ip_3$ and $\bar{b} = b_1 - ib_2$.

To simplify further calculations we will assume that $b_1 + ib_2$ is rotated by a "real" rotation so that $b_2 = 0$, in which case b_1 will be denoted by b. The reader should keep in mind that b is either real or imaginary; the imaginary case corresponding to the hyperbolic complexification. We will now rescale the preceding equations by introducing a rescaled variable $s = \frac{1}{\lambda_3}t = \frac{2}{\lambda}t$ and a new constant $a = \frac{1}{2}\lambda b$. Then equations (15.2) become

$$\frac{dz_1}{ds} = -\frac{1}{2}(z_1z_3 - aw_3), \frac{dw_1}{ds} = z_1w_3 - z_3w_1 + saz_3,$$

$$\frac{dz_2}{ds} = \frac{1}{2}(z_2z_3 - aw_3), \frac{dw_2}{ds} = -z_2w_3 + z_3w_2 - saz_3, \quad (15.3)$$

$$\frac{dz_3}{ds} = \frac{a}{2}(w_2 - w_1), \frac{dw_3}{ds} = \frac{1}{2}(z_1w_2 - z_2w_1) - \frac{1}{2}sa(z_1 - z_2).$$

These equations correspond to the modified Hamiltonian

$$\mathcal{H} = z_1z_2 - \frac{1}{2}z_3^2 + \frac{a}{2}(w_1 + w_2).$$

Then,

$$\frac{d}{ds}(z_1^2 - aw_1 + sa^2) = 2z_1\frac{dz_1}{ds} - a\frac{dw_1}{ds}$$

$$= 2z_1\frac{1}{2}(-z_1z_3 + aw_3) - a(z_1w_3 - z_3w_1 + sa)$$

$$= -2z_3(z_1^2 - aw_1 + sa^2).$$

Similarly, $\frac{d}{ds}(z_2^2 - aw_2 + sa^2) = 2z_3(z_2^2 - aw_2 + sa^2)$. Hence,

$$I_4 = (z_1^2 - aw_1 + sa^2)(z_2^2 - aw_2 + sa^2) \quad (15.4)$$

is an integral of motion for system (15.3).

Let us now investigate the structure of the algebraic variety $V(I_1, I_2, I_3, I_4)$ defined by the constants

$$I_1 = 2\mathcal{H} = 2z_1z_2 + M_3^2 + a(w_1 + w_2), I_2 = \sum_{i=1}^{3} p_i^2 + sM_i^2,$$

$$I_3 = \sum_{i=1}^{3} p_iM_i, \text{ and } I_4 = (z_1^2 - aw_1 + sa^2)(z_2 - aw_2 + sa^2).$$

These equations simplify if we replace w_1 and w_2 by two new variables

$$q_1 = z_1^2 - aw_1 + sa^2, \quad q_2 = z_2^2 - aw_2 + sa^2. \tag{15.5}$$

Then, $I_1 = 2z_1z_2 + M_3^2 + a(w_1 + w_2) = (z_1 + z_2)^2 - (q_1 + q_2) + M_3^2 + 2sa^2$. This equation can be written as

$$(z_1 + z_2)^2 - (q_1 + q_2) + M_3^2 = c_1, \tag{15.6}$$

where $c_1 = I_1 - 2sa^2 = 2\mathcal{H} - 2sa^2$. The remaining constants are also modified according to the following calculations:

$$\begin{aligned}
I_2 &= w_1w_2 + p_3^2 + s(4z_1z_2 + M_3^2) \\
&= w_1w_2 + 2sz_1z_2 + s(I_1 - a(w_1 + w_2)) + p_3^2 \\
&= (w_1 - sa)(w_2 - sa) - sa^2 + 2sz_1z_2 + p_3^2 + sI_1 \\
&= \frac{1}{a^2}(z_1^2 - q_1)(z_2^2 - q_2) + 2sz_1z_2 + p_3^2 + sI_1 - sa^2.
\end{aligned}$$

The above can then be written as

$$z_1^2z_2^2 - q_1z_2^2 - q_2z_1^2 - 2sa^2z_1z_2 + a^2p_3^2 = a^2(I_2 - s\lambda I_1 + sa^2) - I_4 = c_2, \tag{15.7}$$

and finally $aI_3 = az_1w_2 + az_2w_1 + ap_3M_3$ becomes

$$(z_1z_2 + sa^2)(z_1 + z_2) - (q_1z_2 + q_2z_1) + ap_3M_3 = c_3 = aI_3. \tag{15.8}$$

Let $V(c_1, c_2, c_3, c_4)$ denote the algebraic variety defined by equations (15.6), (15.7), (15.8), and $c_4^2 = I_4 = q_1q_2$. We will presently show that each choice of these constants defines an elliptic curve

$$\mathcal{C} = \{(z, u) \in \mathbb{C} \times \mathbb{C} : u^2 = P(z)\}, P(z) = -z^4 + 2\mathcal{H}z^2 - 2c_3z + c_2 \tag{15.9}$$

that is crucial for the subsequent investigations of our Hamiltonian system (equations (15.3)). The precise details are introduced through the following:

Proposition 15.1 *On* $V(c_1, c_2, c_3, c_4)$, *the variables* $\zeta_1 = M_3 z_1 - ap_3, \zeta_2 = M_3 z_2 - ap_3$ *satisfy*

$$\zeta_1^2 = P(z_1) + q_1(z_1 - z_2)^2, \ \zeta_2^2 = P(z_2) + q_2(z_1 - z_2)^2. \tag{15.10}$$

Proof

$$
\begin{aligned}
\zeta_1^2 &= M_3^2 z_1^2 - 2aM_3 p_3 z_1 + a^2 p_3^2 = (c_1 + (q_1 + q_2) - (z_1 + z_2)^2) z_1^2) \\
&\quad - 2(c_3 - (z_1 z_2 + sa^2)(z_1 + z_2) + z_1 q_2 + z_2 q_1) z_1 \\
&\quad + c_2 - z_1^2 z_2^2 - 2sa^2 z_1 z_2 + z_1^2 q_2 + z_2^2 q_1 \\
&= -z_1^4 + (c_1 + 2sa^2) z_1^2 - 2c_3 z_1 + c_2 + q_1(z_1 - z_2)^2 \\
&= P(z_1) + q_1(z_1 - z_2)^2,
\end{aligned}
$$

because

$$(q_1 + q_2) z_1^2 - 2(z_1 q_2 + z_2 q_1) z_1 + z_1^2 q_2 + z_2^2 q_2 = q_1(z_1 - z_2)^2,$$

and

$$-(z_1 + z_2)^2 z_1^2 + 2(z_1 z_2 + sa^2)(z_1 + z_2) z_1 - z_1^2 z_2^2 - 2sa^2 z_1 z_2 = -z_1^4 + 2sa^2 z_1^2.$$

A similar calculation shows that $\zeta_2^2 = P(z_2) + q_2(z_1 - z_2)^2$. $\qquad\square$

Corollary 15.2 *Each point* $(z_1, z_2, z_3, w_1, w_2, w_3)$ *in* $V(c_1, c_2, c_3, c_4)$ *is mapped into the product* $C \times C$ *via the formulas*

$$u_1^2 = (M_3 z_1 - p_3)^2 - q_1(z_1 - z_2)^2, \ u_2^2 = (M_3 z_2 - p_3)^2 - q_2(z_1 - z_2)^2. \tag{15.11}$$

Proof Indeed, $u_1^2 = \zeta_1^2 - q_1(z_1 - z_2)^2 = P(z_1)$ and $u_2^2 = \zeta_2^2 - q_2(z_1 - z_2)^2 = P(z_2)$. $\qquad\square$

Thus every point of $V(c_1, c_2, c_3, c_4)$ corresponds to four points $(z_1, \pm u_1)$ and $(z_2, \pm u_2)$ on the elliptic curve $u^2 = P(z)$. The elliptic curve is related to the solutions of our Hamiltonian system according to the following corollary.

Corollary 15.3 *Solutions of Kirchhoff–Kowalewski system satisfy*

$$-4\left(\frac{dz_1}{ds}\right)^2 = P(z_1) + q_1(z_1 - z_2)^2, \ -4\left(\frac{dz_2}{ds}\right)^2 = P(z_2) + q_2(z_1 - z_2)^2. \tag{15.12}$$

Proof Equations $\frac{dz_i}{ds} = \pm\frac{1}{2}(-z_i z_3 + aw_3)$ are the same as $\frac{dz_i}{dt} = \pm i\frac{1}{2}(M_3 z_i + ap_3), i = 1, 2$, and therefore,

$$-4\frac{dz_i}{ds}^2 = \xi_i^2 = P(z_i) + q_i(z_1 - z_2)^2, i = 1, 2.$$

$\qquad\square$

On $I_4 = 0$ the solutions of (15.13) satisfy $\frac{dz}{ds} = \pm \frac{i}{2}\sqrt{P(z)}$. This equation is essentially the same as the equation

$$\frac{dz}{dt} = \sqrt{P(z)}. \tag{15.13}$$

Their solutions differ only by a reparametrization $s = \pm \frac{i}{2}t$.

Corollary 15.4 *Every solution of $\frac{dz}{ds} = \frac{i}{2}\sqrt{P(z)}$ is the projection of a solution of the Kirchhoff–Kowalewski equation.*

Proof Let $z(s)$ be a solution of the above system. Then choose points z_1, z_2, w_1, w_2 that satisfy $z_2^2 - aw_2 + sa^2 \neq 0$, $z_1 = z(0)$ and $-aw_1 + sa^2 = -z_1^2$ and consider the solution of the Kirchhoff–Kowalewski system that emanates from these points at $s = 0$. Then $q_1(0) = 0$ and $q_2(0) \neq 0$. Since $I_4 = q_1(0)q_2(0) = 0$, and I_4 is constant along the solutions, $q_1(s) = 0$ for all s. Therefore $z_1(t)$ is a solution of $-4(\frac{dz_1}{ds})^2 = P(z_1)$. Hence either $\frac{dz_1}{ds} = \frac{i}{2}P(z_1)$, or $\frac{dz_1}{ds} = -\frac{i}{2}P(z_1)$. In the first case $z_1(s) = z(s)$ and in the second case, $z_1(-s) = z(s)$. □

Differential equation (15.13) has a long history beginning with the work of count Fagnano in 1718 concerning the arc of a lemniscate and its doubling property. Fagnano's discovery led Euler to the investigations of the solutions of (15.14) with an arbitrary fourth degree polynomial. Euler succeeded in getting the general solution in 1753 and in the process discovered addition formulas for elliptic integrals [Sg2]. Since the solutions to Kirchhoff–Kowalewski include the solutions of Euler concerning the arcs of lemniscates and their generalizations, it is not surprising that our investigations must be closely related to the work of Euler. It is somewhat surprising, however, that the path to the solutions initiated by S. Kowalewski sheds new light on the work of Euler and provides an easy access to his solutions.

To explain this in more detail, we will need to introduce additional definitions.

Definition 15.5 A complex valued function $R(z_1, z_2)$ of two complex variables z_1 and z_2 is called a bi-quadratic form if it is symmetric, i.e., if it satisfies $R(z_1, z_2) = R(z_2, z_1)$, and if it can be written as $A(z_1)z_2^2 + B(z_1)z_2 + C(z_1)$ for some functions $A(z), B(z), C(z)$.

It follows that a bi-quadratic form satisfies

$$A(z_1)z_2^2 + B(z_1)z_2 + C(z_1) = A(z_2)z_1^2 + B(z_2)z_1 + C(z_2). \tag{15.14}$$

Therefore, $2A(z_1) = A''(z_2)z_1^2 + B''(z_2)z_1 + C''(z_2)$. This implies that each coefficient A, B, C is a second degree polynomial. Let

$$A = a_2 z^2 + a_1 z + a_0, B = b_2 z^2 + b_1 z + b_0, C = c_2 z^2 + c_1 z + c_0.$$

Then (15.15) implies that $a_1 = b_2, a_0 = c_2$ and $c_1 = b_0$. Hence,

$$R(z_1, z_2) = a_2 z_1^2 z_2^2 + a_1 z_1 z_2 (z_1 + z_2) + a_0 (z_1^2 + z_2^2) + b_1 z_1 z_2 + b_0 (z_1 + z_2) + c_0.$$

The substitution $z_1 z_2 = \frac{1}{2}(z_1^2 + z_2^2) - (z_1 - z_2)^2)$ shows that every bi-quadratic form $R(z_1, z_2)$ is of the form

$$
\begin{aligned}
R_\theta(z_1, z_2) = {} & A + 2B(z_1 + z_2) + 3C(z_1^2 + z_2^2) + 2D z_1 z_2 (z_1 + z_2) \\
& + E z_1^2 z_2^2 + \theta (z_1 - z_2)^2,
\end{aligned}
\tag{15.15}
$$

for some coefficients A, B, C, D, E, and θ.

Remark 15.6 This choice of notation is adopted from the earlier publications [JA; Wl] for easier comparisons.

Each bi-quadratic form (15.15) defines a unique polynomial

$$P(z) = A + 4Bz + 6Cz^2 + 4Dz^3 + Ez^4 \tag{15.16}$$

through the identification $P(z) = R_\theta(z, z)$. Conversely, to each polynomial P described by (15.16) there corresponds a bi-quadratic form

$$R(z_1, z_2) = A + 2B(z_1 + z_2) + 3C(z_1^2 + z_2^2) + 2D z_1 z_2 (z_1 + z_2) + E z_1^2 z_2^2, \tag{15.17}$$

such that $R(z, z) = P(z)$, but the same is true for the form R_θ for any coefficient θ. That is, R_θ is the most general bi-quadratic form that corresponds to a given fourth degree polynomial P.

Proposition 15.7 *Let P be a polynomial of degree four and let R be the associated bi-quadratic form (15.17). Then the equation*

$$R^2(z_1, z_2) + (z_1 - z_2)^2 \hat{R}(z_1, z_2) = P(z_1)P(z_2) \tag{15.18}$$

admits a unique solution

$$
\begin{aligned}
\hat{R} = {} & \hat{A} + 2\hat{B}(z_1 + z_2) + 3\hat{C}(z_1^2 + z_2^2) + 2\hat{D} z_1 z_2 (z_1 + z_2) \\
& + \hat{E} z_1^2 z_2^2 + \hat{\theta}(z_1 - z_2)^2 \ \text{with}
\end{aligned}
$$

$$
\begin{aligned}
& \hat{A} = -4B^2, \hat{B} = 2(AD - 3BC), \hat{C} = \frac{2}{3}(AE + 2BD - 9C^2), \\
& \hat{D} = 2(BE - 3CD), \hat{E} = -4D^2, \hat{\theta} = 9C^2 - 4BD - AE.
\end{aligned}
$$

Proof Let $F(z, w) = P(z)P(w) - R^2(z, w)$. Then, $\frac{\partial F}{\partial z} = P'(z)P(w) - 2R\frac{\partial R}{\partial z}$, which at $w = z$ is equal to zero because $P'(z) = 2\frac{\partial R}{\partial z}$. An analogous argument shows that $\frac{\partial F}{\partial w}$ is equal to zero at $w = z$. Consequently, $P(z)P(w) - R^2(z, w) = F(z, w) = (z - w)^2\hat{R}(z, w)$ for some bi-quadratic form $\hat{R}$. Then,

$$2R(z, w)\frac{\partial R}{\partial z}(z, w) + 2(z - w)\hat{R}(z, w) + (z - w)^2\frac{\partial \hat{R}}{\partial z}(z, w) = P'(z)P(w),$$

and

$$2\left(\frac{\partial R}{\partial w}\frac{\partial R}{\partial z} + R\frac{\partial^2 R}{\partial w \partial z}\right) - 2\hat{R} + 2(z - w)\frac{\partial \hat{R}}{\partial w} - 2(z - w)\frac{\partial \hat{R}}{\partial w} + (z - w)^2\frac{\partial^2 \hat{R}}{\partial w \partial z}$$
$$= P'(z)P'(w).$$

At $w = z$, $2\frac{\partial R}{\partial z} = 2\frac{\partial R}{\partial w} = P'(z)$ and $\frac{\partial^2 R}{\partial z \partial w} = 8Dz + 4Ez^2$. Hence the preceding equality reduces to

$$\frac{1}{2}(P'(z))^2 + 2P(z)(8Dz + 4Ez^2) - 2Q(z) = (P'(z))^2.$$

Therefore,

$$Q = P(z)(8Dz + 4Ez^2) - \frac{1}{4}(P'(z))^2.$$

A straightforward calculation yields

$$Q(z) = -4B^2 + (8AD - 24BC)z + (8BD + 4AE - 36C^2)z^2$$
$$+ (8BE - 24CD)z^3 - 4D^2z^4,$$

which in turn determines the coefficients of $\hat{R}$ except for θ. To calculate θ, let $\hat{R}_0 = \hat{A} + 2\hat{B}(z + w) + 3\hat{C}(z^2 + w^2) + 2\hat{D}zw(z + w) + \hat{E}z^2w^2$, where the $\hat{A}, \hat{B}, \hat{C}, \hat{D}, \hat{E}$ are the coefficients induced by Q. Then $\hat{R} = \hat{R}_0 + \theta(z - w)^2$. It follows that

$$\theta(z - w)^4 = P(z)P(w) - (R^2(z, w) + (z - w)^2\hat{R}_0).$$

So θ is the coefficient of the homogeneous term of degree four in $P(z)P(w) - (R^2(z, w) + (z - w)^2\hat{R}_0$, i.e.,

$$\theta(z - w)^4 = AE(w^4 + z^4) + 4BD(z^3w + zw^3) + 36C^2z^2w^2$$
$$- (2AEz^2w^2 + 9C^2(z^2 + w^2)^2 + 8BDzw(z + w)^2)$$
$$- 2(z - w)^2(AE + 2BD - 9C^2)(z + w)^2.$$

The above then yields $\theta = -(AE + 4BD - 9C^2)$.   □

Corollary 15.8 *Let $R_\theta(z_1, z_2) = R(z_1, z_2) - \theta(z_1 - z_2)^2$. Then $\hat{R}_\theta(z_1, z_2) = \hat{R}(z_1, z_2) + 2R(z_1, z_2)\theta - \theta^2(z_1 - z_2)^2$ is the unique bi-quadratic form that satisfies*

$$R_\theta^2(z_1, z) + (z_1 - z_2)^2 \hat{R}_\theta(z_1, z_2) = P(z_1)P(z_2). \tag{15.19}$$

Let us now return to the Kichhoff–Kowalewski system. Here, $P(z) = c_2 - 3c_3 z + 2\mathcal{H}z^2 - z^4$ and $R(z_1, z_2) = c_2 - c_3(z_1 + z_2) + \mathcal{H}(z_1^2 + z_2^2) - z_1^2 z_2^2$.

Proposition 15.9 *Let $\zeta_i = M_3 z_i - a p_3, i = 1, 2$ be as in Proposition 15.1. On $V(c_1, c_2, c_3, c_4)$*

$$\zeta_1 \zeta_2 = R(z_1, z_2) - (\mathcal{H} - sa^2)(z_1 - z_2)^2.$$

Proof $(\zeta_1 - \zeta_2)^2 = M_3^2(z_1 - z_2)^2 = \left(2\mathcal{H} - 2sa^2 - (z_1 + z_2)^2 + q_1 + q_2\right)(z_1 - z_2)^2$. Therefore,

$$\begin{aligned}
\zeta_1 \zeta_2 &= \frac{1}{2}\left(\zeta_1^2 + \zeta_2^2 - \left(2\mathcal{H} - 2sa^2 - (z_1 + z_2)^2 + q_1 + q_2\right)(z_1 - z_2)^2\right) \\
&= \frac{1}{2}(P(z_1) + P(z_2) + (q_1 + q_2)(z_1 - z_2)^2 \\
&\quad - \left(2\mathcal{H} - 2sa^2 - (z_1 + z_2)^2 + q_1 + q_2\right)(z_1 - z_2)^2) \\
&= \frac{1}{2}(P(z_1) + P(z_2) - \left(2(\mathcal{H} - sa^2) - (z_1 + z_2)^2\right)(z_1 - z_2)^2) \\
&= c_2 - c_3(z_1 + z_2) + \mathcal{H}(z_1^2 + z_2^2) - z_1^2 z_2^2 - (\mathcal{H} - sa^2)(z_1 - z_2)^2 \\
&= R(z_1, z_2) - (\mathcal{H} - sa^2)(z_1 - z_2)^2 .
\end{aligned}$$

$\square$

If we now let $\theta = \mathcal{H} - sa^2$, and use R_θ to denote the form $\zeta_1 \zeta_2$, then $R_\theta(z_1, z_2)$ satisfies equation (15.19). However,

$$\begin{aligned}
R_\theta^2 = \zeta_1^2 \zeta_2^2 &= (P(z_1) + q(z_1 - z_2)^2)(P(z_2) + q_2(z_1 - z_2)^2) \\
&= P(z_1)P(z_2) + (P(z_1)q_2 + P(z_2)q_1)(z_1 - z_2)^2 + q_1 q_2(z_1 - z_2)^4 \\
&= P(z_1)P(z_2) + (z_1 - z_2)^2(P(z_1)q_2 + P(z_2)q_1 + c_4^2(z_1 - z_2)^2),
\end{aligned}$$

and, equation (15.19) reduces to

$$\hat{R}_\theta(z_1, z_2) + P(z_1)q_2 + P(z_2)q_1 + c_4^2(z_1 - z_2)^2 = 0. \tag{15.20}$$

This equation plays a key role in the following proposition.

Proposition 15.10 *The Cartesian product $\mathcal{C} \times \mathcal{C}$ is a 4-fold cover of $V(c_1, c_2, c_3, c_4)$.*

Proof Proposition 15.1 and its corollary show that each point of $V(c_1, c_2, c_3, c_4)$ is mapped into $C \times C$ via the mapping (15.12) and that every point of $V(c_1, c_2, c_3, c_4)$ corresponds to four points $((z_1, \pm u_1), (z_2, \pm u_2))$.

To show the converse, let (u_1, z_1) and (u_2, z_2) denote any points in C. The substitution of $q_2 = \frac{1}{q_1} c_4^2$ into (15.20) yields

$$u_1^2 c_4^2 + u_2^2 q_1^2 + \left(\hat{R}_\theta + c_4^2 (z_1 - z_2)^2 \right) q_1 = 0 .$$

Hence,

$$q_1 = -\frac{\left(\hat{R}_\theta + c_4^2 (z_1 - z_2)^2 \right) + \sqrt{\left(\hat{R}_\theta + c_4^2 (z_1 - z_2)^2 \right)^2 - 4 u_1^2 u_2^2 c_4^2}}{2 u_2} . \quad (15.21)$$

Once q_1 and q_2 are determined, then the remaining variables p_3 and M_3 are obtained from

$$M_3 z_1 - p_3 = \zeta_1 = \sqrt{u_1^2 - q_1 (z_1 - z_2)^2} ,$$

and

$$M_3 z_2 - p_3 = \zeta_2 = \sqrt{u_2^2 - q_2 (z_1 - z_2)^2} .$$

□

The actual form of $\hat{R}_\theta$ is rather cumbersome,

$$\hat{R}_\theta = -c_3^2 + 2 c_2 (\mathcal{H} - sa^2)) + 2 c_3 sa^2 (z_1 + z_2) - 2 (c_2 - \mathcal{H} sa^2)(z_1^2 + z_2^2)$$
$$+ 2 c_3 z_1 z_2 (z_1 + z_2) - 2 (\mathcal{H} - sa^2) z_1^2 z_2^2 + (c_2 + 2 \mathcal{H} - (sa^2))(z_1 - z_2)^2,$$

but fortunately, it is only its property (15.19) that is essential for the calculations that follow.

15.1 Eulers' solutions and addition formulas of A. Weil

Let us now return to the general polynomial $P(z)$ and the fundamental property

$$R_\theta^2(z, w) + (z - w)^2 \hat{R}_\theta(z.w) = P(z) P(w), R_\theta(z, w) = R(z, w) - \theta (z - w)^2. \quad (15.22)$$

Then $\hat{R}_\theta$ can be written as $\hat{R}_\theta = a_\theta(z) w^2 + 2 b_\theta(z) w + c_\theta(z)$ for some quadratic polynomials $a_\theta, b_\theta, c_\theta$. If we write

$$\hat{R} = \hat{A} + 2 \hat{B}(z + w) + 3 \hat{C}(z^2 + w^2) + 2 \hat{D} z w (z + w) + \hat{E} z^2 w^2 - \hat{F}(z - w)^2,$$

then

$$\hat{R}_\theta = 2A\theta + \hat{A} + (4B\theta + 2\hat{B})(z + w) + (6C\theta + 3\hat{C})(z^2 + w^2)$$
$$+ (4D\theta + 2\hat{D})zw(z + w) + (2E\theta + \hat{E})z^2w^2 - (\theta^2 + F)(z - w)^2,$$

and hence,

$$a_\theta = (2E\theta + \hat{E})z^2 + (4D\theta + E\hat{D})z + (6C + 3\hat{C} - (\theta^2 + F)),$$
$$b_\theta = (2D\theta + \hat{D})z^2 + (\theta^2 + F)z + (2B\theta + \hat{B}),$$
$$c_\theta = (6C\theta + 3\hat{C} - (\theta^2 + F))z^2 + (4B\theta + 2\hat{B})z + (2A\theta + \hat{A}).$$

After the substitutions, from Proposition 15.7

$$a_\theta = (2E\theta - 4D^2)z^2 + (4D\theta + 4(BE - 3CD))z + AE - (\theta - 3C)^2,$$
$$b_\theta = (2D\theta + 2(BE - 3CD))z^2$$
$$+ (\theta^2 - 9C^2 + AE + 4BD)z + (2B\theta + 2AD - 3BC),$$
$$c_\theta = (AE - (\theta - 3C)^2)z^2 + (4B\theta + 4(AD - 3BC))z + (2A\theta - 4B^2).$$

Proposition 15.11 *Let G_θ denote the discriminiant $b_\theta^2 - a_\theta c_\theta$. Then $G_\theta(z) = p(\theta)P(z)$, where*

$$p(\theta) = 2\theta(\theta - 3C)^2 + 2\theta(4BD - AE) + 4B^2E + 4AD^2 - 24BCD. \quad (15.23)$$

Proof If $P(z_0) = 0$ then $R_\theta^2(z_0, w) + (z_0 - w)2\hat{R}_\theta(z_0, w) = 0$. If $\hat{R}_\theta(z_0, w) = 0$ for some w then $R(z_0, w) - \theta(z_0 - w)^2 = 0$. Since this expression is quadratic in w it has two roots: either $w = z_0$ is double root of $P(z)$, or $w \neq z_0$. In either case, w is a double root of $\hat{R}_\theta(z_0, w) = 0$ and hence $G_\theta(z_0) = 0$. This shows that $P(z)$ is a factor of G_θ. Since both G_θ and P are polynomials of the same degree, G_θ is a multiple of P, i.e., there exists a function $p(\theta)$ such that $G_\theta(z) = p(\theta)P(z)$. Upon equating the highest powers in $b_\theta^2 - a_\theta c_\theta = p(\theta)P$ we get

$$(2D\theta + 2(BE - 3CD))^2 - (2E\theta - 4D^2)(AE - (\theta - 3C)^2) = p(\theta)E,$$

which easily recovers (15.22). $\qquad\square$

Expression (15.23) is most naturally linked with the cubic curve

$$\eta^2 = 4\xi^3 - i\xi - j,$$
$$i = AE - 4BD + 3C^2, \, j = ACE + 2BCD - AD^2 - EB^2 - C^3. \quad (15.24)$$

For, if $\theta = 2(\xi + C)$ and $\eta = \frac{p}{4}$, then (15.23) becomes (15.24).

The quantities i and j are known as the invariants of a quadratic form

$$f(x, y) = Ay^4 + 4Bxy^3 + 6Cx^2y^2 + 4Dx^3y + Ex^4.$$

They enjoy the following fundamental property: if I and J denote the invariants of

$$g(s, t) = at^4 + 4bt^3 s + 6ct^2 s^2 + 4dts^3 + es^4,$$

where $g(s, t) = f(\alpha s + \beta t, \gamma s + \delta t)$, then $I = (\alpha\delta - \gamma\beta)^4 i$, and $J = (\alpha\delta - \gamma\beta)^6 j$ [Pc, p. 353].

We will now show that curve $\Gamma\{(\xi, \eta) : \eta^2 = 4\xi - i\xi - j\}$ is of fundamental importance not only for the solutions of the lemniscatic equation $\frac{dz}{dt} = \sqrt{P(z)}$ but also for the more general equation $\frac{dx}{\sqrt{P(z)}} \pm \frac{dw}{\sqrt{P(w)}} = 0$. It turns out that the solutions to these two problems pave the way to the solutions of Kirchhoff–Kowalewski system.

Let us first make a few passing remarks about the solutions of $\frac{dz}{ds} = \sqrt{P(z)}$. The next proposition is well known in the theory of elliptic functions (see [Sg2]).

Proposition 15.12 *Assume that P has four distinct roots z_0, z_1, z_2, z_3. Then, there exists a linear fractional transformation $z = \frac{\alpha s + \beta}{\gamma s + \delta}$ with $\Delta = \alpha\delta - \beta\gamma \neq 0$ such that $\int \frac{dz}{\sqrt{P(z)}} = \Delta \int \frac{ds}{\sqrt{4s^3 - g_2 s - g_3}}$, where $g_2 = \Delta^4 i$ and $g_3 = \Delta^6 j$.*

Therefore the solutions of $\frac{dz}{dt} = \sqrt{P(z(t))}$ can be expressed in terms of Weierstrass' $\wp$ function as $z(t) = \frac{\alpha\wp(\Delta t) + \beta}{\gamma\wp(\Delta t) + \delta}$ for a suitable linear fractional transformation $\frac{\alpha s + \beta}{\gamma s + \delta}$. In practice, however, this formula is not convenient, since the search for the right linear fractional transformation involves solving several polynomial equations of degree four. One can bypass the use of fractional transformations by making use of addition formulas for elliptic functions as shown in [Ap]. Let us now come back to the equation

$$\frac{dz}{\sqrt{P(z)}} \pm \frac{dw}{\sqrt{P(w)}} = 0$$

and to its remarkable solutions by Euler [Eu]. The following proposition is a paraphrase of Euler's solutions.

Proposition 15.13 *Let $R_\theta^2(z, w) + (z - w)^2 \hat{R}_\theta(z.w) = P(z)P(w)$,*

$$\hat{R}_\theta(z, w) = \hat{R}(z, w) + 2R(z_1, z_2)\theta - (z - w)^2\theta^2, \text{ and}$$

$$R_\theta(z, w) = R(z, w) - \theta(z - w)^2$$

be as in Corollary 15.8 of Proposition 15.7. Then $\hat{R}_\theta(z, w) = 0$ is a solution for either

$$\frac{dz}{\sqrt{P(z)}} + \frac{dw}{\sqrt{P(w)}} = 0, \ or \ \frac{dz}{\sqrt{P(z)}} - \frac{dw}{\sqrt{P(w)}} = 0.$$

Conversely, for every solution $w(z)$ of either of the above equations there exists a number θ such that $\hat{R}_\theta(w, z) = 0$.

Instead of proving Euler's theorem directly, we will take a slightly more general approach based on the ideas of A. Weil [Wl]. Weil used the results of Euler to show that there is a group structure on $\mathcal{C} \cup \Gamma$, where Γ the cubic curve $\eta^2 = 4\xi^3 - g_2\zeta - g_3$ defined by the covariant invariants $g_2 = AE - ABD + 3C^2$ and $g_3 = ACE + 2BCD - AD^2 - B^2E - C^3$ of the form associated with $P(z)$. While computationally there is very little difference between Euler and Weil, Weil's interpretation of the theorem of Euler demystifies Kowalewski's integration procedure, and hence is very relevant for some formulas down the road.

To get to Weil's group structure, let us first explain how Euler's theorem can be used to show that Γ acts on on $\mathcal{C}$. Note first that $u^2 = P(z)$ corresponds to two points (z, u) and $(z, -u)$ on $\mathcal{C}$. If $M = (z, u)$, then we will write $-M$ for the point $(z, -u)$. For each pair of points $\pm M = (z, \pm u)$ of $\mathcal{C}$, and each point $\Delta = (\xi, \eta)$ of Γ, the corresponding discriminant $G_\theta(z)$ with $\theta = 2(\xi + C)$ is given by $G_\theta(z) = p(\theta)P(z) = 4\eta^2u^2$.

The relation

$$a_\theta(z)w^2 + 2b_\theta(z)w + c_\theta(z) = 0$$

defines two complex numbers w_1, w_2 given by

$$w_{1/2} = -\frac{b_\theta(z) \pm \sqrt{G_\theta(z)}}{a_\theta(z)} = -\frac{b_\theta(z) \pm 2\eta u}{a_\theta(z)}.$$

Let

$$w_1 = -\frac{b_\theta(z) + 2\eta u}{a_\theta(z)}, \quad w_2 = -\frac{b_\theta(z) - 2\eta u}{a_\theta(z)}.$$

We will say that w_1 is induced by M, and that w_2 is induced by $-M$. Since

$$a_\theta(w_1)z^2 + 2b_\theta(w_1)z + c_\theta(w_1) = 0,$$

z is given either by $z = -\dfrac{b_\theta(w_1) - 2\eta\sqrt{P(w_1)}}{a_\theta(w_1)}$ or by $z = -\dfrac{b_\theta(w_1) + 2\eta\sqrt{P(w_1)}}{a_\theta(w_1)}$.

Let $N_1 = (w_1, u_1)$ denote the point of $\mathcal{C}$ for which $z = -\dfrac{b_\theta(w_1) - 2\eta u_1}{a_\theta(w_1)}$. Following A. Weil, we will write

$$\Delta + M = N_1.$$

In an analogous manner, equation $a_\theta(w_2)z^2 + 2b_\theta(w_2)z + c_\theta(w_2) = 0$ defines a point u_2 such that $N_2 = (w_2, u_2)$ is a point of $\mathcal{C}$, and $z = -\dfrac{b_\theta(w_2) - 2\eta u_2}{a_\theta(w_2)}$. In this situation we will write

$$N_2 = \Delta + (-M) = \Delta - M.$$

Suppose now that θ is fixed and that $\hat{R}_\theta(z, w) = 0$ can be expressed as s function $z(w)$ in some open neighborhood of a point (a, b). For this we need to assume that $G_\theta(z) \neq 0$ in the neighborhood. Then

$$\frac{1}{2}\frac{\partial \hat{R}_\theta}{\partial z} = za_\theta(w) + b_\theta(w) = \rho\sqrt{G_\theta(w)}, \quad \frac{1}{2}\frac{\partial \hat{R}_\theta}{\partial w} = wa_\theta(z) + b_\theta(z) = \sigma\sqrt{G_\theta(z)}.$$

On $\hat{R}_\theta(x, w) = 0$

$$\frac{dz}{dw} = -\frac{\partial \hat{R}_\theta}{\partial w}\bigg/\frac{\partial \hat{R}_\theta}{\partial z} = -\frac{2(a_\theta(z)w + 2b_\theta(z))}{2(a_\theta(w)z + 2b_\theta(w))}.$$

At (z, w_1),

$$\frac{dz}{dw} = -\frac{a_\theta(z)w_1 + b_\theta(z)}{a_\theta(w_1)z + b_\theta(w_1)} = \frac{u}{u_2},$$

and at (z, w_2),

$$\frac{dz}{dw} = -\frac{a_\theta(z)w_2 + b_\theta(z)}{a_\theta(w_2)z + b_\theta(w_2)} = -\frac{u}{u_2}.$$

In the first case $z(w)$ is a solution of $\dfrac{dz}{\sqrt{P(z)}} - \dfrac{dw}{\sqrt{P(w)}} = 0$, while in the second case $z(w)$ is a solution of $\dfrac{dz_1}{\sqrt{P(z)}} + \dfrac{dw}{\sqrt{P(w)}} = 0$. This proves the first part of Proposition 15.13.

To show the second part of the theorem, we need to show that any solution of either $\dfrac{dz}{\sqrt{P(z)}} \pm \dfrac{dw}{\sqrt{P(w)}} = 0$ can be realized as the zero set of $\hat{R}_\theta$ for some θ.

This part also lends itself to group theoretic investigations. For we have shown that for any $\Delta \in \Gamma$ and any $\pm M \in \mathcal{C}$, $\Delta \pm M$ are points in $\mathcal{C}$. This implies that the sum $M + N$ and the difference $N - M$ are in Γ for any two points M and N of $\mathcal{C}$ since Γ is a group. The points $\Delta = (\xi, \eta)$ on Γ that correspond to $M + N$ and $N - M$ are determined by the appropriate values of θ such that $\hat{R}_\theta(z_1, z_2) = 0$. Then $\hat{R}_\theta = \hat{R} + 2R\theta - \theta^2(z_1 - z_2)^2 = 0$ gives

$$\theta = \frac{R(z_1, z_2) \pm \sqrt{R^2(z_1, z_2) + (z_1 - z_2)^2 \hat{R}(z_1, z_2)}}{(z_1 - z_2)^2},$$

and $R^2(z_1, z_2) + (z_1 - z_2)^2 \hat{R}(z_1, z_2) = P(z_1)P(z_2)$ yields

$$\theta = \frac{R(z_1, z_2) \pm u_1 u_2}{(z_1 - z_2)^2}. \tag{15.25}$$

Each choice of θ determines complex numbers $\xi = 2(\theta + C)$, which in turn determines the appropriate point on Γ. The value of θ that corresponds to $\Delta = M + N$ must tend to a finite limit when N approaches M, while the value of θ that corresponds to $\Delta = N - M$ tends to infinity, since Δ tends to the group identity on Γ.

It is easy to see that the correct choice of θ is given $\theta = \dfrac{R(z_1, z_2) - u_1 u_2}{(z_1 - z_2)^2}$.

Indeed,

$$
\begin{aligned}
\frac{R(z_1, z_2) - u_1 u_2}{(z_1 - z_2)^2} &= \frac{(R(z_1, z_2) - u_1 u_2)(R(z_1, z_2) + u_1 u_2)}{(R(z_1, z_2) + u_1 u_2)(z_1 - z_2)^2} \\
&= \frac{R^2(z_1, z_2) - u_1^2 u_2^2}{(z_1 - z_2)^2 (R(z_1, z_2) + u_1 u_2)} \\
&= -\frac{(z_1 - z_2)^2 \hat{R}(z_1, z_2)}{(z_1 - z_2)^2 (R(z_1, z_2) + u_1 u_2)} \\
&= -\frac{\hat{R}(z_1, z_2)}{R(z_1, z_2) + u_1 u_2}.
\end{aligned}
$$

As z_2 tends to z_1, the preceding expression tends to $-\dfrac{\hat{P}(z_1)}{P(z_1) + u_1^2} = \dfrac{\hat{P}(z_1)}{2P(z_1)}$, where $\hat{P}$ denotes the polynomial defined by $\hat{R}(z, z)$.

Thus, $\theta_1 = \dfrac{R(z_1, z_2) - u_1 u_2}{(z_1 - z_2)^2}$ determines $\xi_1 = 2(\theta_1 + C)$, while $\theta_2 = \dfrac{R(z_1 z_2) + u_1 u_2}{(z_1 - z_2)^2}$ determines $\xi_2 = 2(\theta_2 + C)$. Then,

$$
M + N = (\xi_1, \eta_1) = \Delta_1, \text{ and } N - M = (\xi_2, \eta_2) = \Delta_2.
$$

The values of η_1 and η_2 are uniquely determined through the relations $\Delta_1 - M = N$ and $\Delta_2 + M = N$.

The transformations that appear in Kowalewski's paper correspond to the infinitesimal version of Weil's addition formulas stated precisely in the following proposition.

Proposition 15.14 *Let $M = (z_1, u_1)$ and $N = (z_2, u_2)$ denote arbitrary points of $\mathcal{C}$ and let $\Delta_1 = (\xi_1, \eta_1)$ and $\Delta_2 = (\xi_2, \eta_2)$ denote the points of Γ such that $-M + N = \Delta_1$ and $M + N = \Delta_2$ then,*

$$
\frac{d\xi_1}{\eta_1} = -\frac{dz_1}{u_1} + \frac{dz_2}{u_2}, \qquad \text{and} \qquad \frac{d\xi_2}{\eta_2} = \frac{dz_1}{u_1} + \frac{dz_2}{u_2}. \tag{15.26}
$$

Proof Points $N + M$ lie on $\hat{R}_\theta(z_1, z_2) = 0$ for $\theta = \frac{R(z_1, z_2) - u_1 u_2}{(z_1 - z_2)^2}$. For this choice of θ,

$$z_1 = -\frac{b_\theta(z_2) - 2\eta_2 u_2}{a_\theta(z_2)}, \; z_2 = -\frac{b_\theta(z_1) - 2\eta_2 u_1}{a_\theta(z_1)} \text{ and } \theta = 2(\xi_2 + C).$$

On $\hat{R}_\theta(z_1, z_2) = 0$, $d\hat{R}_\theta = (-2(z_1 - z_2)^2\theta + 2R)d\theta + \frac{\partial \hat{R}_\theta}{\partial z_1}dz_1 + \frac{\partial \hat{R}_\theta}{\partial z_2}dz_2 = 0$. Then.

$$\frac{1}{2}\frac{\partial \hat{R}_\theta}{\partial z_1} = z_1 a_\theta(z_2) + b_\theta(z_2) = -2\eta_2 u_2, \frac{1}{2}\frac{\partial \hat{R}_\theta}{\partial z_2} = z_2 a_\theta(z_1) + b_\theta(z_1) = -2\eta_2 u_1,$$

$$\frac{\partial \hat{R}_\theta}{\partial \theta} = (-2\theta(z_1 - z_2)^2 + 2R)d\theta = 4u_1 u_2 d\xi_2.$$

It follows that $4u_1 u_2 d\xi_2 - 4\eta_2 u_2 dz_1 - 4\eta_2 u_1 dz_2 = 0$, or $\frac{d\xi_2}{\eta_2} = \frac{dz_1}{u_1} + \frac{dz_2}{u_2}$.

Points $N - M$ lie on $\hat{R}_\theta(z_1, z_2) = 0$ for $\theta = \frac{R(z_1, z_2) + u_1 u_2}{(z_1 - z_2)^2}$. For this θ,

$$z_1 = -\frac{b_\theta(z_2) - 2\eta_1 u_2}{a_\theta(z_2)}, \; z_2 = -\frac{b_\theta(z_1) - 2\eta_1 u_1}{a_\theta(z_1)} \text{ and } \theta = 2(\xi_1 + C),$$

$$\frac{1}{2}\frac{\partial \hat{R}_\theta}{\partial z_1} = z_1 a_\theta(z_2) + b_\theta(z_2) = -2\eta_1 u_2, \frac{1}{2}\frac{\partial \hat{R}_\theta}{\partial z_2} = z_2 a_\theta(z_1) + b_\theta(z_1) = 2\eta_2 u_1,$$

$$\frac{\partial \hat{R}_\theta}{\partial \theta} = (-2\theta(z_1 - z_2)^2 + 2R)d\theta = -4u_1 u_2 d\xi_2.$$

These expressions yield $\frac{d\xi_1}{\eta_1} = -\frac{dz_1}{u_1} + \frac{dz_2}{u_2}$.   □

With this background material at our disposal, we now take up the integration procedure.

15.2 The hyperelliptic curve

Let us return to equations (15.12)

$$-4\left(\frac{dz_1}{dt}\right)^2 = (M_3 z_1 - p_3)^2 = \zeta_1^2 = u_1^2 + q_1(z_1 - z_2)^2,$$

$$-4\left(\frac{dz_2}{dt}\right)^2 = \zeta_2^2 = u_2^2 + q_2(z_1 - z_2)^2,$$

with $u_i^2 = P(z_i) = c_2 - 2c_3 z_i + 2\mathcal{H}z_i^2 - z_i^4, i = 1, 2$.

Rather than integrating these equations on $\mathcal{C} \times \mathcal{C}$ we shall integrate them on $\Gamma \times \Gamma$ through the transformation

$$N - M = \Delta_1 \quad \text{and} \quad M + N = \Delta_2$$

and its infinitesimal analogue (15.26). Let $M = (z_1, u_1)$, $N = (z_2, u_2)$ and let $\Delta_1 = (\xi_1, \eta_1)$, $\Delta_2 = (\xi_2, \eta_2)$. Along the extremal curves $z_1(t)$, $z_2(t)$, equations (15.26) yields the following:

$$\left(\frac{d\xi_1}{dt}\right)^2 \frac{1}{\eta_1^2(t)} = \frac{1}{u_1^2(t)} \left(\frac{dz_1}{dt}\right)^2 + \frac{1}{u_2^2(t)} \left(\frac{dz_2}{dt}\right)^2 - \frac{2}{u_1(t)u_2(t)} \frac{dz_1}{dt} \frac{dz_2}{dt},$$
$$\left(\frac{d\xi_2}{dt}\right)^2 \frac{1}{\eta_2^2(t)} = \frac{1}{u_1^2(t)} \left(\frac{dz_1}{dt}\right)^2 + \frac{1}{u_2^2(t)} \left(\frac{dz_2}{dt}\right)^2 + \frac{2}{u_1(t)u_2(t)} \frac{dz_1}{dt} \frac{dz_2}{dt}.$$

(15.27)

Since

$$\frac{dz_1}{dt} = i\frac{1}{2} (M_3(t)z_1(t) - p_3(t)) = i\frac{1}{2}\zeta_1(t),$$
$$\frac{dz_2}{dt} = -i\frac{1}{2} (M_3(t)z_2(t) - p_3(t)) = -i\frac{1}{2}\zeta_2(t),$$
$$\left(\frac{dz_1}{dt}\right) \left(\frac{dz_2}{dt}\right) = \frac{1}{4}\zeta_1(t)\zeta_2(t) = \frac{1}{4}R_\theta\left(z_1(t), z_2(t)\right),$$

where $\theta = \mathcal{H} - sa^2$ and $R_\theta = R - (z_1 - z_2)^2\theta$.
Hence, equations (15.27) can be rewritten as

$$\frac{-4}{\eta_1^2}\left(\frac{d\xi_1}{dt}\right)^2 = 2 + (z_1(t) - z_2(t))^2 \left(\frac{q_1(t)}{u_1^2(t)} + \frac{q_2(t)}{u_2^2(t)}\right) - \frac{2R_\theta\left(z_1(t), z_2(t)\right)}{u_1(t)u_2(t)},$$
$$\frac{-4}{\eta_2^2}\left(\frac{d\xi_2}{dt}\right)^2 = 2 + (z_1(t) - z_2(t))^2 \left(\frac{q_1(t)}{u_1^2(t)} + \frac{q_2(t)}{u_2^2(t)}\right) + \frac{2R_\theta\left(z_1(t), z_2(t)\right)}{u_1(t)u_2(t)}.$$

Therefore,

$$-\frac{4}{\eta_i^2}\left(\frac{d\xi_i}{dt}\right)^2 = \frac{2u_1^2 u_2^2 + (z_1 - z_2)^2(u_1^2 q_2 + u_2^2 q_1) \mp 2R_\theta u_1 u_2}{u_1^2 u_2^2}, \quad i = 1, 2.$$

With the aid of the key equation (15.20): $u_1^2 q_2 + u_2^2 q_1 + \hat{R}_\theta + c_4^2(z_1 - z_2^2) = 0$, and $R_\theta^2 = u_1^2 u_2^2 - (z_1 - z_2)^2 \hat{R}_\theta$, the right-hand side of the above equation simplifies to

$$-\frac{4}{\eta_i^2}\left(\frac{d\xi_i}{dt}\right)^2 = \left(\frac{u_1 u_2 \mp R_\theta}{u_1 u_2}\right)^2 - \frac{(z_1 - z_2)^4 c_4^2}{u_1^2 u_2^2}.$$

Recall now that $\theta_i = 2(\xi_i + C) = 2(\xi_i + \frac{\mathcal{H}}{3})$, $i = 1, 2$, where

$$\theta_1 = \frac{R + u_1 u_2}{(z_1 - z_2)^2}, \quad \text{and} \quad \theta_2 = \frac{R - u_1 u_2}{(z_1 - z_2)^2}.$$

Then,

$$\left(\frac{R_\theta \mp u_1 u_2}{u_1 u_2}\right)^2 = \left(\frac{R - (\mathcal{H} - sa^2)(z_1 - z_2)^2 \mp u_1 u_2}{u_1 u_2}\right)^2$$

$$= \frac{1}{u_1^2 u_2^2}\left(2\left(\xi_i + \frac{\mathcal{H}}{3}\right)(z_1 - z_2)^2 - (\mathcal{H} - sa^2)(z_1 - z_2)^2\right)$$

$$= 4\frac{(z_1 - z_2)^4}{u_1^2 u_2^2}\left(\xi_i + \frac{\mathcal{H}}{6} - \frac{1}{2}sa^2\right).$$

It follows that

$$-\frac{4}{\eta_i^2}\left(\frac{d\xi_i}{dt}\right)^2 = 4\frac{(z_1 - z_2)^4}{u_1^2 u_2^2}\left(\left(\xi_i - \frac{\mathcal{H}}{6} + \frac{sa^2}{2}\right)^2 - \frac{c_4^2}{4}\right), \quad i = 1, 2.$$

If $k_1 = \dfrac{\mathcal{H}}{6} - \dfrac{sa^2}{2} + \dfrac{c_4}{2}$, and $k_2 = \dfrac{\mathcal{H}}{6} - \dfrac{sa^2}{2} - \dfrac{c_4}{2}$, then

$$-\left(\frac{d\xi_1}{dt}\right)^2 = \eta_1^2 \frac{(z_1 - z_2)^4}{u_1^2 u_2^2}(\xi_1 - k_1)(\xi_1 - k_2),$$

$$-\left(\frac{d\xi_2}{dt}\right)^2 = \eta_2^2 \frac{(z_1 - z_2)^4}{u_1^2 u_2^2}(\xi_2 - k_1)(\xi_2 - k_2).$$

Now $\xi_2 - \xi_1 = \dfrac{1}{2}(\theta_2 - \theta_1) = \dfrac{u_1 u_2}{(z_1 - z_2)^2}$, and therefore the above equations become

$$\left(\frac{d\xi_1}{dt}\right)^2 = \frac{-\eta_1^2}{(\xi_1 - \xi_2)^2}(\xi_1 - k_1)(\xi_1 - k_2),$$

$$\left(\frac{d\xi_2}{dt}\right)^2 = \frac{-\eta_2^2}{(\xi_1 - \xi_2)^2}(\xi_2 - k_1)(\xi_2 - k_2).$$

But $\eta_1^2 = 4\xi_1^3 - g_2\xi_1 - g_3$ and $\eta_2^2 = 4\xi_2^3 - g_2\xi_2 - g_3$. So

$$\left(\frac{d\xi_1}{dt}\right)^2 = \frac{U(\xi_1)}{(\xi_1 - \xi_2)^2} \quad \text{and} \quad \left(\frac{d\xi_2}{dt}\right)^2 = \frac{U(\xi_2)}{(\xi_1 - \xi_2)^2},$$

where $U(\xi) = -(4\xi^3 - g_2\xi - g_3)(\xi - k_1)(\xi - k_2)$.

It follows that $\left(\dfrac{d\xi_1}{d\xi_2}\right)^2 = \dfrac{U^2(\xi_1)}{U^2(\xi_2)}$, and therefore, $\dfrac{d\xi_1}{\sqrt{U(\xi_1)}} \pm \dfrac{d\xi_2}{\sqrt{U(\xi_2)}} = 0$.

On $\xi_1 \neq \xi_2$

$$\frac{d\xi_1}{\sqrt{U(\xi_1)}} + \frac{d\xi_2}{\sqrt{U(\xi_2)}} = 0 \tag{15.28}$$

is the correct equation. This equation coincides with the equation of Kowalewski reported in her celebrated paper of 1889 [Kw]. Equation (15.28) can also be obtained in an independent manner biased on the theory of discriminately separable polynomials initiated by V. Dragovic [Drg]. However, we will not go into these details.

The solutions of the last equation are given by $F(\xi_1) + F(\xi_2) = $ constant, where F is the hyperelliptic integral

$$F(z) = \int_{z_0}^{z} \frac{d\zeta}{\sqrt{U(\zeta)}} .$$

From a theoretical point of view, the above formula provides a complete solution to our Hamiltonian system, although the task of unraveling this answer back to the original variables remains a laborious exercise. In all essential details, this procedure is similar to the one used in the Kirchhoff–Lagrange case and will be omitted.

The Kirchhoff–Kowalewski system, however, stands apart from the other integrable systems, in the sense that is still shrouded in some mystery concerning the geometric origins of its integral of motion. For a while it seemed that the gyrostat in two constant fields held the key to Kowalewskai's mystery due to an outstanding contribution by I. A. Bobenko and his coauthors in 1989 [BR]. It turned out, however, that the Kowalewski's gyrostat can recover only the semi-direct version of the Kowalewski integral, a limitation that further compounds the mystery behind the original discovery of Kowalewski. It is partly for this reason, and partly for its own interest, that we will include the gyrostat in the section below.

15.3 Kowalewski gyrostat in two constant fields

The Hamiltonian that corresponds to the Kowalewski gyrostat in two constant fields contains an extra parameter γ and is of the form

$$\mathcal{H} = \frac{1}{2}(M_1^2 + M_2^2 + 2M_3^2 + 2M_3\gamma) - g_1 - h_2 .$$

This Hamiltonian, when considered as a left-invariant Hamiltonian on $sp_4(\mathbb{R})$, admits a Lax-pair representation $\frac{dL(\lambda)}{dt} = [M(\lambda), L(\lambda)]$ in which the spectral invariants of $L(\lambda)$ yield additional integrals of motion for $\mathcal{H}$ with the integral of motion discovered by S. Kowalewski a particular case when $\gamma = h_2 = 0$ [BR]. This remarkable discovery seemed to validate the belief that behind every integrable system there is an appropriate Lax-pair representation whose spectral invariants account for its integrability, and was offered as a natural explanation for "the peculiar geometry of the Kowalewski top and the origin of its integrability" [BR].

We will subsequently reproduce the results of Bobenko *et al.* by showing that the above integrals of motion, or, more precisely, their holomorphic extensions, are a consequence of certain symmetries inherited from $so_5(\mathbb{C})$. The exposition is based on a brilliant observation of A. Savu that the parameter γ, itself, is an integral of motion in $sp_4(\mathbb{R})$ [IvS].

To explain these symmetries in detail, we shall first obtain the appropriate Hamiltonian on $T^*SO_5(\mathbb{C})$. This Hamiltonian is also induced through the automorphism σ on $SO_5(\mathbb{C})$ given by $\sigma(g) = DgD^{-1}$, where D is a diagonal matrix with the diagonal entries $-1, -1, 1, 1, 1$.

Then the tangent map σ_* at the identity induces a splitting of $so_5(\mathbb{C})$ into the direct sum $\mathfrak{k} \oplus \mathfrak{p}$, where $\mathfrak{k}$ denotes the Lie subalgebra of fixed points of σ_*. Corresponding to this splitting, matrices M in $so_5(\mathbb{C})$ will be written in block form $M = \begin{pmatrix} A & -B^T \\ B & C \end{pmatrix}$ with A a 2×2 block, B a 3×2 block, and C a 3×3 block. The projections $M_{\mathfrak{k}}$ and $M_{\mathfrak{p}}$ of M on $\mathfrak{k}$ and $\mathfrak{p}$ are then given by $M_{\mathfrak{k}} = \begin{pmatrix} A & 0 \\ 0 & C \end{pmatrix}$ and $M_{\mathfrak{p}} = \begin{pmatrix} 0 & -B^T \\ B & 0 \end{pmatrix}$.

It follows that $\mathfrak{k} = \mathfrak{k}_1 \oplus \mathfrak{k}_2$ with $\mathfrak{k}_1$ isomorphic to $so_3(\mathbb{C})$ and $\mathfrak{k}_2$ isomorphic to $so_2(\mathbb{C})$, i.e., $\mathfrak{k}_1 = \{M = \begin{pmatrix} 0 & 0 \\ 0 & C \end{pmatrix}, C \in so_3(\mathbb{C})\}$, and $\mathfrak{k}_2 = \{M = \begin{pmatrix} A & 0 \\ 0 & 0 \end{pmatrix}, A \in so_2(\mathbb{C})\}$; hence, $\mathfrak{k}$ is a four-dimensional subalgebra of $so_5(\mathbb{C})$. The Cartan space $\mathfrak{p}$ is a six-dimensional vector subspace of $so_5(\mathbb{C})$ equal to the sum $V_1 \oplus V_2$ with V_1 and V_2 equal to the projections of B on the first, respectively, the second column of B. Evidently, V_1 and V_2 are three-dimensional complex vector subspaces of $so_5(\mathbb{C})$.

The usual Cartan conditions

$$[\mathfrak{p}, \mathfrak{p}] \subseteq \mathfrak{k}, [\mathfrak{p}, \mathfrak{k}] \subseteq \mathfrak{p}, [\mathfrak{k}, \mathfrak{k}] = \mathfrak{k}$$

break down to more refined conditions

$$[\mathfrak{k}_1, \mathfrak{k}_2] = 0, [\mathfrak{k}_1, V_1] \subseteq V_1, [\mathfrak{k}_1, V_2] \subseteq V_2, [\mathfrak{k}_2, V_1] \subseteq V_2, [\mathfrak{k}_2, V_2] \subseteq V_1,$$
$$[V_1, V_2] \subseteq \mathfrak{k}_2, [V_1, V_1] \subseteq \mathfrak{k}_1, [V_2, V_2] \subseteq \mathfrak{k}_1.$$

$$(15.29)$$

Then A_1, A_2, A_3, A_4 will denote a basis in $\mathfrak{k}$ where A_1, A_2, A_3 denote the standard basis in $so_3(\mathbb{C})$ embedded in $\mathfrak{k}_1$, while $A_4 = \begin{pmatrix} A & 0 \\ 0 & 0 \end{pmatrix}$, with

$A = \begin{pmatrix} 0 & -1 \\ 1 & 0 \end{pmatrix}$ is in $\mathfrak{k}_2$, and B_1, B_2, B_3 and C_1, C_2, C_3 will denote the standard basis for V_1 and V_2, i.e., the bases which coincide with the standard basis e_1, e_2, e_3 in $\mathbb{C}^3$ under the usual identification of each of V_1 and V_2 with $\mathbb{C}^3$.

Then the dual space $so_5^*(\mathbb{C})$ will be considered as the direct sum $\mathfrak{k}_1^* \oplus \mathfrak{k}_2^* \oplus V_1^* + V_2^*$, with $m_1, m_2, m_3, m_4, p_1, p_2, p_3, q_1, q_2, q_3$ the coordinates of a point ℓ in $so_5^*(\mathbb{C})$ relative to the dual basis $A_1^*, A_2^*, A_3^*, A_4^*, B_1^*, B_2^*, B_3^*, C_1^*, C_2^*, C_3^*$.

Finally, $\ell \in so_5^*(\mathbb{C})$ will be identified with $L \in so_5(\mathbb{C})$ via the correspondence $\ell(X) = -\frac{1}{2}Tr(LX)$, $X \in so_5(\mathbb{C})$. It then follows, by an easy calculation, that

$$L = \begin{pmatrix} 0 & -m_4 & -p_1 & -p_2 & -p_3 \\ m_4 & 0 & -q_1 & -q_2 & q_3 \\ p_1 & q_1 & 0 & -m_3 & m_2 \\ p_2 & q_2 & m_3 & 0 & -m_1 \\ p_3 & q_3 & -m_2 & m_1 & 0 \end{pmatrix}.$$

To make easier comparisons with the equations in the previous section, we will identify $so_5(\mathbb{C})$ with with the subalgebra $\mathfrak{g}$ of $sl_4(\mathbb{C})$ via the correspondence

$$\begin{pmatrix} 0 & -m_4 & -p_1 & -p_2 & -p_3 \\ m_4 & 0 & -q_1 & -q_2 & q_3 \\ p_1 & q_1 & 0 & -m_3 & m_2 \\ p_2 & q_2 & m_3 & 0 & -m_1 \\ p_3 & q_3 & -m_2 & m_1 & 0 \end{pmatrix} \rightarrow \begin{pmatrix} M + M_4 & P + iQ \\ P - iQ & M - M_4 \end{pmatrix},$$

where

$$M_4 = \frac{i}{2}\begin{pmatrix} m_4 & 0 \\ 0 & m_4 \end{pmatrix}, M = \frac{1}{2}\begin{pmatrix} im_3 & m_1 + im_2 \\ -m_1 + im_2 & -im_3 \end{pmatrix},$$

$$P = \frac{1}{2}\begin{pmatrix} ip_3 & p_1 + ip_2 \\ -p_1 + ip_2 & -ip_3 \end{pmatrix}, Q = \frac{1}{2}\begin{pmatrix} iq_3 & q_1 + iq_2 \\ -q_1 + iq_2 & -iq_3 \end{pmatrix}.$$

Then $\mathfrak{p}$ is identified with the vector space of matrices $\begin{pmatrix} 0 & P + iQ \\ P - iQ & 0 \end{pmatrix}$, and $\mathfrak{k}$ with the subalgebra of matrices $\begin{pmatrix} M + M_4 & 0 \\ 0 & M - M_4 \end{pmatrix}$. Relative to the decomposition $\mathfrak{p} = \mathfrak{p}_1 \oplus \mathfrak{p}_2$, $\mathfrak{p}_1$ is identified with the space of matrices $\begin{pmatrix} 0 & P \\ P & 0 \end{pmatrix}$, and $\mathfrak{p}_2$ with the space of matrices $\begin{pmatrix} 0 & iQ \\ -iQ & 0 \end{pmatrix}$.

In this representation, the integral group K of $\mathfrak{k}$, i.e., the group generated by $\{e^A, A \in \mathfrak{k}\}$, is identified with the matrices

$$\begin{pmatrix} gh & 0 \\ 0 & gh^{-1} \end{pmatrix}, g \in SL_2(\mathbb{C}), h = e^{i\alpha}, \alpha \in \mathbb{C}.$$

Therefore, K is isomorphic to $SL_2(\mathbb{C}) \times SO_2(\mathbb{C})$

As we have remarked before, K acts on $\mathfrak{p}$ by adjoint action and gives rise to the semi-direct product $G_s = \mathfrak{p} \rtimes K$, which, in turn, endows the matrices in $so_5(\mathbb{C})$ with the semi-direct product Lie algebra $\mathfrak{g}_s$.

As usual, $\langle L, X \rangle$ denotes the quadratic form $-Tr(LX)$ on $\mathfrak{g}$. Then $\langle L, X \rangle = M \cdot N + m_4 n_4 + P \cdot R + Q \cdot S$ for any matrix $X = \begin{pmatrix} N + \frac{in_4}{2}I & R + iS \\ R - iS & N - \frac{in_4}{2}I \end{pmatrix}$, where $X \cdot Y$ denotes the standard inner product $-2Tr(XY)$ in $sl_2(\mathbb{C})$. Hence, $\mathfrak{k}, \mathfrak{p}_1, \mathfrak{p}_2$ are mutually orthogonal relative to $\langle , \rangle$.

We will also rely on the conventions used earlier and identify matrices $X = \frac{1}{2}\begin{pmatrix} ix_3 & x_1 + ix_2 \\ -x_1 + ix_2 & -ix_3 \end{pmatrix}$ in $sl_2(\mathbb{C})$ with the coordinate vector $x = \begin{pmatrix} x_1 \\ x_2 \\ x_3 \end{pmatrix}$

relative to the basis of the skew-Hermitian Pauli matrices

$$A_1 = \frac{1}{2}\begin{pmatrix} 0 & 1 \\ -1 & 0 \end{pmatrix}, A_2 = \frac{1}{2}\begin{pmatrix} 0 & i \\ i & 0 \end{pmatrix}, A_3 = \frac{1}{2}\begin{pmatrix} i & 0 \\ 0 & -i \end{pmatrix}.$$

Recall that the Pauli matrices are orthonormal relative to the inner product $X \cdot Y = -2Tr(XY)$. Therefore, $X \cdot Y = x_1 y_1 + x_2 y_2 + x_3 y_3$ for any matrices X and Y in $sl_2(\mathbb{C})$. In particular, $X \cdot X = x_1^2 + x_2^2 + x_3^2$. Moreover, $X^2 = -\frac{1}{4}(X \cdot X)$ $I = -\frac{1}{4}(x_1^2 + x_2^2 + x_3^2)I$, for any matrix X. Hence, $Tr(X^2) = -\frac{1}{2}X \cdot X$. We will also make use of the correspondence between the cross product $z = y \times x$ and the Lie bracket $Z = [X, Y]$, noted earlier in the text.

With this background behind, let us now come to the Hamiltonian that leads to the gyrostat. The passage is slightly roundabout and begins with the Hamiltonian $\mathcal{H}$ given by

$$\mathcal{H} = \frac{1}{2}\langle L, \Lambda \rangle, \text{ with } \Lambda = \begin{pmatrix} \Omega + \alpha A_3 & B + iC \\ B - iC & \Omega + \alpha A_3 \end{pmatrix},$$

where α is a complex number, $\Omega = \frac{1}{2}\begin{pmatrix} \frac{i}{\lambda_3}m_3 & \frac{1}{\lambda}(m_1 + im_2) \\ -\frac{1}{\lambda}(m_1 - im_2) & -\frac{i}{\lambda_3}m_3 \end{pmatrix}$,

and where $\begin{pmatrix} 0 & B + iC \\ B - iC & 0 \end{pmatrix}$ is a fixed element of $\mathfrak{p}$. This Hamiltonian can be also written as

$$\mathcal{H} = \frac{1}{2}M \cdot \Omega + \alpha M_3 + B \cdot P + C \cdot Q. \tag{15.30}$$

The Hamiltonian equations on $\mathfrak{g} = so_5(\mathbb{C})$ are given by

$$\frac{dM}{dt} = [\Omega, M] + \alpha[A_3, M] + [B, P] + [C, Q],$$

$$\frac{dM_4}{dt} = i(QB + BQ) - i(CP + PC)),$$

$$\frac{dP}{dt} = [\Omega, P] + [B, M] + \alpha[A_3, P] + im_4 C,$$ (15.31)

$$\frac{dQ}{dt} = [\Omega, Q] + [C, M] + \alpha[A_3, Q] + m_4 B,$$

as a consequence of the Lie bracket structure described by (15.29). The corresponding equation on $\mathfrak{g}_s$ take a slightly different form:

$$\frac{dM}{dt} = [\Omega, M] + \alpha[A_3, M] + [B, P] + [C, Q],$$

$$\frac{dM_4}{dt} = i(QB + BQ) - i(CP + PC)),$$ (15.32)

$$\frac{dP}{dt} = [\Omega, P] + \alpha[A_3, P], \frac{dQ}{dt} = [\Omega, Q] + \alpha[A_3, Q],$$

The above equations can be also written in vector form as

$$\frac{dm}{dt} = m \times \omega + \alpha(m \times e_3) + p \times b + q \times c, \frac{dm_4}{dt} = -q \cdot b + p \cdot c,$$

$$\frac{dp}{dt} = p \times \omega + \alpha(p \times e_3) + s(m \times b - m_4 c),$$ (15.33)

$$\frac{dq}{dt} = q \times \omega + \alpha(q \times e_3) + s(m \times c + m_4 b),$$

where m, p, q, ω denote the coordinate vectors of M, P, Q, Ω relative to the Pauli bases. An easy calculation shows that $\dfrac{dm_3}{dt} = p_1 b_2 - p_2 b_1 + q_1 c_2 - q_2 c_1$, and therefore,

$$\frac{d}{dt}(m_3 + m_4) = p_1 b_2 - p_2 b_1 + q_1 c_2 - q_2 c_1$$

$$- q_1 b_1 - q_2 b_2 - q_3 b_3 + p_1 c_1 + p_2 c_2 + p_3 c_3$$

$$= p_1(b_2 + c_1) + p_2(c_2 - b_1) - q_1(b_1 - c_2)$$

$$- q_2(b_2 + c_1) + p_3 c_3 - q_3 b_3.$$

Hence, $\dfrac{d}{dt}(m_3(t) + m_4(t)) = 0$, whenever $c_1 + b_2 = 0$, $c_2 - b_1 = 0$, and $c_3 = b_3 = 0$. That is, $m_3(t) + m_4(t)$ is an integral of motion for both systems (15.31) and (15.32), whenever $b = \begin{pmatrix} b_1 \\ b_2 \\ 0 \end{pmatrix}$ and $c = \begin{pmatrix} -b_2 \\ b_1 \\ 0 \end{pmatrix}$.

Let

$$\gamma = m_3(t) + m_4(t), b = \begin{pmatrix} b_1 \\ b_2 \\ 0 \end{pmatrix}, c = \begin{pmatrix} -b_2 \\ b_1 \\ 0 \end{pmatrix} = b^\perp.$$

Then equations (15.33) reduce to

$$\frac{dp}{dt} = p \times \omega + \alpha(p \times e_3) + s(m \times b - (\gamma - m_3)b^\perp),$$
(15.34)
$$\frac{dq}{dt} = q \times \omega + \alpha(q \times e_3) + s(m \times b^\perp - (\gamma - m_3)b).$$

We will now assume that the Kowalewski conditions $\lambda = 2\lambda_3$ hold, which will be further normalized to $\lambda = 1$ for computational simplicity. Then we have the following proposition.

Proposition 15.15 *Suppose that* $\alpha = -\gamma$. *If* $M(t), M_4(t), P(t), Q(t)$ *is a solution of equation (15.31), then*

$$\bar{M}(t) = M(t), \bar{M}_4(t) = M_4(t), \bar{P}(t) = P(t) + B, \bar{Q}(t) = Q(t) + B^\perp$$

is a solution of (15.32).

Proof Let us first note that

$$b \times \omega + m \times b = b \times (\omega - m) = b \times (m_3 e_3) = -m_3 b^\perp,$$
$$b^\perp \times \omega + m \times b^\perp = b^\perp \times (\omega - m) = b^\perp \times (3_3 e_3) = m_3 b.$$

Therefore,

$$b \times \omega + m \times b + m_3 b^\perp = 0, \ b^\perp \times \omega + m \times b^\perp - m_3 b = 0,$$

or

$$[\Omega, B] + [B, M] + m_3 B^\perp = 0 \text{ and } [\Omega, B^\perp] + [B^\perp, M] - m_3 B = 0. \quad (15.35)$$

Then,

$$\frac{d\bar{P}}{dt} = \frac{dP}{dt} = [\Omega, \bar{P} + B] + [B, M] + \alpha[A_3, \bar{P} + B] + M_3 B^\perp - \gamma B^\perp$$
$$= [\Omega, \bar{P}] + \alpha[A_3, \bar{P}] - \gamma B^\perp + \alpha[A_3, B] = [\Omega, \bar{P}] + \alpha[A_3, \bar{P}] - (\alpha + \gamma)B^\perp,$$

$$\frac{d\bar{Q}}{dt} = [\Omega, \bar{Q} + B^\perp] + [B^\perp, M] + \alpha[A_3, \bar{Q} + B^\perp] - M_3 B + \gamma B$$
$$= [\Omega, \bar{Q}] + \alpha[A_3, \bar{Q}] + \alpha[A_3, B^\perp] + \gamma B = [\Omega, \bar{Q}] + \alpha[A_3, B^\perp] + (\alpha + \gamma)B.$$

When $\alpha + \gamma = 0$,

$$\frac{d\bar{P}}{dt} = [\Omega, \bar{P}] - \gamma[A_3, \bar{P}], \frac{d\bar{Q}}{dt} = [\Omega, \bar{Q}] - \gamma[A_3, \bar{Q}],$$

$$\frac{d\bar{M}}{dt} = [\Omega, \bar{M}] - \gamma[A_3, M] + [B, P] + [B^\perp, Q], \frac{d\bar{m}_4}{dt} = -(\bar{Q} \cdot B) + (\bar{P} \cdot B^\perp).$$

(15.36)

These equations coincide with the Hamiltonian equations of $\mathcal{H}$ on the semi-direct product $\mathfrak{g}_s$. □

Equations

$$\frac{dM}{dt} = [\Omega, M] - \gamma[A_3, M] + [B, P] + [C, Q],$$

$$\frac{dP}{dt} = [\Omega, P] - \gamma[A_3, P], \frac{dQ}{dt} = [\Omega, Q] - \gamma[A_3, Q],$$

(15.37)

are called *the Kowalewski gyrostat in two constant fields* [BR]. In vector form

$$\frac{dm}{dt} = m \times (\omega - \gamma e_3) + p \times b + q \times b^\perp,$$

$$\frac{dp}{dt} = p \times (\omega - \gamma e_3), \frac{dq}{dt} = q \times (\omega - \gamma e_3)$$

these equations formally resemble the Euler–Poisson equations of the heavy top in the presence of two constant vector fields p and q.

Equations (15.37) can be regarded as the projections of (15.32) on the semi-direct product $(\mathfrak{p}_1 \times \mathfrak{p}_2) \rtimes sl_2(\mathbb{C})$ with the Lie bracket induced by the diagonal action $g(P, Q) = (Ad_g(P), Ad_g(Q)), g \in SL_2(\mathbb{C}), P \in \mathfrak{p}_1, Q \in \mathfrak{p}_2$. They correspond to the Lie–Poisson equations on the dual of $(\mathfrak{p}_1 \times \mathfrak{p}_2) \rtimes sl_2(\mathbb{C})$ generated by the Hamiltonian

$$\mathcal{H} = \frac{1}{2}(m_1^2 + m_2^2 + 2m_3^2) + b_1\bar{p}_1 + b_2\bar{p}_2 - b_2\bar{q}_1 + b_1\bar{q}_2 - \gamma m_3.$$

We will presently show that the Lax-pair representation of the Kowalewski gyrostat yields certain integrals of motion which depend on γ that reduce to the Kowalewski integral (in the semi-direct case) for $\gamma = 0$. To show this, we first note that equations (15.38) yield three obvious integrals of motion:

$$I_2 = \bar{Q} \cdot \bar{Q} = \bar{q}_1^2 + \bar{q}_2^2 + \bar{q}_3^2,$$

$$I_3 = \bar{P} \cdot \bar{P} = \bar{p}_1^2 + \bar{p}_2^2 + \bar{3}_3^3,$$

$$I_4 = \bar{P} \cdot \bar{Q} = \bar{p}_1\bar{q}_1 + \bar{p}_2\bar{q}_2 + \bar{p}_3\bar{q}_3.$$

The first two integrals are Casimirs, while the last one is valid only for the Hamiltonians that do not depend explicitly on the variable m_4. On the coadjoint orbit $m_3 + m_4 = \gamma$,

$$L = \begin{pmatrix} M + (\gamma - m_3)I & \bar{P} + B + i(\bar{Q} + B^\perp) \\ \bar{P} + B - i(\bar{Q} + B^\perp) & M - (\gamma - m_3)I \end{pmatrix},$$

and

$$\begin{aligned} \langle L, L \rangle &= m_1^2 + m_2^2 + m_3^2 + (\gamma - m_3)^2 + 2(\bar{P} \cdot B) + 2(\bar{Q} \cdot B^\perp) \\ &\quad + \bar{Q} \cdot \bar{Q} + \bar{P} \cdot \bar{P} + B \cdot B + B^\perp \cdot B^\perp \\ &= 2\mathcal{H} + \bar{Q} \cdot \bar{Q} + \bar{P} \cdot \bar{P} + B \cdot B + B^\perp \cdot B^\perp \\ &= 2\mathcal{H} + I_2 + I_3 + B \cdot B + B^\perp \cdot B^\perp + \gamma^2, \end{aligned}$$

Therefore, the Hamiltonian $\mathcal{H} = I_1$, I_2, I_3 and $\langle L, L \rangle$ are functionally dependent.

We will presently demonstrate that equations for the Kowalewski gyrostat have certain dilational symmetries which account for two extra integrals of motion of which one will yield the integral of Kowalewski as reported in [BR]. On the level surface $\alpha = -\gamma$ the equations

$$\frac{dM}{dt} = [\Omega, M] + [B, \bar{P}] + [B^\perp, \bar{Q}] - \gamma[A_3, M],$$

$$\frac{dM_4}{dt} = i(\bar{Q}B + B\bar{Q}) - i(B^\perp \bar{P} + \bar{P}B^\perp),$$

$$\frac{d\bar{P}}{dt} = [\Omega, \bar{P}] - \gamma[A_3, \bar{P}], \quad \frac{d\bar{Q}}{dt} = [\Omega, \bar{Q}] - \gamma[A_3, \bar{Q}]$$

are invariant under the dilations $\bar{Q} \to \frac{1}{\lambda}\bar{Q}$ $\bar{P} \to \frac{1}{\lambda}\bar{P}$, $B \to \lambda B$, and $B^\perp \to \lambda B^\perp$. This implies that equations (15.32) admit a spectral representation $\frac{dL(\lambda)}{dt} = [\Lambda(\lambda), L(\lambda)]$ with

$$\Lambda(\lambda) = \begin{pmatrix} \Omega + \alpha A_3 & \lambda(B + iB^\perp) \\ \lambda(B - iB^\perp) & \Omega + \alpha A_3 \end{pmatrix},$$

$$L(\lambda) = \begin{pmatrix} M + \frac{1}{2}(\gamma - m_3)iI & \frac{1}{\lambda}\bar{P} + \lambda B + i\left(\frac{1}{\lambda}\bar{Q} + \lambda B^\perp\right) \\ \frac{1}{\lambda}\bar{P} + \lambda B - i\left(\frac{1}{\lambda}\bar{Q} - \lambda B^\perp\right) & M - \frac{1}{2}(\gamma - m_3)iI \end{pmatrix}.$$

It follows that spectral invariants of $L(\lambda)$ are the integrals of motion for the above Hamiltonian system for each λ. Moreover, these spectral invariants are in involution relative the the Poisson bracket induced by the semi-direct product Lie algebra, as a consequence of Proposition 9.12 in Chapter 9 (Indeed, $\lambda L(\lambda)$ is of the form $\lambda L_{\mathfrak{k}} + L_{\mathfrak{p}} + \lambda^2 B$).

Let us now investigate the integrals of motion associated with $I(\lambda) = \langle L(\lambda)^2, L(\lambda)^2 \rangle - \langle L(\lambda, L(\lambda)) \rangle$. We will first need the following:

Lemma 15.16 *Let* $I = Tr(L^4) - \frac{1}{4}(Tr(L^2))^2$. *Then,*

$$I = m_4^2(M \cdot M) + (P \cdot P)(Q \cdot Q) - (P \cdot Q)^2$$
$$+ (M \cdot P)^2 + (M \cdot Q)^2 + 2m_4(M \cdot [P, Q]). \qquad (15.38)$$

Proof First note that

$$L^2 = \begin{pmatrix} (M + \frac{im_4}{2}I)^2 + (P + iQ)(P - iQ) & M(P + iQ) + (P + iQ)M \\ (P - iQ)M + M(P - iQ) & (M - \frac{im_4}{2}I)^2 + (P - iQ)(P + iQ) \end{pmatrix}$$
$$= \begin{pmatrix} M^2 + P^2 + Q^2 - \frac{m_4^2}{2}I + i(m_4 M - [P, Q]) & M(P + iQ) + (P + iQ)M \\ M(P - iQ) + (P - iQ)M & M^2 + P^2 + Q^2 - \frac{m_4^2}{4}I - i(m_4 M - [P, Q]) \end{pmatrix}.$$

Therefore, $Tr(L^4) = 2Tr(-\frac{1}{4}(P \cdot P + Q \cdot Q + M \cdot M + m_4^2)^2 I) - m_4^2 M^2 - 2m_4 M[P, Q] - [P, Q]^2 + \frac{1}{2}(ST + TS))$, where $S = M(P + iQ) + (P + iQ)M$ and $T = M(P - iQ) + (P - iQ)M$. Then,

$$\frac{1}{2}Tr(ST + TS) = -2Tr(M^2(P^2 + Q^2) + (MP)^2 + (MQ)^2)$$
$$= \frac{1}{4}((M \cdot M)(P \cdot P + Q \cdot Q) + \frac{1}{4}(2(M \cdot P)^2 - (M \cdot M)(P \cdot P))$$
$$+ \frac{1}{4}(2(M \cdot Q)^2 - (M \cdot M)(Q \cdot Q))$$
$$= \frac{1}{2}((M \cdot P)^2 + (M \cdot Q)^2).$$

Hence,

$$Tr(L^4) - \frac{1}{4}(Tr(L^2))^2 = -2Tr((m_4^2 M^2 + [P, Q]^2$$
$$+ 2m_4(M[P, Q]) + (M \cdot P)^2 + (M \cdot Q)^2.$$

The substitutions $-2Tr([P, Q]^2) = (P \cdot P)(Q \cdot Q) - (P \cdot Q)^2$, $-2Tr(M^2) = M \cdot M$, and $-2Tr(M[P, Q]) = M \cdot [P, Q]$ into the above equation yield (15.38). $\qquad \square$

Therefore,

$$I(\lambda) = (P(\lambda) \cdot P(\lambda))(Q(\lambda) \cdot Q(\lambda)) - (P(\lambda) \cdot Q(\lambda))^2 + (M \cdot Q(\lambda))^2$$
$$+ (M \cdot P(\lambda))^2 + 2(\gamma - m_3)M \cdot [P(\lambda), Q(\lambda)] + (\gamma - m_3)^2 M \cdot M$$

with $Q(\lambda) = \frac{1}{\lambda}\bar{Q} + \lambda B$, and $P(\lambda) = \frac{1}{\lambda}\bar{P} + \lambda B^{\perp}$.

Hence, $I(\lambda) = \lambda^4 I_5 + \lambda^2 I_6 + I_7 + \frac{1}{\lambda^2}I_8 + \frac{1}{\lambda^4}I_9$ and each of I_5, I_6, I_7, I_8, I_9 is an integral of motion for $\mathcal{H}$. It is easy to see that $I_5 = B \cdot B$ because $B \cdot B^{\perp} = 0$, and $I_9 = (\bar{Q} \cdot \bar{Q})(\bar{P} \cdot \bar{P}) - (\bar{Q} \cdot \bar{P})^2 = I_2 I_3 - I_4^2$. Hence, I_5 does not provide any new information, and I_9 is functionally dependent on I_1, I_2, I_3, I_4. The same applies to I_6 for the following reasons:

$$I_6 = (\bar{Q} \cdot B^\perp)(B \cdot B) + (\bar{P} \cdot B)(B^\perp \cdot B^\perp) + (M \cdot B)^2 + (M \cdot B^\perp)^2$$
$$- 2(\gamma_3 - m_3)M \cdot [B, B^\perp]$$
$$= (B \cdot B)(\bar{Q} \cdot \hat{B}^\perp + \bar{P} \cdot \hat{B} + m_1^2 + m_2^2 + 2m_3^2 - 2\gamma_3 m_3) = 2(B \cdot B)\mathcal{H},$$

since $M \cdot [B, B^\perp] = -m_3 B \cdot B$, and $(M \cdot B)^2 + (M \cdot \hat{B}^\perp)^2 = (m_1^2 + m_2^2)(B \cdot B)$.
Let us now address I_7 and I_8.

$$I_8 = 2(\bar{P} \cdot \bar{P})(\bar{Q} \cdot B^\perp) + 2(\bar{Q} \cdot \bar{Q})(\bar{P} \cdot B) - 2(\bar{P} \cdot \bar{Q})(\bar{P} \cdot B^\perp + \bar{Q} \cdot B)$$
$$+ (M \cdot \bar{P})^2 + (M \cdot \bar{Q})^2 + 2(\gamma - m_3)M \cdot [\bar{P}, \bar{Q}].$$

This integral of motion is essentially the same as the negative of the integral reported in [BR] (provided that the parameter γ is replaced by its negative, b is taken as $b = e_1$, the variables $M, \bar{P}, \bar{Q}$ are renamed ℓ, g, h, and I_8 is renamed I_1).

The remaining integral I_7 is given by the following expression:

$$I_7 = (\bar{Q} \cdot \bar{Q})(B \cdot B) + (\bar{P} \cdot \bar{P})(B^\perp \cdot B^\perp) + 4(\bar{Q} \cdot B^\perp)(\bar{P} \cdot B^\perp)$$
$$- (\bar{Q} \cdot B^\perp + \bar{P} \cdot B)^2 + 2(M \cdot \bar{Q})(M \cdot B^\perp) + 2(M \cdot \bar{P})(M \cdot B)$$
$$+ 2(\gamma - m_3)M \cdot ([\bar{P}, B^\perp] + [B, \bar{Q}]) + (\gamma - m_3)^2(m_1^2 + m_2^2 + m_3^2)$$
$$= I_2(B \cdot B) + I_3(B^\perp \cdot B^\perp) + 4(\bar{Q} \cdot B^\perp)(\bar{P} \cdot B^\perp) - (\bar{Q} \cdot B^\perp + \bar{P} \cdot B)^2$$
$$+ 2(M \cdot \bar{Q})(M \cdot B^\perp) + 2(M \cdot \bar{P})(M \cdot B) + 2(\gamma - m_3)M \cdot ([\bar{P}, B^\perp]$$
$$+ [B, \bar{Q}]) + (\gamma - m_3)^2(m_1^2 + m_2^2 + m_3^2).$$

Let now $I = I_7 - (I_2(B \cdot B) + I_3(B^\perp \cdot B^\perp))$. It follows from above that

$$I = 4(\bar{Q} \cdot B^\perp)(\bar{P} \cdot B^\perp) - (\bar{Q} \cdot B^\perp + \bar{P} \cdot B)^2 + 2(M \cdot \bar{Q})(M \cdot B^\perp)$$
$$+ 2(M \cdot \bar{P})(M \cdot B) + 2(\gamma - m_3)M \cdot ([\bar{P}, B^\perp] + [B, \bar{Q}]) + (\gamma - m_3)^2 M \cdot M$$

The expression for I can be written in a slightly better form in regard to its relation to the Kowalewski integral of motion. To get to this expression, first note that

$$M \cdot ([\bar{P}, B^\perp] + [B, \bar{Q}]) = \bar{p}_3(M \cdot B) + \bar{q}_3(M \cdot B^\perp) - m_3(\bar{P} \cdot B + \bar{Q} \cdot B^\perp).$$

Then, note that

$$(\gamma - m_3)^2(m_1^2 + m_2^2 + m_3^2) + 2(\gamma - m_3)M \cdot ([\bar{P}, B^\perp] +]B, \bar{Q}])$$
$$= (\gamma - m_3)^2(m_1^2 + m_2^2) + (\gamma - m_3)^2 m_3^2 - 2(\gamma - m_3)m_3(\bar{Q} \cdot B^\perp + \bar{P} \cdot B)$$
$$+ 2(\gamma - m_3)(\bar{q}_3(M \cdot B^\perp)) + \bar{p}_3(M \cdot B))$$

$$= (\gamma - M_3)^2(M_1^2 + M_2^2) + \left(\mathcal{H} - \frac{1}{2}(m_1^2 + m_2^2)\right)^2 - (\bar{P} \cdot B + \bar{Q} \cdot B^\perp)^2$$

$$+ 2(\gamma - m_3)\left(\bar{q}_3(M \cdot B^\perp) + \bar{p}_3(M \cdot B)\right).$$

Also note that $-\gamma m_3 + m_3^2 = \mathcal{H} - (\bar{Q} \cdot B^\perp + \bar{P} \cdot B) - \frac{1}{2}(m_1^2 + m_2^2)$.

Therefore,

$$2(\gamma - m_3)(M \cdot ([\bar{P}, B^\perp] + [B, \bar{Q}]) + (\gamma - m_3)^2(M \cdot M)$$
$$= (\gamma^2 - \gamma m_3)(m_1^2 + m_2^2)$$
$$\quad + \left(\mathcal{H} - (\bar{P} \cdot B + Q \cdot B^\perp) - \frac{1}{2}(m_1^2 + m_2^2)\right)(m_1^2 + m_2^2)$$
$$\quad + \left(\mathcal{H} - \frac{1}{2}(m_1^2 + m_2^2)\right)^2 - (\bar{P} \cdot B + \bar{Q} \cdot B^\perp)^2$$
$$\quad - 2(\gamma - m_3)\left(\bar{q}_3(M \cdot B^\perp) + \bar{p}_3(M \cdot B)\right)$$
$$= \gamma(\gamma - m_3)(m_1^2 + m_2^2) + \mathcal{H}^2 - \frac{1}{4}(m_1^2 + m_2^2)$$
$$\quad - (m_1^2 + m_2^2)(\bar{P} \cdot B + \bar{Q} \cdot B^\perp) - (\bar{P} \cdot B + \bar{Q} \cdot B^\perp)^2$$
$$\quad + 2(\gamma - m_3)\left(\bar{q}_3(M \cdot B^\perp) + \bar{p}_3(M \cdot B)\right).$$

Finally note that

$$2(M \cdot \bar{Q})(M \cdot B^\perp) + 2(M \cdot \bar{P})(M \cdot B) - 2m_3\left(\bar{q}_3(M \cdot B^\perp) + \bar{p}_3(M \cdot B)\right)$$
$$= 2(m_1\bar{q}_1 + m_2\bar{q}_2)(M \cdot B^\perp) + (m_1\bar{p}_1 + m_2\bar{p}_2)(M \cdot B)).$$

It follows that $J = I - \mathcal{H}^2$, where J is given by

$$J = 4(\bar{Q} \cdot B^\perp)(\bar{P} \cdot B) - (\bar{Q} \cdot B + \bar{P} \cdot B^\perp)^2 + 2(m_1\bar{q}_1 + m_2\bar{q}_2)(M \cdot B^\perp)$$
$$\quad + 2(m_1\bar{p}_1 + m_2\bar{p}_2)(M \cdot B) - \frac{1}{4}(m_1^2 + m_2^2)^2 - (m_1^2 + m_2^2)(\bar{P}B + \bar{Q}B^\perp)$$
$$\quad - (\bar{P} \cdot B + \bar{Q} \cdot B^\perp)^2 + \gamma(\gamma - m_3)(m_1^2 + m_2^2)$$
$$\quad + 2\gamma\left(\bar{q}_3(M \cdot B^\perp) + \bar{p}_3(M \cdot B)\right).$$

The above integral of motion can be somewhat simplified by noting that

$$4(\bar{Q} \cdot B^\perp)(\bar{P} \cdot B) - (\bar{P} \cdot B^\perp + \bar{Q} \cdot B) - (\bar{P} \cdot B + \bar{Q} \cdot B^\perp)^2$$
$$= -(B \cdot B)\left((\bar{q}_1 - \bar{p}_2)^2 + (\bar{q}_2 - \bar{p}_1)^2\right).$$

Let J_0 denote the part of J which is independent of γ. It follows that

$$J_0 = -\frac{1}{4}(m_1^2 + m_2^2)^2 - (m_1^2 + m_2^2)(\bar{P} \cdot B + \bar{Q} \cdot B^\perp) + 2(m_1\bar{q}_1 + m_2\bar{q}_2)(M \cdot B^\perp)$$
$$+ 2(m_1\bar{p}_1 + m_2\bar{p}_2)(M \cdot B) - (B \cdot B)\left((\bar{q}_1 + \bar{p}_2)^2 + (\bar{q}_2 - \bar{p}_1)^2\right).$$

To relate J_0 to the integral obtained by S. Kowalewski, introduce the following variables:

$$z_1 = \frac{1}{2}(m_1 + im_2), z_2 = \frac{1}{2}(m_1 - im_2),$$
$$w_1 = (\bar{q}_1 + \bar{p}_2) + i(\bar{q}_2 - \bar{p}_1), w_2 = (\bar{q}_1 + \bar{p}_2) - i(\bar{q}_2 - \bar{p}_1),$$
$$c_1 = i(b_1 + ib_2), c_2 = -i(b_1 - ib_2).$$

We leave it to the reader to show that

$$- (m_1^2 + m_2^2)(\bar{P} \cdot B + \bar{Q} \cdot B^\perp) + 2(m_1\bar{q}_1 + m_2\bar{q}_2)(M \cdot B^\perp)$$
$$+ 2(m_1\bar{p}_1 + m_2\bar{p}_2)(M \cdot B) = 2z_1^2 w_1 c_1 + 2z_2^2 w_2 c_2.$$

Hence,

$$J_0 = -4z_1^2 z_2^2 + 2z_1^2 w_1 c_1 + 2z_2^2 w_2 c_2 - c_1 c_2 w_1 w_2$$
$$= -4\left(z_1^2 - \frac{c_1}{2}w_1\right)\left(z_2^2 - \frac{c_2}{2}w_2\right). \tag{15.39}$$

Since

$$\gamma(\gamma - m_3)(m_1^2 + m_2^2) + 2\gamma\left(\bar{q}_3(M \cdot B^\perp) + \bar{p}_3(M \cdot B)\right)$$
$$= 4\gamma(\gamma + m_3)z_1 z_2 + \frac{1}{2}\bar{q}_3(z_2 c_1 + z_1 c_2) + \frac{1}{2i}\bar{p}_3(z_2 c_1 - z_1 c_2)),$$
$$-\frac{1}{4}J = \left(z_1^2 - \frac{c_1}{2}w_1\right)\left(z_2^2 - \frac{c_2}{2}w_2\right)$$
$$- \gamma\left((\gamma + m_3)z_1 z_2 + \frac{1}{2}\bar{q}_3(z_2 c_1 + z_1 c_2) + \frac{1}{2i}\bar{p}_3(z_2 c_1 - z_1 c_2)\right).$$

This integral of motion is a slight generalization of the one reported in [BR], and coincides with the one found by S. Kowalewski when $\bar{q} = 0$ and $\gamma = 0$.

Let us now reflect briefly on the overall picture, at the conclusion of this, somewhat indirect, journey to the integral of Kowalewski. The semi-direct product $\mathfrak{g}_s = \mathfrak{p} \rtimes \mathfrak{k}$ is a ten-dimensional Lie algebra of rank 2. Hence the generic coadjoint orbits in $\mathfrak{g}_s^*$ are eight-dimensional symplectic submanifolds of $\mathfrak{g}_s^*$. They are given by $I_2 = constant$ and $I_3 = constant$. It follows that the maximal number of functionally independent integrals of motion in involution relative to the Poisson structure on $\mathfrak{g}_s^*$, is equal to four. Our findings show that $\mathcal{H}$ is completely integrable on each generic orbit, since

$$F_1 = \mathcal{H}, F_2 = I_4^2, F_3 = I_7, F_4 = J$$

constitute four independent integrals all in involution with each other.

Kowalewski's integral of motion occurs as a particular case of the above situation when $0 = \gamma$ and $\bar{Q} = 0$. For then, $-\frac{1}{4}J = \left(z_1^2 - \frac{c_1}{2}w_1\right)\left(z_2^2 - \frac{c_2}{2}w_2\right)$ agrees with the integral found by S. Kowalewski and F_3 reduces to to the Casimir $(M \cdot \bar{P})^2$.

An alternative way to get to the Kowalewski gyrostat is to begin with the Casimir

$$\langle L, L \rangle = M \cdot M + m_4^2 + P \cdot P + Q \cdot Q$$

on $so_5(\mathbb{C})$ and then seek a Hamiltonian for which

$$m_4 + m_3, (P - B) \cdot (P - B), (Q - C) \cdot (Q - C)$$

are constants of motion, for some constant elements $B \in \mathfrak{p}_1$ and $C \in \mathfrak{p}_2$. For then,

$$\langle L, L \rangle = m_1^2 + m_2^2 + 2m_3^2 - 2\gamma_3 m_3 + 2\bar{P} \cdot B + 2\bar{Q} \cdot C + \bar{P} \cdot \bar{P} + \bar{Q} \cdot \bar{Q} + \gamma^2$$
$$= 2\mathcal{H} + \bar{P} \cdot \bar{P} + \bar{Q} \cdot \bar{Q} + \gamma^2, \text{ where } \bar{P} = P - B, \bar{Q} = Q - C,$$

where $\mathcal{H}$ denotes the Hamiltonian associated with the Kowalewski gyrostat. This ad-hoc procedure gives $c = b^\perp$ immediately, and ultimately leads to the Lax-pair representation as explained above.

However, the Hamiltonian $\mathcal{H} = \frac{1}{2}(m_1^2 + m_2^2 + (m_3 - \gamma)^2) + P \cdot B + Q \cdot C$ obtained by this procedure is very different from the Hamiltonian associated with the elastic problem of Kirchhoff. It is associated with the problem of minimizing the integral $\frac{1}{2}\int(u_1^2 + u_2^2 + (u_3 - \gamma)^2)\,dt$ over the trajectories of

$$\frac{dg}{dt} = g \begin{pmatrix} 0 & 0 & -b_1 & -b_2 & -b_3 \\ 0 & 0 & -c_1 & -c_2 & -c_3 \\ b_1 & c_1 & 0 & -u_3 & u_2 \\ b_2 & c_2 & u_3 & 0 & -u_1 \\ b_3 & c_3 & u_2 & u_1 & 0 \end{pmatrix}.$$

This problem is not affine-quadratic and bears no resemblance to the elastic problem of Kirchhoff. From this perspective, the connection to Kowalewski top is artificial and seems to compound the mystery, rather than explain "the peculiar geometry of the Kowalewski top" as was originally claimed in [BR].

16

Elastic problems on symmetric spaces: the Delauney–Dubins problem

We have already encountered elastic curves in several places, albeit in a somewhat incidental manner, more as a byproduct of our interest in the affine-quadratic Hamiltonians and the theory of mechanical tops, rather than a topic in its own right. The material in this chapter is motivated by a class of geometric problems in which the energy functional depends on second-order derivatives of curves in a Riemannian manifold M with the curvature problem in its forefront.

To contrast and compare with the previous encounters with elastic curves we will begin with some general remarks about Riemannian geometry and control theory.

16.1 The curvature problem

Let us begin with a Riemannian manifold M with its Riemannian structure defined by the inner product $\langle\,,\,\rangle_x$, $x \in M$, T_xM that varies smoothly with the base point x. Then for any parametrized curve $x(t)$ in M, the length of $x(t)$ in an interval $[0, T]$ is given by $L = \int_0^T \left\| \frac{dx}{dt} \right\| dt$, where $\left\| \frac{dx}{dt} \right\| = \sqrt{\left\langle \frac{dx}{dt}, \frac{dx}{dt} \right\rangle_{x(t)}}$. This notion of length is independent of the choice of parametrization. So in talking about the length of curves that connect two given points x_0 and x_1 one can restrict the parameter to a fixed interval, normally to $[0, 1]$.

Riemannian geometry begins with the geodesic problem, a study of "straight lines" or geodesics, curves of shortest length that connect two given points sufficiently near each other. In the differential geometry literature, it is usually assumed that the geodesics are sufficiently regular so that can be found among the solutions of the Euler–Lagrange equation

$$\frac{d^2 x_k}{dt^2} + \sum_{Ij=1}^{n} \Gamma_{ij}^{k} \frac{dx_i}{dt} \frac{dx_j}{dt} = 0, k = 1, \ldots, n, \tag{16.1}$$

where $(x_1, \ldots, x_n)$ denote any system of coordinates on M, and Γ_{ij}^{k} denote the associated Christoffel symbols. The Christoffel symbols are defined through the Levi–Civita connection ∇ by $\nabla_{X_j} X_i = \sum_{k=1}^{n} \Gamma_{ij}^{k} X_k$ for any frame of vector fields $X_1, \ldots, X_n$. The dependence on the Euler–Lagrange equation gives differential geometry a peculiar twist: the theory of connections becomes a principal object of study followed with applications to geodesics, rather than the other way around, with geodesics at the forefront as the basic concept that leads to connections and curvature under a more refined study.

The geodesic problem admits a natural formulation as a time optimal control problem under the assumption that the addmissible curves are extended to the class of Lipschitzian curves on M. For then, any two points that can be connected by an addmissible curve, can be also connected by a regular curve, a curve along which $\frac{dx}{dt} \neq 0$. This implies that in regard to the geodesic problem, all curves can be parametrized by arc length, in which case the length of a curve $x(t)$ from $x_0 = x(0)$ to $x_1 = x(T)$ is equal to the "time" that $x(t)$ reaches the terminal point x_1. Then any orthonormal frame of vector fields $X_1, \ldots, X_n$ induces a "control system"

$$\frac{dx}{dt} = \sum_{i=1}^{n} u_i(t) X_i(x(t)), \sum_{i=1}^{n} u_i^2(t) = 1, \tag{16.2}$$

whose time optimal solutions coincide with the geodesics. In this regard, it is better to convexify the control system, that is, allow the controls to take values in the unit ball $\sum_{i=1}^{n} u_i^2(t) \leq 1$, because the convexified system admits time optimal solutions, whenever the vector fields are complete. More importantly, however, the geodesic control problem leads directly to the right Hamiltonian via the Maximum Principle without any need for the Levi–Civita connection, and that constitutes a major advantage over the classical approach, particularly in regard to the theory of integrable systems, as we have amply demonstrated in the earlier chapters.

Leaving aside the finer justifications for our bias towards the Hamiltonian world, let us now come to the second-order systems and to the curvature problem. Recall first that the geodesic curvature $\kappa(t)$ of a curve $x(t)$ parametrized by its arc length is given by the norm of the covariant derivative $\left\| \left(\frac{D_{x(t)}}{dt} \right) \left(\frac{dx}{dt} \right) \right\|$. If $X_1, \ldots, X_n$ is any frame of vector fields in M, and if $v(t)$ is any curve of tangent vectors along a curve $x(t)$, then $v(t) = \sum_{i=1}^{n} v_i(t) X_i(x(t))$, and the covariant derivative $\frac{D_{x(t)}}{dt}(v(t))$ of $v(t)$ along $x(t)$ is given by the following formula:

$$\frac{D_{x(t)}}{dt}(v(t)) = \sum_{k=1}^{n}\left(\frac{dv_k}{dt} + \sum_{i,j=1}^{n} v_i(t)v_j(t)\Gamma_{ij}^{k}(x(t))\right)X_k(x(t)). \qquad (16.3)$$

In particular, when $v_i(t) = \frac{dx_i}{dt}$, $i = 1, \ldots, n$, then

$$\frac{D_{x(t)}}{dt}\left(\frac{dx}{dt}\right) = \sum_{k=1}^{n}\left(\frac{d^2 x_k}{dt^2} + \sum_{i,j=1}^{n}\frac{dx_i}{dt}\frac{dx_j}{dt}\Gamma_{ij}^{k}(x(t))\right)X_k(x(t)).$$

Hence, a curve is a geodesic if and only if $\frac{D_{x(t)}}{dt}\left(\frac{dx}{dt}\right) = 0$. The latter implies that $\left\|\frac{dx}{dt}\right\| = ct$ for some constant $c > 0$, because

$$\frac{d}{dt}\left\langle\frac{dx}{dt}, \frac{dx}{dt}\right\rangle = 2\left\langle\frac{D_{x(t)}}{dt}\left(\frac{dx}{dt}\right), \frac{dx}{dt}\right\rangle = 0.$$

This implies that geodesics are necessarily regular curves.

To go beyond the geodesics, let

$$\frac{dx}{dt} = \sum_{k=1}^{n} y_k X_k(x(t)), \quad \frac{D_{x(t)}}{dt}\left(\frac{dx}{dt}\right) = \sum_{k=1}^{n} u_k(t)X_k(x(t)),$$

for some functions $u_1(t), \ldots, u_n(t)$. Then,

$$\left\|\frac{dx}{dt}\right\|^2 = \sum_{ij=1}^{n} y_i y_j \langle X_i, X_j\rangle = \sum_{ij=1}^{n} Q_{ij}(x)y_i(t)y_j(t) = (Q(x)y, y),$$

$$\left\|\frac{D_{x(t)}}{dt}\left(\frac{dx}{dt}\right)\right\|^2 = \sum_{ij=1}^{n} u_i u_j \langle X_i, X_j\rangle = (Q(x)u, u),$$

where Q is the matrix with its entries $Q_{ij}(x) = \langle X_i(x), X_j(x)\rangle_x$, and $(\ ,\ )$ the Euclidean product in $\mathbb{R}^n$.

If $x(t)$ is a curve parametrized by arc-length, then $1 = \left\|\frac{dx}{dt}\right\|^2 = (Q(x)y, y)$, and $\kappa^2 = \left\|\frac{D_{x(t)}}{dt}\left(\frac{dx}{dt}\right)\right\|^2$. However, in view of the relation $\langle\frac{dx}{dt}, \frac{dx}{dt}\rangle = 1$,

$$\frac{d}{dt}\left\langle\frac{dx}{dt}, \frac{dx}{dt}\right\rangle = 2\left\langle\frac{D_{x(t)}}{dt}\left(\frac{dx}{dt}\right), \frac{dx}{dt}\right\rangle = (Q(x)u(t), y(t)) = 0.$$

Under these conditions (16.3) defines a control system

$$\frac{dx_k}{dt} = y_k, \frac{dy_k}{dt} = u_k(t) - \sum_{i,j=1}^{n} y_i(t)y_j(t)\Gamma_{ij}^{k}(x(t))), k = 1, \ldots, n, \qquad (16.4)$$

in the cylinder $\{(x, y) : (Q(x)y, y) = 1\}$, where the controls $u(t) = u_1(t), \ldots,$ $u_n(t)$ are subject to the constraint $(Q(x)y, u) = 0$. The optimal control problem defined by $\frac{1}{2} \int_0^T (Q(x(t)), u(t), u(t)) \, dt$ will be called the curvature problem.

Of course, the curvature problem can be defined intrinsically on the unit tangent bundle of M as follows. Let $\pi : TM \to M$ denote the natural projection $v \to x, v \in T_xM$ and let T^1M denote the unit tangent bundle of M. If $v_0 \in T^1_{x_0}M$ and $v_1 \in T^1_{x_1}M$ are two given points with $x_0 \neq x_1$, let $\mathcal{A}(v_0, v_1, T)$ denote the class of differentiable curves $x(t)$ in M subject to the following conditions:

1. $x(0) = x_0, x(T) = x_1$.
2. $\left\| \frac{dx}{dt} \right\| = 1$ for t in $[0, T]$, and $\frac{dx}{dt}(0) = v_0, \frac{dx}{dt}(T) = v_1$.
3. The covariant derivative $\frac{D_{x(t)}}{dt}\left(\frac{dx}{dt}\right)$ is measurable and bounded in the interval $[0, T]$

The curvature problem then can be defined more formally as:

Definition 16.1 the curvature problem Find a curve $v(t) \in \mathcal{A}(v_0, v_1, T)$ that minimizes $\frac{1}{2} \int_0^T \left\| \frac{D_{x(t)}}{dt}\left(\frac{dx}{dt}\right) \right\|^2 dt$ over all other curves in $\mathcal{A}(v_0, v_1, T)$. The curvature problem is said to be *free* if the requirement of fixed length is dropped, and is said to be relaxed if the terminal condition $\frac{dx}{dt}(T) = v_1$ is omitted.

For the sake of consistency with the previous material the solution curves will be called elastic. Hilbert and Cohn-Vossen suggested the solutions to the relaxed curvature problem as an alternative definition for the geodesics [HV]. However, this suggestion went largely unnoticed by the mathematical community. In fact, apart from the spaces of constant curvature, not much is known about the solutions to the curvature problem.

The curvature problem draws attention to several related problems of interest in mechanics and other branches of mathematics. The most immediate of these is to ask for the minimum of $\frac{1}{2} \int_0^T \left\| \frac{D_{x(t)}}{dt}\left(\frac{dx}{dt}\right) \right\|^2 dt$ when the constraint on the choice of parametrization is dropped. This variant of the curvature problem results in a relaxed control system defined on the entire space $\mathbb{R}^n \times \mathbb{R}^n$, with the control functions no longer being subjected to any state dependent constraints [Lt; NP]. We will refer to this problem as *the covariant derivative problem*.

The next class of problems deals with parametrized curves $x(t)$ on a Riemannian manifold subject to the condition that $\left\| \frac{D_{x(t)}}{dt}\left(\frac{dx}{dt}\right) \right\|$ is bounded. When the curves are parametrized by arc length, then the bound on the covariant derivative is the same as the bound on the curvature. In this class of curves with bounded curvature the problem of Delauney, posed in the late

1880s, is the oldest. It asked for the curves of longest and shortest length among all space curves with a given curvature $\kappa(t) = c$ that connect two given line elements in the space [Cr]. In this terminology, a line element is the same as a tangent vector.

Carathéodory ends his book on the calculus of variations with Delauney's problem in which he briefly comment on its history, concluding that the general solution to this problem was not fully known until Weierstrass, who apparently was the first to successfully integrate the associated Euler equation. However, Carathéodory suggested that the Hamiltonian approach is more insightful, and proposed an original method of arriving at the correct Hamiltonian equations. In conclusion, he claimed that the associated Hamiltonian equations are integrable by quadratures in terms of elliptic functions, obtained by solving the fundamental equation

$$\dot{u}^2 = \lambda^2[(w - u^2)(u - 1)^2 - k^2], \tag{16.5}$$

where w and k are constants, and $\lambda = \pm 1$ depending on the sign of u.

Unfortunately, Carathéodory's key equation seems to be incorrect. The correct equation seems to have been obtained earlier by Josepha von Schwarz in 1934 [Sc], and to make the matter even more confusing, Carathéodory does not comment on this discrepancy, even though he cites Schwarz's work in his bibliography. Schwarz's more detailed and more extensive treatment of Delauney's problem provides a solution by quadratures based on the equation

$$(u - \mu_0)^2 \left(\frac{du}{ds}\right)^2 = (u - \mu_0)^2(k^2 - u^2) - h_0^2, \tag{16.6}$$

with $\mu_0 = -1$ for the minimum length and $\mu_0 = 1$ for the maximum length and h_0 a constant.

We will come back to this equation in more detail later on in this chapter. For the time being, however, let us just note that this problem also lends itself to control theoretic formulation as the time optimal problem of finding a trajectory $(x(t), y(t))$ of control system (16.4) that connects (x_0, v_0) to (x_1, v_1) in the least possible time with a control that satisfies $(Q(x(t))u(t), u(t)) = c$.

Control theoretic formulation makes it transparent that the problem of Delauney is not well posed, in the sense that not all points can be reached in a minimum time by the controls that satisfy $(Qu, u) = c$. The reason is simple: control set $\{u : (Qu, y) = 0, (Qu, u) = c\}$ is not convex and the reachable set is not closed. So it is natural to enlarge the set of controls to the convex closure $\{u : (Qu, y) = 0, (Qu, u) \leq c\}$. As we have demonstrated in Chapter 3, the reachable sets by the controls in a set U and the reachable sets by the controls in the convex hull of U have the same topological closure (under the completeness hypothesis).

This extension brings us to the problem of L. Dubins, which asked for the curves of shortest length which join two given line elements in $\mathbb{R}^n$, among the curves whose curvature is less or equal to a given constant c. In his remarkable paper of 1957, Dubins proved that minimizers exist in the class of continuously differentiable functions having integrable second derivatives, and he characterized optimal solutions in the plane as the concatenations of circles of curvature $\pm c$ and straight lines, with at most three switchings from one arc to another [Db]. Apart from proving the existence of optimizers, Dubins did not go further into the the nature of optimal solutions in dimensions greater than two. It is relevant to point out that at the time of Dubins' paper, the calculus of variations had no adequate means to deal with variational problems with inequality constraints, and Dubins, unaware of control theory and its quest for the Maximum Principle, tackled the problem with "bare hands."

In view of the above remarks, the problem of Dubins can be regarded as the convexified version of Delauney's problem. In dimensions greater than two, these two problems overlap, and for that reason we will refer to Dubins' problem as *the Delauney–Dubins problem* (Some authors refer to the above problem as Markov–Dubins problem (for instance, [CK] and [Ss4]), since Markov was the first to pose Dubins's question in 1887 for planar curves, bearing in mind applications to the design of railroad tracks [MK].

One could also consider the analogous time optimal problem for curves constrained by $\left\| \frac{D_{x(t)}}{dt} \left(\frac{dx}{dt} \right) \right\| \leq c$ when the requirement on the parametrization by arc-length is dropped. This problem is referred to as the dynamic Dubins–Markov problem in [CK]. It lends itself to a nice mechanical description of finding the fastest path for a vehicle whose acceleration is bounded. It is a natural extension of the street car problem well known in optimal control literature.

So all of the above problems, geometric or mechanical, lend themselves to control theoretic interpretations, at least locally, when the boundary data lies in the same coordinate system. In this chapter we will consider these problems in more detail, with a particular interest in the effect of the Riemannian curvature on their solutions.

16.2 Elastic problem revisited – Dubins–Delauney
on space forms

Let us begin with $M = \mathbb{E}^n$, where Euclidean space $\mathbb{E}^n$ with its Euclidean metric $\langle \, , \, \rangle$. Then $x = (x_1, \ldots, x_n)$ will denote the coordinates of any point in M relative to an orthonormal basis in M and $y = (y_1, \ldots, y_n)$ will denote the coordinates of a tangent vector at x. Since the tangent bundle of M is isomorphic with $\mathbb{R}^n \times \mathbb{R}^n$ equation (16.3) is global. As usual, TM will be identified with T^*M via the Euclidean product. Then $T^*(TM)$ will be identified with $\mathbb{R}^{4n}$ with its points represented by the quadruples (x, y, p, q).

In this situation the covariant derivative problem reduces to a particularly simple linear-quadratic control problem: minimize the quadratic cost $\frac{1}{2} \int_0^T \|u(t)\|^2 \, dt$ over the solutions of a linear system

$$\frac{dx}{dt} = y, \frac{dy}{dt} = u. \tag{16.7}$$

The Maximum Principle then leads to the Hamiltonian system

$$\frac{dx}{dt} = y, \frac{dy}{dt} = q, \frac{dp}{dt} = 0, \frac{dq}{dt} = -p,$$

generated by the affine Hamiltonian $H = \frac{1}{2}(q, q) + (p, y)$. This system is easily solvable, with

$$x(t) = -\frac{1}{6}at^3 + \frac{1}{2}bt^2 + y_0 t + x_0, y(t) = -\frac{1}{2}at^2 + bt + y_0 \tag{16.8}$$

the general solution. Since $det\begin{pmatrix} -\frac{1}{6}t^3 & \frac{1}{2}t^2 \\ -\frac{1}{2}t^2 & t \end{pmatrix} = \frac{1}{12}t^4$, any terminal state (x_1, y_1) in $\mathbb{R}^n \times \mathbb{R}^n$ can be reached from any initial point (x_0, y_0) by a unique extremal at any time $T > 0$. This extremal curve is known as *the cubic spline function* in the interpolation theory. It is used to interpolate any curve in $\mathbb{R}^n$ up to its first-order jets at a finite number of points $t_0, t_1, \ldots, t_m$. For this reason the covariant derivative problem has been suggested as the interpolating prototype for any Riemannian manifold [Lt; NHP]. It turns out, somewhat surprisingly, that the extremal equations for this problem are difficult to solve beyond the Euclidean setting. We will illustrate this situation in more detail with the extremal equations on the sphere.

Consider now the minimum time transfer in (16.7) when the controls are restricted by $\|u(t)\| \le c$. In this situation the extremal curves are the integral curves of the Hamiltonian $h_{u(t),\mu} = -\mu + p \cdot y + q \cdot u(t), \mu = 0, 1$, that is, the solutions of

$$\frac{dx}{dt} = y(t), \frac{dy}{dt} = u(t), \frac{dp}{dt} = 0, \frac{dq}{dt} = -p. \tag{16.9}$$

generated by the controls $u(t)$ that satisfy

$$-\mu + p(t) \cdot y(t) + q(t)u(t) \ge -\mu + p(t) \cdot y(t) + v \cdot q(t)$$

for all $v \in \mathbb{R}^n$ such that $\|v\| \le c$. It follows that $p(t)$ is constant and $q(t) = at + b$, where $p = -a$ and $b = q(0)$ Hence $q(t) \ne 0$ for all t unless a and b

are colinear, in which case, $q(t) = 0$ exactly once. The preceding inequality implies that every extremal control $u(t)$ is of the form $u(t) = c\frac{q(t)}{||q(t)||}$ on any open interval where $q(t)$ is not equal to zero. Therefore, the projections of the extremal curves are the solutions of the following equations:

$$\frac{dx}{dt} = y, \frac{dy}{dt} = c\frac{a + bt}{\sqrt{||a + bt||^2}} = c\frac{a + bt}{\sqrt{||b||^2(t + \frac{a,b}{||b||^2})^2 + \frac{||a||^2||b||^2 - (a,b)^2}{||b||^2}}}.$$

$$(16.10)$$

To simplify the analysis assume that $x_0 = y_0 = 0$. Then there are two types of solutions depending on a and b. When a and b are not colinear then $\frac{dy}{dt} \neq 0$. In such a case let $t + \frac{(a,b)}{||b||} = \Delta \sinh \theta$, where $\Delta^2 = \frac{1}{||b||^2}(||a||^2||b||^2 - (a,b)^2)$. Then $\frac{dy}{d\theta} = \frac{c}{||b||}\left(a + \left(\Delta \sinh \theta - \frac{(a,b)}{||b||}\right)b\right)$.

Therefore,

$$y(\theta) = \frac{c}{||b||}(a\theta + (\Delta(\cosh \theta - 1) - (a, b)\theta)b).$$

We leave the remaining integration to the reader.

When $a = \lambda b$ for some number λ, then $a + tb = (t \mid \lambda)b$. Hence,

$$\frac{dy}{dt} = c\frac{b}{||b||}\frac{t + \lambda}{|t + \lambda|}.$$

When $\lambda > 0$, then $y(t) = c\frac{b}{||b||}t$ for $t > 0$ and $x(t) = c\frac{b}{||b||}\frac{t^2}{2}$. Hence the vehicle follows a straight line with maximal acceleration. However, when $\lambda < 0$, then $q(t) = 0$ at $t = -\lambda$ and there is jump in the extremal control from $u = -c\frac{b}{||b||}$ to its opposite $u = c\frac{b}{||b||}$. Therefore,

$$y(t) = -c\frac{b}{||b||}t, t \leq -\lambda, \text{ and } y(t) = c\frac{b}{||b||}(t + \lambda) + c\frac{b}{||b||}\lambda, t > -\lambda.$$

This situation will occur when the terminal velocity is zero. For if the vehicle is to start with zero velocity and terminate with zero velocity, then it is obliged to reverse the acceleration somewhere during its journey.

Let us now pass to the situation where the admissible curves are parametrized by arc length. In these situations the state space $N = \mathbb{R}^n \times S^n$ is a proper submanifold of the ambient space $\mathbb{R}^n \times \mathbb{R}^n$ and the controls take values in the tangent bundle of the sphere S^n. In the extraneous coordinates of the ambient space, $N = \{(x, y) \in \mathbb{R}^n \times \mathbb{R}^n : ||y||^2 = 1\}$ and $T^*N = \{(x, y, p, q) : ||y||^2 = 1, (y, q) = 0\}$.

To conform to our earlier notations related to the use of the Maximum Principle in the coordinates of the ambient space, let $G_1 = ||y||^2 - 1 = 0$

and $G_2 = y \cdot q = 0$. For the curvature problem the associated Hamiltonian lift is given by

$$h_{u,\mu} = -\mu \frac{1}{2}||u||^2 + p \cdot y + q \cdot u + \lambda_1 G_1 + \lambda_2 G_2, \ \mu = 0, 1, \quad (16.11)$$

where the multipliers λ_1 and λ_2 are determined by the conditions that $\{h_{u,\mu}, G_1\} = 0$ and $\{h_{u,\mu}, G_2\} = 0$. These conditions are fulfilled if and only if

$$\{g_u, G_2\} + \lambda_1\{G_2, G_1\} = 0, \ \{g_u, G_1\} + \lambda_2\{G_1, G_2\} = 0,$$

where $g_u = -\mu\frac{1}{2}||u||^2 + p \cdot y + q \cdot u$. It follows that $\{G_1, G_2\} = 2$, $\{h_u, G_1\} = -2u \cdot y$, and $\{g_u, G_2\} = p \cdot y - u \cdot q$. Hence,

$$\lambda_1 = -\frac{1}{2}(p \cdot y - u \cdot q), \ \lambda_2 = -2u \cdot y.$$

Suppose now that $(x(t), y(t), p(t), q(t))$ an extremal curve generated by a control $u(t)$, i.e., suppose that

$$\begin{aligned}
\frac{dx}{dt} &= \frac{\partial h_{u,\mu}}{\partial p} = y, \ \frac{dy}{dt} = \frac{\partial h_{u,\mu}}{\partial q} = u, \\
\frac{dp}{dt} &= -\frac{\partial h_{u,\mu}}{\partial x} = 0, \ \frac{dq}{dt} = -\frac{\partial h_{u,\mu}}{\partial y} = -p - \lambda_1 \frac{\partial G_1}{\partial y} - \lambda_2 \frac{\partial G_2}{\partial y},
\end{aligned} \quad (16.12)$$

subject to the condition that

$$h_{u(t),\mu}(x(t), y(t), p(t), q(t)) \geq h_{v,\mu}(x(t), y(t), p(t), q(t)) \quad (16.13)$$

for all $v \in \mathbb{R}^n$ with $(v, y(t)) = 0$.

To determine the extremal control we need to replace $h_{v,\mu}$ with $h_{v,\mu} + \lambda_0(v \cdot y)$, where the multiplier λ_0 is to be determined from the Lagrange multiplier rule. We will leave it to the reader to show that the abnormal extremals project onto the straight lines in N.

In the case of normal extremals (16.13) yields

$$-u(t) + q(t) + \lambda_0 y(t) = 0. \quad (16.14)$$

But, then $-u(t) \cdot y(t) + q(t) \cdot y(t) + \lambda_0||y(t)||^2 = 0$, and, therefore, $\lambda_0 = 0$ and $u(t) = q(t)$. It follows that the normal extremal curves are the solutions of the restricted Hamiltonian system associated with the Hamiltonian

$$H = \frac{1}{2}||q||^2 + p \cdot y + \lambda_1 G_1 + \lambda_2 G_2, \quad (16.15)$$

with $\lambda_1 = \frac{1}{2}(p \cdot y - ||q||^2)$, and $\lambda_2 = \frac{1}{2}(q \cdot y) = 0$. That is, the extremals are the solutions of

$$\frac{dx}{dt} = y, \ \frac{dy}{dt} = q, \ \frac{dp}{dt} = 0, \ \frac{dq}{dt} = -p + (p \cdot y - ||q||^2)y. \quad (16.16)$$

(Straight lines are the projections of both the normal and the abnormal extremals.)

To avoid unnecessary repetitions with the Delauney–Dubins problem, we will postpone our discussion of the solutions in (16.16) until we have both sets of equations at our disposal.

For the Delauney–Dubins problem,

$$h_{u,\mu} = -\mu + p \cdot y + q \cdot u + \lambda_1 G_1 + \lambda_2 G_2 = g_u + \lambda_1 G_1 + \lambda_2 G_2, , \|u\| \leq c.$$

Then $\{h_{u,\mu}, G_1\} = \{h_{u,\mu}, G_2\} = 0$ implies that

$$\lambda_1 = -\frac{\{g_u, G_2\}}{\{G_1, G_2\}} = -\frac{1}{2}(p \cdot y - u \cdot q), \lambda_2 = -\frac{\{g_u, G_1\}}{\{G_2, G_1\}} = \frac{1}{2}(y \cdot q).$$

Each $x(t), y(t), p(t), q(t)$ extremal curve $x(t), y(t), p(t), q(t)$ formally satisfies equation (12), with $u(t) \cdot y(t) = 0$, and

$$-\mu + p(t) \cdot y(t) + q(t)u(t) \geq -\mu + p(t) \cdot y(t) + v \cdot q(t)$$

for all $v \in \mathbb{R}^n$ such that $\|v\| \leq c$ and $v \cdot y(t) = 0$.

On open intervals where $q(t)$ is not equal to zero, $u(t) = c\frac{q(t)}{\|q(t)\|}$ and $\lambda_1 = \frac{1}{2}(p \cdot y(t) - c\|q(t)\|)$. The extremal curves are the solutions of

$$\frac{dx}{dt} = y(t), \frac{dy}{dt} = c\frac{q(t)}{\|q(t)\|}, \frac{dp}{dt} = 0, \frac{dq}{dt} = -p + (p \cdot y(t) - c\|q(t)\|)y(t).$$

$$(16.17)$$

corresponding to the Hamiltonian $h = -\mu + p \cdot y + c\|q(t)\| + \lambda_1 G_1 + \lambda_2 G_2$. In contrast to the dynamic case, where each extremal curve $q(t)$ is equal to zero at most once, here it may happen that $q(t) = 0$ on an interval. Our next proposition describes this phenomenon in more detail.

Proposition 16.2 *If $(x(t), y(t), p(t), q(t))$ is an extremal curve such that $q(t) = 0$ on an interval (t_0, t_1), then $u(t) = 0$ on (t_0, t_1), and $x(t)$ is a straight line on this interval.*

Proof The extremal $(x(t), y(t), p(t), q(t))$ satisfies

$$\frac{dx}{dt} = y(t), \frac{dy}{dt} = u(t), \frac{dp}{dt} = 0, \frac{dq}{dt} = -p + (p \cdot y(t) - u(t) \cdot q(t))y(t).$$

On an open interval where $q(t) = 0$, $p = (p \cdot y(t))y(t) = 0$. Since the Hamiltonian is equal to zero along an extremal curve $p \cdot y(t) = \mu$. Then $\mu = 1$, for if not, then $p = 0$, and both $p = 0$ and $q = 0$ violate the non-degeneracy condition for abnormal extremals. Therefore, $p = y$ and hence $0 = \frac{dy}{dt} = u(t)$. $\square$

16.2.1 Integrals of motion and integrability

All of the above problems are invariant under the isometry group G of the underlying space M, and consequently, the Hamiltonians always have extra integrals of motion generated by the associated moment map. We will show that these integrals of motion lead to elegant solutions in both the curvature and the Dubins–Delauney problem, not just on $\mathbb{E}^n$, but on spaces of constant curvature as well. Paradoxically, the simple solutions in $\mathbb{E}^n$ of either the covariant derivative problem, or the dynamic Dubins problem in $\mathbb{E}^n$ are rather exceptional. It seems that these two problems are not integrable on spaces of non-zero curvature. We will come to this point again in the next section.

In the meantime, let us investigate the symmetries induced by the action of the group of motions of $\mathbb{E}^n$. System (16.7), the constraints $||u|| \leq c$ and $(u, y) = 0$, as well as the appropriate cost functionals are all invariant under the diagonal action $(v, R)(x, y) \rightarrow (v + Rx, Ry)$ of G. That means that each infinitesimal generator

$$V(x, y) = (Ax + a, Ay) = \frac{d}{d\epsilon} \left(e^{A\epsilon}x + \epsilon a, e^{A\epsilon}y \right) |_{\epsilon=0}, \, A \in so_n(R), \, a \in \mathbb{R}^n$$

is a symmetry for all of the above problems. Therefore, the Hamiltonian $h = p \cdot (Ax + a) + y \cdot Ay$ is constant along the extremal curves, that is, $\{H, h\} = 0$, where H denotes the Hamiltonian that generates the extremal curves. The above implies that p is constant (which we already knew) and that $p \cdot Ax + q \cdot Ay$ is constant for each skew-symmetric matrix A. Since

$$p \cdot Ax + q \cdot Ay = \langle A, p \wedge x + q \wedge y \rangle,$$

where $\langle , \rangle$ is the trace product on the space of skew-symmetric matrices

$$\Lambda = p \wedge x + q \wedge y \tag{16.18}$$

is constant along the solutions of each system (16.16) and (16.17). The matrix Λ corresponds to the moment map associated with the above group action.

It follows that the spectral invariants of Λ are constants of motion for any Hamiltonian system whose projection is invariant under G. Since Λ is skew-symmetric with a four-dimensional range, its non-zero spectrum is given by a polynomial of degree 4 of the form

$$\lambda^4 + a\lambda^2 + b = 0. \tag{16.19}$$

A calculation of this spectrum, quite similar to the one in the section on elastic curves in Chapter 13, reveals two integrals of motion

$$I_1 = ||p|| \text{ and } I_2 = ||p||^2||q||^2 - (q \cdot p)^2 - (y \cdot p)^2||q||^2. \tag{16.20}$$

The second integral is the square of the volume spanned by y, p, q. Since

$$I_2 = ||p - (p \cdot y)y - (p \cdot q)\frac{q}{||q^2}||^2||q||^2,$$

I_2 is positive and can be written as $I_2 = h^2$.

With these integrals at our disposal we can easily recover the results of Proposition 13.8 in Chapter 13 for the curvature problem. Since the setting is sufficiently different we will restate the main result and give it an independent proof.

Proposition 16.3 *Let $\kappa(t)$ and $\tau(t)$ denote the curvature and the torsion of an elastic curve. Then,*

1. $\frac{d\xi}{dt}^2 + \xi^3 - 4H\xi^2 - 4(I_1^2 - H^2)\xi + 4I_2 = 0$, *where $\kappa^2(t) = \xi(t)$.*
2. $(\kappa^2\tau)^2 = I_2 = h^2$.
3. *If $T(t), N(t), B(t)$ denote the Serret–Frenet triad defined by*

$$\frac{dx}{dt} = T(t), \frac{dT}{dt} = \kappa N, \frac{dN}{dt} = -\kappa T + \tau B,$$

then $\frac{dB}{dt}(t)$ is contained in the linear span of $T(t), N(t), B(t)$. Hence, the Serret–Frenet frame generated by an elastic curve is at most three dimensional.

Proof Since $\kappa^2 = ||q||^2$, $\frac{d}{dt}k^2 = 2(\frac{dq}{dt} \cdot q) = -2p \cdot q$. Therefore,

$$\frac{1}{4}\left(\frac{d\xi}{dt}\right)^2 = (p \cdot q)^2 = ||q||^2||p||^2 - (y \cdot p)^2||q||^2 - I_2$$

$$= \kappa^2||p||^2 - \left(H - \frac{1}{2}||q||^2\right)^2||q||^2 - I_2^2 = \xi I_1^2 - \left(H - \frac{1}{2}\xi)^2\right)\xi + I_2$$

$$= -\frac{1}{4}\xi^3 + H\xi^2 + (I_1^2 - H^2)\xi - I_2.$$

Therefore (1) holds.

Since, $\frac{dx}{dt} = y$ and $||y|| = 1$, $T(t) = y(t)$. Hence, $\frac{dT}{dt} = \frac{dy}{dt} = q(t)$, and $N = \frac{1}{||q||}q$. Then,

$$\frac{dN}{dt} = -\left(q \cdot \frac{dq}{dt}\right)\frac{1}{||q||^3}q + \frac{1}{||q||}(-p + (p \cdot y - ||q||^2)y)$$

$$= \frac{p \cdot q}{||q||^3}q + \frac{1}{||q||}(-p + (p \cdot y - ||q||^2)y)$$

$$= -\frac{p \cdot q}{||q||^2}N + \frac{1}{||q||}(-p + (p \cdot y)T) - \kappa T = -\kappa T + \tau B.$$

The above yields

$$\kappa \tau B = -p + (p \cdot y)T + \frac{p \cdot q}{||q||}N.$$

This implies that $\frac{dB}{dt}$ is contained in the linear span of T, N, B and moreover, it implies that

$$(\kappa\tau)^2 = \left\| -p + (p \cdot y)y + \frac{p \cdot q}{||q||^2}q \right\|^2$$
$$= \frac{1}{||q||^2}(||p||^2||q||^2 - (p \cdot q)^2 - (p \cdot y)^2||q||^2) = \frac{I_2}{\kappa^2}.$$

$\square$

The proof is now complete.

Remark 16.4 The group symmetries are equally relevant for the covariant derivative problem. For, if $\xi(t) = ||u(t)||^2$, where $u(t) = q(t) = -pt + q_0$, then $u \cdot \frac{du}{dt} = -q(t) \cdot p$. Hence,

$$\frac{1}{4}\left(\frac{d\xi}{dt}\right)^2 = \frac{1}{4}\xi^3 + 4H\xi^2 + (I_1^2 - H^2)\xi - I_2,$$

where $H = \frac{1}{2}||q||^2 + p \cdot y$. Therefore, the optimal cost $\frac{1}{2}\int_0^t ||u(s)||^2\,ds = \frac{1}{2}\int_0^t \xi(s)\,ds$ evolves over the same elliptic curve for both the unconstrained and the curvature problem.

Let us now return to the extremal equations for the Delauney–Dubins problem. On open intervals where $q(t)$ is not equal to zero, the extremals show a great deal of similarity with the equations of the elastic problem. The proposition below summarizes the essential properties.

Proposition 16.5 Let $\kappa(t)$ and $\tau(t)$ denote the curvature and the torsion associated with an extremal curve $(x(t), y(t), p(t), q(t))$. On any interval (t_0, t_1) such that $q(t) \neq 0$, $\kappa = c$, and $||q||$ is a solution of

$$\left(||q||\frac{d}{dt}||q||\right)^2 = -c^2||q||^4 + 2\mu c||q||^3 + (I_1^2 - \mu^2)||q||^2 - I_2. \quad (16.21)$$

Moreover, $(||q||^2\tau)^2 = I_2 = h^2$ and $\frac{dB}{dt}(t)$ in the Serret–Frenet triad defined by

$$\frac{dx}{dt} = T(t), \frac{dT}{dt} = \kappa N, \frac{dN}{dt} = -\kappa T + \tau B,$$

is contained in the linear span of $T(t), N(t), B(t)$. Hence the Serret–Frenet frame generated by an extremal curve is at most three dimensional.

Proof   Since $\kappa = ||u(t)||$ and $u(t) = c\frac{q(t)}{||q(t)||}$, $\kappa = c$. As in the previous case $I_1 = ||p||$ and $I_2 = ||p||^2||q||^2 - (p \cdot q)^2 - (p \cdot y)^2||q||^2$ are constants of motion. Then $\frac{d}{dt}||q|| = \frac{1}{||q||}\left(q \cdot \frac{dq}{dt}\right) = -\frac{1}{||q||}(p \cdot q)$ and therefore,

$$\left(||q||\frac{d}{dt}||q||\right)^2 = (p \cdot q)^2 = ||p||^2||q||^2 - (p \cdot y)^2||q||^2 - I_2$$

$$= I_1^2||q||^2 - (\mu - c||q||)^2||q||^2 - I_2$$

$$= -c^2||q||^4 + 2\mu c||q||^3 + (I_1^2 - \mu^2)||q||^2 - I_2.$$

Now $\frac{dT}{dt} = \frac{dy}{dt} = c\frac{q(t)}{||q(t)||}$ and so $N(t) = \frac{q(t)}{||q(t)||}$. The rest of the proof is the same as in the proof of Proposition 16.3.   □

Let us now consider the special case $I_2 = 0$. Equation (16.21) reduces to

$$\frac{d||q||^2}{dt} = -c^2||q||^2 + 2\mu c||q|| + I_1^2 - \mu^2. \tag{16.22}$$

Apart from the stationary solutions $c||q|| = \mu \pm ||p||$, the solutions of (16.22) are of the form

$$c||q(t)|| = \mu - ||p|| \sin c((t - t_0)), \quad ||q(t)|| \neq 0 \tag{16.23}$$

The stationary solutions result in helices since both the curvature and the torsion are constant. The case $\mu = 0$, $p = 0$ is ruled out by the Maximum Principle. The case $||p|| \geq 1$ is significant, in the sense that it contains solutions $q(t)$ which are zero at isolated instances of time.

Definition 16.6 The hypersurface $S = \{(x, y, p, q) : q = 0\}$ is called the switching surface.

All extremal curves which cross the switching surface are confined to $I_2 = 0$ as a consequence of their continuity. An extremal curve that does not originate on S may cross S either *tangentially or transversally*. If T is the time of crossing then the crossing is tangential if $-||q(T)||^2 + 2\mu||q(T)|| + ||p||^2 - \mu^2 = 0$, otherwise it is transversal. It is transversal when $\lim_{t \to T} \frac{d||q(t)||}{dt} \neq 0$. Otherwise, it is tangential. The critical case $||p|| = 1$ is the only case in which the crossing is tangential. All other crossings are transversal and reside on $||p|| > 1$.

In the normal and non-geodesic case the time interval between two consecutive crossings is larger than π while in the abnormal case this time interval is equal to π. Both of these observations follow from (16.23).

Proposition 16.7 *Let $(x(t), y(t), p, q(t))$ be an extremal curve that crosses the switching surface transversally, i.e., $||p|| > 1$. Then $x(t)$ consists of* *concatenations of arcs of circles of radius $\frac{1}{\sqrt{c}}$ all contained in the plane defined by $x(0), y(0),$ and p.*

Proof Let κ and τ be the curvature and torsion of $x(t)$. Then $\kappa(t) = c$, and $\tau = 0$ on every interval on which $||q|| > 0$. On these intervals $x(t)$ moves along an arc of a circle of radius $\frac{1}{\sqrt{c}}$ centered at some point a, and therefore, $x(t)$ can be represented as

$$x(t) - a = \frac{1}{\sqrt{c}}(A \cos \sqrt{c}t + B \sin \sqrt{c}t), \ ||A||^2 = ||B||^2 = 1, \ (A \cdot B) = 0.$$

Then, $\frac{dx}{dt} = y(t)$ and $\frac{dy}{dt} = -\sqrt{c}(x(t) - a) = \kappa N$. It follows that the normal vector $N(t)$ is equal to $\frac{1}{\sqrt{c}}(a - x)$).

Suppose now that T denoted the time that the two arcs of circles are joined together, that is, suppose that $q(T) = 0$. Let $N_-(T) = \lim_{t < T, t \to T} N(t)$ and $N_+(T) = \lim_{t > T, t \to T} N(t)$. On any open interval I such that $q(t) \neq 0$, $N(t) = \frac{q(t)}{||q(t)||}$ and

$$\frac{dN(t)}{dt} = \frac{1}{||q(t)||}(-p + (p \cdot y - c||q(t)||)y(t))$$

$$- \frac{q(t)}{||q(t)||^2}\frac{d||q(t)||}{dt} - \kappa T - cy(t).$$

Therefore,

$$-p + (p \cdot y)y(t) = N(t)\frac{d||q(t)||}{dt}.$$

It follows from Proposition 16.5 that $\lim_{t \to T} \frac{d||q(t)||}{dt} = \pm\sqrt{||p||^2 - 1}$ depending whether this limit is from the right or from the left. Since the Hamiltonian is equal to zero along an extremal curve, $p \cdot y(T) = \mu$. Hence,

$$N_- = -\frac{-p + \mu y(T)}{\sqrt{||p||^2 - 1}}, \ N_+ = \frac{-p + \mu y(T)}{\sqrt{||p||^2 - 1}}.$$

Since these normals are colinear, the concatenated arcs of the circles are in the same plane. □

Proposition 16.8 *Suppose now that an extremal curve $(x(t), y(t), p, q(t))$ crosses the switching surface tangentially ($||p|| = 1$). Then either there is no switching at the time of the crossing and $x(t)$ is a circle of radius $\frac{1}{\sqrt{c}}$, or $x(t)$ is the concatenation of an arc of a circle of radius $\frac{1}{\sqrt{c}}$ with a straight line, possibly followed by another arc of a circle of radius $\frac{1}{\sqrt{c}}$.*

The proof is simple and will be omitted.

Let us now consider the extremals that reside on $I_2 \neq 0$. It will be convenient to rescale the time variable by $s = ct$. The rescaled variable $||q(s)||$ in equation (16.21) satisfies

$$\left(||q||\frac{d}{ds}||q||\right)^2 = -||q||^4 + 2\frac{\mu}{c}||q||^3 + \frac{(I_1^2 - \mu^2)}{c^2}||q||^2 - \frac{I_2}{c^2}.$$

The preceding equation can be written in the von Schwarz's form [VS] as

$$(u - \mu_0)^2 \left(\frac{du}{ds}\right)^2 = (u - \mu)^2(k^2 - u^2) - h_0^2$$

with $u = ||q|| - \frac{\mu}{c}$, $\mu_0 = -\frac{\mu}{c}$, $k = \frac{I_1}{c}$, $h_0 = \frac{\sqrt{I_2}}{c}$. The latter equation can be further rescaled to

$$(\zeta - l)^2 \left(\frac{d\zeta}{ds}\right)^2 = (\zeta - l)^2(1 - \zeta^2) - h^2 \tag{16.24}$$

with the rescaled variables

$$\zeta = \frac{u}{k}, \; l = \frac{\mu}{k}, \; h^2 = \frac{h_0^2}{k^4}.$$

Then ζ_1 and ζ_2, the roots of the equation $(\zeta - l)^2(1 - \zeta^2) - h^2 = 0$, are the stationary solutions of (16.24). Along them $||q(t)||$ is constant and hence the torsion of the corresponding extremal curve $x(t)$ is constant. Therefore, $x(t)$ is a helix.

Any other solution $\zeta(t)$ satisfies

$$\zeta_1 < \zeta(t) < \zeta_2,$$

and can be expressed in terms of elliptic functions by integrating

$$\int \frac{\zeta - l}{\sqrt{(\zeta - l)^2(1 - \zeta^2) - h^2}} d\zeta = \pm t. \tag{16.25}$$

Then, $\tau^2(t) = \frac{I_2}{||q||^4}$, and the solutions are reduced to solving the Serret–Frenet system

$$\frac{dT}{dt} = cN(t), \; \frac{dN}{dt} = -cT(t) + \tau(t)B(t), \; \frac{dB}{dt} = -\tau(t)N(t).$$

To relate to Dubins' paper of 1957 consider the planar case. Then p, q and y must be linearly dependent, since they lie in the same plane, hence $I_2 = 0$. Therefore every solution is a either an arc of a circle or a line segment, or a concatenation of arcs of circles and line segments.

An optimal solution that involves a line segment cannot have two consecutive circle switchings since such extremals reside on $||p|| = 1$ and the switching interval is 2π. Therefore, optimal solutions that involve a line segment must of the form *CLC* or any sub-path of these.

In the remaining case, optimal solutions are the concatenations of circles. We showed that such solutions reside on $||p|| > 1$, and the time interval between any two consecutive switchings must be equal and greater than π. But, then any path $C_t C_\alpha C_\alpha$ with $t > 0$ and $\alpha > \pi$ cannot be optimal (Monroy's lemma, see [Mi1, p. 141]). Hence, optimal paths along arcs of circles must be of the form $C_\alpha C_\beta C_\gamma$ with $\beta > \pi$, or any sub-path of these arcs.

So apart from Monroy's lemma, the main contents of Dubins' paper can be read directly from our Hamiltonian setup. It would be of interest to investigate which of these optimal planar solutions remain optimal in higher-dimensional spaces.

16.2.2 The sphere and the hyperboloid

Guided by our earlier results, we will consider both cases simultaneously in terms of the parameter $\epsilon = \pm 1$, with $S^n = \{x \in \mathbb{R}^{n+1} : ||x||^2 = 1\}, \epsilon = 1$ for the elliptic case, and the hyperboloid $\mathbb{H}^n = \{x \in \mathbb{R}^{n+1} : x_{n+1}^2 = 1 + \sum_{i=1}^n x_i^2,$ $x_{n+1} > 0\}, \epsilon = -1$ for the hyperbolic case. Then $\mathbb{S}_\epsilon^n$ will denote the unit sphere S^n when $\epsilon = 1$, and $\mathbb{H}^n$ when $\epsilon = -1$, but this time $\mathbb{S}_\epsilon^n$ will be considered as an immersed submanifold of the ambient space $M_\epsilon = \mathbb{R}^{n+1}$ equipped with the quadratic form

$$(v, w)_\epsilon = \sum_{i+1}^n v_i w_i + \epsilon v_0 w_0,$$

and the induced "norm" $||v||_\epsilon^2 = (v \cdot v)_\epsilon$, so that $\mathbb{S}_\epsilon^n = \{x \in M_\epsilon : ||x||_\epsilon^2 - \epsilon = 0\}$.

Then, the tangent bundle TM_ϵ will be identified with points (x, y) in $\mathbb{R}^{n+1} \times \mathbb{R}^{n+1}$ such that $||x||_\epsilon^2 - \epsilon = 0, (x, y)_\epsilon = 0$. The second tangent bundle $T(TM_\epsilon)$ will be identified with points $(x, y, \dot{x}, \dot{y})$ in $\mathbb{R}^{4(n+1)}$ with $(\dot{x}, \dot{y})$ the tangent vectors at (x, y). The cotangent bundle $T^*T(M_\epsilon)$ will be identified with the tangent bundle $T(TM_\epsilon)$ via the formula

$$l((\dot{x}, \dot{y})) = (p, \dot{x})_\epsilon + \epsilon(q, \dot{y})_\epsilon,$$

for each point l in the cotangent space at (x, y). In this correspondence, $(x, y, \dot{x}, \dot{y}) \in TM_\epsilon \iff (x, y, p, \epsilon q) \in T^*M_\epsilon$. In the coordinates (x, y, p, q), the Poisson bracket on T^*M_ϵ is given by

$$\{f, h\} = \left(\frac{\partial f}{\partial x} \cdot \frac{\partial h}{\partial p}\right)_\epsilon + \left(\frac{\partial f}{\partial y} \cdot \frac{\partial h}{\partial q}\right)_\epsilon - \left(\left(\frac{\partial f}{\partial p} \cdot \frac{\partial h}{\partial x}\right)_\epsilon + \left(\frac{\partial f}{\partial q} \cdot \frac{\partial h}{\partial y}\right)_\epsilon\right).$$
$$(16.26)$$

We now return to our second-order systems. If $v(t)$ is a curve of tangent vectors along a curve $x(t)$ on $\mathbb{S}^n$, then the covariant derivative $\frac{D_{x(t)}}{dt}(v(t))$ is equal to the orthogonal projection of $\frac{dv}{dt}$ on the tangent space at $x(t)$, that is,

$$\frac{D_{x(t)}}{dt}(v(t)) = \frac{dv}{dt} - \epsilon\left(\frac{v(t)}{dt}, x(t)\right)_\epsilon x(t).$$

In particular, when $v = \frac{dx}{dt}$ then

$$\frac{D_{x(t)}}{dt}\left(\frac{dx}{dt}\right) = \frac{d^2x(t)}{dt^2} + \epsilon\left(\frac{d^2x(t)}{dt^2}, x(t)\right)_\epsilon = \frac{d^2x(t)}{dt^2} - \epsilon\left(\frac{dx}{dt}, \frac{dx}{dt}\right)_\epsilon x(t).$$

Therefore, the covariant derivative problem can be phrased as the problem of minimizing $\frac{1}{2}\int_0^T \|u(t)\|^2\, dt$ over the solutions $(x(t), y(t))$ of the constrained system

$$\frac{dx}{dt} = y(t),\ \frac{dy}{dt} = u(t) - \epsilon(y(t), y(t))_\epsilon x(t),$$

$$G_1 = \|x\|_\epsilon^2 - \epsilon = 0, G_2 = (x, y)_\epsilon = 0. \tag{16.27}$$

The curvature problem can be phrased similarly over the solutions of

$$\frac{dx}{dt} = y(t),\ \frac{dy}{dt} = u(t) - \epsilon x(t),\ u(t) \cdot x(t) = 0, \tag{16.28}$$

subject to additional constraints

$$(u(t) \cdot x(t))_\epsilon = (u(t) \cdot y(t))_\epsilon = 0,$$

$$G_1 = \|x\|_\epsilon - \epsilon = 0,\ G_2 = \|y\|_\epsilon^2 - 1 = 0,\ G_3 = (x, y)_\epsilon = 0. \tag{16.29}$$

We will reverse the order of our investigations, and begin with the curvature and the Delauney–Dubins problems first. The state space N for these problems is defined by three constraints G_1, G_2, G_3; hence the cotangent bundle T^*N_ϵ is defined by six constraints

$$G_1 = \|x\|_\epsilon - \epsilon = 0,\ G_2 = \|y\|_\epsilon^2 - 1 = 0,\ G_3 = (x, y)_\epsilon = 0,$$

$$G_4 = (x, p)_\epsilon = 0,\ G_5 = (y, q)_\epsilon = 0,\ G_6 = (y, p)_\epsilon + \epsilon(x, q)_\epsilon = 0.$$

An easy calculation using (16.26) shows that these constraints conform to the following Poisson bracket table:

Table 16.1

$\{,\}$	G_1	G_2	G_3	G_4	G_5	G_6
G_1	0	0	0	$2\|x\|_\epsilon^2$	0	$2(x, y)_\epsilon$
G_2	0	0	0	0	$2\|y\|_\epsilon^2$	$2\epsilon(x, y)_\epsilon$
G_3	0	0	0	$(x, y)_\epsilon$	$(x, y)_\epsilon$	$\epsilon\|x\|_\epsilon^2 + \|y\|_\epsilon^2$
G_4	$-2\|x\|_\epsilon^2$	0	$-(x, y)_\epsilon$	0	0	$(y, p)_\epsilon - (x, q)_\epsilon$
G_5	0	$-2\|y\|_\epsilon^2$	$-(x, y)_\epsilon$	0	0	$\epsilon(x, q)_\epsilon - \epsilon(y, p)_\epsilon$
G_6	$-2(x, y)_\epsilon$	$-2\epsilon(x, y)_\epsilon$	$-\epsilon\|x\|_\epsilon^2$ $-\|y\|_\epsilon^2$	$-(y, p)_\epsilon$ $+(x, q)_\epsilon$	$\epsilon(y, p)_\epsilon$ $-\epsilon(x, q)_\epsilon$	0

For the curvature problem, the cost-extended Hamiltonian is given by

$$h_{u,\mu} = -\mu \frac{1}{2}\|u\|_\epsilon^2 + (y,p)_\epsilon + (q, u - \epsilon x)_\epsilon + \sum_{i=1}^{6} \lambda_i G_i, \mu = 0, 1, \quad (16.30)$$

where the multipliers λ_i, are chosen so that $\{h_{u,\mu}, G_i\} = 0$ for any control u that satisfies $(u, x)_\epsilon = (u, y)_\epsilon = 0$. To calculate the above multipliers, it will be convenient first to calculate the intermediate Poisson brackets $\{h_0, G_i\}$ where $h_0 = (p, y)_\epsilon + (q, u - \epsilon x)_\epsilon$. Then,

$$\{h_0, G_1\} = 2(x, y)_\epsilon, \{h_0, G_2\} = -2(y, (u - \epsilon x)_\epsilon,$$
$$\{h_0, G_3\} = -\|y\|_\epsilon^2 - (x, u - \epsilon x)_\epsilon,$$
$$\{h_0, G_4\} = -(q, x)_\epsilon - (p, y)_\epsilon, \{h_0, G_5\} = (p, y)_\epsilon - (q, u - \epsilon x)_\epsilon,$$
$$\{h_0, G_6\} = 2\epsilon((p, x)_\epsilon - (q, y)_\epsilon) - (p, u)_\epsilon.$$

The multipliers are determined from

$$0 = \{h_{u,\mu}, G_i\} = \{h_0, G_i\} + \sum_{j=1}^{6} \lambda_j \{G_j, G_i\}.$$

On T^N, $\{G_1, G_i\} = -2\delta_{i4}$, $\{G_2, G_i\} = -2\delta_{i5}$, $\{G_3, G_i\} - 2\delta_{i6}$, and $\{h_0, G_1\} = \{h_0, G_2\} = \{h_0, G_3\} = 0$. Therefore $\lambda_4 = \lambda_5 = \lambda_6 = 0$. Similar calculations yield

$$\lambda_1 = 0, \lambda_2 = \frac{1}{2}(q, u)_\epsilon, \lambda_3 = \frac{1}{2}(p, u)_\epsilon.$$

Along any extremal curve $x(t), y(t), p(t), q(t)$, the corresponding control $u(t)$ satisfies

$$h_{u(t),\mu}(x(t), y(t), p(t), q(t)) \geq h_{v,\mu}(x(t), y(t), p(t), q(t)),$$

for $v \in \mathbb{R}^n$, such that $(v \cdot x(t))_\epsilon = (v \cdot y(t))_\epsilon = 0$. An easy calculation based on the Lagrange multiplier rule shows that each extremal control in the normal case must be of the form

$$u = q - \epsilon(q, x)_\epsilon x - (q, y)_\epsilon y. \quad (16.31)$$

Subnormal extremals project onto the geodesics and can be ignored, since geodesics are also the projections of normal extremals.

The substitution $u = q - \epsilon(q, x)_\epsilon x - (q, y)_\epsilon y$ into the expression (16.30) results in the Hamiltonian

$$h = \frac{1}{2}(\|q\|^2 - \epsilon(q, x)_\epsilon^2 - (q, y)_\epsilon^2) + (y, p)_\epsilon - \epsilon(q, x)_\epsilon + \lambda_2 G_2 + \lambda_3 G_3,$$
$$(16.32)$$

whose Hamiltonian equations, when restricted to T^*N, are given by

$$\frac{dx}{dt} = y, \quad \frac{dy}{dt} = q - \epsilon(q, x)_\epsilon x - \epsilon x,$$
$$\frac{dp}{dt} = \epsilon q + \epsilon(q, x)_\epsilon q - \lambda_3 y, \quad \frac{dq}{dt} = -p - 2\lambda_2 y - \lambda_3 x. \tag{16.33}$$

If H denotes the restriction of h to T^*N, then

$$H = \frac{1}{2}(\|q\|^2 - (x, q)_\epsilon^2) + 2(y, p)_\epsilon = \frac{1}{2}\kappa^2 + 2(y, p)_\epsilon, \tag{16.34}$$

with $\kappa = \|u\|$ the geodesic curvature of $x(t)$. Equations (16.33), together with H in (16.34) constitute the Hamiltonian system generated by the curvature problem.

The Hamiltonian for the Delauney–Dubins problem is obtained in a similar manner. The reader may readily verify that in the case that $q(t)$ is neither equal to zero, nor colinear with $x(t)$ on an open interval I, the extremal control must be of the form

$$u(t) = c\frac{q(t) - \epsilon(q, x)_\epsilon x(t)}{\|q(t) - \epsilon(q, x)_\epsilon x(t)\|} \tag{16.35}$$

and

$$\frac{dx}{dt} = y, \quad \frac{dy}{dt} = c\frac{q(t) - \epsilon(q, x)_\epsilon x(t)}{\|q(t) - \epsilon(q, x)_\epsilon x(t)\|} - \epsilon x,$$
$$\frac{dp}{dt} = c\frac{\epsilon(q, x)_\epsilon q}{\|q(t) - \epsilon(q, x)_\epsilon x(t)\|} - \lambda_3 y, \quad \frac{dq}{dt} = -p - 2\lambda_2 y - \lambda_3 x \tag{16.36}$$

are the extremal equations for the Dubins–Delauney problem corresponding to the Hamiltonian

$$H = -\mu + 2(y, p)_\epsilon + c\|q - \epsilon(q, x)_\epsilon x\|. \tag{16.37}$$

The reader may also verify that an extremal curve in which $q(t)$ is colinear with $x(t)$ projects onto a geodesic on the base space $\mathbb{S}_\epsilon^n$.

16.2.3 Integrability

Let SO_ϵ denote the isometry group of $\mathbb{S}_\epsilon^n$. Then

$$S = x \wedge_\epsilon p + y \wedge_\epsilon q$$

is the moment matrix associated with the geodesic symmetries induced by the action of SO_ϵ. Hence S is constant along the solutions of either Hamiltonian system (16.33) or (16.37). As before, $v \otimes_\epsilon w$ is the rank one marix defined by $(v \otimes_\epsilon w)x = (w, x)_\epsilon v$, and $v \wedge_\epsilon w = v \otimes_\epsilon w - w \otimes_\epsilon v$.

The spectral invariants of S provide the appropriate integrals of motion in terms of which the extremal equations can be integrated. We leave it to the reader to show that the characteristic polynomial of the restriction of S to the vector space spanned by x, y, p, q is given by

$$\lambda^4 + a\lambda^2 + b = 0,$$

where

$$a = ||q||_\epsilon^2 - 2(y,p)_\epsilon (x,q)_\epsilon + \epsilon ||p||_\epsilon^2,$$
$$b = \epsilon(||p||_\epsilon^2 ||q||_\epsilon^2 - (q,x)_\epsilon^2 - ||q||^2 (y,p)_\epsilon^2) + (y,p)_\epsilon^2 (q,x)_\epsilon^2 - ||p||_\epsilon^2 (q,x)_\epsilon^2.$$

On T^*N_ϵ, $(y,p)_\epsilon = -\epsilon(q,x)_\epsilon$, hence,

$$a = \epsilon(||p||_\epsilon^2 + 2(y,p)_\epsilon^2 + \epsilon ||q||_\epsilon^2),$$
$$b = \epsilon((||p||_\epsilon^2 - (y,p)_\epsilon^2)(||q||_\epsilon^2 - \epsilon(x,q)_\epsilon^2) - (p,q)_\epsilon^2).$$

It follows that

$$I_1 = ||p||_\epsilon^2 + 2(y,p)_\epsilon^2 + \epsilon ||q||_\epsilon^2,$$
$$I_2 = (||p||_\epsilon^2 - (y,p)_\epsilon^2)(||q||_\epsilon^2 - \epsilon(x,q)_\epsilon^2) - (p,q)_\epsilon^2$$

are integrals of motion for each Hamiltonian system (16.33) and (16.37).

Together with H, these integrals of motion describe the geometric invariants of the elastic curves, exactly as in Proposition 13.8 of Chapter 13.

Proposition 16.9   Let $(x(t), y(t), p(t), q(t))$ denote any solution of the Hamiltonian system associated with the curvature problem (equations (16.33)). Let $\kappa(t)$ and $\tau(t)$ denote the curvature and the torsion of the projected curve $x(t)$ and let $\xi(t) = \kappa^2(t)$. Then

$$\frac{d\xi}{dt}^2 = -\xi^3 + 4(H - \epsilon)\xi^2 + 4(I_1 - H^2)\xi - 4I_2, \text{ and } (\kappa^2\tau)^2 = I_2. \quad (16.38)$$

If $T(t), N(t), B(t)$ denote the Serret–Frenet triad defined by

$$\frac{dx}{dt} = T(t), \frac{D_x}{dt}T(t) = \kappa N(t), \frac{D_x}{dt}N(t) = -\kappa T(t) + \tau B(t),$$

then $\frac{D_x}{dt}B(t)$ is contained in the linear span of $T(t), N(t), B(t)$. Hence, the Serret–Frenet frame generated by an elastic curve is at most three dimensional.

Proof   We will only verify the equation for the curvature. Other statements of the proposition follow by the identical arguments used in the Euclidean case. Since $\xi = \kappa^2 = ||q||^2 - \epsilon(q,x)_\epsilon^2$,

$$\frac{d\xi}{dt} = 2((q,\dot{q})_\epsilon - \epsilon((q,x)_\epsilon((\dot{q},x_\epsilon) + (q,\dot{x})_\epsilon)) = -2(p,q)_\epsilon.$$

Therefore,

$$\frac{1}{4}\left(\frac{d\xi}{dt}\right)^2 = -I_2 + (\|p\|_\epsilon^2 + 2(y,p)_\epsilon^2)\xi = -I_2 + (I_1 - 3(y,p)_\epsilon^2 - \epsilon\|q\|_\epsilon^2)\xi$$

$$= -I_2 + (I_1 - 4(y,p)_\epsilon^2 - \epsilon\xi)\xi = -I_2 + \left(I_1 - \left(H - \frac{1}{2}\xi\right)^2 - \epsilon\xi\right)\xi$$

$$= -\frac{1}{4}\xi^3 + (H - \epsilon)\xi^2 + (I_1 - H^2)\xi - I_2^2.$$

$\square$

The solutions to the Delauney–Dubins problem, obtained by essentially the same arguments as in the Euclidean situation, are described by the following proposition.

Proposition 16.10 *Let $\kappa(t)$ and $\tau(t)$ denote the curvature and the torsion associated with an extremal curve $(x(t), y(t), p(t), q(t))$ for the Delauney–Dubins problem (equations (16.36). Let $\xi(t) = \|q(t) - \epsilon(x(t), q(t))_\epsilon x(t)\|$. Then $\xi(t)$ is a solution of*

$$\left(\xi\frac{d\xi}{dt}\right)^2 = -(c^2 + \epsilon)\xi^4 + 2c\mu\xi^3 + (I_1 - \mu^2)\xi^2 - I_2. \tag{16.39}$$

Moreover, $(\xi^2\tau)^2 = I_2$ and $\frac{dB}{dt}(t)$ in the Serret–Frenet triad is contained in the linear span of $T(t), N(t), B(t)$. Hence, the Serret–Frenet frame generated by an extremal curve is at most three dimensional.

Proof In this situation $I_2 = (\|p\|_\epsilon^2 - (y,p)_\epsilon^2)\xi^2 - (p,q)^2$ and $(y,p)_\epsilon^2 = \frac{1}{4}(\mu - c\xi)^2$. Moreover, $2\xi\frac{d\xi}{dt} = \frac{d}{dt}(\|q\|^2 - \epsilon(x,q)_\epsilon^2 p) = -2(p,q)_\epsilon$.

Therefore,

$$\left(\xi\frac{d\xi}{dt}\right)^2 = (p,q)^2 = (\|p\|^2 - (p,y)^2)\xi^2 - I_2$$

$$= (I_1 - 2(y,p)_\epsilon^2 - \epsilon\|q\|_\epsilon^2 - (y,p)_\epsilon^2)\xi^2 - I_2$$

$$= (I_1 - 3(y,p)_\epsilon^2 - \epsilon\|q\|^2)\xi^2 - I_2$$

$$= (I_1 - 4(y,p)_\epsilon^2 - \epsilon\xi^2)\xi^2 - I_2 = (I_1 - (\mu - c\xi)^2 - \epsilon\xi^2)\xi^2 - I_2$$

$$= -(c^2 + \epsilon)\xi^4 + 2c\mu\xi^3 + (I_1 - \mu^2)\xi^2 - I_2.$$

The rest of the proof is completely analogous as in the Euclidean case and will be omitted. $\square$

The content of this proposition is essentially the same as that found in [Mo], with the exception that in the hyperbolic case one should take $c > 1$, for otherwise, the control system is not controllable [Mt].

The hypersurface $S = \{(x,y,p,q) : \xi = ||q - \epsilon(x,q_\epsilon)x|| = 0\}$ is the switching surface. If an extremal curve crosses the switching surface at some time T, then either $q(T)$ is equal to zero, or it is colinear with $x(T)$. In either case, $(p(T),q(T))_\epsilon$ is equal to zero, and $I_2 = 0$. Therefore, extremal curves that cross the switching surface reside on the hypersurface $I_2 = 0$.

The extremals that reside on $I_2 = 0$ are the solutions of

$$\left(\frac{d\xi}{dt}\right)^2 = -(c^2 + \epsilon)\xi^2 + 2\mu c\xi + I_1 - \mu^2. \tag{16.40}$$

Analogous to the Euclidean case, the stationary solutions project onto non-Euclidean helices (curves having both the curvature and the torsion constant). Otherwise, the solutions are of the form

$$(c^2 + \epsilon)\xi(t) = a - b\sin\sqrt{c^2 + \epsilon}\,(t - t_0),\ a = \mu c,\ b = \sqrt{I_1(c^2 + \epsilon) - \mu^2\epsilon}.$$

The associated extremal curve does not cross the switching surface when $a > b$. It crosses the switching surface tangentially when $a = b$, that is, when $I_1 = \mu^2$. The crossing is transversal for all other values of a and b. In the normal and non-geodesic case, the time interval between two consecutive crossings is larger than $\dfrac{\pi}{\sqrt{c^2+\epsilon}}$, while, in the abnormal case, this time interval is equal to $\dfrac{\pi}{\sqrt{c^2+\epsilon}}$.

On hypersurfaces $I_2 \neq 0$, the solutions of (16.32) can be expressed in terms of elliptic functions much in the same manner as in the Euclidean case. They are generic, and non-switching. It would be nice to know their optimality status. Some switching extremals project onto optimal solutions. For instance, the concatenation of three circles is optimal in any dimension, but some extremals which are optimal in two dimensions may lose their optimality in higher dimensions. These considerations, however, require a separate study, which will not be pursued here. Instead, we will now turn our attention to the Hamiltonian system associated with the covariant derivative problem.

16.2.4 Non-Euclidean covariant derivative problem

In this situation the cotangent bundle T^*N of the tangent bundle N of $\mathbb{S}^n_\epsilon$ is realized as the set of all quadruples (x,y,p,q) in $\mathbb{R}^{4(n+1)}$ subject to four constraints

$$G_1(x) = ||x||_\epsilon^2 - \epsilon = 0, G_2(x, y) = (x, y)_\epsilon = 0,$$
$$G_3 = (x, p)_\epsilon = 0, G_4 = (p, y)_\epsilon + \epsilon(x, q)_\epsilon = 0.$$

The Hamiltonian lift of the cost-extended system

$$\frac{dx_0}{dt} = \frac{1}{2}||u(t)||_\epsilon^2, \frac{dx}{dt} = y(t), \frac{dy}{dt} = u(t) - \epsilon(y(t), y(t))_\epsilon x(t)$$

is given by

$$h_{u,\mu} = -\frac{\mu}{2}||u||_\epsilon^2 + (p, y)_\epsilon + (q, u - \epsilon||y||_\epsilon^2 x)_\epsilon + \sum_{i=1}^{4} \lambda_i G_i.$$

To ensure that T^*N is invariant for the flow of $\vec{h}_{u,\mu}$, we require that

$$\{h_{u,\mu}, G_i\} = 0, i = 1, 2, 3, 4 \tag{16.41}$$

on T^*N. For our purposes we need to consider only the multipliers which do not vanish on T^*N, therefore, it suffices to solve (16.41) only on T^*N. To do so, let $h_0 = (p, y)_\epsilon + (q, u - \epsilon||y||_\epsilon^2 x)_\epsilon$. Then,

$$\{h_0, G_1\} - 2(x, y)_\epsilon, \{h_0, G_2\} = \quad (x, u)_\epsilon, \{h_0, G_3\} = \epsilon||y||_\epsilon^2 (q, x)_\epsilon - (p, y)_\epsilon,$$
$$\{h_0, G_4\} = -\epsilon(q, y)_\epsilon(1 + ||y||_\epsilon^2) + \epsilon(p, x)_\epsilon(1 + ||y||_\epsilon^2) - (p, u)_\epsilon.$$
$$\tag{16.42}$$

The preceding equalities imply that $\{h_0, G_1\} = \{h_0, G_2\} = 0$ on T^*N. Furthermore,

$$\{G_i, G_1\} = -2\epsilon\delta_{i3}, \{G_i, G_2\} = -(1 + ||y||_\epsilon^2)\delta_{i4},$$
$$\{G_i, G_3\} = 2\epsilon\delta_{i1}, \{G_i, G_4\} = (1 + ||y||_\epsilon^2)\delta_{i2},$$

as can be readily verified from Table 16.1. Together with (16.41) and (16.42) these relations imply that the restrictions of $\lambda_i, i = 1, 2, 3, 4$ to T^*N are given by

$$\lambda_1 = \frac{1}{2}(\epsilon(p, y)_\epsilon - ||y||_\epsilon^2 (q, x)_\epsilon),$$

$$\lambda_2 = \epsilon(q, y)_\epsilon + \frac{1}{1 + ||y||_\epsilon^2}(p, q)_\epsilon, \lambda_3 = 0, \lambda_4 = 0. \tag{16.43}$$

In the normal case, the extremal controls are of the form $u = q - \epsilon(q, x)_\epsilon x$, which results in a single Hamiltonian

$$h = \frac{1}{2}(||q||^2 - \epsilon(x, q)_\epsilon^2) + (p, y)_\epsilon - \epsilon||y||^2(q, x)_\epsilon$$
$$+ \lambda_1 G_1 + \lambda_2 G_2 + \lambda_3 G_3 + \lambda_4 G_4.$$

The restriction of h to T^*N can be written as

$$H = \frac{1}{2}(||q||_\epsilon^2 - \epsilon(q,x)_\epsilon^2) + (1 + ||y||_\epsilon^2)(p,y)_\epsilon,$$

because $-\epsilon(q,x)_\epsilon = (p,y)_\epsilon$.

The corresponding Hamiltonian equations on T^*N are then given by

$$\frac{dx}{dt} = y, \frac{dy}{dt} = q - \epsilon(x,q)_\epsilon x - \epsilon||y||_\epsilon^2 x = u - \epsilon||y||_\epsilon^2 x,$$

$$\frac{dp}{dt} = \epsilon(q,x)_\epsilon q + \epsilon||y||_\epsilon^2 q - 2\lambda_1 x - \lambda_2 y, \frac{dq}{dt} = -p - \lambda_2 x + 2\epsilon(q,x)_\epsilon y.$$

$$(16.44)$$

The covariant derivative problem is also SO_ϵ invariant, and therefore $I_1 = ||p||_\epsilon^2 + 2(y,p)_\epsilon^2 + \epsilon||q||_\epsilon^2$ and $I_2 = (||p||_\epsilon^2 - (y,p)_\epsilon^2)(||q||_\epsilon^2 - \epsilon(x,q)_\epsilon^2) - (p,q)_\epsilon^2$ are integrals of motion. However, in contrast to the previous problems, the existence of these two integrals of motion is insufficient to describe the solutions, and without an extra integral of motion, it is difficult to say anything about the solutions of (16.44), apart from the obvious geodesic solution $u = 0, p = q = 0$. The Hamiltonian equations for the dynamic Dubins problem can be obtained along similar lines. This problem is also subject to the same difficulties.

16.3 Curvature problem on symmetric spaces

Let us now derive the Hamiltonian equations associated with the curvature problem on a symmetric space $M = G/K$ associated with a semi-simple and connected Lie group G with an involutive automorhism σ. Recall first the basic ingredients from Chapter 8:

1. Cartan decomposition $\mathfrak{g} = \mathfrak{p} \oplus \mathfrak{k}$ of the Lie algebra $\mathfrak{g}$ induced b the tangent map σ subject to the conditions

$$[\mathfrak{p},\mathfrak{p}] \subseteq \mathfrak{k}, \ [\mathfrak{p},\mathfrak{k}] \subseteq \mathfrak{p}, \ [\mathfrak{k},\mathfrak{k}] \subseteq \mathfrak{k}.$$

2. Horizontal distribution $\mathcal{D}$ defined by the left-invariant vector fields with values in $\mathfrak{p}$.
3. Riemannian pair (G,K), and Ad_K-invariant, positive-definite, quadratic form $\langle\,,\,\rangle$ on $\mathfrak{p}$ with the associated norm $||\ ||$.
4. Horizontal curves $g(t)$ and their relations to the projected curves $x(t) = \pi(g(t))$

$$\frac{dx}{dt} = \pi_*(G(t)A(t)), \ A(t) = g^{-1}(t)\frac{dg}{dt} \text{ and } \left|\left|\frac{dx}{dt}(t)\right|\right| = ||A(t)||, \quad (16.45)$$

where $||\frac{dx}{dt}||$ is the Riemannian length in the base manifold.

Also recall that the covariant derivative $\frac{D_{x(t)}}{dt}(v(t))$ of a curve of tangent vectors $v(t)$ along $x(t)$ is given by the projection $\pi_*(g(t)\frac{dB}{dt}(t))$, where $g(t)$ is a horizontal curve that projects onto $x(t)$, and $B(t)$ is the horizontal lift of $v(t)$.

It follows from above that the geodesic curvature $\kappa(t)$ of a curve $x(t)$ parametrized by arc length is given by $\kappa(t) = \|\frac{dA}{dt}(t)\|$, $\|A(t)\| = 1$, where $\pi(g(t)) = x(t)$, and $A(t) = g^{-1}(t)\frac{dg}{dt}(t)$.

We will now consider the curvature problem on G/K over the curves $x(t)$ that satisfy the given boundary conditions

$$\left\|\frac{dx}{dt}(t)\right\| = 1, x(0) = x_0, \frac{dx}{dt}(0) = v_0, \ x(T) = x_1, \frac{dx}{dt}(T) = v_1$$

and have a bounded and measurable covariant derivative $\frac{D_{x(t)}}{dt}\left(\frac{dx}{dt}(t)\right)$ in the interval $[0, T]$. We will denote this set by $\gamma(v_0, v_1, T)$.

To take advantage of the left-invariant symmetries, we will consider the lifted version on the tangent bundle of $G \times S_{\mathfrak{p}}$, where $S_{\mathfrak{p}}$ denotes the unit sphere in $\mathfrak{p}$. Each curve $x(t) \in \gamma(v_0, v_1, T)$ lifts to a curve $(g(t), \Lambda(t)) \in G \times S_{\mathfrak{p}}$ such that $g(t)$ is horizontal curve that projects onto $x(t)$ and the derivative $\frac{dg}{dt}$ projects onto $\frac{dx}{dt}$. That is, the lifted curve $g(t)$ satisfies (16.45). Note also that the boundary data lift to the initial manifold $S_0 = \{(g_0h, h^{-1}A_0h) : h \in K\}$, and the terminal manifold $S_1 = \{(g_1h, h^{-1}A_1h) : h \in K\}$.

We will let $\Gamma(S_0, S_1, T)$ denote the set of curves $(g(t), \Lambda(t)) \in G \times S_\epsilon$, $t \in [0, T]$, such that $(g(0), \Lambda(0)) \in S_0$ and $(g(T), \Lambda(T)) \in S_1$. This data then defines the lifted curvature problem:

Definition 16.11 The lifted curvature problem Find a curve $(g(t), \Lambda(t))$ in $\Gamma(S_0, S_1, T)$ such that $(g(t), \Lambda(t))$ is a solution of

$$\frac{dg}{dt}(t) = g(t)\Lambda(t), \ \frac{d\Lambda}{dt} = U(t), \langle\Lambda(t), U(t)\rangle = 0, \ t \in [0, T] \qquad (16.46)$$

that attains the minimum of $\frac{1}{2}\int_0^T \|U(t)\|^2 dt$ among all other solutions in $\Gamma(S_0, S_1, T)$.

The proposition below states precisely the connections between the curvature problem on the base manifold and the lifted curvature problem on T^*G.

Proposition 16.12 *A curve $x(t)$ in $\gamma(v_0, v_1, T)$ is a solution to the curvature problem if and only if $x(t)$ is the projection of an optimal solution $(g(t), \Lambda(t))$ of the lifted curvature problem.*

Proof The projection $x(t) = \pi(g(t))$ of every solution curve $(g(t), \Lambda(t))$ of equation (16.46) satisfies $\frac{dx}{dt} = \pi_*(g(t)\Lambda(t))$. Since $||\Lambda(t)|| = 1$, $\left|\left|\frac{d\Lambda}{dt}\right|\right| = ||U(t)||$ is equal to the geodesic curvature of $x(t)$. Moreover, $x(0) = \pi(g(0))$, $\frac{dx}{dt}(0) = \pi_*(g(0)\Lambda(0)) = \pi_*(g_0 h(h^{-1} A_0 h)) = v_0$, and similarly $x(T) = \pi(g(T))$, $\frac{dx}{dt}(T) = \pi_*(g(T)\Lambda(T)) = \pi_*(g_1 h(h^{-1} A_1 h)) = v_1$. Therefore, the projections of solutions of equation (16.46) that belong to $\Gamma(S_0, S_1, T)$ are in $\gamma(v_0, v_1, T)$, and conversely, every curve in $\gamma(v_0, v_1, T)$ is the projection of a solution curve of (16.46) in $\Gamma(S_0, S_1, T)$. Hence optimal solutions of the curvature problem on the base manifold coincide with the projections of optimal solutions to the lifted curvature problem.   □

16.3.1 Extremal curves of the lifted problem

We will follow the formalism developed in the previous section. The tangent bundle T^*G will be realized as the product $G \times \mathfrak{g}$ via the left translations. For simplicity of exposition we will assume that the quadratic form $\langle\,,\,\rangle$ is a scalar multiple of the Killing form Kl on $\mathfrak{p}$. This assumption means that we may consider $\langle\,,\,\rangle$ to be the restriction of an invariant, non-degenerate quadratic form on $\mathfrak{g}$, relative to which $\mathfrak{k}$ and $\mathfrak{p}$ are orthogonal. Then the cotangent bundle $T^*(G \times S_\mathfrak{p})$ will be identified with $T^*G \times T^*(S_\mathfrak{p})$, and T^*G will be identified with $G \times \mathfrak{g}^*$ via the left translations, and finally, $\mathfrak{g}^*$ will be identified with $\mathfrak{g}$ via the quadratic form $\langle\,,\,\rangle$. Similarly, the cotangent bundle $T^*(S_\mathfrak{p})$ will be identified with the tangent bundle $\{(\Lambda, X) \in \mathfrak{p} \times \mathfrak{p} : ||\Lambda|| = 1, \langle\Lambda, X\rangle = 0\}$.

The Poisson structure $\{\,,\,\}$ on the cotangent bundle of $G \times S_\mathfrak{p}$ is the direct sum of the Poisson structures $\{\,,\,\}_1$ on T^*G and $\{\,,\,\}_2$ on $T^*S_\mathfrak{p}$. Having identified T^*G with $G \times \mathfrak{g}^*$, then the Poisson structure on T^*G, restricted to the left-invariant functions f and h on $\mathfrak{g}^*$, coincides with the Lie–Poisson structure on $\mathfrak{g}^*$. After the identification of $\ell \in \mathfrak{g}^*$ with $L \in \mathfrak{g}$, the first Poisson structure is given by the familiar formula $\{f, h\}_1(L) = \langle[df, dh], L\rangle$.

The second Poisson structure $\{\,,\,\}_2$ coincides with the restriction ot the canonical Poisson structure on $\mathfrak{p}^*$ to $T^*S_\mathfrak{p}$ and is given by $\{f, h\}_2(\Lambda, X) = \left\langle\frac{\partial f}{\partial \Lambda}, \frac{\partial h}{\partial X}\right\rangle - \left\langle\frac{\partial f}{\partial X}, \frac{\partial h}{\partial \Lambda}\right\rangle$. So the configuration space $T^*(G \times S_\mathfrak{p})$ is identified with the quadruples (g, L, Λ, X) subject to the constraints $G_1 = ||\Lambda|| - 1 = 0$ and $G_2 = \langle\Lambda, X\rangle = 0$.

We will now obtain the equations for the extremal curves in the coordinates of the ambient space $G \times \mathfrak{g} \times \mathfrak{g} \times \mathfrak{g}$ much in the same way as in the previous section. We begin with the Hamiltonian lift

$$\mathcal{H}_{\mu, U} = -\frac{\lambda}{2}||U||^2 + \langle L, \Lambda\rangle + \langle X, U\rangle + \lambda_1 G_! + \lambda_2 G_2, \mu = 0, 1, \quad (16.47)$$

where the multipliers λ_1 and λ_2 are to be chosen so that the integral curves of the corresponding Hamiltonian vector field $\vec{\mathcal{H}}_{\mu,U}$ that originate in $T^*(G \times S_{\mathfrak{p}})$ remain there for all t for any control $U(t)$ that satisfies the constraint $\langle U(t), \Lambda \rangle = 0$. This condition will be fulfilled whenever $\{\mathcal{H}_{\mu,U}, G_1\} = \{\mathcal{H}_{\mu,U}, G_2\} = 0$. The above Hamiltonian can be rewritten as

$$\mathcal{H}_{\mu,U} = -\frac{\lambda}{2}||U||^2 + \langle P, \Lambda \rangle + \langle X, U \rangle + \lambda_1 G_1 + \lambda_2 G_2, \mu = 0, 1,$$

where $L = P + Q$, with $P \in \mathfrak{p}$ and $Q \in \mathfrak{k}$. Remember that $\mathfrak{p}$ and $\mathfrak{k}$ are orthogonal relative to $\langle , \rangle$, and therefore, $\langle L, \Lambda \rangle = \langle P, \Lambda \rangle$.

Since both G_1 and G_2 are functions on $T^*S_{\mathfrak{p}}$ the preceding equations reduce to

$$\{\mathcal{H}_0, G_1\}_2 + \lambda_2 \{G_2, G_1\}_2 = 0, \text{ and } \{\mathcal{H}_0, G_2\}_2 + \lambda_1 \{G_1, G_2\}_2 = 0,$$

where $\mathcal{H}_0 = -\lambda\frac{1}{2}||U||^2 + \langle P, \Lambda \rangle + \langle X, U \rangle$.

It follows that

$$\{\mathcal{H}_0, G_1\} = \left\langle \frac{\partial \mathcal{H}_0}{\partial \Lambda}, \frac{\partial G_1}{\partial X} \right\rangle - \left\langle \frac{\partial \mathcal{H}_0}{\partial X}, \frac{\partial G_1}{\partial \Lambda} \right\rangle = -2\langle U, \Lambda \rangle = 0,$$

$$\{\mathcal{H}_0, G_2\} = \left\langle \frac{\partial \mathcal{H}_0}{\partial \Lambda}, \frac{\partial G_2}{\partial X} \right\rangle - \left\langle \frac{\partial \mathcal{H}_0}{\partial X}, \frac{\partial G_2}{\partial \Lambda} \right\rangle = \langle P, \Lambda \rangle - \langle U, X \rangle,$$

$$\{G_1, G_2\} = 2||\Lambda||^2 = 2.$$

Hence,

$$\lambda_1 = \frac{1}{2}(\langle U, X \rangle - \langle P, \Lambda \rangle), \lambda_2 = 0.$$

Indeed, the integral curves of $\vec{\mathcal{H}}$ are given by

$$\frac{dg}{dt} = g\left(\frac{\partial \mathcal{H}}{\partial L}\right) = g\Lambda, \frac{dL}{dt} = \left[\frac{\partial \mathcal{H}}{\partial L}, L\right] = [\Lambda, L],$$

$$\frac{d\Lambda}{dt} = \frac{\partial \mathcal{H}}{\partial X} = U, \frac{dX}{dt} = -\frac{\partial \mathcal{H}}{\partial \Lambda} = -P - 2\lambda_1 \Lambda. \tag{16.48}$$

The Maximum Principle states that each optimal trajectory $(g(t), \Lambda(t), U(t))$ of (16.46), is the projection of an extremal curve $(g(t), \Lambda(t), L(t), X(t))$, a solution of system (16.48), subject to the condition that

$$\mu\frac{1}{2}||U(t)||^2 + \langle L(t), \Lambda(t) \rangle + \langle U(t), X(t) \rangle$$

$$\geq \mu\frac{1}{2}||V||^2 + \langle L(t), \Lambda(t) \rangle + \langle V, X(t) \rangle, \tag{16.49}$$

for every $V \in \mathfrak{p}$ such that $\langle V, \Lambda(t) \rangle = 0$, and also subject to the transversality conditions at both ends.

Transversality means that the extremal must annihilate the tangent spaces of the initial and the terminal manifolds at its end points. Let us begin with the initial manifold $S_0 = \{g_0 h, h^{-1} A_0 h), h \in K\}$, which we can take at $g_0 = e$ without any loss of generality. Then $(g(0), \Lambda(0)) \in S_0$ means that $g(0) = h_0, \Lambda(0) = h_0^{-1} A_0 h_0$ for some $h_0 \in K$. The tangent space at this point consists of vectors $(h_0 \dot{h}, [\Lambda(0), \dot{h}]), \dot{h} \in \mathfrak{k}$. After the idenitifications of the cotangent spaces with the tangent spaces via the quadratic form, the annihilation of the tangent vectors by the extremals translates into the orthogonality condition $\langle L(0), \dot{h} \rangle + \langle X(0), [\dot{h}, \Lambda(0)] \rangle = 0, \dot{h} \in \mathfrak{k}$. The above is the same as

$$\langle Q(0) + [\Lambda(0), X(0)], \dot{h} \rangle = 0, \dot{h} \in \mathfrak{k}.$$

Hence,

$$Q(0) + [\Lambda(0), X(0)] = 0. \tag{16.50}$$

A similar condition can be derived at the terminal point.

Let us now return to the inequality (16.49). It implies that $U = X(t)$ in the normal case and that $X(t) = 0$ in the abnormal case. So in the normal case, the Maximum Principle implies that each optimal trajectory $(g(t), \Lambda(t))$ is the projection of an integral curve of a single Hamiltonian

$$H = \frac{1}{2} \|X\|^2 + \langle P, \Lambda \rangle + \lambda_1 G_1, \tag{16.51}$$

while in the abnormal case, the Maximum Principle simply says that the abnormal extremal is confined to the submanifold $X = 0$.

Proposition 16.13 *Every abnormal extremal is generated by $U(t) = 0$, hence projects onto a geodesic in the base manifold M.*

Proof Let $(g(t), \Lambda(t), L(t), X(t))$ be an abnormal extremal. When $X(t) = 0$, equations (16.48) reduce to $P(t) = 2\lambda_1 \Lambda(t)$. Therefore, $Q(t)$ is constant, since $\frac{dQ}{dt} = [\Lambda(t), P(t)]$. The transversality condition (16.50) implies that $Q(t) = 0$. But then $P(t)$ is constant since $\frac{dP}{dt} = [\Lambda, Q]$. It follows that $\lambda_1 \Lambda(t)$ is constant. Since $P = 0$ is ruled out by the Maximum Principle, $\lambda_1 \neq 0$. Therefore, $\Lambda(t)$ is constant, and hence $U(t) = 0$. $\square$

Since the geodesics are also the projections of normal extremal curves, the abnormal curves will be ignored. Normal extremals are the integral curves of H in (16.51). They are the solutions of

$$\frac{dP}{dt} = [\Lambda, Q], \quad \frac{dQ}{dt} = [\Lambda, P], \quad \frac{d\Lambda}{dt} = X, \quad \frac{dX}{dt} = -P - (\|X\|^2 - \langle P, \Lambda \rangle)\Lambda .$$
$$\tag{16.52}$$

subject to the transversality conditions. To integrate these conditions with the symmetries of the problem, note that

$$\frac{d}{dt}(Q + [\Lambda, X]) = [\Lambda, P] + [\Lambda, -P - \alpha\Lambda] = 0.$$

Therefore, $Q(t) + [\Lambda(t), X(t)]$ is constant.

The above integral of motion is a consequence of the symmetries

$$g \rightarrow gh, \ \Lambda \rightarrow h^{-1}\Lambda h, \ h \in K,$$

for the lifted curvature problem. The infinitesimal generator of one parameter group of symmetries $(g e^{\epsilon Y}, e^{-\epsilon Y} \Lambda e^{\epsilon Y}), Y \in \mathfrak{k}$ is equal to $(gY, [Y, \Lambda])$. Therefore, the Hamiltonian lift H_Y, given by $H_Y = \langle Y, Q \rangle + \langle [Y, \Lambda], X \rangle$ is constant along the flow of (16.52). It follows that

$$Q(t) + [\Lambda(t), X(t)]$$

is the moment map associated with the above symmetries. The moment map is equal to zero because of transversality condition (16.50). Therefore $Q(t) = -[\Lambda(t), X(t)]$ and (16.52) can be rewritten as

$$\frac{dP}{dt} = -[\Lambda, [\Lambda, X]], \ \frac{dQ}{dt} = [\Lambda, P],$$

$$\frac{d\Lambda}{dt} = X, \ \frac{dX}{dt} = -P - (||X||^2 - \langle P, \Lambda \rangle)\Lambda \tag{16.52a}$$

16.3.2 The Euler–Lagrange equation

At this point it may be instructive to relate the extremal equations (16.52) to the Euler–Lagrange equation

$$2(\nabla_T)^3 T + \nabla_T(3\kappa^2 + \lambda)T + 2R(\nabla_T T, T)T = 0 \tag{16.53}$$

cited by Langer and Singer in [LS, p. 3]. In this notation, $T(t)$ is the unit tangent vector of a critical curve, and R is the Riemannian curvature tensor defined by $R(X, Y)Z = \nabla_X \nabla_Y Z - \nabla_Y \nabla_X Z - \nabla_{X,Y} Z$. The Lie bracket is subject to the convention $\nabla_X Y - \nabla_Y X = [X, Y]$.

To translate to our notation, $T(t) = g(t)\Lambda(t)$, and $(\nabla_T)^k T = g(t)\frac{d^k\Lambda}{dt^k}$. The Riemannian curvature tensor is given by $R(X, Y)Z = g[[A, B], C]$ for any left-invariant vector fields $X = gA$, $Y = gB$, and $Z = gC$ [HI, p. 215]. The sign is altered to conform to our definition of the Lie bracket.

Therefore, the Euler–Lagrange equation (16.53) can be reduced to the following equation:

$$\frac{d^3\Lambda}{dt^3} + \frac{1}{2}\frac{d}{dt}((3\kappa^2 + \lambda)\Lambda) + \left[\left[\frac{d\Lambda}{dt}, \Lambda, \right]\Lambda\right] = 0. \qquad (16.54)$$

This equation could be easily obtained from (16.52): since $\frac{d\Lambda}{dt} = X$, $\frac{d^2\Lambda}{dt^2} = \frac{dX}{dt} = -P + (\langle P, \Lambda\rangle - ||X||^2)\Lambda$. The quantity $\langle P, \Lambda\rangle - ||X||^2$ is equal to $H - \frac{3}{2}\kappa^2$, because $H = \frac{1}{2}||X||^2 + \langle P, \Lambda\rangle = \frac{1}{2}\kappa^2 + \langle P, \Lambda\rangle$. Therefore,

$$\frac{d^3\Lambda}{dt^3} = -\frac{dP}{dt} = -[\Lambda, [\Lambda, X]] - \frac{d}{dt}\left(\frac{3}{2}\kappa^2 - H)\Lambda\right)$$

$$= -\left[\Lambda, \left[\Lambda, \frac{d\Lambda}{dt}\right]\right] - \frac{d}{dt}\left(\frac{3}{2}\kappa^2 - H\right)\Lambda,$$

which agrees with (16.54) for $\lambda = -2H$.

Conversely one may begin with equation (16.54) and arrive at the equations (16.52) by defining $X = \frac{d\Lambda}{dt}$, $Q = -[\Lambda, \frac{d\Lambda}{dt}]$, and $P = \frac{d^2\Lambda}{dt^2} - \frac{d}{dt}\left(\frac{3}{2}\kappa^2 - H\right)\Lambda$. We leave the verification to the reader.

16.3.3 Riemannian curvature and the extremal equations

Let us now investigate the effect of the Riemannian curvature tensor on the solvability of the extremal equations. Recall that the Riemannian curvature of the base space $M = G/K$ is given by the bilinear form $\kappa : \mathfrak{p} \times \mathfrak{p} \to \mathbb{R}$ with

$$\kappa(A, B) = \langle[[A, B], A], B\rangle \qquad (16.55)$$

(Definition 8.37, Chapter 8). The Riemannian curvature can be also expressed as $\kappa(A, B) = \langle R(X, Y)X, Y\rangle$, where R the Riemannian curvature tensor $R_g(X, Y)Z = g[[A, B], C]$ over the left-invariant vector fields $X(g) = gA$, $Y(g) = gB$ and $Z(g) = gC$.

Also recall that sectional curvature of any two-dimensional linear subspace S in $\mathfrak{p}$ is equal to $\kappa(A, B)$, where A and B is any orthonormal basis in S. For symmetric spaces of constant curvature the sectional curvatures are all equal.

Let us now investigate the advantage of constant curvature on the solutions of the curvature problem. So assume that $ad^2A(B) = -\epsilon B$, $\epsilon \neq 0$ for any A and B in $\mathfrak{p}$ that satisfy $||A|| = 1$ and $\langle B, A\rangle = 0$. Then equations (16.52a) become

$$\frac{dP}{dt} = \epsilon X, \quad \frac{dQ}{dt} = [\Lambda, P], \quad \frac{d\Lambda}{dt} = X, \quad \frac{dX}{dt} = -P - (||X||^2 - \langle P, \Lambda\rangle)\Lambda.$$

But then, $-\frac{dP}{dt} + \epsilon\frac{d\Lambda}{dt} = 0$, and therefore, $\epsilon\Lambda(t) - P(t) = A$ for some element $A \in \mathfrak{p}$. Furthermore, (16.50) implies that

$$[\Lambda(t), Q(t)] + [\Lambda(t), [\Lambda(t), X(t)]] = [\Lambda(t), Q(t)] - \epsilon X(t) = 0.$$

We therefore have

Proposition 16.14   *On spaces of constant curvature ϵ, extremal equations (16.52a) satisfy*

$$X(t) = \frac{1}{\epsilon}[\Lambda(t), Q(t)] \text{ and } \epsilon\Lambda(t) - P(t) = A, \tag{16.56}$$

for some constant element $A \in \mathfrak{p}$. Consequently,

$$\begin{aligned}\frac{dP}{dt} &= \epsilon\frac{d\Lambda}{dt} = \epsilon X = [\Lambda, Q] = \frac{1}{\epsilon}[P + A, Q], \\ \frac{dQ}{dt} &= [\Lambda, P] = \frac{1}{\epsilon}[P + A, P] = \frac{1}{\epsilon}[A, P].\end{aligned} \tag{16.57}$$

Corollary 16.15   *Let $\bar{P} = \frac{1}{\epsilon}P, \bar{Q} = \frac{1}{\epsilon}Q, \bar{A} = \frac{1}{\epsilon}A$, then $X(t) = [\Lambda(t), \bar{Q}(t)]$ and*

$$\frac{d\bar{P}}{dt} = [\bar{P} + \bar{A}, \bar{Q}], \frac{d\bar{Q}}{dt} = [\bar{A}, \bar{P}], \Lambda = \bar{P} + \bar{A}. \tag{16.58}$$

The information provided by Proposition 16.14 is easily transported to the material on elastic curves in Chapter 13 via the following proposition.

Proposition 16.16   *Suppose that $(g(t), \Lambda(t))$ is the projection of an extremal curve $(X(t), \bar{P}(t), \bar{Q}(t))$. There exists a curve $h(t) \in K$ such that $h(t)\bar{A}h^1(t) = \Lambda(t)$. Then,*

$$\bar{g}(t) = g(t)h(t), L_{\mathfrak{k}} = h^{-1}\bar{Q}(t)h(t), L_{\mathfrak{p}} = h^{-1}(t)\bar{P}h(t)$$

satisfy the following equations

$$\frac{d\bar{g}}{dt} = \bar{g}(\bar{A} + L_{\mathfrak{k}}), \frac{dL_{\mathfrak{k}}}{dt} = [\bar{A}, L_{\mathfrak{p}}],$$

$$\frac{dL_{\mathfrak{p}}}{dt} = [L_{\mathfrak{k}}, L_{\mathfrak{p}}] + [\bar{A}, L_{\mathfrak{k}}]. \text{ Moreover, } L_k = -[\bar{A}, h_1 Xh].$$

Proof   Let $\bar{P}(t)$ and $\bar{Q}(t)$ be as in equation (16.58). If $h(t)$ denotes the solution of $\frac{dh}{dt} = Q(t)h(t)$ with $h_0 \in K$ defined by $h_0\bar{A}h_0^{-1} = \Lambda(\theta)$, then $\Lambda(t) = h(t)\bar{A}h^{-1}(t)$, since they both satisfy the same differential equation with the same initial conditions. then

$$\frac{dL_{\mathrm{p}}}{dt} = h^{-1}([Q,P] + [\Lambda,Q])h = [L_{\mathfrak{k}}, L_{\mathrm{p}}] + [\bar{A}, L_{\mathfrak{k}}],$$

$$\frac{dL_{\mathfrak{k}}}{dt} = h^{-1}[\Lambda,P]h = [\bar{A}, L_{\mathrm{p}}], \qquad (16.59)$$

$$\frac{d\bar{g}}{dt} = g\Lambda h + gQh = \bar{g}(h^{-1}\Lambda h + h^{-1}Qh) = \bar{g}(\bar{A} + L_{\mathfrak{k}})$$

The remaining equality $L_{\mathfrak{k}} = -[\bar{A}, h^{-1}Xh]$ follows from $Q + [\Lambda, X] = 0$. $\square$

Corollary 16.17 *Let* $\mathfrak{k}_{\bar{A}} = \{X \in \mathfrak{k} : [\bar{A}, X] = 0\}$ *and let* $\mathfrak{k}_{\bar{A}}^{\perp}$ *denote its orthogonal complement in* $\mathfrak{k}$. *Then* $L_{\mathfrak{k}}(t)$ *in the above proposition belongs to* $\mathfrak{k}_{\bar{A}}^{\perp}$

Proof If $Y \in \mathfrak{k}_{\bar{A}}$ then

$$\langle L_{\mathfrak{k}}, Y \rangle_{\epsilon} = -\langle Y, [\bar{A}, h^{-1}Xh] \rangle = \langle [\bar{A}, Y], h^{-1}Xh \rangle = 0.$$

$\square$

Corollary 16.18

$$I = ||P||^2 + \epsilon ||Q||^2 I_2 = ||Q||^2 ||P||^2 - ||Q||^2 \langle A, P \rangle^2 - \langle Q, P \rangle^2$$

are constants of motion for system (16.57)

Proof Equation (16.59) is the same as equation (13.4) in Chapter 13. Hence it yields the same isospectral integrals. $\square$

Corollary 16.19 *Suppose that* $\epsilon = \pm 1$. *Then the curvature and the torsion of an elastic curve are subject to the same conditions as stated in Proposition 13.7 of Chapter 13.*

16.4 Elastic curves and the rolling sphere problem

Equations (16.58) can be also interpreted as the extremal equations associated with a certain sub-Riemannian problem on $SO_{\epsilon} \times SO_{\epsilon}$ that is intimately connected with the rolling sphere geodesics. The rolling sphere geodesics are curves of shortest length on a Riemannian manifold M that can be traced by the point of contact of an oriented sphere as it rolls on M from a given initial configuration to a fixed final configuration. Below we will show a remarkable fact, that the elastic curves can be obtained entirely through the rolling sphere problems, and, along the way, we will be able to show that the rolling sphere geodesics trace elastic curves on all two-dimensional stationary manifolds of constant curvature [JZ].

Let us begin this brief detour into the rolling sphere problems by first identifying equations (16.58) with a sub-Riemannian problem on $G_\epsilon \times G_\epsilon$ where $G_\epsilon = SO_\epsilon$. Let us first establish some notations. The Lie algebra $\mathfrak{g}$ of G is equal to $\mathfrak{g}_\epsilon \times \mathfrak{g}_\epsilon$ and can be identified with $\mathfrak{g}_1 \oplus \mathfrak{g}_2$, where $\mathfrak{g}_1 = \mathfrak{g}_\epsilon \times \{0\}$, and $\mathfrak{g}_2 = \{0\} \times \mathfrak{g}_\epsilon$. Then $\mathfrak{g}_\epsilon = \mathfrak{p}_\epsilon \oplus \mathfrak{k}_\epsilon$ induces factors $\mathfrak{p}_1 = \mathfrak{p}_\epsilon \times (0)$, $\mathfrak{p}_2 = (0) \times \mathfrak{p}_\epsilon$. It also induces complementary subalgebras $\mathfrak{k}_1$ and $\mathfrak{k}_2$ defined similarly. The resulting decomposition $\mathfrak{g} = \mathfrak{p} \oplus \mathfrak{k}$, with $\mathfrak{p} = \mathfrak{p}_1 \oplus \mathfrak{p}_2$, and $\mathfrak{k} = \mathfrak{k}_1 \oplus \mathfrak{k}_2$ conforms to the usual Cartan conditions. Let us remind the reader that the Lie bracket on $\mathfrak{g}$ obeys the following rule:

$$[A + B, C + D] = [A, C] + [B, D]$$

for any A, C in $\mathfrak{g}_1$ and B, D in $\mathfrak{g}_2$.

We are now ready to introduce the rolling sphere distribution. For simplicity of exposition, assume that $\epsilon = \pm 1$ and then assume that α and β are any numbers such that $\alpha - \beta = 1, \alpha \neq \beta$. Then consider the left-invariant distribution $\mathcal{D}$ on G defined by

$$\mathcal{D}((g_1, g_2)) = \{(\alpha X(g_1), \beta X(g_2)) : X \in \mathfrak{p}_\epsilon\}. \tag{16.60}$$

We will subsequently relate $\mathcal{D}$ to the rolling spheres, but first let us consider $\mathcal{D}$ in its own right. To begin with, we have the following fundamental property.

Proposition 16.20 *Let $\Gamma = \{(\alpha X, \beta X) : X \in \mathfrak{p}_\epsilon\}$ and let $Lie(\Gamma)$ denote the Lie algebra generated by Γ. Then $Lie(\Gamma)$ is equal to $\mathfrak{g}$ whenever $\alpha \neq \beta$. If $\alpha = \beta$, then $Lie(\Gamma)$ is isomorphic to so_ϵ.*

Proof If $(\alpha X, \beta X)$ and $(\alpha Y, \beta Y)$ are any elements of Γ then their Lie bracket $(\alpha^2[X, Y], \beta^2[X, Y])$ is in $Lie(\Gamma)$. Therefore, $\{(\alpha^2 Z, \beta^2 Z) : Z \in \mathfrak{k}_\epsilon\} \subseteq Lie(\Gamma)$ since $[\mathfrak{p}_\epsilon, \mathfrak{p}_\epsilon] = \mathfrak{k}_\epsilon$. The Lie brackets of order two show that $(\alpha^3 ad^2(X)(Y), \beta^3 ad^2(X)(Y)) \in Lie(\Gamma)$. If furthermore, X and Y are orthogonal, and $||X|| = 1$, then $ad^2(X)(Y) = -\epsilon Y$. Then any linear combination $\lambda_1(\alpha Y, \beta Y) - \lambda_2 \epsilon(\alpha^3 Y, \beta^3 Y)$ is in $Lie(\Gamma)$. In particular, when $\lambda_1 = \lambda_2 \epsilon \alpha^2$, then

$$\lambda_1(\alpha Y, \beta Y) - \lambda_2 \epsilon(\alpha^3 Y, \beta^3 Y) = \lambda_2 \epsilon \beta(\alpha - \beta)(0, Y).$$

Since Y is arbitrary, $(0, \mathfrak{p}_\epsilon)) \in Lie(\Gamma)$, whenever $\alpha \neq \beta$. Similar argument with $\lambda_1 = \lambda_2 \epsilon \beta^2$ shows that $(\mathfrak{p}_\epsilon, 0) \in Lie(\Gamma)$. But then $[(\mathfrak{p}_\epsilon, 0), (\mathfrak{p}_\epsilon, 0)] = (\mathfrak{k}_\epsilon, 0)$ and $[(0, \mathfrak{p}_\epsilon), (0, \mathfrak{p})_\epsilon] = (0, \mathfrak{k}_\epsilon)$. Hence, $Lie(\Gamma) = \mathfrak{g}$. In the remaining case, $\alpha = \beta, Lie(\Gamma) = \{(X, X) : X \in so_\epsilon\}$. □

Consider now the horizontal curves in G whose tangent vectors belong to $\mathcal{D}$. Horizontal curves $g(t) = (S(t), T(t))$ are the solutions of

$$\frac{dS}{dt} = \alpha S(t) U_\epsilon(t), \quad \frac{dT}{dt} + \beta T(t) U_\epsilon(t), \tag{16.61}$$

for some control curve $U_\epsilon(t) \in \mathfrak{p}_\epsilon$. Proposition 16.20 implies that pair of points in G can be connected by a horizontal curve, a consequence of the Orbit theorem.

We will now introduce a left-invariant metric on the space of horizontal curves through the quadratic form $\langle X, Y \rangle_\epsilon = -\frac{\epsilon}{2} Tr(XY)$ on $\mathfrak{g}_\epsilon$. More precisely,

$$\langle (X_1, X_2), (Y_1, Y_2) \rangle = -\frac{\epsilon}{2} Tr(X_1 Y_1) - \frac{\epsilon}{2} Tr(X_2 Y_2) = \langle X_1, Y_1 \rangle_\epsilon + \langle X_2, Y_2 \rangle_\epsilon,$$

for $(X_1, X_2) \in \mathfrak{g}_\epsilon \times \mathfrak{g}_\epsilon$, and $(Y_1, Y_2) \in \mathfrak{g}_\epsilon \times \mathfrak{g}_\epsilon$. This quadratic form is positive-definite on $\mathfrak{p}$ and defines a natural metric $||X||^2 = \langle X, X \rangle, X \in \mathfrak{p}$. We will then use $\int_0^T \sqrt{||U_\epsilon||^2} \, dt$ to denote the length of a horizontal curve $g(t)$ on $[0, T]$ (instead of the other natural choice $\int_0^T \sqrt{\alpha^2 ||U_\epsilon||^2 + \beta^2 ||U_\epsilon||} \, dt = \sqrt{\alpha^2 + \beta^2} \int_0^T \sqrt{||U_\epsilon||^2} \, dt$).

It then follows that the associated sub-Riemannian problem of finding a horizontal curve of shortest length that connects two given points in G is well defined. As usual, we will apply the Maximum Principle to obtain the equations for the extremal curves. But first, let us note that this sub-Riemannian problem is invariant under the diagonal adjoint action of K on G defined by

$$h \in K \rightarrow (hSh^{-1}, hTh^{-1}), \tag{16.62}$$

where K is the isotropy group associated with the base space $S_\epsilon = G_\epsilon/K$.

After the usual identification of $\mathfrak{g}$ with $\mathfrak{g}^*$, the Maximum Principle leads to the Hamiltonian

$$H = \frac{1}{2} ||\alpha P_1 + \beta P_2||^2, \tag{16.63}$$

induced by the extremal controls $U_\epsilon = \alpha P_1 + \beta P_2$, whose integral curve constitute the normal extremals associated with the above sub-Riemannian problem. Since the optimal trajectories are the projections of normal extremals, abnormal extremals will be ignored [JZ]. It follows that the extremals are the solutions of the following system of equations:

$$\frac{dS}{dt} = \alpha S U_\epsilon, \quad \frac{dP_1}{dt} = \alpha [U_\epsilon, Q_1], \quad \frac{dQ_1}{dt} = \alpha [U_\epsilon, P_1],$$
$$\frac{dT}{dt} = \beta T U_\epsilon, \quad \frac{dP_2}{dt} = \beta [U_\epsilon, Q_2], \quad \frac{dQ_2}{dt} = \beta [U_\epsilon, P_2]. \tag{16.64}$$

Then, $\frac{d}{dt}(Q_1 + Q_2) = 0$, hence $Q_1(t) + Q_2(t)$ is constant. It is easy to check that this constant of motion is a consequence of the symmetry (16.62).

On surface $Q_1 + Q_2 = 0$, equations (11.64) admit another integral of motion $\beta P_1(t) + \alpha P_2(t)$, because

$$\beta \frac{dP_1}{dt} + \alpha \frac{dP_2}{dt} = \alpha \beta [U_\epsilon, Q_1 + Q_2] = 0.$$

Let $\beta P_1 + \alpha P_2 = -\frac{1}{\alpha}A$. Together with $U_\epsilon = \alpha P_1 + \beta P_2$, these equations yield

$$P_1 = \frac{1}{\beta^2 - \alpha^2}\left(-\frac{\beta}{\alpha}A - \alpha U_\epsilon\right), \; P_2 = \frac{1}{\beta^2 - \alpha^2}(A + \beta U_\epsilon).$$

Introduce now new variables P, Q and Λ defined by

$$P = (\alpha^2 - \beta^2)P_2, \; Q = (\beta^2 - \alpha^2)Q_2, \; \Lambda = -\beta U_\epsilon.$$

Then $\Lambda = A + P$, and

$$\begin{aligned}
\frac{dP}{dt} &= (\alpha^2 - \beta^2)\frac{dP_2}{dt} = (\alpha^2 - \beta^2)\beta[U_\epsilon, Q_2] = [\Lambda, Q], \\
\frac{dQ}{dt} &= (\beta^2 - \alpha^2)\frac{dQ_2}{dt} = [\Lambda, P].
\end{aligned} \tag{16.65}$$

In addition,

$$\frac{dS}{dt} = \alpha S U_\epsilon = -\frac{\alpha}{\beta}S\Lambda, \frac{dT}{dt} = \beta T U_\epsilon = -T\Lambda. \tag{16.65a}$$

Equations (11.65) have the same form as equations (11.58).

Let us now connect these equations to the rolling spheres. In this context, a sphere of radius ρ, denoted by $\mathbb{S}^n_{\rho,\epsilon}$, is the Euclidean sphere $\{x \in \mathbb{R}^{n+1} : \|x\|^2 = \rho^2\}$ when $\epsilon = 1$, and the hyperboloid $\{x \in \mathbb{R}^{n+1} : x_0^2 - \sum_{i-1}^{n} x_i^2 = \rho^2, x_0 > 0\}$ when $\epsilon = -1$. We are interested in the situations where an the oriented sphere $\mathbb{S}^n_{\rho,\epsilon}$ rolls without slipping and twisting on the stationary sphere $\mathbb{S}^n_{\sigma,\epsilon}$, with $\rho \neq \sigma$. Such rollings are described by the pairs of curves $(x(t), R(t))$, where $x(t)$ denotes the path of the point of contact with the stationary sphere, and $R(t)$ the rotation of a fixed frame on the moving sphere relative to a stationary frame in the ambient space. A rolling $(x(t), R(t))$ is said to be isometric with no slipping if it satisfies the following equations:

$$\frac{dx}{dt} = \alpha \Omega_\epsilon(t)x(t), \; \frac{dR}{dt} = \Omega_\epsilon(t)R(t),$$

for some curve $\Omega_\epsilon(t) \in \mathfrak{p}_\epsilon$, where $\alpha = \frac{\rho}{\rho+\sigma}$ ([JZ]).

Let $S(t)$ denote a horizontal curve in SO_ϵ such that $x(t) = \sigma S(t)e_0$. Then, $\frac{dS}{dt} = S(t)A_\epsilon(t)$ with $A_\epsilon(t) \in \mathfrak{p}_\epsilon$. It follows that

$$\alpha \Omega_\epsilon S e_0 = \frac{dx}{dt} = SA_\epsilon e_0.$$

Hence,

$$A_\epsilon = \alpha S^{-1}\Omega_\epsilon S.$$

If $U(t) = S^{-1}(t)\Omega_\epsilon(t)S(t)$, then $\frac{dS}{dt} = \alpha S(t)U_\epsilon(t)$. If we now define $T(t) = R^{-1}(t)S(t)$, then

$$\frac{dT}{dt} = -R^{-1}\Omega_\epsilon S + \alpha R^{-1}SU_\epsilon = (\alpha - 1)T(t)U_\epsilon(t) = \beta T(t)U_\epsilon(t).$$

Hence, $(S(t), T(t))$ is an itegral curve of the rolling distribution $\mathcal{D}$ (equation (16.61)).

It follows that the Riemannian length of curve $x(t)$ traced by the point of contact is equal to the sub-Riemannian length of the lifted curve $(S(t), T(t)$. The rolling sphere problem is concerned with the paths $(x(t), R(t))$ that connect two given configurations (x_0, R_0) and (x_1, R_1) in such a way that the length of the curve traced by the point of contact is minimal. Such curves are said to be optimal. It follows that each optimal curve $(x(t), R(t))$ is the projection of a horizontal curve $(S(t), T(t))$ whose sub-Riemannian length is minimal among the horizontal curves that connect the initial manifold $N_0 = \{(S_0, T_0) : S_0e_0 = x_0, T_0 = R_0^{-1}S_0\}$ to the terminal manifold $N_1 = \{(S_0, T_0) : S_0e_0 = x_1, T_0 = R_1^{-1}S_0\}$.

Therefore optimal curves are the projections of sub-Riemannian extremals that satisfy the transversality conditions. It is easy to verify that an extremal $L(t) = P_1(t) + P_2(t) + K_1(t) + K_2(t)$ satisfies the above transversality conditions if and only if $K_1(t) + K_2(t) = 0$. It follows that optimal solutions $(x(t), R(t))$ of the rolling sphere problem conform to the following set of equations:

$$x(t) = \sigma S(t)e_0, R(t) = S(t)T^{-1}(t),$$

$$\frac{dS}{dt} = \alpha S(t)U_\epsilon(t) = -\frac{\alpha}{\beta}S\Lambda(t) = \frac{\rho}{\sigma}S(t)\Lambda(t), \Lambda(t) = A + P(t), \quad (16.66)$$

$$\frac{dP}{dt} = [\Lambda(t), Q(t)], \frac{dQ}{dt} = [\Lambda(t), P(t)].$$

16.4.1 Elastic curves and rolling sphere geodesics

The rolling sphere extremals and the curvature extremals share the same isospectral matrix

$$L_\mu = P + \mu Q + (1 - \mu^2)A.$$

In the curvature problem the matrix Q is of special form, due to the condition that $Q = \frac{1}{\epsilon}[\Lambda, [\Lambda, Q]]$, while this condition is absent in the rolling sphere problem. As a result the spectral invariants of L_μ for the rolling sphere cannot be calculated in the same manner as in the case of the curvature problem.

We will show below that the relevant integrals of motion can be extracted from the functions $\phi_{\mu,k} = Tr((L_\mu)^k$ as was done in [JZ]. The following lemma will be useful.

Lemma 16.21 *Let* $L = \begin{pmatrix} 0 & -\epsilon p^T \\ p & M \end{pmatrix}$ *with M an antisymmetric n×n matrix, and p a column vector in* $\mathbb{R}^n$, *then*

$$C_1 = ||p||^2 + \epsilon||M||^2, C_2 = ||M||^2||p||^2 - ||Mp||^2 + \epsilon(||M||^4 - \frac{1}{2}Tr(M^4))$$

are the spectral invariants of L. Here $||p||$ *denotes the Euclidean norm in* $\mathbb{R}^n$, *and* $||M||$ *the trace norm in* $so_n(R)$.

Proof

$$L^2 = \begin{pmatrix} -\epsilon||p||^2 & -\epsilon p^T M \\ Mp & M^2 - \epsilon(p \otimes p) \end{pmatrix},$$

and therefore, $C_1 = -\frac{\epsilon}{2}Tr(L^2) = ||p||^2 + \epsilon||M||^2$.

An easy calculation shows that

$$Tr(L^4) = ||p||^4 - 2\epsilon Mp \cdot p^T M + Tr(M^2 - \epsilon(p \otimes p))^2,$$

where $x \cdot y$ denotes the Euclidean inner product in $\mathbb{R}^n$. Since the matrix M is antisymmetric, $Mp \cdot p^T M = -||Mp||^2$ and the above can be rewritten as

$$Tr(L^4) = 2||p||^4 + 4\epsilon||Mp||^2 + Tr(M^4) .$$

But then

$$C_2 = \frac{\epsilon}{2}\left(C_1^2 - \frac{1}{2}Tr(L^4)\right)$$

$$= \frac{\epsilon}{2}\left(\left(||p||^4 + 2\epsilon||M||^2||p||^2 + \epsilon^2||M||^4\right)\right.$$

$$\left. -(||p||^4 + 2\epsilon||Mp||^2 + \frac{1}{2}Tr(M)^4)\right)$$

$$= ||M||^2||p||^2 - ||Mp||^2 + \epsilon\left(||M||^4 - \frac{1}{2}Tr(M^4)\right).$$

$\square$

The above invariants can be written as

$$C_1 = ||P||^2 + \epsilon||Q||^2, C_2 = ||Q||^2||P||^2 - ||[P,Q]||^2 + \epsilon(||Q||^4 - \frac{1}{2}Tr(Q^4)),$$

$$(16.67)$$

where $P = \begin{pmatrix} 0 & -\epsilon p^T \\ p & 0 \end{pmatrix}$ and $Q = \begin{pmatrix} 0 & 0 \\ 0 & M \end{pmatrix}$. When L is replaced by the

spectral matrix L_μ, P is replaced by $\epsilon P + (1 - \mu^2)\epsilon A$ and Q by $\mu\epsilon Q$. Then

$$C_1(\mu) = ||A + P||^2 - 2\mu^2\epsilon(\langle A, \Lambda \rangle_\epsilon + \epsilon||Q||^2) + \mu^4||A||^4.$$

The above shows that

$$I_0 = ||A + P||^2, \; I_1 = ||P||^2 + \epsilon||Q||^2, I_3 = \frac{1}{2}||Q||^2 - \langle A, \Lambda \rangle_\epsilon$$

are integrals of motion for both systems. Only two of the above integrals are functionally independent, since

$$I_3 = \frac{\epsilon}{2}(\epsilon||Q||^2 - 2\epsilon\langle A, \Lambda \rangle_\epsilon) = \frac{\epsilon}{2}(I_1 - (||P||^2 + 2\langle A, P \rangle_\epsilon + 2||A||^2))$$

$$= \frac{\epsilon}{2}(I_1 - I_0 - ||A||^2).$$

Then $C_2(\mu)$ yields two new integrals of motion

$$I_2 = ||Q||^2||P||^2 - ||[Q, P]||^2 + \epsilon(||Q||^4 - \frac{1}{2}Tr(Q^4)),$$

$$I_4 = ||Q||^2||A||^2 - ||[A, Q]||^2.$$

Remark 16.22 The numbering of the above integrals is the same as in [JZ] for easier comparisons.

The first integral $||A + P||^2 = ||A||^2$ is equal to $\frac{\beta^2}{2}H_r$, with H_r the Hamiltonian for the rolling sphere problem, while the second integral I_1 is a Casimir on $\mathfrak{g}_\epsilon$. In the curvature problem, the Hamiltonian H_c is equal to $\frac{1}{2}||\Lambda, Q]||^2 + \langle \Lambda, P \rangle$, and there is an additional constraint $Q = -\epsilon[\Lambda, [\Lambda, Q]]$. The latter constraint implies that $Q = q \wedge \lambda$, where q is a vector orthogonal to λ. Then, $[\Lambda, Q] = q \wedge_\epsilon e_0$ and hence, $||[\Lambda, Q]||^2 = ||q||^2 = ||Q||^2$. Moreover, $\langle \Lambda, P \rangle = \langle \Lambda, \epsilon\Lambda - A \rangle_\epsilon = \epsilon||\Lambda||^2 - \langle \Lambda, A \rangle_\epsilon = \epsilon - \langle A, \Lambda \rangle_\epsilon$. Hence,

$$H_c = \frac{1}{2}||Q||^2 + \langle \Lambda, P \rangle = \frac{1}{2}||Q||^2 - \langle A, \Lambda \rangle_\epsilon + \epsilon = I_3 + \epsilon.$$

Thus the Hamiltonian for the curvature problem and I_3 are functionally dependent. The following proposition is essential for further comparisons between these two problems.

Proposition 16.23 $I_4 = 0$ if and only if $Q = [\Lambda, X]$ for some $X \in \mathfrak{p}_\epsilon$ orthogonal to Λ. On $I_4 = 0$, $I_2 = ||Q||^2||P||^2 - ||[Q, P]||^2$.

Proof If $Q = [\Lambda, X]$ for some $X \in \mathfrak{p}_\epsilon$ orthogonal to Λ, then $Q = \begin{pmatrix} 0 & 0 \\ 0 & x \wedge_\epsilon \lambda \end{pmatrix}$, where $X = x \wedge_\epsilon e_0$ and $\Lambda = \lambda \wedge_\epsilon e_0$. It follows that

$||Q||^2 = ||\lambda||^2 ||x||^2$, and $[\Lambda, Q] = ||\lambda||^2 x \wedge_\epsilon e_0$. Therefore, $||[\Lambda, Q]||^2 = ||\Lambda||^4 ||x||^2 = ||\Lambda||^2 ||Q||^2$. Evidently, $I_4 = 0$.

To prove the converse, let $\mathfrak{k}_\Lambda = \{Q : [\Lambda, Q] = 0\}$, and let $\mathfrak{k}_\Lambda^\perp$ denote its orthogonal complement in $\mathfrak{k}$. Let $Q(t) = Q_0(t) + Q_1(t)$ with $Q_0 \in \mathfrak{k}_\Lambda$ and $Q_1 \in \mathfrak{k}_\Lambda^\perp$. Then $ad\Lambda$ maps $\mathfrak{k}_\Lambda^\perp$ in $\mathfrak{p}_\epsilon$ onto $\mathfrak{k}_\Lambda^\perp$ by Proposition 13.6 in Chapter 13. Hence $Q_1 = ad\Lambda(X)$ for some $X \in \mathfrak{p}_\epsilon$ such that $\langle \Lambda, X \rangle_\epsilon = 0$. If $X = x \wedge_\epsilon e_0$ for $x \in \mathbb{R}^n$, then $ad\Lambda(X) = \begin{pmatrix} 0 & 0 \\ 0 & x \wedge_\epsilon \lambda \end{pmatrix}$. Then $[\Lambda, Q_1] = ||\lambda||^2 x \wedge_\epsilon e_0$. It follows that

$$||[\Lambda, Q]||^2 = ||[\Lambda, Q_1]||^2 = ||\Lambda||^2 ||Q_1||^2.$$

If $I_4 = 0$, then $0 = ||\Lambda||^2 (||Q_0||^2 + ||Q_1||^2) - ||\Lambda||^2 ||Q_1||^2$, which yields $Q_0 = 0$.

On $I_4 = 0$, $Q = \begin{pmatrix} 0 & 0 \\ 0 & x \wedge_\epsilon \lambda \end{pmatrix}$. Therefore,

$$Q^2 = -\begin{pmatrix} 0 & 0 \\ 0 & ||x||^2 \lambda \otimes \lambda + ||\lambda||^2 x \otimes x \end{pmatrix} \text{ and}$$

$$Q^4 = \begin{pmatrix} 0 & 0 \\ 0 & ||x||^2 ||\lambda||^2 (||x||^2 \lambda \otimes \lambda + ||\lambda||^2 x \otimes q \end{pmatrix}.$$

Hence, $||Q||^4 - \frac{1}{2} Tr(Q^4) = 0$. $\qquad\square$

Proposition 16.24 *A curve $x(t)$ on the stationary unit sphere is elastic if and only if it is the projection of a rolling extremal on $I_0 = ||A + P||^2 = \frac{1}{\rho^2}$ and $I_4 = 0$.*

Proof Suppose that $x(t)$ is an elastic curve on the stationary sphere G_ϵ/K with $\epsilon = \pm 1$. Then $x(t)$ is the projection of an extremal curve $(g(t), \Lambda(t)), P(t), Q(t), X(t))$ with $x(t) = \pi(g(t)), \frac{dg}{dt} = g(t)\Lambda(t), ||\Lambda(t)|| = 1, \Lambda(t) = A + P$ for some constant matrix $A \in \mathfrak{p}_\epsilon$. Additionally, $Q + [\Lambda, X] = 0$, and P and Q are solutions of $\frac{dP}{dt} = [\Lambda, Q], \frac{dQ}{dt} = [\Lambda, P]$ (Corollary 16.18). To make the correspondence easier, assume that $\pi(g) = ge_0$, that is, take $K = \{1\} \times SO_n(\mathbb{R})$. Let

$$\Lambda_r(t) = \frac{1}{\rho} \Lambda(t), P_r(t) = (1 - \rho^2)\Lambda_r(t) + \rho P(t), Q_r(t) = Q(t), A_r = \rho A.$$

$$(16.68)$$

Let $S(t)$ denote the solution of $\frac{dS}{dt} = \rho S(t)\Lambda_r(t)$ with $g(0) = S(0)$. Since $S(t)$ and $g(t)$ satisfy the same equation and have the same initial values, $g(t) = S(t)$. Then, $x_r(t) = x(t)S(b)e_0 = g(t)e_0 = x(t)$. Furthermore,

$$\Lambda_r - P_r = \rho^2\Lambda_r - \rho P = \rho(\epsilon\Lambda - P) = \rho A = A_r,$$

$$\frac{dP_r}{dt} = \frac{1-\rho^2}{\rho}[\Lambda, Q] + \rho[\Lambda, Q] = \frac{1}{\rho}[\Lambda_r, \epsilon Q] = [\Lambda_r, Q_r], \qquad (16.69)$$

$$\frac{dQ_r}{dt} = \epsilon[\Lambda, P] = \left[\frac{1}{\rho}\Lambda, \rho P\right] = [\Lambda_r, P_r].$$

Hence, $x(t)$ is the point of contact associated with the extremal curve of the rolling sphere problem associated with (16.69), and $\|A_r + P_r\| = \|\Lambda_r\| = \frac{1}{\rho}\|\Lambda\| = \frac{1}{\rho}$ and $I_4(\Lambda_r, Q_r) = 0$.

To prove the converse, assume that $x(t)$ is a rolling geodesic and a projection of an extremal on $I_0 = \frac{1}{\rho^2}$ and $I_4 = 0$. Proposition 16.23 implies that $Q_r = [\Lambda_r, X]$ for some X in $\mathfrak{p}$ orthogonal to Λ_r. Now use (16.68) and (16.69) in reversed order to show that the point of contact is elastic. □

Corollary 16.25 *Let $\kappa(t)$ and $\tau(t)$ denote the curvature and the torsion of a curve $x(t)$ that is a projection of a rolling problem extremal that satisfies $\|\Lambda_r\| = 1$ and $I_4(\Lambda_r, Q_r) = 0$. Let $\xi(t) = \kappa^2(t)$. Then*

$$\left(\frac{d\xi}{dt}\right)^2 = -\xi^3 + 4I_3\xi^2 + 4(\|A\|^2 - I_3^2)\xi - \frac{4}{\rho^2}I_2,$$

$$(\kappa^2\tau)^2 = \frac{1}{\rho^2}I_2. \qquad (16.70)$$

Moreover, if $T(t), N(t), B(t)$ denote the Serret–Frenet triad along $x(t)$ then $\frac{dB}{dt}(t)$ is contained in the linear span of $T(t), N(t), B(t)$. Consequently, all the remaining curvatures along $x(t)$ are equal to zero.

Proof Since the point of contact $x(t)$ on $\|A + P\| = \frac{1}{\rho}$ and $I_4 = 0$ is elastic, its curvature $\kappa(t)$ and torsion $\tau(t)$ satisfy the equations in Proposition 13.7 in Chapter 13. That is, $\xi(t) = \kappa^2(t)$ is a solution of

$$\left(\frac{d\xi}{dt}\right)^2 = -\xi^3 + 4(H - \epsilon)\xi^2 + 4(I_1 - H^2)\xi - 4I_2$$

and $(\kappa(t)^2\tau(t))^2 = I_2$. These equations are the same as equations (16.70) because $H - \epsilon = (I_3 + \epsilon) - \epsilon = I_3$. In addition,

$$I_1 = \|P\|^2 + \epsilon\|Q\|^2 = \frac{1}{\rho^2}\|P_r - (1 - \rho^2)\Lambda_r\|^2 + \epsilon\|Q\|^2$$

$$= \frac{1}{\rho^2}\| - A_r + \rho^2\Lambda_r\|^2 + \epsilon\|Q\|^2$$

$$= \frac{1}{\rho^2}||A_r||^2 + \rho^2||\Lambda_r||^2 - 2\langle A_r, \Lambda_r\rangle_\epsilon + \epsilon||Q||^2$$
$$= ||A||^2 + 1 + 2\epsilon I_3.$$

Hence $I_1 - H^2 = ||A||^2 + 1 + 2\epsilon I_3 + (I_3 + \epsilon)^2 = ||A||^2 - I_3^2$. The reader can easily verify that

$$||P||^2||Q||^2 - ||[P, Q]||^2 = \frac{1}{\rho^2}(||P_r||^2||Q_r||^2 - ||[P_r, Q_r]||^2.$$

$\square$

Corollary 16.26   *On two-dimensional spheres the projection of an extremal on the stationary manifold is elastic.*

Proof   When the rolling takes place on two-dimensional spheres, $\mathfrak{k} = so_3(R)$ and every element $Q \in so_3(R)$ can be written as $Q = a \wedge b$ for some vectors a and b in $\mathbb{R}^3$. Hence, the integrals of motion for the two problems coincide. $\square$

17

The non-linear Schroedinger's equation and Heisenberg's magnetic equation–solitons

The material in this chapter is largely inspired by another spectacular property of elastic curves – they appear as soliton solutions in the non-linear Schroedinger's equation. We will be able to demonstrate this fact by introducing a symplectic structure on an infinite-dimensional Fréchet manifold of framed curves of fixed length over a three-dimensional space form; the symplectic structure will then identify some partial differential equations of mathematical physics with the Hamiltonian flows generated by the functionals defined by the geometric invariants of curves, such as the curvature or the torsion functional.

The passage to infinite dimensions will be based on familiar geometric notions developed in the previous chapters. The fundamental space consists of framed curves, anchored at the initial point and further constrained by the condition that the tangent vector of the projected curve coincides with the first leg of the orthonormal frame. Such class of curves are called anchored Darboux curves, and in particular include the Serret–Frenet framed curves.

The symplectic form ω is defined on the space of horizontal curves of fixed length in the universal covers of the orthonormal frame bundles of the underlying manifolds: $SL_2(C)$ for the hyperboloid $\mathbb{H}^3$, and $SU_2 \times SU_2$ for the sphere S^3. The form ω is left-invariant and is induced by the Poisson–Lie bracket in the appropriate Lie algebra. More precisely, the form ω in each of the above cases is defined over the curves whose tangents take values in the Cartan space $\mathfrak{p}$ corresponding to the decomposition $\mathfrak{g} = \mathfrak{p} + \mathfrak{k}$ of the Lie algebra $\mathfrak{g}$ subject to the usual Lie algebraic relations

$$\mathfrak{p}, \mathfrak{p}] = \mathfrak{k}, \ [\mathfrak{p}, \mathfrak{k}] = \mathfrak{p}, \ [\mathfrak{k}, \mathfrak{k}] = \mathfrak{k}.$$

In the case of the hyperboloid $\mathfrak{g}$ is equal to $sl_2(C)$ and the Cartan space $\mathfrak{p}$ is equal to the space of Hermitian matrices, while in the case of the sphere $\mathfrak{g}$ is equal to $su_2 \times su_2$ and the Cartan space is isomorphic to the space of skew-Hermitian matrices $\mathfrak{h}$. The symplectic forms in each of these two cases are isomorphic to each other, as a consequence of the isomorphism between $\mathfrak{p}$ and $\mathfrak{h}$ given by $i\mathfrak{h} = \mathfrak{p}$.

The symplectic structure associates a Hamiltonian vector field $\vec{f}$ to each function f. We will show that the Lie algebra projection of the Hamiltonian vector field associated with $f = \frac{1}{2}\int_0^L \kappa^2(s)\,ds$ satisfies the Heisenberg magnetic equation (HME)

$$\frac{\partial \Lambda}{\partial t}(s,t) = \frac{1}{i}\left[\Lambda(s), \frac{\partial^2 \Lambda}{\partial s^2}(s,t)\right]. \qquad (17.1)$$

We will also show that the horizontal-Darboux curves are parametrized by the curves in SU_2, which, along the solutions of (17.1), satisfy Schroedinger's non-linear equation (NLS)

$$-i\frac{\partial \psi}{\partial t}(t,s) = \frac{\partial^2 \psi}{\partial s^2}(t,s) + i\frac{1}{2}(|\psi(t,s)|^2 + c(t))\psi(t,s). \qquad (17.2)$$

In this remarkable triangle, the elastic curves generate the soliton solutions for (17.2). Finally, we will show that the modified Korteweg–de Vries equation and the curve shortening equation are Hamiltonian, generated by $\int_0^L \kappa^2(s)\tau(s)\,ds$ and $\int_0^L \tau(s)\,ds$, and are all in involution with each other as well as with $\frac{1}{2}\int_0^L \kappa^2(s)\,ds$.

17.1 Horizontal Darboux curves

Let us now recall complex quaternions and $SL_2(\mathbb{C})$ discussed in Section 14.1 of Chapter 14. We will resume the notations used in that chapter, and identify points $z = (z_0, z_1, z_2, z_3)$ in $\mathbb{C}^4$ with the matrices Z in $SL_2(C)$ through

$$Z = \begin{pmatrix} z_0 + iz_1 & z_2 + iz_3 \\ -z_2 + iz_3 & z_0 - iz_1 \end{pmatrix}, z_0^2 + z_1^2 + z_2^2 + z_3^2 = Det(Z) = 1.$$

Then the sphere $S^3 = \{x \in \mathbb{R}^4 : x_0^2 + x_1^2 + x_2^2 + x_3^2 = 1\}$ is identified with matrices $X = \begin{pmatrix} u & v \\ -\bar{v} & \bar{u} \end{pmatrix}$ in SU_2 when z is restricted to $\mathbb{R}^4$, while the hyperboloid $\mathbb{H}^3 = \{x \in \mathbb{R}^4 : x_0^2 - x_1^2 - x_2^2 - x_3^2 = 1, x_0 > 0\}$ is identified with the positive-definite Hermitian matrices

$$P = \begin{pmatrix} x_0 + x_1 & x_2 + ix_3 \\ x_2 - ix_3 & x_0 - x_1 \end{pmatrix}, Det(P) = 1, \qquad (17.3)$$

by setting $z_0 = x_0, z_1 = -ix_1, z_2 = ix_3, z_3 = -ix_2$.

For notational convenience G will denote the ambient space $SL_2(C)$ and $\mathfrak{g}$ will denote its Lie algebra. Once more we will make use of the fact that $\mathfrak{g}$ admits a Cartan decomposition $\mathfrak{g} = \mathfrak{p} \oplus \mathfrak{h}$, with $\mathfrak{p}$ the space of Hermitian matrices of trace zero, and $\mathfrak{h}$ the subalgebra of skew-Hermitian matrices $\mathfrak{h}$ subject to the usual rules

$$[\mathfrak{p}, \mathfrak{p}] = \mathfrak{h}, [\mathfrak{p}, \mathfrak{h}] = \mathfrak{p}, [\mathfrak{h}, \mathfrak{h}] = \mathfrak{h}, i\mathfrak{p} = \mathfrak{h}.$$

Matrices

$$B_1 = \frac{1}{2}\begin{pmatrix} 1 & 0 \\ 0 & -1 \end{pmatrix}, B_2 = \frac{1}{2}\begin{pmatrix} 0 & -i \\ i & 0 \end{pmatrix}, B_3 = \frac{1}{2}\begin{pmatrix} 0 & 1 \\ 1 & 0 \end{pmatrix},$$

will be referred to as the Hermitian Pauli matrices. They form a basis for $\mathfrak{p}$, while the skew-Hermitian Pauli matrices $A_1 = iB_1, A_2 = iB_2, A_3 = iB_3$ form a basis for $\mathfrak{h}$. Together, these matrices form a basis for $\mathfrak{g}$ and conform to the following Lie bracket table:

Table 17.1

$\{,\}_\epsilon$	A_1	A_2	A_3	B_1	B_2	B_3
A_1	0	$-A_3$	A_2	0	$-B_3$	B_2
A_2	A_3	0	$-A_1$	B_3	0	$-B_1$
A_3	$-A_2$	A_1	0	$-B_2$	B_1	0
B_1	0	$-B_3$	B_2	0	$-A_3$	A_2
B_2	B_3	0	$-B_1$	A_3	0	$-A_1$
B_3	$-B_2$	B_1	0	$-A_2$	A_1	0

The quadratic form on $\mathfrak{g}$ defined by $\langle A, B \rangle = 2Trace(AB)$ will be called the trace form. The Hermitian Pauli matrices form an orthonormal basis relative to the trace form, which is the main reason for choosing it, rather than the usual choice $\pm\frac{1}{2}Tr(AB)$. Then

$$\langle A, B \rangle = a_1 b_1 + a_2 b_2 + a_3 b_3$$

for any Hermitian matrices $A = \sum_{i=1}^{3} a_i B_i$ and $B = \sum_{i=1}^{3} b_i B_i$. The trace form is invariant in the usual sense, $\langle A, [B, C] \rangle = \langle [A, B], C \rangle$ for any matrices A, B, C in $\mathfrak{g}$. It is also SU_2 invariant, in the sense that $\langle gAg^*, gBg^* \rangle = \langle A, B \rangle$ for any g in SU_2, where g^* denotes the Hermitian transpose $\bar{g}^T$ of g.

Since $\langle iA, iB \rangle = -\langle A, B \rangle$ similar formula holds on $\mathfrak{h}$ with the sign reversed. Then $\langle , \rangle_h$ will denote the restriction of the trace form to $\mathfrak{p}$, and $\langle , \rangle_s$ will denote the negative of the restriction of the trace form to $\mathfrak{h}$. $\| \ \|_h$, and $\| \ \|_s$ will denote the induced norms on $\mathfrak{p}$ and $\mathfrak{h}$.

We will now pass to the universal covers $SU_2 \times SU_2$ and $SL_2(C)$ of the orthonormal frame bundles $SO_4(R)$ and $SO(1,3)$ of the underlying sphere and the hyperboloid. Recall that $SU_2 \times SU_2$ is a double cover of $SO_4(R)$, and that $SL_2(\mathbb{C})$ is a double cover of $SO(1,3)$. (Propositions 14.2 and 14.4.)

We will make use of the fact that both $SU_2 \times SU_2$ and $SL_2(\mathbb{C})$ are principal SU_2-bundles over S^3 and $\mathbb{H}^3$, respectively, via the following realizations: in the case of the sphere, the action is diagonal $R(p,q) = (pR^*, qR^*)$ for (p,q) in $SU_2 \times SU_2$ and $R \in SU_2$, with the projection map given by $X = \pi(p,q) = pq^*$, while in the case of the hyperboloid, the action is given by $(R,g) \rightarrow gR^*$ for all $g \in G$ and $R \in SU_2$ and the projection map π is given by $\pi(g) = gg^*$.

Curves $g(t) = (p(t), q(t))$ in $SU_2 \times SU_2$ will be called spherical horizontal if $p^* \frac{dp}{dt}(t) = P(t)$, $q(t)^* \frac{q}{dt}(t) = -P(t)$ for some curve $P(t)$ in $\mathfrak{h}$, and the left-invariant distribution $\mathcal{H}_s((p,q)) = \{(pP, q(-P)) : P \in \mathfrak{h}\}$ in $SU_2 \times SU_2$ will be called the spherical connection. Curves $g(t)$ in $SL_2(C)$ will be called hyperbolic horizontal if $g^{-1}(t)\frac{dg}{dt}(t) = B(t)$ for some curve of matrices $B(t)$ that take values in $\mathfrak{p}$. Analogously, the left-invariant distribution $\mathcal{H}_h(g) = \{gB : B \in \mathfrak{p}\}$ will be called the hyperbolic connection.

Remark 17.1 This terminology is consistent with the terminology used earlier in the text. In the realizations $SO_4(R)/SO_3(R)$ and $SO(1,3)/SO_3(R)$ used earlier for the space forms, horizontal distributions were defined as the distributions spanned by the left-invariant vector fields with values in $\mathfrak{p}_\epsilon = \left\{ \begin{pmatrix} 0 & -\epsilon p^T \\ p & 0 \end{pmatrix} : p \in R^3 \right\}$. In the isomorphism $so_4(R) \cong su_2 \times su_2$ each matrix $\begin{pmatrix} 0 & -p^T \\ p & 0 \end{pmatrix}$ is identified with $(P, -P)$ for $P = \frac{1}{2}\begin{pmatrix} ip_1 & p_2 + ip_3 \\ -p_2 + ip_3 & -ip_1 \end{pmatrix}$, while in the case of the hyperboloid, horizontal curves are identified with the left-invariant distributions in $SL_2(\mathbb{C})$ that take values in the Hermitian matrices in $sl_2(\mathbb{C})$.

Definition 17.2 The length of any spherical horizontal curve $g(t) = (p(t), q(t))$ in an interval $[0, T]$ is equal to $\int_0^T \|p^*(t)\frac{dp}{dt}(t)\|_s \, dt = \int_0^T \|P(t)\|_s \, dt$ and the length of a hyperbolic horizontal curve $g(t)$ in $[0, T]$ is equal to $\int_0^T \|g^{-1}(t)\frac{dg}{dt}(t)\|_h \, dt = \int_0^T \|B(t)\|_h \, dt$.

If $X(t) = \begin{pmatrix} x_0 + ix_1 & x_2 + ix_3 \\ -x_2 + ix_3 & x_0 - ix_1 \end{pmatrix}$ is the projection of a spherical horizontal curve $(p(t), q(t))$ then $\frac{dX}{dt}(t) = X(t)(2q(t)P(t)q(t)^*)$. Then

$$\int_0^T \sqrt{\frac{dx_0}{dt}^2 + \frac{dx_1}{dt}^2 + \frac{dx_2}{dt}^2 + \frac{dx_3}{dt}^2} \, dt = \int_0^T \sqrt{\frac{1}{2} Tr\left(X\frac{dX}{dt}\right)(X\frac{dX}{dt})^*} \, dt$$

$$= \int_0^T \sqrt{2Tr(P(t)P(t)^*)} \, dt$$

$$= \int_0^T \|P(t)\|_s \, dt.$$

Similarly the length of a hyperbolic horizontal curve coincides with the Riemannian length $\int_0^T \sqrt{-\frac{dx_0}{ds}^2 + \frac{dx_1}{ds}^2 + \frac{dx_2}{ds}^2 + \frac{dx_3}{ds}^2} \, dt$ of the projected curve $X(t) = g(t)g^*(t)$.

The above calculation recovers a general fact that we encountered earlier, namely that the Riemannian metric of a symmetric space is induced by the left-invariant sub-Riemannian metric on a horizontal distribution. In this setting then every curve in the base manifold can be lifted to a horizontal curve and any two liftings differ by an element in SU_2, consistent with the general theory of principal bundles.

On the sphere SU_2 each pair (p,q) in $SU_2 \times SU_2$ defines an orthonormal frame (v_1, v_2, v_3) at $X = pq^*$, where

$$v_1 = 2pA_1q^* = 2pq^*(qA_1q^*), v_2 = 2pA_2q^* = p(qA_2q^*), v_3 = 2pA_3q^*$$
$$= 2pq^*(qA_3q^*).$$

Conversely, every orthonormal frame at a point $X \in SU_2$ can be represented by the tangent vectors $v_1 = 2XU_1, v_2 = 2XU_2, v_3 = 2XU_3$ for some matrices U_1, U_2, U_3 in $\mathfrak{h}$ that are orthonormal relative to the trace form. There are exactly two matrices $\pm q \in SU_2$ such that $U_1 = qA_1q^*, U_2 = qA_2q^*, U_3 = qA_3q^*$. Having found q, p is uniquely defined by $p = Xq$.

In the case of the hyperboloid each g in $SL_2(C)$ defines an orthonormal frame

$$v_1 = 2gB_1g^*, \quad v_2 = 2gB_2g^*, \quad v_3 = 2gB_3g^*$$

at $X = gg^*$, and conversely every orthonormal frame v_1, v_2, v_3 at a point $X \in \mathbb{H}^3$ can be identified with exactly two matrices $\pm g \in SL_2(C)$ via the above relations.

Definition 17.3 Curves $g(t)$ in the universal covers of the orthonormal frame bundle will be called framed curves. Framed curves defined on a fixed interval $[0, L]$ over a base curve $X(s)$ in the underlying symmetric space will be called Darboux if $\frac{dX}{ds} = v_1(s)$, where v_1 denotes the first leg of the frame. Darboux curves which satisfy $g(0) = I$ will be called anchored.

Condition $\frac{dX}{ds} = v_1(s)$ implies that $X(s)$ is parametrized by arc length and therefore L is the length of $X(s)$. The fact that the orthonormal bundles are replaced by their universal covers does not matter in the subsequent exposition since all Darboux curves will be anchored.

In the spherical case anchored Darboux curves $g(s) = (p(s), q(s))$ are the solutions of

$$\frac{dp}{ds}(s) = p(s)P(s), \frac{dq}{ds}(s) = q(s)Q(s), P(s) - Q(s) = 2A_1, \qquad (17.4)$$

satisfying the initial conditions $p(0) = I, q(0) = I$. Condition $P(s) - Q(s) = 2A_1$ can also be expressed as

$$P(s) = U(s) + A_1, Q(s) = U(s) - A_1, \text{ where } U(s) = \frac{1}{2}(P(s) + Q(s)).$$
$$(17.5)$$

Definition 17.4 Anchored spherical Darboux curves are said to be reduced if the curve $U(s)$ in (17.5) is of the form $U(s) = \begin{pmatrix} 0 & u(s) \\ -\bar{u}(s) & 0 \end{pmatrix}$ for some complex curve $u(s)$.

Alternatively, curves $U(s)$ are reduced if and only if they lie in the orthogonal complement $A_1^\perp$ of A_1.

Every anchored Darboux curve $(p(s), q(s))$ can be transformed into a reduced Darboux curve $(\tilde{p}(s), \tilde{q}(s))$ without altering the base curve $X(s)$ by taking $\tilde{p} = ph, \tilde{q} = qh$, where $h(s)$ the solution of $\frac{dh}{ds} = -h(s)\begin{pmatrix} iu_1(s) & 0 \\ 0 & -iu_1(s) \end{pmatrix}, h(0) = I$, with $\begin{pmatrix} iu_1(s) & 0 \\ 0 & -iu_1(s) \end{pmatrix}$ the diagonal part of $U(s)$. Consequently, every base curve $X(s)$ can be lifted to a reduced Darboux curve. However, reduced Darboux framed curves exclude the Serret–Frenet frames, a fact that will be of some relevance later on.

Definition 17.5 Curves $(p(s), q(s))$ in $SU_2 \times SU_2$ which are the solutions of

$$\frac{d}{ds}(p(s), q(s)) = (p(s), q(s))(\Lambda(s), -\Lambda(s)),$$
$$\Lambda(0) = A_1, \|\Lambda(s)\| = 1, p(0) = q(0) = I \qquad (17.6)$$

will be called spherical horizontal-Darboux curves.

Every anchored spherical Darboux curve $g(s) = (p(s), q(s))$ can be transformed into a spherical horizontal-Darboux curve $\tilde{p}(s) = p(s)R^*(s), \tilde{q}(s) = q(s)R^*(s)$ for some matrix $R(s) \in SU_2, R(0) = I$ without altering the projected curve $X(s) = p(s)q^*(s)$. In fact, $R(s)$ is a solution of $\frac{dR}{ds} = \frac{1}{2}R(s)(P(s)+Q(s))$, and $\tilde{p}$ and $\tilde{q}$ are the solutions of

$$\frac{d\tilde{p}}{ds} = \tilde{p}\left(\frac{1}{2}R(P-Q)R^*\right) = \tilde{p}(RA_1R^*), \frac{d\tilde{q}}{ds} = \tilde{q}\left(\frac{1}{2}R(Q-P)R^*\right) \quad (17.7)$$

$$= \tilde{q}(-RA_1R^*). \quad (17.8)$$

Conversely, every curve $\Lambda(s) \in \mathfrak{h}$ with $\|\Lambda(s)\| = 1, \Lambda(0) = A_1$ can be written as $\Lambda(s) = R(s)A_1R^*(s)$ for some curve $R(s)$ in SU_2 with $R(0) = I$ because SU_2 acts transitively on the sphere $\|\Lambda\| = 1$ by conjugations. The correspondence between Λ and R is not bijective: if $R_0 \rightarrow \Lambda$ then $R_0h \rightarrow \Lambda$ for any

$$h = \begin{pmatrix} z & 0 \\ 0 & \bar{z} \end{pmatrix}, \|z\| = 1.$$

Curves $R(s)$ defined by $\Lambda(s) = R(s)A_1R^*(s)$ with $R(0) = I$ define spherical Darboux curves $(p(s), q(s))$ via the relations (17.6) where $U(s) = R^*(s)\frac{dR}{ds}(s)$. If the diagonal part of $U(s)$ is equal to zero then $(p(s), q(s))$ is a reduced Darboux curve. It follows that such curves set up a bijective correspondence between the horizontal-Darboux curves and the reduced Darboux curves. Thus every curve $X(s)$ parametrized by arc length on the interval $[0, L]$ with boundary conditions $X(0) = I$ and $\frac{dX}{ds}(0) = 2A_1$ can be lifted to a unique spherical horizontal-Darboux curve and also to a unique reduced anchored spherical Darboux curve.

In the subsequent exposition we will be less formal and refer to the spherical horizontal-Darboux curves as the solutions of the initial value problem

$$\frac{dp}{ds}(s) = p(s)\Lambda(s), \|\Lambda(s)\| = 1, p(0) = I, \quad (17.9)$$

since then the second factor $q(s)$ is defined by of $\frac{dq}{ds} = q(s)(-\Lambda(s)), q(0) = I$.

Definition 17.6 Spherical horizontal-Darboux curves $p(s)$ for which $\Lambda(s) = R(s)A_1R^*(s)$ for some curve $R(s) \in SU_2$ such that $R(L) = R(0) = I$ are called frame-periodic.

Frame-periodicity implies not only that $\Lambda(s)$ is periodic but also implies that the corresponding Darboux curve $(p(s), q(s))$ is a solution of an equation with periodic right-hand side, since the matrix $U(s)$ is periodic. However, if $U(s)$ is periodic then its diagonal part $D = \begin{pmatrix} iu_1 & 0 \\ 0 & -iu_1 \end{pmatrix}$ is periodic, and therefore $h(s)$, the solution of $\frac{dh}{ds} = -h(s)D(s), h(0) = I$ satisfies $h(L) = I$, from which it follows that the reduced Darboux curve has periodic right-hand side as well.

Definition 17.7 We will use $\mathcal{D}_s(L), \mathcal{HD}_s(L)$, and $\mathcal{PHD}_s(L)$ respectively to denote the space of anchored spherical Darboux curves, the space of horizontal-Darboux curves, and the space of frame-periodic horizontal curves on the interval $[0, L]$.

Hyperbolic Darboux curves will be defined analogously. Every anchored Darboux curve $g(s) \in SL_2(C)$ satisfies $g(0) = I$ and defines frames $v_1(s) = 2g(s)B_1g^*(s), v_2 = 2g(s)B_2g^*(s), v_3(s) = 2g(s)B_3g^*(s)$ over the projected curve $X(s) = g(s)g^*(s)$ such that $\frac{dX}{ds} = 2g(s)B_1g^*(s)$. It then follows that

$$\frac{dg}{ds} = g(s)(B_1 + A(s))$$

for some curve $A(s)$ in $\mathfrak{h}$. For if $\frac{dg}{ds} = g(s)(B(s) + A(s))$ with $B(s) \in \mathfrak{p}$ and $A(s) \in \mathfrak{h}$, and if $\tilde{g}(s) = g(s)R^{-1}(s)$ for some $R(s) \in SU_2$ then both g and $\tilde{g}$ project onto the same curve $X(s)$. In particular, if $\frac{dR}{ds} = R(s)A(s)$ then $\frac{d\tilde{g}}{ds} = \tilde{g}(s)(R(s)B(s)R^*(s))$. Hence

$$\frac{dX}{ds} = 2\tilde{g}(s)(R(s)B(s)R^*(s))\tilde{g}^* = 2g(s)B(s)g^*(s) = 2g(s)B_1g^*(s),$$

and therefore $B(s) = B_1$. Similar to the spherical case, the hyperolic Darboux curves for which the diagonal part of the matrix A is equal to zero will be called reduced. It follows that any base curve $X(s)$ of an anchored Darboux curve is initially fixed at $X(0) = I$ and has a fixed initial tangent vector $\frac{dX}{ds}(0) = 2B_1$. Furthermore, it follows from above that $X(s)$ is the projection of a horizontal curve $\tilde{g}(s)$ such that

$$\tilde{g}^*\frac{d\tilde{g}}{ds}(s) = \Lambda(s) = R(s)B_1R^*(s)$$

for some curve $R(s)$ in SU_2.

Hyperbolic horizontal curves $g(s)$ will be called hyperbolic horizontal-Darboux if $g(0) = I$ and

$$g^{-1}(s)\frac{dg}{ds}(s) = \Lambda(s) = R(s)B_1R * S, \tag{17.10}$$

for some curve $R(s)$ in SU_2 with $R(0) = I$.

It follows that every curve $X(s)$ on the hyperboloid parametrized by arc length on $[0, L]$ that satisfies $X(0) = I$ and $\frac{dX}{ds}(0) = 2B_1$ is the projection of a unique hyperbolic horizontal-Darboux curve $g(s)$. Moreover, the relation $\Lambda(s) = R(s)B_1R^*(s), R(0) = I$ defines an anchored hyperbolic curve $\tilde{g} = gR$ over X. As in the spherical case, the correspondence between hyperbolic horizontal-Darboux curves and reduced hyperbolic anchored Darboux curves if bijective.

The above implies that the horizontal-Darboux curves in both the spherical and the hyperbolic case are parametrized by matrices $R(s)$ in SU_2 which are the solutions of $\frac{dR}{ds} = R(s)U(s), R(0) = I$, with $U(s) = \begin{pmatrix} 0 & u(s) \\ -\bar{u}(s) & 0 \end{pmatrix}$ for some complex curve $u(s)$ through the relations

$$\frac{dg}{ds} = \Lambda(s) = R(s)C_1R^*(s), R(0) = I, \qquad (17.11)$$

where $C_1 = A_1$ in the spherical case and $C_1 = B_1$ in the hyperbolic case. Hyperbolic horizontal-Darboux curves $g(s)$ are frame-periodic if $\Lambda(s)$ in (17.10) satisfies $\Lambda(s) = R(s)B_1R^*(s)$ for some curve $R(s) \in \mathfrak{h}$ such that $R(0) = R(L) = I$.

Analogous to the spherical case, $\mathcal{D}_h(L)$, $\mathcal{HD}_h(L)$, and $\mathcal{PHD}_h(L)$ will respectively denote the anchored hyperbolic Darboux curves, the horizontal-Darboux curves, and the frame-periodic Darboux curves. In both the spherical and the hyperbolic case frame-periodicity implies that the matrix $U(s) = R(s)^* \frac{dR}{ds}$ is smoothly periodic. The same applies to the matrix $\Lambda(s) = R(s)A_1R^*(s)$ (respectively $\Lambda(s) = R(s)B_1R^*(s)$). This implies that the projections of frame-periodic curves necessarily have periodic curvature and torsion, but need not be closed.

17.2 Darboux curves and symplectic Fréchet manifolds

We will make use of the general theory developed by R. S. Hamilton in [Hm] and regard each space of anchored or frame-periodic Darboux curves and their horizontal projections as an infinite-dimensional Fréchet manifold. Recall that a topological Hausdorff vector space V is called a Fréchet space if its topology is induced by a countable family $\{p_n\}$ of semi-norms and if it is complete relative to the semi-norms in $\{p_n\}$ (see, for instance [Yo]). A Fréchet manifold is a topological Hausdorff space equipped with an atlas whose charts take values in open subsets of a Fréchet vector space V such that any change of coordinate charts is smooth.

The paper of Hamilton [Hm] singles out an important class of Fréchet manifolds, called tame, in which the implicit function theorem is true. One of the main theorems in [Hm] is that the set of smooth mappings from a compact manifold interval into a finite-dimensional Riemannian manifold M is a tame Fréchet manifold. It therefore follows from the implicit function theorem that closed subsets of tame Fréchet manifolds $\mathcal{M}$, defined by the zero sets of finitely many smooth functions on $\mathcal{M}$ are tame submanifolds of $\mathcal{M}$. Since the anchored Darboux curves are particular cases of the above situation, they are tame Fréchet manifolds and the same applies to their horizontal projections.

Tangent vectors and tangent bundles of Fréchet manifolds are defined in the same manner as for finite-dimensional manifolds. In particular, tangent vectors at a point x in a Fréchet manifold $\mathbb{M}$ are the equivalence classes of curves $\sigma(t)$ in $\mathcal{M}$ all emanating from x (i.e., $\sigma(0) = x$), and all having the same tangent vector $\frac{d\sigma}{dt}(0)$ in each equivalence class. The set of all tangent vectors at x denoted by $T_x\mathbb{M}$ constitutes the tangent space at x.

The tangent bundle of a Fréchet manifold $\mathbb{M}$ is a Fréchet manifold. A vector field X on $\mathcal{M}$ is a smooth mapping from $\mathcal{M}$ into the tangent bundle $T\mathcal{M}$ such that $X(x) \in T_x\mathbb{M}$ for each $x \in \mathbb{M}$. On tame Fréchet manifolds vector fields can be defined as the derivations in the space of smooth functions on $\mathbb{M}$.

However, the dual of a Fréchet space is not a Fréchet space, hence it is difficult to speak of the cotangent bundle of a Fréchet space. We will circumvent this difficulty by introducing a closed, non-degenerate 2-form on the tangent bundle which then will enable us to speak of Hamiltonian fields associated with functions. We will then demonstrate the benefits of this formalism by linking the geometry of curves to the equations of mathematical physics. As in finite-dimensional situations, a skew-symmetric, non-degenerate, and closed 2-form will be called symplectic.

Let us first address the structure of the tangent bundle.

Proposition 17.8   *(a) The tangent space $T_p(\mathcal{HD}_s)(L)$ at a spherical horizontal-Darboux curve $p(s)$ with $\Lambda(s) = p^*(s)\frac{dp}{ds}(s)$ consists of curves $v(s) = X(s)V(s)$ with $V(s)$ the solution of*

$$\frac{dV}{ds}(s) = [\Lambda(s), V(s)] + U(s) \tag{17.12}$$

such that $V(0) = 0$, where $U(s)$ is a curve in $\mathfrak{h}$ subject to the conditions that $U(0) = 0$ and $\langle \Lambda(s), U(s)\rangle_s = 0$.

(b) Tangent vectors $v(s) = X(s)V(s)$ at frame-periodic horizontal-Darboux curves $X(s)$ are generated by smoothly periodic curves $U(s)$ whose period is equal to L.

The calculations of tangent vectors make use of the covariant derivative. On the hyperboloid the formula is the same as used in previous chapters; the covariant derivative $\frac{D_{\pi(g)}}{ds}(v)$ of a curve of tangent vectors $v(s) = \pi_* g(s)U(s)$, $U(s) \in \mathfrak{p}$, along a horizontal curve $g(s)$ in $SL_2(C)$ given by

$$\frac{D_{\pi(g)}}{ds}(v)(s) = \pi_*\left(g(s)\frac{dU}{ds}(s)\right).$$

On the sphere, however, a representation different from the one used before is more convenient: the covariant derivative of a curve of tangent vectors

$v(s) = X(s)U(s)$ along a curve $X(s)$ in SU_2 is given by

$$\frac{D_X}{ds}(v)(s) = X(s)\left(\frac{dU}{ds} + \frac{1}{2}[U(s), \Lambda(s)]\right),$$

where $\Lambda(s) = X^*(s)\frac{dX}{ds}(s)$. The reader can easily verify that the covariant derivative on SU_2 is equal to the orthogonal projection of the ordinary derivative in $\mathbb{R}^4$ onto the tangent space of the sphere when the sphere is considered a submanifold of $\mathbb{R}^4$ (see also [DC]).

The subsequent calculations will also make use of the following:

Lemma 17.9   *Suppose that $X(s, t)$ is a field of curves in SU_2 with its infinitesimal directions*

$$A(s, t) = X^*(s, t)\frac{\partial X}{\partial s}(s, t) \text{ and } B(s, t) = X^*(s, t)\frac{\partial X}{\partial t}(s, t).$$

Then

$$\frac{\partial A}{\partial t} - \frac{\partial B}{\partial s} + [A, B] = 0. \tag{17.13}$$

Proof   On any Riemannian manifold the mixed partial derivatives $\frac{D_X}{ds}\left(\frac{\partial X}{\partial t}\right)$ and $\frac{D_X}{ds}\left(\frac{\partial X}{\partial s}\right)$ are equal. Hence,

$$X\left(\frac{\partial B}{\partial s} + \frac{1}{2}[B, A]\right) = X\left(\frac{\partial A}{\partial t} + \frac{1}{2}[A, B]\right)$$

and therefore,

$$\frac{\partial A}{\partial t} - \frac{\partial B}{\partial s} + [A, B] = 0.$$

$\square$

Equation (17.13) is also known as the zero-curvature equation [Fa]. Return now to the proof of the proposition.

Proof   Let $Y(s, t)$ denote a family of anchored horizontal-Darboux curves such that $Y(s, 0) = p(s)$. Then, $v(s) = \frac{\partial Y}{\partial t}(s, t)_{t=0}$ is a tangent vector at $X(s)$, where $v(0) = 0$, since the curves $Y(s, t)$ are anchored.

Let $Z(s, t)$ and $W(s, t)$ denote the matrices defined by

$$Z(s, t) = Y(s, t)^*\frac{\partial Y}{\partial s}(s, t), \ W(s, t) = Y(s, t)^*\frac{\partial Y}{\partial t}(s, t).$$

It follows that $\Lambda(s) = Z(s, 0)$, $V(s) = W(s, 0)$. Then

$$\frac{\partial Z}{\partial t} - \frac{\partial W}{\partial s} + [Z, W] = 0$$

for $t = 0$ reduces to

$$\frac{dV}{ds}(s) = [\Lambda(s), V(s)] + U(s),$$

where $U(s) = \frac{\partial Z}{\partial t}(s, 0)$.

Since the curves $s \to Y(s, t)$ are Darboux curves for each t, $\langle Z(s, t), Z(s, t)\rangle_s = 1$ and $Z(0, t) = A_1$. Therefore,

$$\langle Z(s, t), \frac{\partial Z}{\partial t}(s, t)\rangle = 0, \text{ and } \frac{\partial Z}{\partial t}(0, t) = 0,$$

which implies that $\langle \Lambda(s), U(s)\rangle_s = 0$ and $U(0) = 0$.

It remains to show that any curve $V(s)$ in $\mathfrak{h}$ that satisfies (17.12) can be realized by the perturbations $Y(s, t)$ used above. So assume that $V(s)$ be any solution of (17.12) generated by a curve $U(s)$ with $U(0) = 0$ that satisfies $\langle \Lambda(s), U(s)\rangle_s = 0$.

Let $\phi(t)$ denote any smooth function such that $\phi(0) = 0$ and $\frac{d\phi}{dt}(0) = 1$. Define

$$Z(s, t) = \frac{1}{1 + \phi^2(t)\langle U(s), U(s)\rangle_s}(\Lambda(s) + \phi(t)U(s)).$$

Evidently $Z(0, t) = A_1$ for all t, and an easy calculation shows that $\langle Z(s, t), Z(s, t)\rangle_s = 1$. Therefore, $Y(s, t)$, the solution of $\frac{\partial Y}{\partial s}(s, t) = Y(s, t)Z(s, t)$ with $Y(0, t) = I$, corresponds to an anchored horizontal-Darboux curve for each t. Since $U(s) = \frac{\partial Z}{\partial t}(s, 0)$, our proof of part (a) is finished.

To prove part (b) assume that the above curves $Y(s, t)$ belong to $\mathcal{PHD}_s(L)$. Then, the curves $s \to Z(s, t)$ are L-periodic for each t, and therefore, $U(s) = \frac{\partial Z}{\partial t}(s, 0)$ is also L-periodic. □

Tangent spaces of the manifolds $\mathcal{HD}_h(L)$ and $\mathcal{PHD}_h(L)$ are obtained along similar lines as in the spherical case. They are described by the following proposition.

Proposition 17.10 (a) *The tangent space $T_g(\mathcal{HD}_h)(L)$ at a hyperbolic horizontal-Darboux curve $g(s)$ with $\Lambda(s) = g^{-1}(s)\frac{dg}{ds}(s)$ consists of curves $v(s) = g(s)V(s)$ such that*

$$\frac{dV}{ds} = U(s), V(0) = 0, \tag{17.14}$$

where $U(s)$ is a Hermitian curve subject to the following conditions:

$$U(0) = 0, \langle \Lambda(s), U(s)\rangle_h = 0.$$

(b) *For frame-periodic horizontal-Darboux curves, the curve $\frac{dV}{ds}(s)$ must be smoothly periodic having the period equal to L.*

Proof   Let $h(s, t)$ denote a family of anchored horizontal-Darboux curves such that $h(s, 0) = g(s)$. Then $v(s) = \frac{\partial h}{\partial t}(s, t)_{t=0}$ is a tangent vector at $g(s)$ such that $v(0) = 0$, since the curves $h(s, t)$ are anchored.

Let $Z(s, t)$ and $W(s, t)$ denote the matrices defined by

$$Z(s, t) = h(s, t)^{-1}\frac{\partial h}{\partial s}(s, t), \ W(s, t) = h(s, t)^{-1}\frac{\partial h}{\partial t}(s, t).$$

It follows that $\Lambda(s) = Z(s, 0)$ and $v(s) = g(s)V(s)$ with $V(s) = W(s, 0)$. Then,

$$\frac{\partial Z}{\partial t}(s, t) = \frac{\partial W}{\partial s}(s, t)$$

is the hyperbolic analogue of the zero-curvature equation (17.13). For $t = 0$ the above equation reduces to

$$\frac{dV}{ds}(s) = \frac{\partial W}{\partial s}(s, 0) = U(s),$$

where $U(s) = \frac{\partial Z}{\partial t}(s, 0)$. Constraints $\langle Z(s, t), Z(s, t)\rangle_h = 1$ and $Z(0, t) = B_1$ imply that $\langle \Lambda(s), U(s)\rangle_h = 0$, and $U(0) = 0$.

Conversely, any curve $V(s)$ in $\mathfrak{h}$ that satisfies (17.14) can be realized by the perturbations $h(s, t)$ defined in the first part of the proof. The argument is the same as in the spherical case and will be omitted. The same applies to the proof of part (b).   □

Having now the description of the tangent spaces, we will pass to the symplectic form ω. This 2-form will be defined first for the spherical horizontal-Darboux curves. In the process it will become clear how to adapt the results to the hyperbolic horizontal-Darboux curves.

Definition 17.11   Let ω be defined by

$$\omega_\Lambda(V_1, V_2) = -\int_0^T \left\langle \Lambda(s), \left[\frac{dV_1}{ds}, \frac{dV_2}{ds}\right]\right\rangle_s ds = -\int_0^L \langle \Lambda(s), [U_1(s), U_2(s)]\rangle_s ds$$

$$(17.15)$$

for any pair of tangent vectors (V_1, V_2) at a horizontal-Darboux curve $p(s)$ that is defined by $\frac{dp}{ds}(s) = p(s)\Lambda(s)$. Here $U_1(s)$ and $U_2(s)$ are as in equation (17.12) that is, $U_i(0) = 0 \ \langle \Lambda(s), U_i(s)\rangle_s = 0$, and

$$U_i(s) = \frac{dV_i}{ds}(s) - [\Lambda(s), V_i(s)], \ i = 1, 2.$$

As in finite-dimensional situations the choice of sign is a matter of convention.

Proposition 17.12 ω *is skew-symmetric, non-degererate, and closed on each of $\mathcal{HD}_s(L)$ and $\mathcal{PHD}_s(L)$.*

Proof Evidently ω is skew-symmetric. To show that it is non-degenerate assume that $\omega_\Lambda(V_1, V) = 0$ for some tangent vector gV_1 and all other tangent vectors gV. Let $U_1(s)$ correspond to $V_1(s)$ defined by equations (9). Then $U(s) = [\Lambda(s), U_1(s)]$ satisfies $U(0) = 0$, and $\langle \Lambda(s), U(s) \rangle_s = 0$. Therefore the corresponding vector gV with V the solution of equation (17.12) belongs to the tangent space at g. Since $[A, [B, C]] = \langle A, C \rangle_s B - \langle A, B \rangle_s C$ for any elements A, B, C in $\mathfrak{h}$,

$$[U_1, [\Lambda, U_1]] = \langle U_1, U_1 \rangle_s \Lambda = \|U_1\|_s^2 \Lambda.$$

Therefore,

$$\langle \Lambda(s), [U_1(s), U(s)] \rangle_s = \|\Lambda(s)\|_s^2 \|U_1(s)\|_s^2 = \|U_1(s)\|_s^2,$$

which implies that $U_1(s) = 0$ since $0 = \omega_\Lambda(V_1, V) = \int_0^L \|U_1(s)\|^2 \, ds$. But then $V_1(s) = 0$. because $V_1(s)$ is a solution of a linear differential equation with $V(0) = 0$. Thus ω is non-degenerate.

To show that ω is closed, let $v_i(s) = g(s)V_i(s)$, $1 \le i \le 3$ denote any three tangent vectors at a fixed Darboux curve $g(s)$. It is required to show that

$$d\omega(X_1, X_2, X_3) = \sum_{cyclic} X_i(\omega(X_j, X_k)) + \sum_{cyclic} \omega([X_i, X_j], X_k) = 0, \quad (17.16)$$

where X_i denote any vector fields such that $X_i(g) = v_i$ for each $i = 1, 2, 3$.

Let $X_i(z) = zZ_i, i = 1, 2, 3$ denote any such vector fields. At each Darboux curve $z(s)$, $Z_i(s)$ is the solution of

$$\frac{dZ_i}{ds}(s) = [\Lambda_z(s), Z_i(s)] + U_i(s), \text{ where } \Lambda_z = z^* \frac{dz}{ds}.$$

Since the mapping $C \to [\Lambda, C]$ is surjective on the orthogonal complement to Λ there exists a unique curve $C_i(s)$ such that $U_i(s) = [\Lambda(s), C_i(s)], i = 1, 2, 3$. Then, $\frac{dV_i}{ds} = [\Lambda, V_i + C_i]$, and an easy calculation based on Jacobi's identity yields

$$\frac{d}{ds}([V_i, V_j]) = [\Lambda, [V_i, V_j]] + [[\Lambda, C_i], V_j] + [V_i, [\Lambda, C_j]].$$

Therefore,

$$\sum_{cyclic} \omega([X_i, X_j], X_k) = \sum_{cyclic} \int_0^L \langle \Lambda, [[[\Lambda, C_i], V_j] + [V_i, [\Lambda, C_j]], [\Lambda, C_k]] \rangle_s ds$$

$$= -\sum_{cyclic} \int_0^L \langle \Lambda, [((\langle V_i, \Lambda \rangle_s C_j - \langle V_j, \Lambda \rangle_s C_i) + (\langle V_j, C_i \rangle_s \Lambda$$

$$- \langle V_i, C_j \rangle_s \Lambda), [\Lambda, C_k]] \rangle_s \, ds$$

$$= \sum_{cyclic} \int_0^L \langle V_j, \Lambda \rangle_s \langle C_i, C_k \rangle_s - \langle V_i, \Lambda \rangle_s \langle C_j, C_k \rangle_s \, ds = 0.$$

The calculations involving $X_i(\omega(X_j, X_k))$ in (17.16) require additional notations. Let $t \to z_i(s, t)$ denote the integral curves of the vector field X_i that originate at $g(s)$ for $t = 0$, and let

$$\frac{\partial z_i}{\partial t}(s, t) = z_i(s, t) Z_i((z_i(s, t)) \text{ and } \frac{\partial z_i}{\partial s}(s, t) = z_i(s, t) \Lambda_i(z_i(s, t)).$$

For simplicity of notation let $Z_i(z_i(s, t))$ and $\Lambda_i(z_i(s, t))$ be denoted by $Z_i(s, t)$ and $\Lambda_i(s, t)$. Then

$$\frac{\partial \Lambda_i}{\partial t} - \frac{\partial Z_i}{\partial s} + [\Lambda_i, Z_i] = 0,$$

which at $t = 0$ reduce to

$$U_i - \frac{dV_i}{ds} + [\Lambda, V_i] = 0.$$

As in the preceeding calculation, U_i will be represented by $U_i = [\Lambda, C_i]$. Then,

$$X_i(\omega(X_j, X_k)) = \frac{\partial}{\partial t} \int_0^L \langle \Lambda_i(s, t), [[\Lambda_j(s, t), C_j], [\Lambda_k(s, t), C_k]] \rangle_s \, ds|_{t=0}$$

$$= \int_0^L \left\langle \frac{\partial \Lambda_i}{\partial t}(s, t), [[\Lambda_j(s, t), C_j], [\Lambda_k(s, t), C_k]] \right\rangle_s ds|_{t=0}$$

$$+ \int_0^L \left\langle \Lambda_i(s, t), \left[\left[\frac{\partial \Lambda_j}{\partial t}(s, t), C_j \right], [\Lambda_k(s, t), C_k] \right] \right\rangle_s ds|_{t=0}$$

$$+ \int_0^L \left\langle \Lambda_i(s, t), \left[[\Lambda_j(s, t), C_j], \left[\frac{\partial \Lambda_k}{\partial t}(s, t), C_k \right] \right] \right\rangle_s ds|_{t=0}$$

$$= \int_0^L \langle U_i, [U_j, U_k] \rangle_s \, ds + \int_0^L \langle \Lambda, ([[U_j, C_j], [\Lambda, C_k]]$$

$$+ [[\Lambda, C_j], [U_k, C_k]] \rangle_s \, ds.$$

The first integral $\int_0^L \langle U_i, [U_j, U_k] \rangle_s \, ds$ is equal to zero because $\langle U_i, [U_j, U_k] \rangle_s$ is the volume of the parallelepiped with sides U_i, U_j, U_k each of which is in the plane orthogonal to Λ.

The second integral $\int_0^L \langle \Lambda, ([[U_j, C_j], [\Lambda, C_k]] + [[\Lambda, C_j], [U_k, C_k]]) \rangle_s \, ds$ is also equal to zero because

$$\langle \Lambda, ([[U_j, C_j], [\Lambda, C_k]] + [[\Lambda, C_j], [U_k, C_k]]) \rangle_s$$

$$= -\langle \Lambda, (\langle [U_j, C_j], \Lambda \rangle_s C_k + \langle [U_j, C_j], C_k \rangle_s \Lambda + \langle [U_k, C_k], \Lambda \rangle_s C_j$$

$$- \langle [U_k, C_k], C_j \rangle_s \Lambda \rangle_s)$$
$$= - \langle [U_k, C_k], C_j \rangle_s + \langle [U_j, C_j], C_k \rangle_s.$$

Since U_k, C_k, C_j are all in the plane orthogonal to Λ the preceding expression is equal to zero.

Thus ω is closed, and hence symplectic. $\qquad\square$

Corollary 17.13 *The form ω defined on the space of anchored hyperbolic-Darboux curves given by*

$$\omega_\Lambda(V_1, V_2) = \frac{1}{i} \int_0^L \left\langle \Lambda(s), \left[\frac{dV_1}{ds}(s), \frac{dV_2}{ds}(s) \right] \right\rangle_h ds \qquad (17.17)$$

for any tangent vectors $V_1(s), V_2(s)$ at a horizontal curve $g(s)$ is symplectic.

Proof Tangent vectors at a horizontal curve $g(s)$ are of the form $v(s) = g(s)V(s)$ with $V(s)$ a Hermitian curve that satisfies the following conditions:

$$V(0) = 0, \quad \frac{dV}{ds}(0) = 0, \quad \left\langle \Lambda(s), \frac{dV}{ds}(s) \right\rangle_h = 0.$$

Let

$$\tilde{\Lambda} = i\Lambda, \quad \tilde{U}_1 = i\frac{dV_1}{ds}, \quad \tilde{U}_2 = i\frac{dV_2}{ds}.$$

Since $[\mathfrak{p}, \mathfrak{p}] = \mathfrak{h}$ and $i\mathfrak{p} = \mathfrak{h}$, matrices $\tilde{\Lambda}, \tilde{U}_1, \tilde{U}_2$ belong to $\mathfrak{h}$ and satisfy $\langle \tilde{\Lambda}, \tilde{U}_i \rangle_h = 0$ for $i = 1, 2$. Therefore,

$$\omega_\Lambda(V_1, V_2) = \frac{1}{i} \int_0^L \left\langle \Lambda(s), \left[\frac{dV_1}{ds}(s), \frac{dV_2}{ds}(s) \right] \right\rangle_h ds$$
$$= - \int_0^L \langle \tilde{\Lambda}(s), [\tilde{U}_1(s), \tilde{U}_2(s)] \rangle_s ds$$

coincides with the form given in Definition 17.11. $\qquad\square$

Corollary 17.14 $\mathcal{HD}_h(L)$ *and* $\mathcal{PHD}_h(L)$ *are symplectic manifolds relative to the form ω defined by (17.17).*

It may be appropriate to point out that both the spherical and the hyperbolic symplectic form defined above are isomorphic to the symplectic structure of anchored loops on the sphere S^2 given explicitly by

$$\omega_\lambda(u_1, u_2) = \int \lambda(s) \cdot (u_1(s) \times u_2(s)) ds,$$

where $u_1(s)$ and $u_2(s)$ are tangent vectors at $\lambda(s)$ on S^2. For when λ, u_1 and u_2 are identified with the coordinates vectors of Λ, U_1, and U_2 relative to the Pauli matrices, then the coordinate vector of $[U_1, U_2]$ is given by the cross product $u_1 \times u_2$. Therefore,

$$\int \langle \Lambda, [U_1, U_2] \rangle_s \, ds = \int \lambda(s) \cdot (u_1(s) \times u_2(s)) \, ds.$$

To correlate the findings of this paper with the related existing literature, which almost exclusively deals with curves in $\mathbb{R}^3$, it seems appropriate to include a discussion of the remaining simply connected three-dimensional symmetric space, the Euclidean space.

17.2.1 Euclidean Darboux curves

The most convenient way to pass to Euclidean Darboux curves is to realize the Euclidean group of motions as the semi-direct product $S_H(\mathfrak{p}) = \mathfrak{p} \rtimes H$, relative to the adjoint action of $H = SU_2$ on the space of Hermitian matrices $\mathfrak{p}$.

Recall that the semi direct product of a vector space V and a group H which acts linearly on V consists of pairs (x, R) with $x \in V$ and $R \in H$ with the group operation given by $(x, R)(y, T) = (x + Ry, RT)$. In this setting the Lie algebra of $S_H(V)$ consists of pairs (a, A) with $a \in V$ and $A \in \mathfrak{h}$ with the Lie bracket

$$[(a, A), (b, B)] = (A(b) - B(a), [A, B]).$$

In our specific situation the Lie bracket is given by

$$[(a, A), (b, B)] = ([b, A] - [a, B], [A, B]).$$

As a vector space, $\mathfrak{p} \rtimes \mathfrak{h}$ can be identified with $sl_2(C)$ via the embedding $(a, A) \to a + A$ for any (a, A) in $\mathfrak{p} \ltimes su_2$. With this identification, $s_H(V) = \mathfrak{p} \oplus \mathfrak{h}$ and

$$[\mathfrak{p}, \mathfrak{p}] = 0, \quad [\mathfrak{p}, \mathfrak{h}] = \mathfrak{p}, \quad [\mathfrak{h}, \mathfrak{h}] = \mathfrak{h}.$$

The group $S_H(\mathfrak{p})$ acts on $\mathfrak{p}$ by $(x, R)(y) = R(y) + x$ for each $(x, R) \in S_H(\mathfrak{p})$ and each $y \in \mathfrak{p}$. The action is transitive, and H is equal to the isotropy group of the orbit through the origin $y = 0$. Then $\mathfrak{p}$, when identified with the orbit through the origin, becomes the coset space $S_H(\mathfrak{p})/H$.

The space of Hermitian matrices endowed with the metric induced by the trace form becomes a three-dimensional Euclidean space $\mathbb{E}^3$. The preceeding action extends to an action on the tangent bundle of $\mathbb{E}^3$ in which a tangent vector v at y is taken to the tangent vector $R(v)$ at x under the action by an element $(x, R) \in S_H(\mathbb{E}^3)$. The action on the tangent bundle extends further to an action on the orthonormal frame bundle of $\mathbb{E}^3$ such that a frame (v_1, v_2, v_3) at a point $y \in \mathbb{E}^3$ is taken to the frame $(R(v_1), R(v_2), R(v_3))$ at x under the action by an element $(x, R) \in S_H(\mathbb{E}^3)$. The kernel of this action consists of $\pm I$, and hence $S_H(\mathbb{E}^3)/\{\pm I\}$ can be identified with the oriented orthonormal frame bundle of $\mathbb{E}^3$ as the orbit through the standard frame (B_1, B_2, B_3) at the origin.

In the left-invariant representation of the tangent bundle of $S_H(\mathbb{E}^3)$, tangent vectors at a point (x, R) are represented by the pairs $(R(a), RA)$ with $a \in \mathbb{E}^3$ and $A \in \mathfrak{h}$. Hence curves $(x(s), R(s))$ in $S_H(\mathbb{E}^3)$ are represented by the differential equations

$$\frac{dx}{ds}(s) = R(s)(a(s)), \quad \frac{dR}{ds}(s) = R(s)A(s).$$

The terminology concerning Darboux curves in non-Euclidean cases extends naturally to the Euclidean setting. In particular, curves $(x, R) \in S_H(\mathbb{E}^3)$ are Euclidean anchored Darboux curves if $\frac{dx}{ds}(s) = R(s)(B_1)$, i.e., whenever $a(s) = B_1$, subject to further boundary conditions $x(0) = 0$, and $R(0) = I$. Euclidean horizontal-Darboux curves are the projections of Euclidean anchored Darboux curves, i.e., they are the solutions of

$$\frac{dx}{ds}(s) = R(s)(B_1), \quad x(0) = 0$$

with $R(s)$ an arbitrary curve in H such that $R(0) = I$. Frame-periodic Darboux curves (x, R) conform to the periodicity of $R(s)$ with its period equal to the length of $x(s)$.

For any horizontal-Darboux curve $x(s)$

$$\frac{d^2x}{ds^2}(s) = R(s)([B_1, A(s)]),$$

and therefore

$$\kappa^2(s) = \left\| \frac{d^2x}{ds^2}(s) \right\|^2 = u_2^2(s) + u_3^2(s),$$

where $A(s) = \sum u_i(s)A_i$. The frame $R(s)$ is a Serret–Frenet frame if $A(s) = \tau(s)A_1 + \kappa(s)A_3$, in which case the frame vectors $T(s), N(s), B(s)$ are given by $T(s) = R(s)(B_1)$, $N(s) = R(s)(B_2)$, $B(s) = R(s)(B_3)$.

The reader may easily verify that the tangent space at each anchored horizontal-Darboux curve $x(s)$ consists of curves $v(s)$ such that:

(a) $v(0) = \frac{dv}{ds}(0) = 0$, and
(b) $\langle \frac{dx}{ds}(s), \frac{dv}{ds}(s) \rangle = 0$.

The space of frame-periodic Euclidean horizontal-Darboux curves inherits the symplectic structure given by Corollary 17.13 to Proposition 17.12. This symplectic structure is isomorphic to the structure used by J. Millson and B. A. Zombro in [MZ] since su_2 and $so_3(R)$ are isomorphic

17.3 Geometric invariants of curves and their Hamiltonian vector fields

Each function f on $\mathcal{PHD}(L)$ induces a Hamiltonian vector field $\mathcal{X}_f$ defined by

$$df_\Lambda(V) = \omega_\Lambda(\mathcal{X}_f, V), \tag{17.18}$$

where V ranges over all tangent vectors at a horizontal curve $g(s)$ with $g^{-1}\frac{dg}{ds} = \Lambda(s)$. We will now investigate the Hamiltonian vector fields associated with some geometric invariants of curves in the base space, starting with the function $f(\Lambda) = \frac{1}{2}\int_0^L \left\langle \frac{d\Lambda(s)}{ds}, \frac{d\Lambda(s)}{ds} \right\rangle ds$ that corresponds to the elastic energy $\frac{1}{2}\int_0^L \kappa^2(s)\,ds$ of the base curve.

Proposition 17.15 *Suppose that $f(\Lambda) = \frac{1}{2}\int_0^L \left\| \frac{d\Lambda}{ds}(s) \right\|^2 ds$ is a function over the anchored periodic Darboux curves $g(s)$ with $g^{-1}\frac{dg}{ds} = \Lambda(s)$. In the spherical case integral curves $g(s,t)$ of the associated Hamiltonian flow X_f evolve according to*

$$\frac{\partial g}{\partial s}(s,t) = g(s,t)\Lambda(s,t), \quad \frac{\partial \Lambda}{\partial t}(s,t) = \left[\frac{\partial^2 \Lambda}{\partial s^2}, \Lambda(s,t)\right], \tag{17.19}$$

and in the hyperbolic case they evolve according to

$$\frac{\partial g}{\partial s}(s,t) = g(s,t)\Lambda(s,t), \quad \frac{\partial \Lambda}{\partial t}(s,t) = i\left[\frac{\partial^2 \Lambda}{\partial s^2}, \Lambda(s,t)\right], \tag{17.20}$$

Proof Although conceptually alike, the calculations in the spherical setting are different in several aspects from those in the hyperbolic setting, and will be done separately. For notational simplicity the subscripts in the quadratic form will be dropped and all inner products will be denoted by the same symbol $\langle\,,\,\rangle$, and the same will apply to the induced norms.

Let us begin with hyperbolic case. To calculate the directional derivative $df_\Lambda(V)$, let $\hat{g}(s,t)$ be a family of anchored horizontal-Darboux curves that are the solutions of $\frac{\partial \hat{g}}{\partial s} = \hat{g}(s,t)\hat{\Lambda}(s,t)$ such that $\hat{g}(s,0) = g(s)$, $\hat{\Lambda}(s,0) = \Lambda(s)$, $\frac{\partial \hat{\Lambda}}{\partial t}(s,0) = \frac{dV}{ds}(s)$. The directional derivative $df_\Lambda(V)$ is given by

$$df_\Lambda(V) = \frac{1}{2}\frac{\partial}{\partial t}\int_0^L \left\langle \frac{\partial \hat{\Lambda}}{\partial s}(s,t), \frac{\partial \hat{\Lambda}}{\partial s}(s,t) \right\rangle ds|_{t=0},$$

Then,

$$\frac{1}{2}\frac{\partial}{\partial t}\int_0^L \left\langle \frac{\partial \hat{\Lambda}}{\partial s}(s,t), \frac{\partial \hat{\Lambda}}{\partial s}(s,t) \right\rangle ds|_{t=0} = \int_0^L \left\langle \frac{\partial \hat{\Lambda}}{\partial s}(s,t), \frac{\partial}{\partial s}\frac{\partial \hat{\Lambda}}{\partial t}(s,t) \right\rangle ds|_{t=0} =$$

$$\int_0^L \left\langle \frac{d\Lambda}{ds}, \frac{d}{ds}\left(\frac{dV}{ds}\right)\right\rangle ds = -\int_0^L \left\langle \frac{d^2\Lambda}{ds^2}, \frac{dV}{ds}\right\rangle ds + \left\langle \frac{d\Lambda}{ds}, \frac{dV}{ds}\right\rangle \Big|_{s=0}^{|s=L}.$$

The boundary terms $\left\langle \frac{d\Lambda}{ds}, \frac{dV}{ds}\right\rangle |_{s=0}^{s=L}$ are equal to 0 because of periodicity, consequently, $df_\Lambda(V) = -\int_0^L \left\langle \frac{d^2\Lambda}{ds^2}, \frac{dV}{ds}\right\rangle ds$.

The Hamiltonian vector field is of the form $\mathcal{X}_f(g) = gF$ for some Hermitian matrix $F(s)$ that satisfies $df_\Lambda(V) = \frac{1}{i}\int_0^L \left\langle \Lambda(s), \left[\frac{dF}{ds}, \frac{dV}{ds}\right]\right\rangle ds$ for an arbitrary tangential direction $V(s)$. The above is equivalent to

$$\int_0^L \left\langle \frac{d^2\Lambda}{ds^2} + \frac{1}{i}\left[\Lambda(s), \frac{dF}{ds}\right], \frac{dV}{ds}\right\rangle ds = 0. \tag{17.21}$$

Since the mapping $U \rightarrow i[\Lambda, U]$ is bijective on the space of Hermitian matrices orthogonal to $\Lambda(s)$, there exists a Hermitian matrix $U(s)$ such that $U(0) = 0$ and $\frac{dV}{ds} = i[\Lambda(s), U(s)]$ Then equation (17.21) becomes

$$\int_0^L \left\langle i\left[\frac{d^2\Lambda}{ds^2}, \Lambda(s)\right] + \left[\left[\Lambda(s), \frac{dF}{ds}\right], \Lambda(s)\right], U(s)\right\rangle ds = 0.$$

It follows that

$$i\left[\frac{d^2\Lambda}{ds^2}, \Lambda(s)\right] + \left[\left[\Lambda(s), \frac{dF}{ds}\right], \Lambda(s)\right] = 0$$

because $U(s)$ is arbitrary. Since $\left[\Lambda(s), \left[\Lambda(s), \frac{dF}{ds}\right]\right] = \frac{dF}{ds}$, the above becomes

$$i\left[\frac{d^2\Lambda}{ds^2}, \Lambda(s)\right] - \frac{dF}{ds} = 0.$$

It follows that

$$\mathcal{X}_f(g) = g(s)F(s) \text{ with } F(s) = i\int_0^s \left[\frac{d^2\Lambda}{dx^2}(x), \Lambda(x)\right] dx.$$

The integral curves $t \rightarrow g(s, t)$ of $\mathcal{X}_f$ are the solutions of the following equations:

$$\frac{\partial g}{\partial t}(s, t) = g(s, t)i\int_0^s \left[\frac{d^2\Lambda}{dx^2}(x, t), \Lambda(x, t)\right] dx, \quad \frac{\partial g}{\partial s}(s, t) = g(s, t)\Lambda(s, t).$$

$$\tag{17.22}$$

The equality of mixed partial derivatives $\frac{D_g}{ds}\left(\frac{\partial g}{\partial t}\right) = \frac{D_g}{dt}\left(\frac{\partial g}{\partial s}\right)$ implies that the matrices $\Lambda(s,t)$ evolve according to

$$\frac{\partial \Lambda}{\partial t}(s,t) = i\left[\frac{\partial^2 \Lambda}{\partial s^2}, \Lambda(s,t)\right].$$

To prove the analogous formula in the spherical case, let $Y(s,t)$ denote a family of anchored horizontal-Darboux curves such that $Y(s,0) = X(s)$ and $v(s) = \frac{\partial Y}{\partial t}(s,t)_{t=0} = X(s)V(s)$.

Denote by $Z(s,t)$ the matrices defined by $\frac{\partial Y}{\partial s}(s,t) = Y(s,t)Z(s,t)$. It follows that $\Lambda(s) = Z(s,0)$, and that $V(s)$ is the solution of $\frac{dV}{ds} = [\Lambda(s), V(s)] + U(s)$ with $U(s) = \frac{\partial Z}{\partial t}(s,0)$.

Then,

$$df_\Lambda(V) = \frac{1}{2}\frac{\partial}{\partial t}\int_0^L \left\langle\frac{\partial Z}{\partial s}(s,t), \frac{\partial Z}{\partial s}(s,t)\right\rangle ds|_{t=0}$$

$$= \int_0^L \left\langle\frac{d\Lambda}{ds}(s), \frac{dU}{ds}(s)\right\rangle ds$$

$$= -\int_0^L \left\langle\frac{d^2\Lambda}{ds^2}(s), U(s)\right\rangle ds + \left\langle\frac{d\Lambda}{ds}(s), U(s)\right\rangle\Big|_{s=0}^{s=L}.$$

Analogous to the hyperbolic case, the boundary terms vanish, and therefore

$$df_X(V) = -\int_0^L \left\langle\frac{d^2\Lambda}{ds^2}, U(s)\right\rangle ds.$$

The Hamiltonian vector field $\mathcal{X}_f$ that corresponds to f is of the form $\mathcal{X}_f(X)(s) = X(s)F(s)$ for some curve $F(s) \in \mathfrak{h}$. Since $\mathcal{X}_f(X) \in T_X\mathcal{PHD}_s(L)$, $F(s)$ is the solution of

$$\frac{dF}{ds}(s) = [\Lambda(s), F(s)] + U_f(s), F(0) = 0$$

for some curve $U_f(s) \in \mathfrak{h}$ that satisfies

$$U_f(0) = 0 \text{ and } \langle\Lambda(s), U_f(s)\rangle = 0.$$

The curve $U_f(s)$ is determined by the symplectic form ω:

$$df_\Lambda(U) = -\int_0^L \langle\Lambda(s), [U_f(s), U(s)]\rangle ds, \tag{17.23}$$

where $U(s)$ is an arbitrary curve in $\mathfrak{h}$ that satisfies $U(0)=0$ and $\langle\Lambda(s), U(s)\rangle = 0$. Equation (17.23) yields

$$\int_0^L \left\langle\frac{d^2\Lambda}{ds^2} - [\Lambda(s), U_f(s)], U(s)\right\rangle ds = 0. \tag{17.24}$$

If we write $U(s)$ as $U(s) = [\Lambda(s), C(s)]$ for some curve $C(s)$ that satisfies $C(0) = 0$, equation (17.24) becomes

$$\int_0^L \left\langle \left[\frac{d^2\Lambda}{ds^2}(s), \Lambda(s) \right] - [[\Lambda(s), U_f(s)], \Lambda], C(s) \right\rangle ds = 0.$$

Since $C(s)$ is arbitrary,

$$\left[\frac{d^2\Lambda}{ds^2}, \Lambda \right] - [[\Lambda, U_f], \Lambda] = 0.$$

But then $[[\Lambda, U_f], \Lambda] = U_f$, and therefore $U_f = -\left[\Lambda, \frac{d^2\Lambda}{ds^2}\right]$. The integral curves $t \to X(s, t)$ of $\mathcal{X}_f$ are the solutions of

$$\frac{\partial X}{\partial t}(s, t) = X(s, t)F(s, t), \text{ and } \frac{\partial X}{\partial s} = X(s, t)\Lambda(s, t),$$

where $F(s, t)$ is the solution of $\frac{\partial F}{\partial s}(s, t) = [\Lambda(s, t), F(s, t)] - \left[\Lambda(s, t), \frac{\partial^2\Lambda}{\partial s^2}(s, t)\right]$. Matrices $F(s, t)$ and $\Lambda(s, t)$ satisfy the zero-curvature equation $\frac{\partial\Lambda}{\partial t} - \frac{\partial F}{\partial s} + [\Lambda, F] = 0$. Combined with the above, this equation yields

$$\frac{\partial\Lambda}{\partial t}(s, t) = \left[\frac{\partial^2\Lambda}{\partial s^2}(s, t), \Lambda(s, t) \right].$$

Thus in both the hyperbolic and the spherical case $\Lambda(s, t)$ evolves according to the same equation: in the hyperbolic case Λ is Hermitian, while in the spherical case Λ is skew-Hermitian. To pass from the hyperbolic case to the spherical case multiply $\Lambda(s, t)$ in equation (17.19) by i. We leave it to the reader to show that in the Euclidean case the integral curves $g(s, t)$ of the Hamiltonian flow X_f evolve according to

$$\frac{\partial g}{\partial s}(s, t) = g(s, t)\Lambda(s, t), \frac{\partial\Lambda}{\partial t}(s, t) = i \left[\frac{\partial^2\Lambda}{\partial s^2}, \Lambda(s, t) \right],$$

exactly as in the hyperbolic case, except that the curves $g(s, t)$ now evolve in the semi-direct product $S_H(\mathfrak{p})$. $\qquad\square$

Equations (17.19) and (17.20) will be referred to as *Heisenberg's magnetic equation* (HME) (17.1).

Equation (17.20), when expressed in terms of the coordinates $\lambda(s, t)$ of $\Lambda(s, t)$ relative to the basis of Hermitian Pauli matrices, takes on the following form:

$$\frac{\partial\lambda}{\partial t}(s, t) = \lambda(s, t) \times \frac{\partial^2\lambda}{\partial s^2}(s, t), \qquad (17.25)$$

while (17.19) becomes

$$\frac{\partial \lambda}{\partial t}(s,t) = \frac{\partial^2 \lambda}{\partial s^2}(s,t) \times \lambda(s,t)$$

when Λ is expressed in the coordinates relative to the skew Hermitian basis A_1, A_2, A_3.

The reason is simple: if $X = x_1 B_1 + x_2 B_2 + x_3 B_3$ and $Y = y_1 B_1 + y_2 B_2 + y_3 B_3$ then

$$i[X,Y] = z_1 B_1 + z_2 B_2 + z_3 B_3 \text{ with } z = y \times x.$$

But if $X = x_1 A_1 + x_2 A + _2 + x_3 A_3$ and $Y = y_1 A_1 + y_2 A_2 + y_3 A_3$ then

$$[X,Y] = z_1 A_1 + z_2 A_2 + z_3 A_3 \text{ with } z = x \times y.$$

Both of these assertions can be verified through the relations in Table 17.1.

Equation (17.22) is referred to as the continuous isotropic Heisenberg ferromagnetic model in [Fa, Part II, Ch.1]. This equation is related to the filament equation [AKh]

$$\frac{\partial \gamma}{\partial t}(s,t) = \kappa(s,t) B(s,t). \tag{17.26}$$

For, when the solution curves of the filament equation are restricted to curves parametrized by arc-length, i.e., to curves $\gamma(s,t)$ such that $\left\| \frac{\partial \gamma}{\partial s}(s,t) \right\| = 1$, then

$$T(t,s) = \frac{\partial \gamma}{\partial s}(t,s), \text{ and } \frac{\partial T}{\partial s}(s,t) = \kappa(s,t) N(s,t) = \frac{\partial^2 \gamma}{\partial s^2}(s,t). \tag{17.27}$$

It then follows that in the space of arc-length parametrized curves the solutions of the filament equation coincide with the solutions of

$$\frac{\partial \gamma}{\partial t} = \frac{\partial \gamma}{\partial s} \times \frac{\partial^2 \gamma}{\partial s^2}, \tag{17.28}$$

because $B(s,t) = T(s,t) \times N(s,t)$. For each solution curve $\gamma(s,t)$ of (17.28) the tangent vector $T(s,t)$ satisfies

$$\frac{\partial T}{\partial t} = T \times \frac{\partial^2 T}{\partial s^2}$$

as can be easily verified by differentiating with respect to s. But then $T(s,t)$ may be interpreted as the coordinate vector of $\Lambda(s,t)$ relative to an orthonormal basis in either $\mathfrak{h}$ or $\mathfrak{p}$, which brings us back to the Heisenberg's magnetic equation.

To link the HME (17.1) to the non-linear Schroedinger's equation, we will now turn our attention to the matrices $R(s,t)$ in SU_2 defined by the relations

$\Lambda(s,t) = R(s,t)B_1R^*(s,t)$ in the hyperbolic, and $\Lambda(s,t) = R(s,t)A_1R^*(s,t)$ in the spherical case. The field of curves $R(s,t)$ then evolve according to

$$\frac{\partial R}{\partial s}(s,t) = R(s,t)U(s,t), \text{ and } \frac{\partial R}{\partial t} = R(s,t)V(s,t)$$

for some matrices $U(s,t)$ and $V(s,t)$ in $\mathfrak{h}$. Since the matrices $U(s,t)$ and $V(s,t)$ are generators of the same field of curves, they must satisfy the zero-curvature equation

$$\frac{\partial U}{\partial t}(s,t) - \frac{\partial V}{\partial s}(s,t) + [U(s,t), V(s,t)] = 0, \qquad (17.29)$$

as stated by Lemma 17.9 following Proposition 17.8. Moreover, $V(0,t) = 0$ for all t, because the horizontal-Darboux curves are anchored at $s = 0$.

Proposition 17.16 *Let* $R^{-1}(s,t)\frac{\partial R}{\partial s}(s,t) = U(s,t) = \sum u_j A_j$ *be such that either* $\Lambda(s,t)) = R^*(s,t)B_1R(s,t)$ *or* $\Lambda(s,t) = R^*(s,t)A_1R(s,t)$ *is a solution of the approprite HME. If* $u(s,t) = u_2(s,t) + iu_3(s,t)$, *then,*

$$\psi(s,t) = u(s,t)\exp\left(i\int_0^s u_1(x,t)\,dx\right)$$

is a solution of the non-linear Schroedinger's equation

$$\frac{\partial}{\partial t}\psi(s,t) = i\frac{\partial^2\psi}{\partial s^2}(s,t) + i\left(\frac{1}{2}|\psi(s,t)|^2 + c\right)\psi(s,t) \text{ with } c(t) = -\frac{1}{2}|u(0,t)|^2.$$
$$(17.30)$$

In the proof below we will make use of the following formulas:

$$[A, [A, B]] = \langle A, B\rangle A - \langle A, A\rangle B, A \in \mathfrak{p}, B \in \mathfrak{p},$$
$$[[A, B], B] = \langle B, B\rangle A - \langle A, B\rangle B, A \in \mathfrak{h}, B \in \mathfrak{p}. \qquad (17.31)$$

They can be easily verified by direct calculation.

Proof The proof of the theorem will be done for the hyperbolic case first. Then we will point to the modifications required for the proof in the spherical case.

Since $\Lambda(s,t) = R(s,t)B_1R^*(s,t)$, then

$$\frac{\partial\Lambda}{\partial t} = \frac{\partial}{\partial t}(R(s,t)B_1R^*(s,t)) = R[B_1, V]R^*, \quad \frac{\partial\Lambda}{\partial s} = R[B_1, U]R^*$$

and

$$\frac{\partial^2\Lambda}{\partial s^2} = R\left([[B_1, U], U] + \left[B_1, \frac{\partial U}{\partial s}\right]\right)R^*.$$

The fact that $\Lambda(s,t)$ evolves according to Heisenberg's magnetic equation implies that

$$[B_1, V] = i\left([[[B_1, U], U], B_1] + \left[\left[B_1, \frac{\partial U}{\partial s}\right], B_1\right]\right) \qquad (17.32)$$

Relations (17.31) imply that

$$[[B_1, U], U] = \langle U, B_1\rangle U - \langle U, U\rangle B_1 = -(u_2^2 + u_3^2)B_1 + u_1 u_2 B_2 + B_3 u_1 u_3,$$

hence

$$[[[B_1, U], U], B_1] = u_1 u_3 A_2 - u_1 u_2 A_3.$$

Similarly,

$$\left[B_1, \frac{\partial U}{\partial s}\right] = \frac{\partial u_3}{\partial s}B_2 - \frac{\partial u_2}{\partial s}B_3, \text{ and } \left[\left[B_1, \frac{\partial U}{\partial s}\right], B_1\right] = -\frac{\partial u_3}{\partial s}A_3 - \frac{\partial u_2}{\partial s}A_2.$$

Therefore equation (17.32) reduces to

$$[B_1, V] = i\left(u_1(u_3 A_2 - u_2 A_3) - \frac{\partial u_3}{\partial s}A_3 - \frac{\partial u_2}{\partial s}A_2\right)$$
$$= -u_1(u_3 B_2 - u_2 B_3) + \frac{\partial u_3}{\partial s}B_3 + \frac{\partial u_2}{\partial s}B_2.$$

If $V = v_1 A_1 + v_2 A_2 + v_3 A_3$, then $[B_1, V] = v_3 B_2 - v_2 B_3$, which, when combined with the above, yields $v_2 = -u_1 u_2 - \frac{\partial u_3}{\partial s}$, and $v_3 = -u_1 u_3 + \frac{\partial u_2}{\partial s}$. These relations can be written more succinctly as

$$v(s,t) = -u_1(s,t)u(s,t) + i\frac{\partial u}{\partial s}(s,t),$$

where $u = u_2 + iu_3$ and $v = v_2 + iv_3$.

The zero curvature equation implies that

$$\frac{\partial u_1}{\partial t} = \frac{\partial v_1}{\partial s} + \frac{1}{2}\frac{\partial}{\partial s}\left(u_2^2 + u_3^2\right),$$
$$\frac{\partial u}{\partial t} = i\frac{\partial^2 u}{\partial s^2} - 2u_1\frac{\partial u}{\partial s} - \frac{\partial u_1}{\partial s}u - i(v_1 + u_1^2)u. \qquad (17.33)$$

The first equation in (17.33) implies that

$$\frac{\partial}{\partial t}\int_0^s u_1(x,t)dx = v_1(s,t) + \frac{1}{2}(u_2^2(s,t) + u_3^2(s,t)) + c(t),$$

where $c(t) = -v_1(0,t) - \frac{1}{2}(u_2^2(0,t) + u_3^2(0,t)) = -\frac{1}{2}(u_2^2(0,t) + u_3^2(0,t))$, since $V(0,t) = 0$. Substituting $v_1(s,t) = \frac{\partial}{\partial t}\int_0^s u_1(x,t)dx - \frac{1}{2}|u(s,t)|^2 - c$ into the second equation in (17.33) leads to

$$\frac{\partial u}{\partial t} + iu\frac{\partial}{\partial t}\int^s u_1(t,x)\,dx = i\frac{\partial^2 u}{\partial s^2} - 2u_1\frac{\partial u}{\partial s} - u\frac{\partial u_1}{\partial s} - i\left(-\frac{1}{2}|u|^2 - c + u_1^2\right)u.$$

$$(17.34)$$

After multiplying by $\exp\left(i\int_0^s u_1(x,t)\,dx\right)$, equation (17.34) can be expressed as

$$\frac{\partial}{\partial t}\psi(s,t) = \left(i\frac{\partial^2 u}{\partial s^2} - 2u_1\frac{\partial u}{\partial s} - u\frac{\partial u_1}{\partial s} - i\left(u_1^2 - \frac{1}{2}|u|^2 - c\right)u\right)$$

$$\times \exp\left(i\int_0^s u_1(x,t)\,dx\right),$$

$$(17.35)$$

where $\psi(s,t) = u(s,t)\exp\left(i\int_0^s u_1(x,t)\,dx\right)$.

It follows that $\frac{\partial\psi}{\partial s} = \left(\frac{\partial u}{\partial s} + iuu_1\right)\exp\left(i\int_0^s u_1(x,t\,dx)\right)$, and therefore,

$$\frac{\partial^2\psi}{\partial s^2} = \left(\frac{\partial^2 u}{\partial s^2} + 2iu_1\frac{\partial u}{\partial s} + iu\frac{\partial u_1}{\partial s} - u_1^2 u\right)\exp\left(i\int_0^s u_1(x,t)\,dx\right),$$

or

$$i\frac{\partial^2\psi}{\partial s^2} = \left(i\frac{\partial^2 u}{\partial s^2} - 2u_1\frac{\partial u}{\partial s} - u\frac{\partial u_1}{\partial s} - iu_1^2 u\right)\exp\left(i\int_0^s u_1(x,t\,dx)\right), \quad (17.36)$$

Combined with (17.35), equation (17.36) yields

$$\frac{\partial}{\partial t}\psi(t,s) = i\frac{\partial^2\psi}{\partial s^2} + i\left(\frac{1}{2}|\psi|^2 + c(t)\right)\psi.$$

In the spherical case equation (17.32) is replaced by

$$[A_1, V] = \left([[[A_1, U], U], A_1] + \left[\left[A_1, \frac{\partial U}{\partial s}\right], A_1\right]\right).$$

The substitution $A_1 = iB_1$ brings us back to equation (17.32). Therefore the calculations that led to the non-linear Schroedinger equation in the hyperbolic case are equally valid in the spherical case, with the same end result. □

The steps taken in the passage from the Heisenberg equation to the Schroedinger equation are reversible. Any solution $\psi(s,t)$ of (17.2) generates matrices

$$U = \frac{1}{2}\begin{pmatrix} 0 & \psi \\ -\bar{\psi} & 0 \end{pmatrix} \quad \text{and} \quad V = \frac{1}{2}\begin{pmatrix} -i(\frac{1}{2}|\psi|^2 + c(t)) & i\frac{\partial\psi}{\partial s} \\ i\frac{\partial\bar{\psi}}{\partial s} & i(\frac{1}{2}|\psi|^2 + c(t)) \end{pmatrix}$$

that satisfy the zero-curvature equation. Therefore, there exist unique curves $R(s,t)$ in SU_2 with boundary conditions $R(0,t)=I$ that evolve according to the differential equations:

$$\frac{\partial R}{\partial s}(s,t) = R(s,t)U(s,t), \quad \frac{\partial R}{\partial t}(s,t) = R(s,t)V(s,t).$$

Such curves define $\Lambda(s,t)$ through the familiar formulas $\Lambda(s,t)=R(s,t)B_1R^*$ (s,t) or $\Lambda(s,t)=R(s,t)A_1R^*(s,t)$ depending on the case. In the first case,

$$i\left[\Lambda, \frac{\partial^2 \Lambda}{\partial s^2}\right] = R\left(\left[B_1, [[B_1, U], U] + \left[B_1, \frac{\partial U}{\partial s}\right]\right]\right)R^* =$$

$$R\left(\left[B_1, \left[B_1, \frac{\partial U}{\partial s}\right]\right]\right)R^* = \frac{1}{2}R\left(\begin{pmatrix} 0 & iu_s \\ i\bar{u}_s & 0 \end{pmatrix}\right)R^*.$$

Therefore $i\left[\Lambda, \frac{\partial^2 \Lambda}{\partial s^2}\right]$ is equal to $\frac{\partial \Lambda}{\partial t} = R[B_1, V]R^* = \frac{1}{2}\begin{pmatrix} 0 & v \\ -\bar{v} & 0 \end{pmatrix}$ because $v = iu_s$. The spherical case follows along similar lines.

Remark 17.17 The above proposition reveals that $SO_2 = \left\{ \begin{pmatrix} z & 0 \\ 0 & \bar{z} \end{pmatrix}, \right.$ $\left. |z| = 1 \right\}$ is a symmetry group for the non-linear Schroedinger equation, since $\psi(s,t) = u(s,t)\exp\left(i\int_0^s u_1(x,t)\,dx\right)$ remains a solution for any u_1, independently of the symmetric space.

   This observation removes some mystery behind an ingenious observation by H. Hasimoto [H1] that $\psi = \kappa \exp\left(i\int \tau\,dx\right)$ of a curve $\gamma(s,t)$ that satisfies the filament equation is a solution of the non-linear Schroedinger equation, for when $R(s,t)$ is a Serret–Frenet frame then $u_1 = \tau, u_2 = 0, u_3 = \kappa$ in the Euclidean and the hyperbolic case, while $u_1 = \tau + \frac{1}{2}, u_2 = 0, u_3 = \kappa$ in the spherical case.

   Hasimoto's function $\kappa(s,t)\exp i\int_0^s \tau(x,t)\,dx)$ coincides with $u(s,t)$ $\exp\left(i\int_0^s u_1(x,t)\,dx\right)$ in the hyperbolic and the Euclidean case, but not in the spherical case. Of course, the most natural frame is the reduced frame $u_1 = 0$ which bypasses these inessential connections with the torsion.

   The correspondence between the HME (17.1) and NLS (17.2) described by the above proposition strongly suggests that (17.1) is integrable, in the sense that there exist infinitely many functions in involution with each other that are constant along the solutions of (17.1), for it is well known that (17.2) is integrable (see, for instance [Fa; Mg; LP]). Its conserved quantities can be obtained either by an iterative procedure based on isospectral methods due to

Shabat and Zakharov [ShZ], or alternatively, through a procedure known as the Magri scheme.

The Magri scheme is applicable in situations in which there are two compatible Poisson structures $\{\,,\,\}_1$ and $\{\,,\,\}_2$ relative to which a given vector field X is Hamiltonian. This means that there exist two functions f and g such that $X = \vec{f}_1 = \vec{g}_2$, where $\vec{f}_1$ is the Hamiltonian vector field of f relative to the first Poisson structure and $\vec{g}_2$ is the Hamiltonian vector field of g relative to the second Poisson structure. Then $\{f, g\}_1$ and $\{f, g\}_2$ are the constants of motion for X. Then the Hamiltonian vector field induced by these two function is also bi-Hamiltonian and generates two new integrals of motion. The iteration of this process produces a sequence of conserved quantities in involution with each other for the flow of X.

It is entirely possible that Magri's scheme could be used to prove the integrability of (17.1) by introducing another symplectic form on the space of anchored curves given by

$$\Omega_\Lambda(V_1, V_2) = \int_0^L \langle \Lambda, [V_1, V_2] \rangle \, ds.$$

Such a form is mentioned elsewhere in the literature (see, for instance [AKh; Br; LP]) but never used for this specific purpose. We will not undertake this study here. Instead, we will just mention some functions that might be in this hierarchy of conserved quantities.

Proposition 17.18 *The Hamiltonian flow of* $f = -i \int_0^L \left\langle \left[\Lambda, \frac{d\Lambda}{ds} \right], \frac{d^2\Lambda}{ds^2} \right\rangle ds$ *is given by*

$$\frac{\partial \Lambda}{\partial t} = 2 \left(\frac{\partial^3 \Lambda}{\partial s^3} - \left\langle \frac{\partial^3 \Lambda}{\partial s^3}, \Lambda \right\rangle \Lambda \right) - 3 \left\langle \Lambda, \frac{\partial^2 \Lambda}{\partial s^2} \right\rangle \frac{\partial \Lambda}{\partial s}. \tag{17.37}$$

Function f *Poisson commutes with* $f_0 = \frac{1}{2} \int_0^L \kappa^2(s) \, ds$.

Proof Let $V(s)$ be an arbitrary tangent vector at a frame-periodic horizontal-Darboux curve $g(s)$. Then the directional derivative of f in direction V is given by the following expression:

$$df(V) = -i \frac{\partial}{\partial t} \int_0^L \left\langle \left[Z(s, t), \frac{\partial Z}{\partial s}(s, t) \right], \frac{\partial^2 Z}{\partial s^2}(s, t) \right\rangle ds|_{t=0},$$

where $Z(s, t)$ denotes a field of Hermitian matrices such that

$$Z(s, 0) = \Lambda(s) \text{ and } \frac{\partial Z}{\partial t}(s, 0) = \frac{dV}{ds}(s).$$

It follows that

$$
\begin{aligned}
df(V) &= -i \int_0^L \langle \dot{V}, [\Lambda, \dot{\Lambda}] \rangle + \langle \ddot{\Lambda}, [\dot{V}, \dot{\Lambda}] \rangle + \langle \ddot{\Lambda}, [\Lambda, \ddot{V}] \rangle \, ds \\
&= -i \int_0^L 2\langle [\ddot{\Lambda}, \Lambda], \dot{V} \rangle - \langle [\ddot{\Lambda}, \dot{\Lambda}], V \rangle \, ds \\
&= i \int_0^L \left\langle 2\left(\frac{d}{ds}([\ddot{\Lambda}, \Lambda]) + [\ddot{\Lambda}, \dot{\Lambda}] \right), V \right\rangle \, ds \\
&= i \int_0^L \langle 2[\dddot{\Lambda}, \Lambda]) + 3[\ddot{\Lambda}, \dot{\Lambda}], V \rangle \, ds,
\end{aligned}
$$

where the dots indicate derivatives with respect to s. In the preceding calculations periodicity of Λ is implicitly assumed to eliminate the boundary terms in the integration by parts.

Let now $V_1(s)$ denote the Hermitian matrix such that $df(V) = \omega_\Lambda(V_1, V)$ for all tangent vectors V.

Then,

$$
i \int_0^L \langle 2[\dddot{\Lambda}, \Lambda] + 3[\ddot{\Lambda}, \dot{\Lambda}], V \rangle \, ds = \frac{1}{i} \int_0^L \langle [\Lambda, \dot{V_1}], V \rangle \, ds,
$$

which implies

$$
\int_0^L \langle [\Lambda, \dot{V_1}] + 2[\dddot{\Lambda}, \Lambda] + 3[\ddot{\Lambda}, \dot{\Lambda}], V \rangle \, ds = 0.
$$

When $\dot{V} = [\Lambda, C]$ the above becomes

$$
\int_0^L \langle [[\Lambda, \dot{V_1}], \Lambda] + 2[[\dddot{\Lambda}, \Lambda], \Lambda] + 3[[\ddot{\Lambda}, \dot{\Lambda}], \Lambda], C(s) \rangle \, ds = 0.
$$

Since $C(s)$ is an arbitrary curve with $C(0) = 0$ the preceeding integral equality reduces to

$$
[[\Lambda, \dot{V_1}], \Lambda] + 2[[\dddot{\Lambda}, \Lambda], \Lambda] + 3[[\ddot{\Lambda}, \dot{\Lambda}], \Lambda] = 0.
$$

The Lie bracket relations (17.31) imply that

$$
\dot{V_1} + 2(\langle \dddot{\Lambda}, \Lambda \rangle \Lambda - \dddot{\Lambda}) + 3\langle \ddot{\Lambda}, \Lambda \rangle \dot{\Lambda} = 0.
$$

Now it follows by the arguments used earlier in the paper that the Hamiltonian flow X_f satisfies

$$
\frac{\partial \Lambda}{\partial t} = 2(\dddot{\Lambda} - \langle \dddot{\Lambda}, \Lambda \rangle \Lambda) - 3\langle \Lambda, \ddot{\Lambda} \rangle \dot{\Lambda}.
$$

To prove the second part, we need to show that the Poisson bracket of $\{f_0, f_1\}$, given by the formula $\omega_\Lambda(V_0(\Lambda), V_1(\Lambda)) = \frac{1}{i}\int_0^L \langle \Lambda(s), [\dot{V}_0(s), \dot{V}_1(s)] \rangle \, ds$ with $\dot{V}_0(\Lambda) = i[\ddot{\Lambda}, \Lambda]$ and $\dot{V}_1 = -(2(\langle \ddot{\Lambda}, \Lambda \rangle \Lambda - \ddot{\Lambda}) + 3\langle \ddot{\Lambda}, \Lambda \rangle \dot{\Lambda})$, is equal to 0.

An easy calculation shows that

$$[\dot{V}_0, \dot{V}_1] = i(2(\langle \dddot{\Lambda}, \ddot{\Lambda} \rangle) - \langle \Lambda, \ddot{\Lambda} \rangle \langle \dddot{\Lambda}, \Lambda \rangle) - 3\langle \ddot{\Lambda}, \Lambda \rangle \langle \ddot{\Lambda}, \dot{\Lambda} \rangle \Lambda.$$

Hence,

$$\{f_0, f\} = \int_0^L (2\langle \dddot{\Lambda}, \ddot{\Lambda} \rangle - 2\langle \Lambda, \dddot{\Lambda} \rangle \langle \Lambda, \ddot{\Lambda} \rangle - 3\langle \Lambda, \ddot{\Lambda} \rangle \langle \dot{\Lambda}, \ddot{\Lambda} \rangle) \, ds.$$

The integral of the first term is zero because $2\langle \dddot{\Lambda}, \ddot{\Lambda} \rangle = \frac{d}{ds}\langle \ddot{\Lambda}, \ddot{\Lambda} \rangle$.

Since

$$2\langle \Lambda, \ddot{\Lambda} \rangle \langle \Lambda, \dddot{\Lambda} \rangle = \frac{d}{ds}\langle \Lambda, \ddot{\Lambda} \rangle^2 - 2\langle \Lambda, \ddot{\Lambda} \rangle \langle \dot{\Lambda}, \ddot{\Lambda} \rangle,$$

the remaining integrand reduces to one term $-\langle \Lambda, \ddot{\Lambda} \rangle \langle \dot{\Lambda}, \ddot{\Lambda} \rangle$. But then $\frac{1}{4}\frac{d}{ds}\langle \dot{\Lambda}, \dot{\Lambda} \rangle^2 = \langle \Lambda, \ddot{\Lambda} \rangle \langle \dot{\Lambda}, \ddot{\Lambda} \rangle$ because $\langle \dot{\Lambda}, \dot{\Lambda} \rangle = -\langle \Lambda, \ddot{\Lambda} \rangle$. □

The above functional can be restated in terms of the curvature and the torsion of the base curve as $\int_0^L \kappa^2(s)\tau(s) \, ds$ for the following reasons.

If $T(s) = \Lambda(s)$, then $g(s)T(s)$ projects onto the tangent vector of a base curve $\gamma \in \mathbb{H}^3$. Then $N(s)$ and $B(s)$, the matrices that correspond to the normal and the binormal vectors, are given by

$$N = \frac{1}{\kappa}\frac{d\Lambda}{ds} \quad \text{and} \quad B(s) = \frac{1}{i}[T(s), N(s)] = -\frac{i}{\kappa}\left[\Lambda, \frac{d\Lambda}{ds}\right].$$

According to the Serret–Frenet equations $\frac{dN}{ds} = -k\Lambda + \tau B$. Therefore,

$$\tau = \left\langle \frac{dN}{ds}, B \right\rangle = -i\left\langle -\frac{1}{\kappa^2}\frac{d\kappa}{ds}\frac{d\Lambda}{ds} + \frac{1}{\kappa}\frac{d^2\Lambda}{ds^2}, \frac{1}{\kappa}\left[\Lambda, \frac{d\Lambda}{ds}\right]\right\rangle$$

$$= -i\frac{1}{\kappa^2}\left\langle \left[\Lambda, \frac{d\Lambda}{ds}\right], \frac{d^2\Lambda}{ds^2}\right\rangle,$$

and hence, $\kappa^2 \tau = -i\left\langle \left[\Lambda, \frac{d\Lambda}{ds}\right], \frac{d^2\Lambda}{ds^2}\right\rangle$.

Proposition 17.19 *Suppose that $\Lambda(s, t) = R(s, t)B_1R^*(s, t)$ evolves according to the equation (17.37), where $R(s, t)$ is the solution of*

$$\frac{\partial R}{\partial s}(s, t) = R(s, t)\begin{pmatrix} 0 & u(s, t) \\ -\bar{u}(s, t) & 0 \end{pmatrix}, \quad R(0, t) = I.$$

Then, $u(s, t)$ is a solution of

$$\frac{\partial u}{\partial t} - 3|u|^2\frac{\partial u}{\partial s} - 2\frac{\partial^3 u}{\partial s^3} = 0. \tag{17.38}$$

This proposition is proved by a calculation similar to the one used in the proof of Proposition 17.16, the details of which will be omitted. Equation (17.38) is known as the modified Korteweg-de Vries equation [Mg].

Functions $f_0 = \int_0^L k^2 ds$ and $f_1 = \int_0^L k^2 \tau ds$ also appear in the paper of C. Shabat and V. Zacharov [ShZ], but in a completely different context. The first two integrals of motion in the paper of Shabat and Zacharov are up to constant factors given by the following integrals:

$$C_1 = \int_{-\infty}^{\infty} |u(s,t)|^2 \, ds, \; C_2 = \int_{-\infty}^{\infty} (u(s,t))\dot{\bar{u}}(s,t)) - \bar{u}(s,t)\dot{u}(s,t)) \, ds,$$

where C_1 interpreted as the number of particles in the wave packet and C_2 is interpreted as their momentum. To see that C_1 and C_2 are in exact correspondence with functions f_0 and f_1 assume that the Darboux curves are expressed by the reduced frames $R(s)$, i.e., as the solutions of

$$\frac{dR}{ds}(s) = R(s)U(s) \text{ with } U(s) = u_2(s)A_2 + u_3(s)A_3.$$

Then,

$$f_0 = \frac{1}{2} \int_0^L ||\dot{\Lambda}(s)||^2 \, ds = \frac{1}{2} \int_0^L ||[B_1, U(s)]||^2 \, ds = \frac{1}{2} \int_0^L |u(s)|^2 \, ds.$$

Hence, C_1 corresponds to $\int_0^L k^2(s) \, ds$. Furthermore,

$$\int_0^L k^2 \tau \, ds = \frac{1}{2i} \int_0^L (u(s)\dot{\bar{u}}(s) - \bar{u}(s)\dot{u}(s)) \, ds,$$

because

$$i\langle \Lambda, [\dot{\Lambda}, \ddot{\Lambda}] \rangle = \langle [[B_1, U], [B_1, \dot{U}]], B_1 \rangle = \text{Im}(\bar{u}\dot{u}),$$

where $\text{Im}(z)$ denotes the imaginary part of a complex number z. Therefore f_1 corresponds to C_2.

In the language of mathematical physics, vector $\int_0^L \Lambda(s) \, ds$ is called the total spin [Fa]. Here it appears as the moment map associated with the adjoint action of SU_2. It is a conserved quantity since the Hamiltonian is invariant under this action. This fact can be verified directly:

$$\frac{\partial}{\partial t} \int_0^L \Lambda(t,s) \, ds = \int_0^L \frac{\partial \Lambda}{\partial t}(t,s) \, ds = i \int_0^L \left[\frac{\partial^2 \Lambda}{\partial s^2}, \Lambda \right] ds$$

$$= i \int_0^L \frac{\partial}{\partial s}[\Lambda, \dot{\Lambda}] \, ds = 0.$$

The third integral of motion C_3 in [ShZ], called the energy, is given by

$$C_3 = \int_{-\infty}^{\infty} \left(\left| \frac{\partial u}{\partial s}(s,t) \right|^2 - \frac{1}{4} |u(s,t)|^4 \right) ds.$$

It corresponds to the function

$$f_2 = \int_0^L \left(||\ddot{\Lambda}(s)||^2 - \frac{5}{4} ||\dot{\Lambda}(s)||^4 \right) ds$$

$$= \int_0^L \left(\frac{\partial \kappa}{\partial s}(s)^2 + \kappa^2(s)\tau^2(s) - \frac{1}{4}\kappa^4(s) \right) ds.$$

Functions C_0, f_1, f_2 are in involution relative to the Poisson bracket induced by the symplectic form. There seems to be a hierarchy of functions that contains f_0, f_1, f_2 such that any two functions in the hierarchy Poisson commute. For instance, it can be also shown that $f_3 = \int_0^L \tau(s)\, ds$ is in this hierarchy and that its Hamiltonian vector field generates the curve shortening equation [Ep]

$$\frac{\partial \Lambda}{\partial t}(s,t) = \frac{\partial \Lambda}{\partial s}(s,t) = \kappa(s,t)N(s,t).$$

These findings are in exact correspondence with the ones discovered by J. Langer and R. Perline over the rapidly decreasing functions in $\mathbb{R}^3$ [LP].

17.4 Affine Hamiltonians and solitons

For mechanical systems the Hamiltonian function represents the total energy of the system and its critical points correspond to the equilibrium configurations. In an infinite-dimensional setting the behavior of a Hamiltonian system at a critical point of a Hamiltonian does not lend itself to such simple characterizations.

We will now show that the critical points of the Hamiltonian $f = \frac{1}{2}\int_0^L k^2\, ds$ over the horizontal Darboux curves are naturally identified with the affine Hamiltonian

$$H = \frac{1}{2}(H_2^2 + H_3^2) + h_1 \qquad (17.39)$$

generated by

$$\frac{dg}{ds} = g(s)(B_1 + u_2(s)A_2 + u_3(s)A_3) \qquad (17.40)$$

and the cost functional $\frac{1}{2}\int_0^L (u_2^2(s) + u_3^2(s))\, ds$.

To explain, we will need to go back to the parallel frames discussed earlier in the text. For simplicity of exposition we will confine our discussion to the hyperbolic Darboux curves.

Each anchored Darboux curve $g_0(s)$ is a solution of $\frac{dg_0}{ds} = g_0(s)\Lambda(s)$, where $\Lambda(s)$ is a curve of Hermitian matrices such that $||\Lambda(s)|| = 1$ and $\Lambda(0) = B_1$. Every such matrix $\Lambda(s)$ can be represented by a curve $R(s)$ in SU_2 that satisfies $\Lambda(s) = R(s)B_1R^*(s)$ and $R(0) = I$. Then $g(s) = g_0R(s)$ is a solution of

$$\frac{dg}{ds} = g(s)((B_1 + u_1(s)A_1 + u_2A_2 + u_3(s)_3)A_3$$

in $SL_2(\mathbb{C})$, for some real functions $u_1(s), u_2(s), u_3(s)$, that satisfies $g(0) = I$. It projects onto the same base curve $x(s) = g(s)g^*(s) = g_0(s)g_0^*s)$. Each reduced Darboux curve $g_0(s)$ defines a parallel frame

$$v_1(s) = 2g_0(s)B_1g_0^*(s), v_2(s) = g_0(s)B_2g_0^*s, v_3(s) = g_0(s)B_3g_0^*(s)$$

over the base curve $x(s) = g_0(s)g_0^*(s)$, since $v_1(s) = \frac{dx}{ds}$ and

$$\frac{D_x}{ds}(v_2) = g_0(s)[B_2, u_2(s)A_2 + u_3(s)A_3]g_0^*(s) = u_3(s)v_1(s),$$

$$\frac{D_x}{ds}(v_3) = g_0(s)[B_3, u_2(s)A_2 + u_3(s)A_3]g_0^*(s) = -u_2(s)v_1(s).$$

Therefore, the cost functional $\frac{1}{2}\int_0^L (u_2^2(s) + u_3^*(s))\, ds$ associated with (17.40) is equal to $\frac{1}{2}\int_0^L \kappa^2(s)\, ds$ where $\kappa(s)$ denotes the curvature of the base curve $x(s)$.

Therefore the affine Hamiltonian (17.39) is the Hamiltonian associated with the optimal control problem of finding the minimum of $\frac{1}{2}\int_0^L (u_2^2(s) + u_3^2(s))\, ds$ over the trajectories in (34) that conform to the given boundary conditions (the Euler–Griffiths problem in Chapter 13). The variables $h_1, h_2, h_3, H_1, H_2, H_3$ are the coordinates of an element $\ell \in \mathfrak{g}^*$ relative to the dual basis $B_1^*, B_2^*, B_3^*, A_1^*, A_2^*, A_3^*$.

Then,

$$\frac{dg}{ds} = g(s)(B_1 + H_2(s)A_2 + H_3(s)A_3), \frac{dL}{ds} = [dH(s), L(s)] \qquad (17.41)$$

are the equations of the integral curves of H having identified $\mathfrak{g}^*$ with $\mathfrak{g}$ via the trace form. Again, we will take advantage of the Cartan decomposition $\mathfrak{g} = \mathfrak{p} \oplus \mathfrak{h}$ and write $U = \frac{1}{2}\begin{pmatrix} 0 & u(s) \\ -\bar{u}(s) & 0 \end{pmatrix}$, where $u(s) = H_2(s) + iH_3(s)$

so that $dH = B_1 + U$. Then $P = \sum_{i=1}^{3} h_i B_1$ and $Q = \sum_{i=1}^{3} H_i A_i$ are the Hermitian and skew-Hermitian parts of L. It follows that

$$P = \frac{1}{2} \begin{pmatrix} h_1 & iw \\ -i\bar{w} & -h_1 \end{pmatrix}, Q = \frac{1}{2} \begin{pmatrix} -iH_1 & -u \\ \bar{u} & iH_1 \end{pmatrix},$$

where $w = h_2 + ih_3$. Then equations (17.41) take on the familiar form

$$\frac{dQ}{ds} = [U(s), Q(s)] + [B_1, P(s)], \frac{dP}{ds} = [U(s), P(s)] + \epsilon[B_1, Q(s)], \epsilon = -1.$$

$$(17.42)$$

This equation is formally the same as equation (8.7) in Chapter 8 and can be written in expanded form as

$$\frac{dH_1}{ds} = 0, \frac{du}{ds} = iH_1 u(s) - iw(s), \frac{dh_1}{ds}(s) = iRe(w(s)u(s)), \frac{dw}{ds}$$
$$= i(h_1 - \epsilon)u(s). \qquad (17.43)$$

Except for some minor notational details this equation is valid for the spherical and the Euclidean cases as well.

We are now in the position to relate these this material to the solutions of (17.2).

Proposition 17.20 *Let $u(s)$ and H_1 be as in equation (37) and let $H = \frac{1}{2}\|u(s)\|^2 + h_1$. Then $\psi(s,t) = u(s + \xi t)$ is a solution of the non-linear Schroedinger's equation $\frac{\partial \psi}{\partial t} = -i\left(\frac{\partial^2 \psi}{\partial s^2} + \frac{1}{2}|\psi|^2 \psi\right)$ precisely when $\xi = -H_1$ and $H = \epsilon$.*

Proof If $\psi(s,t) = u(s + \xi t)$ then

$$\frac{\partial \psi}{\partial t} = i\xi(H_1\psi - w) \text{ and } \frac{\partial^2 \psi}{\partial s^2} = -H_1^2\psi + H_1 w + (h_1 - \epsilon)\psi.$$

Therefore,

$$0 = -i\frac{\partial \psi}{\partial t} - \left(\frac{\partial^2 \psi}{\partial s^2} + \frac{1}{2}|\psi|^2 \psi\right)$$

$$= \xi(H_1\psi - w) - \left(-H_1^2\psi + H_1 w + (h_1 - \epsilon)\psi + \frac{1}{2}|\psi|^2\right)\psi$$

$$= \xi(H_1\psi - w) - (-H_1^2\psi + H_1 w + (h_1 - \epsilon)\psi + \psi(H - h_1))$$

$$= -(\xi + H_1)w + (\xi H_1 + H_1^2 + \epsilon - H)\psi,$$

whenever $\xi = -H_1$ and $H = \epsilon$. $\qquad \square$

Thus the extremals which reside on energy level $H = \epsilon$ generate soliton solutions of the non-linear Schroedinger's equation traveling with speed equal to the level surface $H_1 = -\xi$. These soliton solutions degenerate to the stationary solution when $H_1 = 0$, that is, when the projections of the extremals on the base curve are elastic. To show that periodic solutions exist on energy level $H = \epsilon$ requires an explicit formula for $u(s)$, which is tantamount to solving the extremal equations explicitly.

The integration procedure is essentially the same as in the Kirchhoff–Lagrange equation. To begin note that $(H_2 h_3 - H_3 h_2)^2 + (H_2 h_2 + H_3 h_3)^2 = (H_2^2 + H_3^2)(h_2^2 + h_3^2)$. Then,

$$
\begin{aligned}
\left(\frac{d}{ds}h_1\right)^2 &= (H_2 h_3 - H_3 h_2)^2 = (H_2^2 + H_3^2)(h_2^2 + h_3^2) - (H_2 h_2 + H_3 h_3)^2 \\
&= (H_2^2 + H_3^2)(I_1 - \epsilon(H_1^2 + H_2^2 + H_3^2) - h_1^2) - (I_2 - h_1 H_1)^2 \\
&= 2(H - h_1)(I_1 - \epsilon H_1^2 - 2\epsilon(H - h_1) - h_1^2) - (I_2 - h_1 H_1)^2 \\
&= 2h_1^3 + c_1 h_1^2 + c_2 h_1 + c_3,
\end{aligned}
$$

where $I_1 = h_1^2 + h_2^2 + h_3^2 + \epsilon(H_1^2 + H_2^2 + H_3^2)$ and $I_2 = h_1 H_1 + h_2 H_2 + h_3 H_3$ denote the universal integrals on $\mathfrak{g}_\epsilon$, and c_1, c_2, c_3 are the constants given by the following expressions:

$$
c_1 = -\left(H_1^2 - 2H - 4\epsilon\right), c_2 = \left(2 I_2 H_1 - 2\epsilon H_1^2 + 4\epsilon H - 2 I_1\right),
$$

$$
c_3 = 2H\left(I_1 - \epsilon H_1^2 - 2\epsilon H\right) - I_2^2.
$$

It follows that $h_1(s)$ is solvable in terms of elliptic functions, and since $k^2 = H_2^2 + H_3^2 = 2(H - h_1)$, the same can be said for the curvature of the projected elastic curve. The remaining variables $u = H_2 + i H_3$ and $w = h_2 + i h_3$ can be integrated in terms of two angles θ and ϕ on the sphere

$$
(h_1 - \epsilon)^2 + |w|^2 = J^2, \text{ where } J^2 = I_1 - \epsilon H_1^2 - 2\epsilon H + \epsilon^2. \tag{17.44}
$$

This new constant J is obtained as follows:

$$
\begin{aligned}
I_1 &= h_1^2 + |w|^2 + \epsilon\left(H_1^2 + |u|^2\right) = h_1^2 + |w|^2 + \epsilon\left(H_1^2 + 2(H - h_1)\right) \\
&= (h_1 - \epsilon)^2 + |w|^2 + \epsilon H_1^2 + 2\epsilon H - \epsilon.
\end{aligned}
$$

Then angles θ and ϕ correspond to the spherical coordinates on the above sphere. They are defined by

$$
(h_1(s) - \epsilon) = J \cos\theta(s) \text{ and } w(s) = J \sin\theta(s) e^{i\phi(s)}. \tag{17.45}
$$

It follows that

$$\frac{dh_1}{ds} = -J \sin\theta \frac{d\theta}{ds}, \text{ and } \frac{dw}{ds} = w\left(\frac{\cos\theta}{\sin\theta}\frac{d\theta}{ds} + i\frac{d\phi}{ds}\right).$$

Furthermore,

$$\frac{u}{w} = \frac{u\bar{w}}{|w|^2} = \frac{H_2 h_2 + H_3 h_3 + i(H_3 h_2 - H_2 h_3)}{J^2 - (h_1 - \epsilon)^2}$$

$$= \frac{I_2 - h_1 H_1 + i\frac{dh_1}{ds}}{J^2 \sin^2\theta}$$

$$= \frac{I_2 - h_1 H_1}{J^2 \sin^2\theta} - \frac{i}{J\sin\theta}\frac{d\theta}{ds}.$$

This formula shows that

$$u(s) = w(s)\left(\frac{I_2 - h_1 H_1}{J^2 \sin^2\theta} - \frac{i}{J\sin\theta}\frac{d\theta}{ds}\right) \tag{17.46}$$

Equations (17.43) combined with (17.45) yield

$$w\left(\frac{\cos\theta}{\sin\theta}\frac{d\theta}{ds} + i\frac{d\phi}{ds}\right) = \frac{dw}{ds} = i(h_1 - \epsilon)u$$

$$= \left(iJ\cos\theta\frac{(I_2 - h_1 H_1)}{J^2 \sin^2\theta} + \frac{\cos\theta}{\sin\theta}\frac{d\theta}{ds}\right)w.$$

Hence,

$$\frac{d\phi}{ds} = J\cos\theta\frac{(I_2 - h_1 H_1)}{J^2 \sin^2\theta} = \frac{J\cos\theta(I_2 - \epsilon H_1 - H_1 J\cos\theta)}{J^2 \sin^2\theta}. \tag{17.47}$$

Now, the equation

$$\left(\frac{dh_1}{ds}\right)^2 = (H_2 h_3 - H_3 h_2)^2$$

$$= (H_2^2 + H_3^2)(h_2^2 + h_3^2) - (H_2 h_2 + H_3 h_3)^2$$

$$= 2(H - h_1)|w|^2 - (I_2 - H_1 h_1)^2$$

is the same as

$$\left(\frac{d\theta}{ds}\right)^2 = 2\,(H - \epsilon - J\cos\theta) - \frac{(I_2 - H_1(\epsilon + J\cos\theta))^2}{J^2 \sin^2\theta}, \tag{17.48}$$

after the substitutions $h_1 - \epsilon - J\cos\theta$ and $|w|^2 = J^2 \sin^2\theta$. The solutions of (17.48) parametize the extremal curves: for then ϕ is given by equation (17.47) and u and w by equations (17.45) and (17.46).

We now return to the question of periodicity. Evidently, both u and w are periodic whenever $\phi(0) = \phi(L)$ and $\theta(0) = \theta(L)$. Soliton solutions propagate with speed $-\xi = H_1$ on energy level $H = \epsilon$. On this energy level, $\phi(0) = \phi(L)$ and $\theta(0) = \theta(L)$, if and only if

$$\int_0^L \frac{J \cos\theta (I_2 + H_1 - H_1 J \cos\theta)}{J^2 \sin^2\theta} \, ds = 0,$$

where θ denotes a closed solution of the equation

$$\left(\frac{d\theta}{ds}\right)^2 = -2J\cos\theta - \frac{(I_2 - H_1(\epsilon + J\cos\theta))^2}{\sin^2\theta}.$$

The paper of T. Ivey and D. Singer demonstrates that there are infinitely many closed solutions generated by the constants I_1, I_2, H_1 [IvS].

Each such solution generates a solution of the non-linear Schroedinger equation that travels with speed equal to the corresponding value of the integral $-I_1$.

Concluding remarks

Well into the writing of this book I became increasingly aware that some of the topics that I initially planned to include in the text had to be left out. To begin with, I realized that integrable systems, even though linked through common symmetries, somehow stood apart from each other, and each required more space and attention than I initially had imagined. Secondly, as the number of pages grew, I was constantly reminded of that old saying that a big book is a big nuisance. So, some inital expectations had to be curtailed.

Initially, I wanted to include a chapter on optimal control of finite dimensional quantum systems. This class of systems fits perfectly our theoretic framework since it is described by left-invariant affine control systems of $G = SU_n$ subordinate to the natural Cartan decomposition on the Lie algebra su_n [BGN]. In that context, it would have been natural to extend the isospectral representations to optimal problems on SU_n and investigate the relevance of the associated integrals of motion for the problems of magnetic resonance imaging.

Secondly, I wanted to make a more extended study of isospectral systems on manifolds whose contangent bundles can be represented as coadjoint orbits on Lie algebras and compute some of the spectral invariants associated with the spectral matrix $-L_{\mathfrak{p}} + \lambda L_{\mathfrak{h}} + (\lambda^2 - s)B$.

I had also wanted to include a study of the elastic problem on the Heisenberg group and compare its integrability properties with the elastic problem on the space forms.

Along with the above, I would have liked to present a more detailed study of infinite dimensional Hamiltonian systems associated with the symplectic form described in the last chapter and investigate its possible relevance for the Korteweg–de Vries equation. The end of the last chapter also calls for a more detailed investigation of the hierarchy of the Poisson commuting functions and of its relation to the results found in [LP] and [ShZ].

Having said this, let me add that these topics would make worthwhile research projects, and I invite the interested reader to carry out the required details.

References

[AM] R. Abraham and J. Marsden, *Foundations of Mechanics*, Benjamin-Cummings, Reading, MA, 1978.

[Ad] J. F. Adams, *Lectures on Lie Groups*, Mathematics Lecture Notes Series, W. A. Benjamin, New York, 1969.

[Al] M. Adler, On a trace functional for formal pseudo-differential operators and symplectic structure of the Korteweg-de Vries equation, *Invent. Math.* **50** (1979), 219–249.

[AB] A. Agrachev, B. Bonnard, M. Chyba and I. Kupka, Sub-Riemannian sphere in Martinet flat case, *ESAIM: Control, Optimization and Calculus of Variations* **2** (1997), 377–448.

[AS] A. Agrachev and Y. Sachkov, *Control Theory from the Geometric Point of View*, Encyclopedia of Mathematical Sciences, vol. **87**, Springer-Verlag, New York, 2004.

[Ap] R. Appell E. and Lacour, *Principes de la theorie des fonctions elliptiques et applications*, Gauthier-Villars, Paris, 1922.

[Ar] V. I. Arnold, *Mathematical Methods of Classical Mechanics*, Graduate Texts in Mathematics, vol. **60**, Springer-Verlag, New York, 1978.

[AKh] V. I. Arnold and B. Khesin, *Topological Methods in Hydrodynamics*, App. Math. Sci. (125), Springer-Verlag, New York, 1998.

[AKN] V. I. Arnold, V. V. Kozlov, and A. I. Neistadt, Mathematical aspects of classical and celestial mechanics, *In Encyclopedia of Mathematical Sciences*, Vol 3, Springer-Verlag, Berlin-Heidelberg, 1988.

[Bl] A. G. Bliss, *Calculus of Variations*, American Mathematical Society, LaSalle, IL, 1925.

[BR] A. I Bobenko, A. G. Reyman and M. A. Semenov-Tian Shansky, The Kowalewski top 99 years later: a Lax pair, generalizations and explicit solutions, *Comm. Math. Phys* **122** (1989), 321–354.

[Bg1] O. Bogoyavlenski, New integrable problem of classical mechanics, *Comm. Math. Phys.* **94** (1984), 255–269.

[Bg2] O. Bogoyavlenski, Integrable Euler equations on SO_4 and their physical applications, *Comm. Math. Phys.* **93** (1984), 417–436.

[Bv] A.V. Bolsinov, A completeness criterion for a family of functions in involution obtained by the shift method, *Soviet Math. Dokl.* **38** (1989), 161–165.

[BJ] B. Bonnard, V. Jurdjevic, I. Kupka and G. Sallet, Transitivity of families of invariant vector fields on semi-direct products of Lie groups, *Trans. Amer. Math. Soc.* **271** (1982), 525–535.

406

[BC] B. Bonnard and M. Chyba, Sub-Riemannian geometry: the Martinet case (Conference on Geometric Control and Non-Holonomic Mechanics, Mexico City, 1996, edited by V. Jurdjevic and R.W. Sharpe), *Canad. Math. Soc. Conference Proceedings*, **25** (1998), 79–100.

[Bo] A. Borel, Kählerian coset spaces of semi-simple Lie groups, *Proc. Nat. Acad. USA*, **40** (1954), 1147–1151.

[BT] R. Bott and W. Tu, *Differential Forms in Algebraic Topology*, Graduate Texts in Mathematics, Springer-Verlag, New York, (1982).

[Br1] R. Brockett, Control theory and singular Riemannian geometry, in *New Directions in Applied Mathematics* (P. Hilton and G. Young, eds.), Springer-Verlag, New York, 1981, 11–27.

[Br2] R. Brockett, Nonlinear control theory and differential geometry, *Proceedings of the International Congress of Mathematicians*, Aug. 16–24 (1983), Warszawa, 1357–1368.

[BGN] R. Brockett, S. Glazer, and N. Khaneja, Sub-Riemannian geometry and time optimal control of three spin systems: quantum gates and coherence transfer, *Phys. Rev. A* **65**(3), 032301, (2002).

[BG] R. Bryant and P. Griffiths, Reductions for constrained variational problems and $\frac{1}{2} \int \kappa^2 \, ds$, *Amer. Jour. Math.* **108** (1986), 525–570.

[Br] J.P Brylinski, *Loop Spaces, Characteristic Classes and Geometric Quantization*, Progress in Math. (108), Birkhauser, Boston, Mass, 1993.

[Cr] C. Carathéodory, *Calculus of Variations*, reprinted by Chelsea in 1982, Teubner, 1935.

[C1] E. Cartan, Sur une classe remarquable d'espaces de Riemann, *Bull. Soc. Math. France* **54** (1927), 114–134.

[C2] E. Cartan, *Leçons sur les invariants integraux*, Herman, Paris, 1958.

[CK] A. l. Castro and J. Koiller, On the dynamic Markov–Dubins problem: from path planning in robotics and biolocomotion to computational anatomy, *Regular and Chaotic Dynamics* **18**, No. 1–2 (2013), 1–20.

[Ch] J. Cheeger and D. Ebin, *Comparison Theorems in Reimannian Geometry*, North-Hollond, Amsterdam, 1975.

[CL] E. A. Coddington and N. Levinson, *Theory of Ordinary Differential Equations*, McGraw Hill, New York, 1955.

[Dr] G. Darboux, *La théory général des surfaces, Vol III*, Translation of the Amer. Math. Soc. original publication Paris 1894, Chelsea, New York, 1972.

[D] P. A. M. Dirac, Generalized Hamiltonian Dynamics, *Can. J. Math.* **1** (1950), 129–148.

[DC] M. P. DoCarmo, *Riemannian Geometry*, Birkhäuser, Boston, MA, 1992.

[Drg] V. Dragovic, Geometrization and generalization of the Kowdewski top, *Commun. Math. Phys.* **298** (2010), 37–64.

[Db] L. E. Dubins, On curves of minimal ength with a constraint on the average curvature and with prescribed initial positions and tangents, *Amer. J. Math.* **79** (1957), 497–516.

[DKN] B. A. Dubrovin, I. M Kirchever and S. P. Novikov, Integrable systems I, In *Dynamical systems IV* (V. I. Arnold and S. P. Novikov, eds.), *Encylopaedia of Mathematical Sciences*, Vol 4. (1990), Springer-Verlag, 174–271.

[Eb] P. B. Eberlein, *Geometry of Nonpositively Curved Manifolds*, Chicago Lectures in Mathematics, The University of Chicago Press, Chicago, IL, 1997.

[Ep] C. K. Epstein and M. I. Weinstein, A stable manifold theorem for the curve shortening equation, *Comm. Pure Appl. Math.* **XL** (1987), 119–139.

[E] L. Euler, Methodus inveniendi lineas curvas maximi minimive proprieatate gaudentes, sive solutio, problematis isoperimetrici lattisimo sensu accepti, *Opera Omnia Ser. I^a* Lausannae **24** (1744).

[Eu] L. Euler, Evolutio generalior formularum comparationi curvarum inserventium, *Opera Omnia Ser I^a* **20** (E347/1765), 318–354.

[Fa] L. Faddeev L. and L.Takhtajan, *Hamiltonian Methods in the Theory of Solitons*, Springer-Verlag, Berlin, 1980.

[FJ] Y. Fedorov and B. Jovanovic, Geodesic flows and Newmann systems on Steifel varieties: geometry and integrability, *Math. Z.* **270**, (3–4) (2012), 659–698.

[Fl1] H. Flaschka, Toda latice, I: Existence of integrals, *Phys. Rev.* B3(9) (1974), 1924–1925.

[Fl2] H. Flaschka, Toda latice, II: Inverse scattering solution, *Progr. Theor. Phys.* **51** (1974), 703–716.

[Fk] V. A. Fock, The hydrogen atom and non-Euclidean geometry, *Izv. Akad. Nauk SSSR, Ser. Fizika* **8** (1935).

[FM] A. T. Fomenko and A. S. Mischenko, Euler equation on finite-dimensional Lie groups, *Math USSR Izv.* **12** (1978), 371–389.

[FT] A. T. Fomenko and V. V. Trofimov, Integrability in the sense of Louville of Hamiltonian systems on Lie algebras, (in Russian), *Uspekhi Mat. Nauk* 2(236) (1984), 3–56.

[FT1] A. T. Fomenko and V. V. Trofimov, Integrable systems on Lie algebras and symmetric spaces, in *Advanced Studies in Contemporary Mathematics*, Vol. 2, Gordon and Breach, 1988.

[GM] E. Gasparim, L. Grama and A.B. San Martin, Adjoint orbits of semisimple Lie groups and Lagrangian submanifolds, to appear, *Proc. of Edinburgh Mat. Soc.*

[GF] I. M. Gelfand and S. V. Fomin, *Calculus of Variations*, Prentice-Hall, Englewood Cliffs, NJ, (1963).

[Gb] V. V. Golubev, *Lectures on the Integration of the Equations of Motion of a Heavy Body Around a Fixed Point*, (in Russian), Gostekhizdat, Moscow, 1977.

[Gr] P. Griffiths, *Exterior Differential Systems and the Calculus of Variations*, Birkhäuser, Boston, MA, 1983.

[GH] P. Griffiths and J. Harris, *Principles of Algebraic Geometry*, Wiley-Interscience, New York, 1978.

[GS] V. Guillemin and S. Sternberg, *Variations on a Theme by Kepler*, vol. 42 Amer. Math. Soc., 1990.

[Gy] G. Györgi, Kepler's equation, Fock Variables, Baery's equation and Dirac. brackets, $N_{yo}V_o$ *cimertio A* **53** (1968), 717–736.

[Hm] R. S. Hamilton, The inverse function theorem of Nash and Moser, *Bull. Amer. Math. Soc.* **7** (1982), 65–221.

[H1] H. Hasimoto, Motion of a vortex filament and its relation to elastica, *J. Phys. Soc. Japan* (1971), 293–294.

[H2] H. Hasimoto, A soliton on the vortex filament, *J. Fluid Mech.* **51** (1972), 477–485.

[Hl] S. Helgason, *Differential Geometry, Lie Groups and Symmetric Spaces*, Academic Press, New York, 1978.

[HV] D. Hilbert and S. Cohn-Vossen, *Geometry and Imagination*, Amer. Math. Soc., 1991.

[Hg] J. Hilgert, K. H. Hofmann and J. D. Lawson, *Lie Groups, Convex Cones ans Semigroups*, Oxford Univesity Press, Oxford, 1989.

[HM] E. Horozov and P. Van Moerbeke, The full geometry of Kowalewski's top and (1, 2) Abelian surfaces, *Comm. Pure and App. Math.* **XLII** (1989), 357-407.

[IS] C. Ivanescu and A. Savu, The Kowalewski top as a reduction of a Hamiltonian system on $Sp(4, \mathbb{R})$, *Proc. Amer. Math. Soc.* (to appear).

[IvS] T. Ivey T and D.A. Singer, Knot types, homotopies and stability of closed elastic curves, *Proc. London Math. Soc* **79** (1999), 429–450.

[Jb] C. G. J. Jacobi, *Vorlesungen uber Dynamik*, Druck und Verlag von G. Reimer, Berlin, 1884.

[J1] V. Jurdjevic, The Delauney-Dubins Problem, in *Geometric Control Theory and Sub-Riemannian Geometry*. (U. Boscain, J. P. Gauthier, A. Sarychev and M. Sigalotti, eds.) INDAM Series 5, Springer Intl. Publishing 2014, 219–239.

[Ja] V. Jurdjevic, The symplectic structure of curves in three dimensional spaces of constant curvature and the equations of mathematical physics, *Ann. I. H. Poincaré* (2009), 1843–1515.

[Jr] V. Jurdjevic, Optimal control on Lie groups and integrable Hamiltonian systems, *Regular and Chaotic Dyn.* (2011), 514–535.

[Je] V. Jurdjevic, Non-Euclidean elasticae, *Amer. J. Math.* **117** (1995), 93–125.

[Jc] V. Jurdjevic, *Geometric Control Theory*, Cambridge Studues in Advanced Mathematics, vol. 52, Cambridge University Press, New York, 1997.

[JA] V. Jurdjevic, Integrable Hamiltonian systems on Lie groups: Kowalewski type, *Annals Math.* **150** (1999), 605–644.

[Jm] V. Jurdjevic, Hamiltonian systems on complex Lie groups and their Homogeneous spaces, *Memoirs AMS*, **178**, (838) (2005).

[JM] V. Jurdjevic and F. Perez-Monroy, Variational problems on Lie groups and their homogeneous spaces: elastic curves, tops and constrained geodesic problems, in *Non-linear Control Theory and its Applications* (B. Bonnard *et al.*, eds.) (2002), World Scientific Press, 2002, 3–51.

[JZ] V. Jurdjevic and J. Zimmerman, Rolling sphere problems on spaces of constant curvature, *Math. Proc. Camb. Phil. Soc.* **144** (2008), 729–747.

[Kl] R. E. Kalman, Y. C. Ho and K. S. Narendra, Controllability of linear dynamical systems, *Contrib. Diff. Equations*, **1** (1962), 186–213.

[Kn] H. Knörrer, Geodesics on quadrics and a mechanical problem of C. Newmann, *J. Riene Angew. Math.* **334** (1982) 69–78.

[Ki] A. A. Kirillov, *Elements of the Theory of Representations*, Springer, Berlin, 1976.

[KN] S. Kobayashi and K. Nomizu, *Foundations of Differential Geometry, Vol I*, Interscience Publishers, John Wiley and Sons, New York, 1963.

[Km] I. V. Komarov, Kovalewski top for the hydrogen atom, *Theor. Math. Phys* **47**(1) (1981), 67–72.

[KK] I. V. Komarov and V. B. Kuznetsov, Kowalewski top on the Lie algebras $o(4)$, $e(3)$ and $o(3, 1)$, *J. Phys. A* **23**(6) (1990), 841–846.

[Ks] B. Kostant, The solution of a generalized Toda latrtice and representation theory, *Adv. Math.* **39** (1979), 195–338.

[Kw] S. Kowalewski, Sur le problème de la rotation d'un corps, solide autor d'un point fixé *Acta Math.* **12** (1889), 177–232.

[LP] J. Langer and R. Perline, Poisson geometry of the filament equation, *J. Nonlinear Sci.*, **1** (1978), 71–93.

[LS] J. Langer and D. Singer, The total squared curvature of closed curves, *J. Diff. Geometry*, **20** (1984), 1–22.

[LS1] J. Langer and D. Singer, Knotted elastic curves in $\mathbb{R}^3$, *J. London Math. Soc.* **2**(30) (1984), 512–534.

[Lt] F. Silva Leite, M. Camarinha and P. Crouch, Elastic curves as solutions of Riemannian and sub-Riemannian control problems, *Math. Control Signals Sys.* **13** (2000), 140–155.

[LL] L. D. Landau and E. M. Lifshitz, *Mechanics*, 3rd edition, Pergamon Press, Oxford, 1976.

[Li] R. Liouville, Sur le movement d'un corps solide pesant suspendue autour d'un point fixé, *Acta Math.* **20** (1896), 81–93.

[LSu] W. Liu and H. J. Sussmann, *Shortest Paths for Sub-Riemannian Metrics on Rank 2 Distributions*, Amer. Math. Soc. Memoirs 564, Providence, RI, (1995).

[Lv] A. E. Love, *A Treatise on the Mathematical Theory of Elasticity*, 4th edition, Dover, New York, 1927.

[Lya] A. M. Lyapunov, On a certain property of the differential, equations of a heavy rigid body with a fixed point *Soobshch. Kharkov. Mat. Obshch. Ser. 2* **4** (1894), 123–140.

[Mg] F. A. Magri, A simple model for the integrable Hamiltonian equation, *J. Math. Phys.* **19** (1978), 1156–1162.

[Mn] S. V. Manakov, Note on the integration of Euler's equations of the dynamics of an n dimensional rigid body, *Funct. Anal. Appl.* **10** (1976), 328–329.

[MK] A. A. Markov, Some examples of the solution of a special kind of problem in greatest and lowest quantities (in Russian), *Soobsch. Karkovsk. Mat. Obshch.* (1887), 250–276.

[MZ] J. Millson and B. A. Zambro, A Kähter structure on the moduli spaces of isometric maps of a circle into Euclidean spaces, *Invert. Math.* **123**(1) (1996), 35–59.

[Ml] J. Milnor, Curvature of left invariant metrics on Lie groups, *Adv. Math.* **21**(3), (1976), 293–329.

[Mi1] D. Mittenhuber, Dubins' problem in hyperbolic spaces, in *Geometric Control and Non-holonomic Mechanics* (V. Jurdjevic and R. W. Sharpe, eds.), vol. 25, CMS Conference Proceedings American Mathematical Society Providence, RI, 1998, 115–152.

[Mi2] D. Mittenhuber, *Dubins' Problem is Intrinsically Three Dimensional*, ESAIM: Contrôle, Optimisiation et Calcul des Variations, vol. 3 (1998), 1–22.

[Mo] F. Monroy-Pérez, Three dimensional non-Euclidean Dubins' problem, in *Geometric Control and Non-Holonomic Mechanics* (V. Jurdjevic and R. W. Sharpe, eds.), vol. 25, CMS Conference Proceedings Canadian Math. Soc., 1998, 153–181.

[Mt] R. Montgomery, Abnormal minimizers, *SIAM J. Control Opt.* **32**(6) (1994), 1605-1620.

[Ms1] J. Moser, Regularization of Kepler's problem and the averaging method on a manifold, *Comm. Pure Appl. Math.* **23**(4) (1970), 609–623.

[Ms2] J. Moser, *Integrable Hamiltonian Systems and Spectral Theory*, Lezioni Fermiane, Academia Nazionale dei Lincei, Scuola Normale, Superiore Pisa, 1981.

[Ms4] J. Moser, *Finitely Many Mass Points on the Line Under the Influence of an Exponential Potential-an Integrable System*, in Lecture notes in Physics, vol 38 Springer, Berlin, 1975, 97–101.

[Ms3] J. Moser, Various aspects of integrable Hamiltonian systems, in *Dynamical Systems*, C.I.M.E. Lectures, Bressanone, Italy, June, 1978, Progress in Mathematics, vol 8, Birkhauser, Boston, MA, 1980, 233–290.

[Na] I. P. Natanson, *Theory of Function of a Real Variable,* Urgar, 1995.

[Nm] C. Newmann, De probleme quodam mechanico, quod ad primam, integralium ultra-ellipticoram classem revocatum *J. Reine Angew. Math.* **56** (1856).

[NP] L. Noakes L. and T. Popiel, Elastica on $SO(3)$, J. Australian Math. Soc. **83**(1) (2007), 105–125.

[NHP] L. Noakes, G. Heizinger and B. Paden, Cubic spines of curved spaces, *IMA J. Math. Control Info.* **6** (1989), 465–473.

[O1] Y. Osipov, Geometric interpretation of Kepler's problem, *Usp. Mat. Nauk* **24**(2), (1972), 161.

[O2] Y. Osipov, The Kepler problem and geodesic flows in spaces of constant curvature, *Celestial Mechanics* **16** (1977), 191–208.

[Os] R. Osserman, From Schwarz to Pick to Ahlfors and beyond, *Notices of AMS* **46**(8) (1997), 868–873.

[Pr] A. M. Perelomov, *Integrable Systems of Classical Mechanics and Lie Algebras, vol. I*, Birkhauser Verlag, Basel, 1990.

[Pc] E. Picard, *Oeuvres de Charles Hermite*, Publiées sous les auspices de l'Academie des sciences, Tome I, Gauthier-Villars, Paris, 1917.

[Po] H. Poincaré, *Les méthodes nouvelles de la mechanique celeste, Tome I*, Gauthier-Villars, Paris, 1892.

[Pt] L. S. Pontryagin, V. G. Boltyanski, R. V. Gamkrelidze and E. F. Mishchenko, *The Mathematical Theory of Optimal Processes*, Translated from Russian by D. E. Brown, Pergamon Press, Oxford, 1964.

[Rt1] T. Ratiu, The motion of the free n-dimensional Rigid body, *Indiana Univ. Math. J.* **29**(4), (1980), 602–629.

[Rt2] T. Ratiu, The C. Newmann problem as a completely integrable system on a coadjoint orbit, *Trans. Amer. Mat. Soc.* **264**(2) (1981), 321–329.

[Rn] C. B. Rauner, The exponential map for the Lagrange problem on differentiable manifolds, *Phil. Trans. Royal Soc. London, Ser. A* **262** (1967), 299–344.

[Rm] A. G. Reyman, Integrable Hamiltonian systems connected with graded Lie algebras, *J. Sov. Math.* **19** (1982), 1507–1545.

[RS] A. G. Reyman and M. A. Semenov-Tian Shansky, Reduction of Hamiltonian systems, affine Lie algebras and Lax equations I., *Invent. Math.* **54** (1979), 81–100.

[RT] A. G. Reyman and M. A. Semenov-Tian Shansky, Group-theoretic methods in the theory of finite-dimensional integrable systems, *Encyclopaedia of Mathematical Sciences* (V. I. Arnold and S. P. Novikov, eds.), Part 2, Chapter 2, Springer-Verlag, Berlin Heidelberg, 1994.

[Ro] H. L. Royden, *Real Analysis*, 3rd Edition, Prentice Hall, Englewood Cliffs, 1988.

[Sc] J. Schwarz, von, Das Delaunaysche Problem der Variationsrechung in kanon-ischen Koordinaten, *Math. Ann.* **110** (1934), 357–389.

[ShZ] C. Shabat and V. Zakharov, Exact theory of two dimensional self-focusing, and one dimensional self-modulation of waves in non-linear media, *Sov. Phys. JETP* **34** (1972), 62–69.

[Sh] R. W. Sharpe, *Differential Geometry: Cartan's Generalization of Klein's Erlangen Program*, Graduate Texts in Mathematics, Springer, New York, 1997.

[Sg1] C. L. Siegel, Symplectic Geometry, *Amer. J. Math.* **65** (1943), 1–86.

[Sg2] C. L. Siegel, *Topics in Complex Function Theory, vol. 1: Elliptic Functions and Uniformization Theory*, Tracts in Pure and Appl. Math. vol. 25, Wiley-Interscience, John Wiley and Sons, New York, 1969.

[Sg] E. D. Sontag, *Mathematical Control Theory, Deterministic Finite-Dimensional Systems*, Springer-Verlag, Berlin, 1990.

[So] J. M. Souriau, *Structure des Systemes Dynamiques*, Dunod, Paris, 1970.

[Sf] P. Stefan, Accessible sets, orbits and foliations with singularities, *London Math. Soc. Proc., Ser. 3*, **29** (1974), 699–713.

[St] S. Sternberg, *Lectures on Differential Geometry*, Prentice-Hall Inc, Englewood Cliffs, NJ, 1964.

[Sz] R. Strichartz, Sub-Riemannian Geometry, *J. Diff. Geometry* **24** (1986), 221–263.

[Sg] M. Sugiura, Conjugate classes of Cartan subalgebras in real semisimple Lie algebras, *J. Math. Soc. Japan*, **11**(4) (1959).

[Ss] H. J. Sussmann, Orbits of families of vector fields and integrability of distributions, *Trans. Amer. Math. Soc.* **180** (1973), 171–188.

[Ss1] H. J. Sussmann, A cornucopia of four-dimensional abnormal sub-Riemannian minimizers, in *Sub-Riemannian Geometry* (A. Bellaïche and J. J. Risler, eds), Progress in Mathematics, vol. 144, Birkhäuser, 1996, 341–364.

[Ss3] H. J. Sussmann, Shortest 3-dimensional paths with a prescribed curvature bound, in *Proc. 34th Conf. on Decision and Control*, New Orleans, 1995, 3306–3311.

[Ss4] H. J. Sussmann, The Markov-Dubins problem with angular acceleration, in *Proc. 36th IEEE Conf. on Decision and Control*, San Diego, CA, 1997, 2639–2643.

[Sy] W. Symes, Systems of Toda type, inverse spectral problems and representation theory, Invent. Math. 59 (1980), 13–53.

[To] M. Toda, *Theory of Nonlinear Lattices*, Springer, New York, 1981.

[Tr] E. Trelat, Some properties of the value function and its level sets for affine control systems with quadratic cost, J. Dyn. Contr. Syst. **6**(4) (2000), 511–541.

[VS] J. von Schwarz, Das Delaunische Problem der Variationsrechung in kanonischen Koordinaten, *Math. Ann.* **110** (1934), 357–389.

[W] K. Weierstrass, *werke, Bd 7, Vorlesungen über Variationsrechung*, Akadem. Verlagsges, Leipzig, 1927.

[Wl] A. Weil, Euler and the Jacobians of elliptic curves, in Arithmetic and Geometry (M. Artin and J. Tate, eds.), Progress in Mathematics, vol. 35, Birkhäuser, Boston, MA, 353–359.

[Wn] A. Weinstein, The local sructure of Poisson manifolds, *J. Diff. Geom.* **18** (1983), 523–557.

[Wh] E. T. Whittaker, *A Treatise on the Analytical Dynamics of Particles and Rigid Bodies*, 3rd edition, Cambridge Press, Cambridge, 1927.

[Wf] J. A. Wolf, *Spaces of Constant Curvature*, 4th edition, Publish or Perish, Inc, Berkeley, CA, 1977.

[Yo] K. Yosida, *Functional Analysis*, Springer-Verlag, Berlin, 1965.

[Yg] L. C. Young, *Lectures in the Calculus of Variations and Optimal Control Theory*, W.B. Saunders, Philadelphia, PA, 1969.

[Zg1] S. L. Ziglin, Branching of solutions and non-existence of first, integrals in Hamiltonian mechanics I, *Funkts. Anal. Prilozhen* **16**(3) (1982), 30–41.

[Zg2] S. L. Ziglin, Branching of solutions and non-existence of first, integrals in Hamiltonian mechanics II, *Funkts. Anal. Prilozhen* **17**(1) (1981), 8–23.

Index

413

Printed in the United States
by Baker & Taylor Publisher Services